J. Carpenter
Rm 8361
Dept. of Chem.
UW - Madison

Kinetics and dynamics of elementary gas reactions

Butterworths Monographs in Chemistry and Chemical Engineering

Butterworths Monographs in Chemistry and Chemical Engineering is a series of occasional texts by internationally acknowledged specialists, providing authoritative treatment of topics of current significance in chemistry and chemical engineering.

Forthcoming titles:

Silicon in Organic Synthesis
Alkaloid Biosynthesis
Particulate Systems
Fluidized Bed Reactors
Crystallisation Process Design

Butterworths Monographs in Chemistry and Chemical Engineering

Kinetics and dynamics of elementary gas reactions

Ian W. M. Smith, MA, PhD

University Lecturer in Physical Chemistry
and Fellow of Christ's College, Cambridge

Butterworths

LONDON · BOSTON

Sydney · Wellington · Durban · Toronto

THE BUTTERWORTH GROUP

United Kingdom London — **Butterworth & Co [Publishers] Ltd**
88 Kingsway, WC2B 6AB

Australia Sydney — **Butterworths Pty Ltd**
586 Pacific Highway, Chatswood, NSW 2067
Also at Melbourne, Brisbane, Adelaide and Perth

Canada Toronto — **Butterworth & Co [Canada] Ltd**
2265 Midland Avenue,
Scarborough, Ontario, MIP 4S1

New Zealand Wellington — **Butterworths of New Zealand Ltd**
T & W Young Building,
77–85 Customhouse Quay, 1

South Africa Durban — **Butterworth & Co [South Africa] [Pty] Ltd**
152–154 Gale Street

USA Boston — **Butterworth [Publishers] Inc**
10 Tower Office Park, Woburn, Massachusetts 01801

First published 1980

ISBN 0 408 70790 9

British Library Cataloguing in Publication Data

Smith, Ian W M
Kinetics and dynamics of elementary gas reactions. – (Butterworth's monographs in chemistry and chemical engineering).
1. Chemical reaction, Rate of
2. Gases, Kinetic theory of
I. Title
541'.39 QD501 79–40533

ISBN 0–408–70790–9

Typeset by The Macmillan Co. of India Ltd, Bangalore
Printed and bound in England by the Camelot Press, Southampton

Preface

Chemists seek to understand natural phenomena in terms of *molecular* behaviour. In the field of chemical kinetics, this ensures that elementary gas-phase reactions occupy a special place: because each molecular event in which chemical change takes place is relatively simple and occurs in isolation, these reactions offer the most hopeful meeting-ground for full scale theory and detailed experiment. This book surveys the state of modern knowledge on elementary reactions in two Parts. Part 1 describes the theoretical and conceptual background to the subject, emphasizing the assumptions and limitations of each theoretical approach, as well as its strengths. In Part 2, selected experimental results are considered in order, first, to demonstrate the scope of present day techniques and, secondly, to illustrate the application of the theoretical ideas which were introduced in Part 1.

The book is intended primarily for working kineticists, especially graduate students. If it encourages them to think deeply about the fundamentals of our subject, it will have fulfilled one of its main aims. However, little that is not taught in the first two years of University chemistry courses is assumed, so I hope that senior undergraduates will also find much to interest them.

Too many people have helped, directly or indirectly, with this book for me to mention all of them by name. However, special thanks must go to Dr. Tony Callear and Professor John Polanyi for guiding my first faltering steps in research and for continuing, despite that experience, to offer me friendly encouragement. I would also like to thank Professor Bradley Moore and his colleagues for their hospitality during a period of sabbatical leave at Berkeley, longer ago than I would care to admit, when my first plans for this book were made. Mrs Nancy Hurley's superb typing of the manuscript has only been matched by the excellence of Eric Smith and Roy Hammans' artwork, and to all three I am extremely grateful. Finally, I come to the members of my research group and my family. Both have suffered, I presume!, from my preoccupation, and they deserve my heartfelt thanks.

Ian W. M. Smith
Cambridge

To my families: the one I grew up with
and the one I am growing old with!

Contents

Part 1:

The theoretical background

Chapter 1

Macroscopic and microscopic kinetics

1.1 Elementary reactions

When chemical change occurs it usually does so through a network of interacting steps which is known as the *reaction mechanism.* The identification of mechanism in this sense is one of two major problems in the field of chemical kinetics. The other is to investigate and attempt to understand the details of chemical reactions which cannot be further subdivided into processes of lesser molecular complexity. These processes, which necessarily involve the participation of a small integral number of molecules, are termed *elementary reactions* and they can be studied in greatest detail in the gas phase, since in solution the role played by nearby solvent molecules cannot be neglected.

Elementary chemical reactions can usually be classified as either *collisional* or *decay* processes. The former are generally *bimolecular*, that is two species collide in each microscopic event that leads to reaction. Reactions of this type in which a single atom is transferred have been studied in greatest detail and they are featured very prominently in this book. An example is

$$F + H_2 \rightarrow HF + H \tag{1.1}$$

Decay processes are unimolecular; they involve a single species that changes to a different form. A familiar example is provided by the spontaneous radiative decay which follows photochemical excitation of an atom or molecule. In unimolecular chemical processes, a molecule with high internal energy may fragment, as in

$$C_2H_6 \rightarrow CH_3 + CH_3 \tag{1.2}$$

in which case the reverse reaction is bimolecular, or it may isomerize, as in

$$CH_3NC \rightarrow CH_3CN \tag{1.3}$$

when the reverse reaction, as well as the forward reaction, is unimolecular.

The separation of elementary reactions into these two categories is useful but must be carried out with care. Thus, neither the dissociation of ethane nor the isomerization of methyl isocyanide are as simple as one might be led to believe by equations (1.2) and (1.3) and the statement that these reactions are

unimolecular. As well as the unimolecular chemical step represented by these equations, collisions in which energy is transferred but no chemical change takes place play a vital role in the kinetics. A somewhat similar situation arises when a bimolecular encounter is 'sticky', that is, it leads to the formation of a *collision complex* which survives for a time which is longer than the characteristic periods of its vibrations and rotations. In these circumstances, the molecular event may be thought of as a bimolecular collision followed by a unimolecular decay process. An example is provided by the reaction of methyl and trifluoromethyl radicals at low pressure. This proceeds via the formation of a 1,1,1-trifluoroethane molecule with considerable internal energy, i.e.,

$$CH_3 + CF_3 \rightarrow (CH_3{-}CF_3)^{\dagger} \rightarrow CH_2{=}CF_2 + HF \qquad (1.4)$$

In recent years, new techniques have enabled the experimentalist to study the molecular collisions that are so important in elementary chemical reactions in increasingly fine detail. Theoretical developments have kept pace—and occasionally outstripped—the experimental progress. The purpose of this book is to introduce its readers to some of this exciting research into *reaction dynamics*, and to encourage them to think about the molecular level events that underlie any kinetic observation.

1.2 Macroscopic kinetics[1]

The past 20 years have seen a huge improvement in the quality and quantity of kinetic data. This situation has been brought about by the invention of a whole armoury of new laboratory techniques that have enabled the experimentalist both to study elementary reactions in isolation and also to observe the rapid changes which frequently accompany them. Although this advance has been spectacular, the aim of much of this work has, in one important sense, remained conventional: to measure the progress of reaction in terms of the rate of change in concentration of a particular chemical entity (e.g., the disappearance of a reagent) over as wide a range of temperature as possible. The definition of a temperature implies that the constituents of the reacting mixture are thermally equilibrated, and this is the usual situation in what are often called 'bulk' or 'bulb' experiments.

1.2.1 The rate constant and the Arrhenius equation

When a temperature is defined for a reacting system, the rate of elementary reactions—as well as those of more complex ones—can be expressed in terms of the concentrations of the chemical species present. To illustrate this, let us consider a generalized bimolecular reaction

$$A + B \rightleftarrows C + D \qquad (1.5)$$

where A, B, C and D all represent different atomic or molecular species. Allowing for the possibility of reaction in both directions, then the rate

equation is given by

$$-\mathrm{d}[\mathrm{A}]/\mathrm{d}t\ (= -\mathrm{d}[\mathrm{B}]/\mathrm{d}t = +\mathrm{d}[\mathrm{C}]/\mathrm{d}t = +\mathrm{d}[\mathrm{D}]/\mathrm{d}t)$$
$$= k_{\mathrm{f}}(T)[\mathrm{A}][\mathrm{B}] - k_{\mathrm{r}}(T)[\mathrm{C}][\mathrm{D}] \quad (1.6)$$

The square brackets denote concentrations, and $k_{\mathrm{f}}(T)$ and $k_{\mathrm{r}}(T)$ are coefficients which relate the rates of the bimolecular reactions in the forward and reverse directions to the concentrations of reactants (that is the species written on the left-hand side of the chemical equation) and of products (i.e., the species on the right-hand side). Clearly, $k_{\mathrm{f}}(T)$ and $k_{\mathrm{r}}(T)$ must have the dimensions of concentration^{-1} time^{-1}. Because a major preoccupation in this book is to relate macroscopic observables, such as $k_{\mathrm{f}}(T)$ and $k_{\mathrm{r}}(T)$, to molecular properties, *molecule* cm^{-3} will be used as the unit of concentration, so bimolecular rate constants will be expressed in units[a] of cm^3 molecule^{-1} s^{-1}. Although $k_{\mathrm{f}}(T)$ and $k_{\mathrm{r}}(T)$ are temperature-dependent, emphasized here by the symbols used, we shall follow the customary practice and refer to them as *rate constants.*

In the majority of cases, the variation of an observed rate constant with temperature can be described, within the accuracy of the experimental measurements, by the Arrhenius relationship

$$k(T) = A\exp(-E_{\mathrm{act}}/RT) \quad (1.7)$$

Differentiating this equation with respect to $(1/RT)$ leads to the following definition for the *activation energy*, E_{act}

$$-\frac{\mathrm{d}\ln k(T)}{\mathrm{d}(1/RT)} = E_{\mathrm{act}} \quad (1.8)$$

These two equations—and these equations alone—should be used to define E_{act} and A; A possesses the same dimensions as $k(T)$ and is termed the *pre-exponential factor.*

The purpose of conventional kinetic studies is to measure $k(T)$ over a wide range of temperature and thus to determine A, E_{act}, and whether these parameters themselves vary with T. Partly to stress the macroscopic nature of the thermal rate constant, the relationship between the Arrhenius parameters and their thermodynamic counterparts is considered next.

1.2.2 Kinetics and thermodynamics

A system at chemical equilibrium is not microscopically static but no overall change is observed because each elementary reaction is proceeding at an equal rate in the forward and reverse directions. Therefore, if reaction (1.5) is again taken as an example, at equilibrium

$$k_{\mathrm{f}}[\mathrm{A}]_{\mathrm{e}}[\mathrm{B}]_{\mathrm{e}} = k_{\mathrm{r}}[\mathrm{C}]_{\mathrm{e}}[\mathrm{D}]_{\mathrm{e}} \quad (1.9)$$

[a] A summary of the units used in this book is provided in Appendix 1, along with values of universal constants and conversion factors.

where the subscripts are used to denote values at equilibrium, and k_f and k_r replace $k_f(T)$ and $k_r(T)$ for the sake of simplicity. Rearrangement of this equation yields

$$k_f/k_r = [C]_e[D]_e/[A]_e[B]_e \quad (1.10)$$

where the right-hand side is simply K_c, the equilibrium constant expressed in terms of concentrations. Hence

$$k_f/k_r = K_c \quad (1.11)$$

This equation, relating the rate constants of chemical kinetics to the equilibrium constant of chemical 'statics' or thermodynamics, gives algebraic form to the principle of *detailed balancing* which states that at equilibrium every molecular process proceeds on average equally fast in both directions. The microscopic basis for this law will be considered in Section 1.3.2.

The connection between kinetic and thermodynamic quantities can be carried further by differentiating the equation

$$\ln k_f - \ln k_r = \ln K_c \quad (1.12)$$

with respect to $(1/RT)$. Remembering equation (1.8), it is clear that the left-hand side of this equation is the difference in activation energies for the forward and reverse reactions, i.e., $-(E_{act,f} - E_{act,r})$. The right-hand side can be evaluated using the appropriate form of the van't Hoff equation, i.e.,

$$\frac{d \ln K_c}{d(1/RT)} = -\Delta E_c^\circ \quad (1.13)$$

Actually the standard state symbols are unnecessary for ideal gases, since E depends only on temperature and not on concentration or pressure. Therefore, one can write simply

$$E_{act,f} - E_{act,r} = \Delta E \quad (1.14)$$

A similar derivation can be used to relate the pre-exponential factors to the standard entropy change for the reaction. For a reaction such as (1.5), in which there are equal numbers of molecules on both sides of the chemical equation, this leads to

$$\ln (A_f/A_r) = \Delta S_c^\circ/R \quad (1.15)$$

and once again the definition chosen for the standard state is unimportant. However, the general form of this equation is

$$\ln (A_f/A_r) = (\Delta S_c^\circ/R) - \Delta v \quad (1.16)$$

It allows for a change, Δv, in the number of molecules on going from left to right in the chemical equation: thus, $\Delta v = 1$ in reaction (1.2). For reactions of order higher than one, the standard state must be specified in the same concentration units as are used in expressing the rate constants and pre-exponential factors. Further details of these relationships can be found elsewhere.[1]

Since accurate thermodynamic data are available for most species, equations (1.14) to (1.16) can be useful in one of two related ways. When values of the Arrhenius parameters for a reaction in both directions have been determined independently, the equations provide an important check on the measurements. Alternatively, when only A_f and $E_{act,\,f}$ have been obtained directly, they can be used to calculate A_r and $E_{act,\,r}$. The equations are not, however, much help in predicting rate data in the absence of any kinetic measurements, although, as the activation energy is never both negative and large in magnitude, equation (1.14) does lead to the conclusion that where $\Delta E > 0$ the activation energy must be $\gtrsim \Delta E$.

These relationships between kinetic and thermodynamic quantities *suggest* that elementary reactions proceeding in opposite directions pass through a common critical state or configuration. Where E_{act} is positive for both reactions, then this activated state must lie at an energy higher than those of the separated reactants or separated products. *Figure 1.1* illustrates this, and emphasizes diagrammatically that it is possible for the activated state to have any energy and still satisfy equation (1.14).

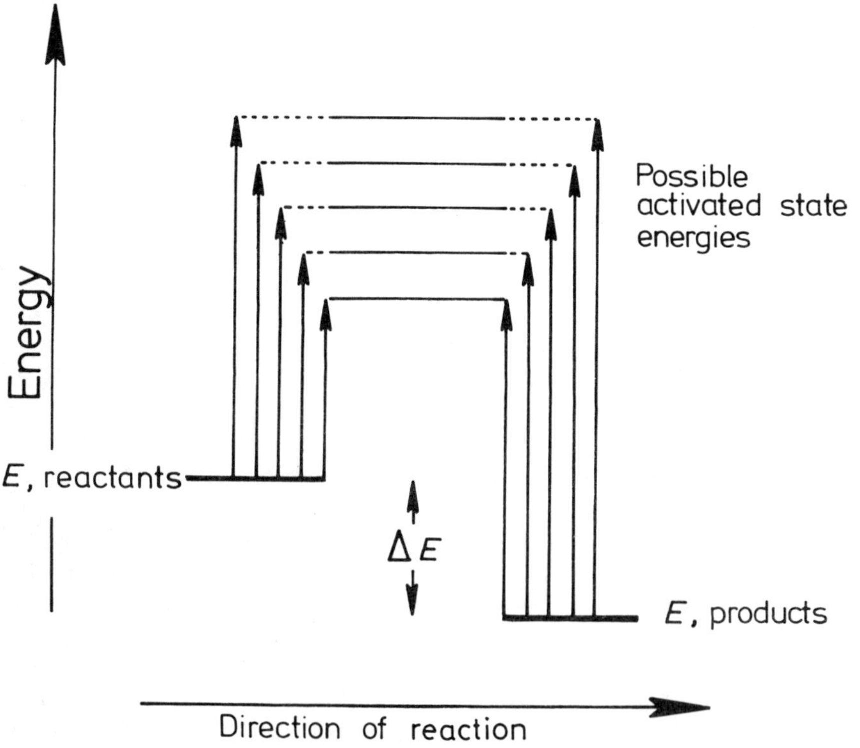

Figure 1.1. Relationship between ΔE and the activation energies for forward and reverse reactions. Possible activation energies are represented by the vertical arrows ($E_{act,\,f}$ on the left; $E_{act,\,r}$ on the right), and the diagram shows that the equation $\Delta E = (E_{act,\,f} - E_{act,\,r})$ holds for each pair of values of $E_{act,\,f}$ and $E_{act,\,r}$

Comparison of equations (1.14) and (1.16) suggests that, as well as energies of activation, it may be possible to consider entropies of activation, such that

$$(S_{\mathrm{act,f}} - S_{\mathrm{act,r}}) = \Delta S_c^\circ \qquad (1.17)$$

and

$$\ln A_{\mathrm{f}} - \ln A_{\mathrm{r}} = (S_{\mathrm{act,f}} - S_{\mathrm{act,r}})/R - (\Delta v_{\mathrm{act,f}} - \Delta v_{\mathrm{act,r}}) \qquad (1.18)$$

Since the activated state is always a single molecular conglomerate,

$$\Delta v_{\mathrm{act,f}} = 1 - \sum_{\mathrm{reactants}} v_{\mathrm{i}} \quad \text{and} \quad \Delta v_{\mathrm{act,r}} = 1 - \sum_{\mathrm{products}} v_{\mathrm{i}} \qquad (1.18a)$$

It then becomes possible to formulate general expressions for the rate constants which include the entropies of activation:

$$k_{\mathrm{f}} = \rho \exp\left(\sum_{\mathrm{reactants}} v_{\mathrm{i}}\right) \exp(S_{\mathrm{act,f}}/R) \exp(-E_{\mathrm{act,f}}/RT) \qquad (1.19)$$

$$k_{\mathrm{r}} = \rho \exp\left(\sum_{\mathrm{products}} v_{\mathrm{i}}\right) \exp(S_{\mathrm{act,r}}/R) \exp(-E_{\mathrm{act,r}}/RT) \qquad (1.20)$$

where ρ is a constant for all reactions.

These expressions, which relate rate constants to differences in thermodynamic quantities for the system (*a*) as separated reactants and (*b*) in the activated state, imply that in both configurations the species are in internal thermal equilibrium; that is, they are distributed among the available energy states according to the Boltzmann laws of statistical thermodynamics. This type of equilibrium is usually maintained by collisions in which energy is transferred. The assumption of thermal equilibration of the *reactants* will be valid as long as molecules in every state making a significant contribution to the reaction rate react over a time which is longer than the 'relaxation' time characterizing the energy transfer processes. For bimolecular reactions this is likely to be so under any but the most extreme conditions, but for unimolecular reactions, as was indicated in Section 1.1, the energy transferring processes must be included in any general discussion of the kinetics. The question of whether species in the *activated state* are in internal thermal equilibrium is much more difficult to answer. This problem will be considered at length in Section 4.2.2 where the fundamental assumptions of *transition state theory* are examined.

In the past few pages, some emphasis has been placed on the macroscopic nature of the thermal rate constant. Clearly, even for an elementary reaction, $k(T)$ measures the average result of a multitude of individual molecular events which differ in their detailed characteristics—for example, the collision energy or the orientation of the species at 'impact'. Any fully detailed theory of reaction rates should start at the *molecular* level, and derive an expression for $k(T)$ by performing correctly weighted summations or integrations over the

variables that characterize the molecular events. Since the statistical laws governing the distributions of these variables are well known, this averaging is not difficult. Unfortunately, the reverse process of deriving the microscopic probabilities of reaction from the measured rate constant and its variation with temperature is impossible. Consequently, any experiment which determines the probability of reaction in events which are, at least to some extent, selected provides a more useful test of a detailed theory than the rate data provided by conventional kinetic experiments. It is this kind of approach which is introduced next.

1.3 Microscopic kinetics

The realization that conventional rate data provide little information about elementary reactions at the molecular level has prompted a growing number of kineticists to devise experiments under non-equilibrium conditions. Those which are potentially most powerful, and which differ most dramatically from 'bulb' experiments, use *crossed molecular beams.*

In these experiments, each reactant is formed into a molecular beam characterized by a well-defined direction. The velocities of the species in each beam and their internal states can—at least, in principle—be selected before the beams intersect in a well-defined crossing zone. The spatial distribution of the products of the collisions that occur in the crossing region, and possibly their translational velocity and internal energy states, are measured. The method is powerful, because the results of single well-defined collisions are observed, and because the results are more closely related to the dynamics in the collisions and the forces that control them than the data from conventional 'bulk' experiments.

1.3.1 The total and differential reaction cross-sections[2,3]

In order to specify the rate or probability of reaction in the kind of experiment that has just been described, it is necessary to define the *reaction cross-section*, a quantity which occupies a position in 'microscopic' kinetics that is equivalent to that of the rate constant in 'macroscopic' kinetics. The meaning of the reaction cross-section can be made clear by reference to an idealized experiment in which a stream of species A, all with the same velocity[a] v, move through a collection of B molecules having a uniform concentration [B].

If the concentration in the beam is [A], its intensity is given by

$$I_A = v[A] \tag{1.21}$$

As the beam passes through the volume containing B some A particles will be

[a] Strictly, the symbol v denotes the *magnitude* of the velocity, a vector quantity. When the *direction* of the velocity is stressed, the symbol $\mathbf{v}$ is used.

scattered and the beam intensity reduced. Assuming that only single two-body collisions are important, the change in intensity of the beam on passing through a thickness $\mathrm{d}l$ of the scattering medium is

$$\mathrm{d}I_{\mathrm{A}} = -S(v)I_{\mathrm{A}}(l)[\mathrm{B}]\,\mathrm{d}l \tag{1.22}$$

This equation is exactly analogous to Beer's law describing the attenuation of light (i.e., a 'beam' of photons) by an absorbing medium, and it defines $S(v)$, the total scattering cross-section, which generally depends on v and corresponds to an effective area presented by a molecule of B to the approaching molecules of A (or vice versa).

When chemical reaction between A and B can occur, the total reduction in I_{A} results from reactive and non-reactive collisions, and $S(v)$ can be considered as the sum of two independent quantities: $S_{\mathrm{r}}(v)$, the reaction cross-section, defined by

$$\mathrm{d}I_{\mathrm{A,r}} = -S_{\mathrm{r}}(v)I_{\mathrm{A}}(l)[\mathrm{B}]\,\mathrm{d}l \tag{1.23}$$

where $\mathrm{d}I_{\mathrm{A,r}}$ is the change in beam intensity due only to reactive collisions, and $S(v) - S_{\mathrm{r}}(v)$, the cross-section for non-reactive scattering. Since $I_{\mathrm{A}} = v[\mathrm{A}]$ and $v = \mathrm{d}l/\mathrm{d}t$,

$$-(\mathrm{d}[\mathrm{A}]/\mathrm{d}t)_{\mathrm{r}} = vS_{\mathrm{r}}(v)[\mathrm{A}][\mathrm{B}] \tag{1.24}$$

and this equation shows that $vS_{\mathrm{r}}(v)$ defines a rate coefficient for reaction in collisions between A + B occurring at a specified relative velocity[a], v.

Measuring the overall decrease in incident beam intensity does not distinguish reactive from non-reactive scattering and cannot by itself yield a value for S_{r}. To obtain S_{r}, it is necessary to detect one of the scattered reaction *products*. Further discussion requires the introduction of the *differential* reaction cross-section. This is related to the flux of product molecules reaching a detector that is positioned at an angle θ relative to $\boldsymbol{v}$ and that subtends a solid angle $\mathrm{d}\omega$ at the point where the beams intersect. Alternatively, the differential reaction cross-section, $\sigma_{\mathrm{r}}(v, \theta)$, can be defined by

$$\mathrm{d}I_{\mathrm{A,r}}(\theta) = -\sigma_{\mathrm{r}}(v, \theta)I_{\mathrm{A}}(l)[\mathrm{B}]\,\mathrm{d}l \tag{1.25}$$

It is that part of the total reaction cross-section which results in product being scattered at an angle θ into an element of solid angle $\mathrm{d}\omega$. Clearly

$$S_{\mathrm{r}}(v) = \int_0^{4\pi} \sigma_{\mathrm{r}}(v, \theta)\,\mathrm{d}\omega \tag{1.26}$$

The relationships between θ and $\mathrm{d}\theta$, and ω and $\mathrm{d}\omega$, are explained fully in Section 3.3.

Thus far, no mention has been made of the internal states of A, B, C and D. If

[a] In this chapter, we proceed without comment, from collisions in the idealized experiment to collisions between A and B at a *relative* velocity defined by $\boldsymbol{v} = \boldsymbol{v}_{\mathrm{A}} - \boldsymbol{v}_{\mathrm{B}}$, where $\boldsymbol{v}_{\mathrm{A}}$ and $\boldsymbol{v}_{\mathrm{B}}$ are the vectorial velocities of A and B, and a collision energy of $\frac{1}{2}\mu v^2$, where $\mu = m_{\mathrm{A}}m_{\mathrm{B}}/(m_{\mathrm{A}} + m_{\mathrm{B}})$ is the reduced mass. This will be justified in Chapter 3.

it is supposed that n and n′ denote sets of quantum numbers specifying the states of A + B and of C + D, respectively, then $\sigma_r(\mathrm{n}'|\mathrm{n}; v, \theta)$ represents the differential cross-section for reactive scattering, it being assumed that both selection of the internal states of A and B and the specific detection of the states of C and D are possible. The final relative speed of the reaction products, i.e., v', need not be specified explicitly, as to conserve energy it is necessary that the difference between the relative translational energy before and after the collision, $\varepsilon_T = p^2/2\mu$ and $\varepsilon_{T'} = p'^2/2\mu'$, is equal to the difference in energy of the states specified by n and n′, i.e.,

$$p'^2/2\mu' = p^2/2\mu - \Delta\varepsilon_{int} \tag{1.27}$$

where $\Delta\varepsilon_{int} = \varepsilon_{n'} - \varepsilon_n + \Delta\varepsilon_0$, as shown later in *Figure 1.3*[a].

1.3.2 Microscopic reversibility[3]

A stage has now been reached at which it is possible to consider the *principle of microscopic reversibility*. This principle arises from the invariance of the Schrödinger equation—and the classical equations of motion—under time reversal. Because of this, the probability of a transition per unit time between a fully specified state of the reactants and a fully specified state of the products is independent of the direction in which time is chosen to move, i.e.

$$P(\mathrm{n}', \theta \leftarrow \mathrm{n}, v) = P(\mathrm{n}, \theta \leftarrow \mathrm{n}', v') \tag{1.28}$$

Because of this relationship, the ratio of the differential detailed rate coefficients [*see* equation (1.24)] connecting any pair of fully specified reactant and product states must be equal to the ratio of the 'phase space volumes' associated with the relative momenta of the reactants and products which are connected through equation (1.27). Further discussion of the principle of statistical mechanics on which this assertion is founded is given in Section 4.1. It leads to the equation

$$\frac{v\sigma_r(\mathrm{n}'|\mathrm{n}; v, \theta)}{v'\sigma_r(\mathrm{n}|\mathrm{n}'; v', \theta)} = \frac{p'^2\,\mathrm{d}p'}{p^2\,\mathrm{d}p} \tag{1.29}$$

which is the microscopic equivalent of equation (1.11). As $v\sigma_r(\mathrm{n}'|\mathrm{n}; v, \theta)$ and $v'\sigma_r(\mathrm{n}|\mathrm{n}'; v', \theta)$ are 'differential detailed rate coefficients' for the reaction in the forward and reverse directions involving state-specified species, this equation may be thought of as defining a 'differential detailed equilibrium coefficient'. Equation (1.29) can be simplified by noting that differentiation of (1.27) yields $v'\,\mathrm{d}p' = v\,\mathrm{d}p$ so that

$$p^2\sigma_r(\mathrm{n}'|\mathrm{n}; v, \theta) = p'^2\sigma_r(\mathrm{n}|\mathrm{n}'; v', \theta) \tag{1.30}$$

[a] E and ε both represent energy in this volume. The symbol ε is used where *molecular* properties and processes are under discussion and is therefore used in conjunction with the Boltzmann constant, k_B. On the other hand, E is used where *molar* quantities are being considered and with the gas constant, R. A summary of the symbols used in this book is provided in Appendix 2.

These expressions require some modification when the species involved possess internal angular momentum. Equation (1.30) becomes

$$p^2 g_{\mathrm{n}} \sigma_{\mathrm{r}}(\mathrm{n}'|\mathrm{n}; v, \theta) = p'^2 g_{\mathrm{n}'} \sigma_{\mathrm{r}}(\mathrm{n}|\mathrm{n}'; v', \theta) \quad (1.31)$$

where g_{n} and $g_{\mathrm{n}'}$ are the total degeneracies associated with the angular momentum quantum states of A + B and of C + D. The σ_{r} are cross-sections averaged over such states, and defined by

$$\sigma_{\mathrm{r}}(\mathrm{n}'|\mathrm{n}; v, \theta) = (1/g_{\mathrm{n}}) \sum_{\mathrm{m_n m_n}} \sigma_{\mathrm{r}}(\mathrm{n}'\mathrm{m}_{\mathrm{n}'}|\mathrm{n}\mathrm{m}_{\mathrm{n}}; v, \theta) \quad (1.32)$$

where $\mathrm{m_n}$ and $\mathrm{m_{n'}}$ are quantum numbers or combination of quantum numbers which differ only in the projection of angular momentum vectors along a chosen axis. Now n and n′ indicate specification of all quantum numbers other than these m quantum numbers.

Equations (1.28) to (1.31) incapsulate the principle of microscopic reversibility. In addition, this section has introduced the parameters which could describe the results of experiments at the ultimate microscopic level. Next we consider how these parameters are related to data which it is possible to obtain from experiments of different types through to those in which only the thermal rate constant is determined.

1.3.3 The relationship between the rate constant and the reaction cross-section

The ultimate goal of experiments in reaction dynamics is to determine sets of differential cross-sections for processes connecting specified reactant and product states in collisions of well-defined relative velocity. At the present time, it remains a distant objective, since the present generation of experiments only yield quantities that are averaged over the parameters that define individual collisions to a greater extent than $\sigma_{\mathrm{r}}(\mathrm{n}'|\mathrm{n}; v, \theta)$.

Only crossed molecular beam experiments yield differential reaction cross-sections. It is now fairly standard in such experiments (Section 7.2) to employ velocity selection of the reactants and velocity analysis of the products, but rarer to select the internal reactant states or to measure the distribution over product states directly. However, the distribution of products with respect to total internal energy can be obtained by making use of the energy conservation equation

$$\varepsilon_{\mathrm{T}'} + \varepsilon_{\mathrm{n}'} = \varepsilon_{\mathrm{T}} + \varepsilon_{\mathrm{n}} - \Delta\varepsilon_0 \quad (1.33)$$

where ε_{T} and ε_{n} are the relative translational and internal energies of the reactants (unprimed) and products (primed), and $\Delta\varepsilon_0$ is the energy difference between the zero-point levels of products and reactants, as shown in *Figure 1.3*[a]. Because, even in a thermal distribution, the spread of ε_{n} values is usually

[a] For an atom-transfer reaction, such as $A + BC \rightarrow AB + C$, $\Delta\varepsilon_0$ corresponds to ΔD_0, the difference in dissociation energies of BC and AB. The sign of $\Delta\varepsilon_0$ determines

small compared with $\Delta\varepsilon_0$, this equation yields the distribution of $\varepsilon_{n'}$ values if ε_T is selected and the distribution of $\varepsilon_{T'}$ values is measured directly.

If differential reaction cross-sections are determined over a sufficiently wide range of scattering angles, the integration in equation (1.26) can be carried out, and a *total* reaction cross-section obtained. Where no selection or analysis of internal states is performed, the value of $S_r(v)$ is one averaged over the distribution of reactant states and includes equally weighted contributions from all the processes in which products are formed independent of their states, i.e.,

$$S_r(v) = \int_0^{4\pi} \sum_{n'} \sum_{n} f_n \sigma_r(n'|n; v, \theta)\, d\omega \tag{1.34}$$

Here f_n is the fraction of the total number of collisions that occurs with A and B in the combination of states denoted by n. If the distributions over these internal states are equilibrated at temperature T,

$$f_n = g_n \exp(-\varepsilon_n/k_B T)/Q_{int,A} Q_{int,B} \tag{1.35}$$

where $Q_{int,A}$ and $Q_{int,B}$ are the partition functions associated with the internal states of A and B (*see* Appendix 3), and g_n and ε_n are the degeneracies and energies for the specified combination of internal states, the latter being referred to the zero-point level of the reactants.

The form of the variation of $S_r(v)$ with v is usually termed the (state averaged) *excitation function.* These have been determined for a few reactions in experiments using either velocity selected crossed molecular beams or the photolytic production of reactive 'hot' atoms with defined amounts of excess translational energy.

Next we can consider the results of experiments where either the internal states of the reactants are selected, or those of the products are observed, but collisions occur with the full thermal spread of relative velocities defined by a temperature T. Then the *detailed rate constant* for formation of products in states denoted by n′ from reactants in states n is given by

$$k(n'|n; T) = \int v S_r(n'|n; v) f(v; T)\, dv \tag{1.36}$$

where $f(v; T)\,dv$ is the fraction of collisions which, in a thermal distribution of relative velocities defined by the temperature T, occur with a relative translational velocity between v and $v + dv$. This distribution function can be derived by straightforward application of the methods of statistical mechanics (Appendix 3). Substitution of the result into equation (1.36) yields

---

whether a reaction is *exoergic* ($\Delta\varepsilon_0$ negative) or *endoergic* ($\Delta\varepsilon_0$ positive). Here, the terms *exothermic* and *endothermic* will be used according to whether there is an overall release or consumption of translational energy; they can therefore be applied to state-selected processes.

$$k(\mathrm{n}'\,|\,\mathrm{n};\,T) = 4\pi(\mu/2\pi k_{\mathrm{B}}T)^{\frac{3}{2}} \int_0^{\infty} S_{\mathrm{r}}(\mathrm{n}'\,|\,\mathrm{n};\,v)v^3 \exp(-\mu v^2/2k_{\mathrm{B}}T)\mathrm{d}v \tag{1.37}$$

The equation for the detailed rate constant for the reverse reaction connecting this combination of reactant and product states has an identical form

$$k(\mathrm{n}\,|\,\mathrm{n}';\,T) = 4\pi(\mu/2\pi k_{\mathrm{B}}T)^{\frac{3}{2}} \int_0^{\infty} S_{\mathrm{r}}(\mathrm{n}'\,|\,\mathrm{n};\,v)v^3 \exp(-\mu v^2/2k_{\mathrm{B}}T)\mathrm{d}v \tag{1.38}$$

Now, equation (1.27) and its derivative form, $\mu v\,\mathrm{d}v = \mu' v'\,\mathrm{d}v'$, together with the microscopic reversibility equation (1.29), can be used to derive the relationship between $k(\mathrm{n}'\,|\,\mathrm{n};\,T)$ and $k(\mathrm{n}\,|\,\mathrm{n}';\,T)$:

$$k(\mathrm{n}'\,|\,\mathrm{n};\,T)/k(\mathrm{n}\,|\,\mathrm{n}';\,T) = (\mu'/\mu)^{\frac{3}{2}}\,(g_{n'}/g_{\mathrm{n}})\exp(-\Delta\varepsilon_{\mathrm{int}}/k_{\mathrm{B}}T) \tag{1.39}$$

This equation is, of course, analogous to equation (1.29) on the one hand and equation (1.11) on the other.

Some experiments, such as those where infrared chemiluminescence is observed (Section 6.2.1), yield detailed rate constants for the formation of products in specific states but from reactants that are totally equilibrated. Consequently, values of $k(\mathrm{n}'\,|;\,T)$ are obtained where

$$k(\mathrm{n}'\,|;\,T) = \sum_{\mathrm{n}} f_{\mathrm{n}} k(\mathrm{n}'\,|\,\mathrm{n};\,T) \tag{1.40}$$

and f_{n} was defined by equation (1.35).

The final connection to the thermal rate constant of macroscopic kinetics then simply requires a further summation over all the states of the products, i.e.,

$$k_{\mathrm{f}}(T) = \sum_{\mathrm{n}'} k(n'\,|;\,T) \tag{1.41}$$

Similar equations to (1.40) and (1.41) exist for $k(\mathrm{n}\,|;\,T)$ and $k_{\mathrm{r}}(T)$, that is for the reverse reaction, and combining all of these with equation (1.39) leads to

$$k_{\mathrm{f}}(T)/k_{\mathrm{r}}(T) = (\mu'/\mu)^{\frac{3}{2}}\left(\prod Q'_{\mathrm{int}}/\prod Q_{\mathrm{int}}\right)\exp(-\Delta\varepsilon_0/k_{\mathrm{B}}T) \tag{1.42}$$

where $\prod Q_{\mathrm{int}}$ is the product of partition functions associated with internal states of the reactants, i.e., $Q_{\mathrm{int,A}}\,Q_{\mathrm{int,B}}$, and $\prod Q'_{\mathrm{int}}$ is the equivalent quantity for the products. From statistical mechanics (*see* Section 4.1), the expression on the right-hand side of this equation is simply K_{c}, the equilibrium constant for the reaction, so this relationship consolidates the connection that has been established between the parameters of microscopic and macroscopic kinetics.

Finally, it should be noted that if the state-specified reaction cross-section is

first averaged over states according to equation (1.34) and then the averaging over $f(v; T)$ is carried out, one obtains

$$k(T) = 4\pi(\mu/2\pi k_B T)^{\frac{3}{2}} \int_0^\infty \{vS_r(v)\} v^2 \exp(-\mu v^2/2k_B T)dv \tag{1.43}$$

This equation shows clearly that $k(T)$ is the average over the relative velocity distribution of the rate coefficient $\{vS_r(v)\}$ that was introduced in equation (1.24). Thus

$$k(T) = \langle \{vS_r(v)\} \rangle \tag{1.44}$$

Frequently, a (mean) cross-section is quoted following the measurement of a rate constant in a bulk experiment. This is calculated from an *approximate* form of this expression:

$$k(T) = \langle S_r(v) \rangle . \langle v \rangle \tag{1.45}$$

where $\langle v \rangle = (8k_B T/\pi\mu)^{\frac{1}{2}}$ is the average relative velocity.

1.4 Towards a detailed rate theory for collisional reactions[2–4]

The links between the microscopic and macroscopic descriptions of kinetics having been firmly established, it is now possible to consider how a fully detailed theory of molecular collision processes may be formulated. Such a theory should first involve the calculation of fully detailed differential cross-sections, i.e., values of $\sigma_r(n'|n; v, \theta)$, and then use the equations derived in Section 1.3 to determine values of the parameters—for example, excitation functions, detailed rate constants, total rate constants—that are required.

In calculating the molecular dynamics of fully specified collisions use is made of the famous *Born–Oppenheimer principle*. This states that the electrostatic forces acting in a molecular system can be assumed to be independent of the nuclear *motions*, and can therefore be calculated with the nuclear *positions* fixed. The simplified justification of this assumption is that electrons are much lighter than nuclei and can therefore adjust virtually instantaneously to any change in position of the nuclei. If then the appropriate Schrödinger equations for the electrons and fixed nuclei are solved for a large number of nuclear frameworks, the potential energy $\mathscr{V}$ can be mapped as a function of the coordinates describing the positions of the nuclei. This would be the first stage in a fully detailed rate theory. The second would be to solve the Schrödinger equation for the motion of the nuclei—or, as a second best, the classical equations of motion—controlled by the potential $\mathscr{V}$.

1.4.1 The potential hypersurface

Because $\mathscr{V}$ for a system of $N(>2)$ atoms depends on $3N-6$ coordinates

defining the relative positions of the nuclei, the potential cannot be represented graphically, unless the number of independent variables is reduced. For an atom-transfer reaction that involves only three atoms, for example

$$A + BC \rightarrow AB + C \tag{1.46}$$

where A, B and C now each represent a single atom, this might be done by just considering collinear configurations. Then can be plotted as a contour diagram of the type shown in *Figure 1.2*, and reference can be made to a *potential energy surface* for the reaction. By analogy, the unrestricted function $\mathscr{V}(r_{AB}, r_{BC}, r_{CA})$ is termed a potential energy *hypersurface*.

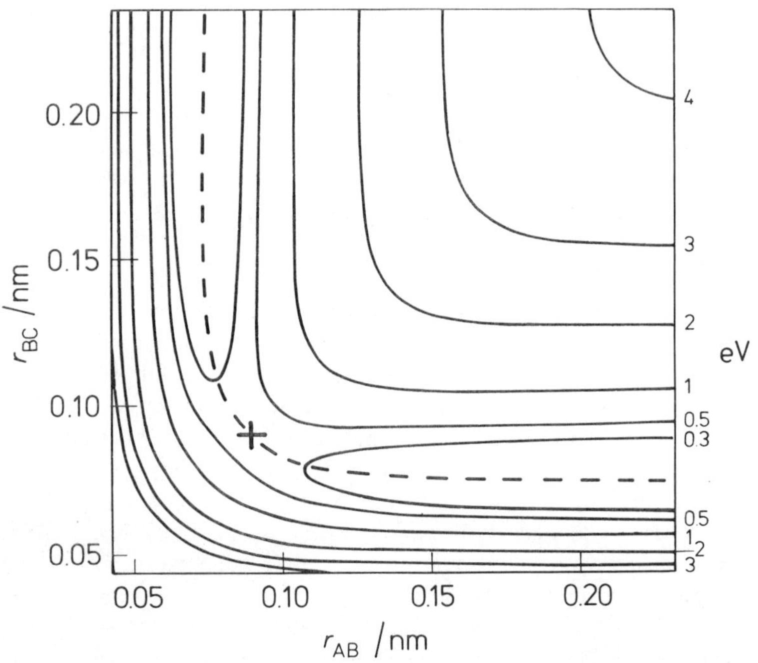

Figure 1.2. Potential-energy surface for the collinear reaction $H_A + H_BH_C \rightarrow H_AH_B + H_C$, *as calculated by Porter and Karplus* [Journal of Chemical Physics, **40**, 1105 (1964)]. *The contour lines join points of equal potential energy* (1 eV $\equiv$ 96.5 kJ mol^{-1}). *The dashed line indicates the reaction path of minimum energy and the cross the highest point on this path*

In general, there are a number of such hypersurfaces, each corresponding to a different electronic state of the molecular system. Except where explicitly stated we shall be considering reactions which proceed 'across' a single hypersurface—usually the lowest. Such reactions are termed '*electronically adiabatic*'. Where hypersurfaces intersect, it is necessary to consider the influence of additional quantum mechanical effects (Sections 2.4 and 4.4).

For atom-transfer reactions involving neutral species, one frequently finds that any 'route' from reactants to products must pass through configurations of higher electronic potential energy. As an example of this, the dashed line in *Figure 1.2* traces out the path of minimum energy between separated reactants and separated products and the cross marks the highest point on this curve. The difference in energy between this point and the separated reagents constitutes the classical *energy barrier* for collinear reaction. There are, of course, an infinite number of potential energy surfaces (each with a different restriction on the independent variables) on which barriers of varying height are located, but it is common practice to assume that the expression 'energy barrier' refers to the minimum potential barrier between reactants and products, and there are indications that for reactions like (1.46) this barrier is often associated with a collinear conformation.

The barrier is defined as 'classical', because no reference has yet been made to the quantized behaviour of nuclei moving under the influence of the potential. As $r_{AB} \rightarrow \infty$, the potential becomes that of an unperturbed BC molecule, i.e., $\mathscr{V}(r_{BC})$; solving the Schrödinger equation for the nuclei yields the vibrational – rotational states of BC, and the lowest of these lies some way above the minimum of the potential curve. Similar quantum effects are present at all points on the hypersurface, but the magnitude of the 'zero-point energy' varies. One result of this is illustrated in *Figure 1.3*: the difference between the

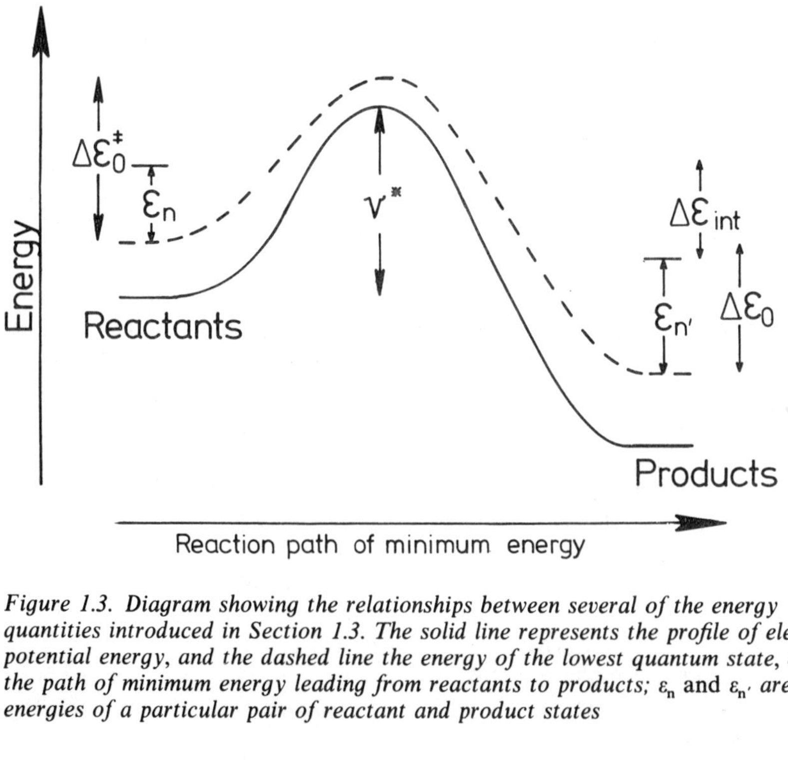

Figure 1.3. Diagram showing the relationships between several of the energy quantities introduced in Section 1.3. The solid line represents the profile of electronic potential energy, and the dashed line the energy of the lowest quantum state, along the path of minimum energy leading from reactants to products; ε_n and $\varepsilon_{n'}$ *are energies of a particular pair of reactant and product states*

lowest energy states of the separated reactants and that at the energy barrier, $\Delta\varepsilon_0^{\ddagger}$, does not correspond to $\mathscr{V}^*$, the difference in the potential energies at these points.

1.4.2 The relationship between the threshold energy and the activation energy[3]

The discussion in the last two paragraphs suggests that, for an individual collision to have any chance of leading to reaction when there is an energy barrier separating reactants and products, the sum of the internal and relative translational energies, $\varepsilon_n + \varepsilon_T$, will need to exceed some critical value. This introduces the concept of a *threshold energy*, ε_T^0, which is taken to be the value of the collision energy below which $S_r(n'|n; \varepsilon_T)$ is effectively zero, i.e.,

for

$$\varepsilon_T < \varepsilon_T^0, \quad S_r(n'|n; \varepsilon_T) = 0$$

and for

$$\varepsilon_T > \varepsilon_T^0, \quad S_r(n'|n; \varepsilon_T) > 0 \tag{1.47}$$

Consequently, if equation (1.37) for the detailed rate constant is rewritten in terms of ε_T, making the substitutions $v^2 = 2\varepsilon_T/\mu$ and $v\,dv = d\varepsilon_T/\mu$,

$$k(n'|n; T) = (1/\pi\mu)^{\frac{1}{2}}(2/k_B T)^{\frac{3}{2}} \int_{\varepsilon_T^0}^{\infty} S_r(n'|n; \varepsilon_T)\,\varepsilon_T \exp(-\varepsilon_T/k_B T)\,d\varepsilon_T \tag{1.48}$$

ε_T^0 can be written as the lower limit of the integration. The threshold energy for collisions without state selection can clearly be defined in an exactly analogous way.

It should be emphasized that the value of ε_T^0 may not be determined solely by energetic considerations, because the molecular collision dynamics, i.e., the detailed path taken by the collision, may restrict how the energy possessed by the reactants can be used in enabling the system to reach the activated state. Consequently, the value of ε_T^0 will not be equal to, or necessarily directly related to, either $\Delta\varepsilon_0^{\ddagger}$ or $\mathscr{V}^*$. A further complication arises from the fact that atoms obey quantum, rather than classical, mechanics. This means that the energy barrier, which cannot be surmounted by classical systems possessing kinetic energy less than the barrier height, can be penetrated with low probability by quantum mechanical systems (*see* Sections 3.6 and 4.4). As a result of this quantum mechanical 'tunnelling' effect, $S_r(\varepsilon_T)$ will never be *exactly* zero for reactions where $\Delta\varepsilon_{int}$ is negative.

Ignoring tunnelling, equation (1.48) or the corresponding equation for $k(T)$ can be used to derive the relationship between the activation energy and the

threshold energy. Remembering that the activation energy is defined by

$$-\frac{\mathrm{d}\ln k(T)}{\mathrm{d}(1/RT)} = E_{\mathrm{act}} \tag{1.8}$$

a general expression for E_{act} for a state-specified reaction can be obtained by operating on equation (1.48) according to (1.8). This yields

$$E_{\mathrm{act}} = \frac{N_{\mathrm{A}} \int_{\varepsilon_{\mathrm{T}}^{0}}^{\infty} \varepsilon_{\mathrm{T}} [\varepsilon_{\mathrm{T}} S_{\mathrm{r}} (\mathrm{n}' \,|\, \mathrm{n}; \varepsilon_{\mathrm{T}})] \exp(-\varepsilon_{\mathrm{T}}/k_{\mathrm{B}} T)\, \mathrm{d}\varepsilon_{\mathrm{T}}}{\int_{\varepsilon_{\mathrm{T}}^{0}}^{\infty} [\varepsilon_{\mathrm{T}} S_{\mathrm{r}} (\mathrm{n}' \,|\, \mathrm{n}; \varepsilon_{\mathrm{T}})] \exp(-\varepsilon_{\mathrm{T}}/k_{\mathrm{B}} T)\, \mathrm{d}\varepsilon_{\mathrm{T}}} - \tfrac{3}{2} RT \tag{1.49}$$

where N_{A} is Avogadro's constant. The first term on the right-hand side of this equation is the average collision energy of those collisions that actually lead to reaction, whereas $\frac{3}{2}RT$ is, of course, the mean collision energy of *all* collisions. Therefore,

$$E_{\mathrm{act}} = \langle E_{\mathrm{r,T}} \rangle - \langle E_{\mathrm{T}} \rangle \tag{1.50}$$

The corresponding treatment of $k(T)$, which is algebraically somewhat more complex, leads to the result that the activation energy is the difference between the mean collision energy plus internal energy for those collisions leading to reaction and the same quantity for all collisions, i.e.,

$$E_{\mathrm{act}} = (\langle E_{\mathrm{r,T}} \rangle + \langle E_{\mathrm{r,int}} \rangle) - (\langle E_{\mathrm{T}} \rangle + \langle E_{\mathrm{int}} \rangle) \tag{1.51}$$

The nature of the relationship between the activation energy and the threshold energy for some different forms of the excitation function is discussed in Section 3.4.4.

1.5 Summary

In this introductory chapter, emphasis has been placed on precisely defining the 'language' that is employed to describe the results of microscopic and macroscopic experiments on elementary bimolecular reactions and on establishing the relationship between various quantities that have been defined during this process. Thus, in the last section, four interrelated energy quantities were considered: the classical energy barrier to reaction ($\mathscr{V}^*$), the difference between the zero-point quantum levels of reactants and the activated state ($\Delta\varepsilon_0^{\ddagger}$), the threshold energy ($\varepsilon_{\mathrm{T}}^0$), and the activation energy (E_{act}). Despite the connection between them, these quantities are neither conceptually nor usually quantitatively the same, and therefore the terms should be used with care.

The second aim of this first chapter has been to provide an outline sketch of how a fully detailed approach to reactive collisions might be developed. In *Figure 1.4*, this is summarized by means of a flow diagram.

The difficulty of carrying out the calculations required in the first two stages

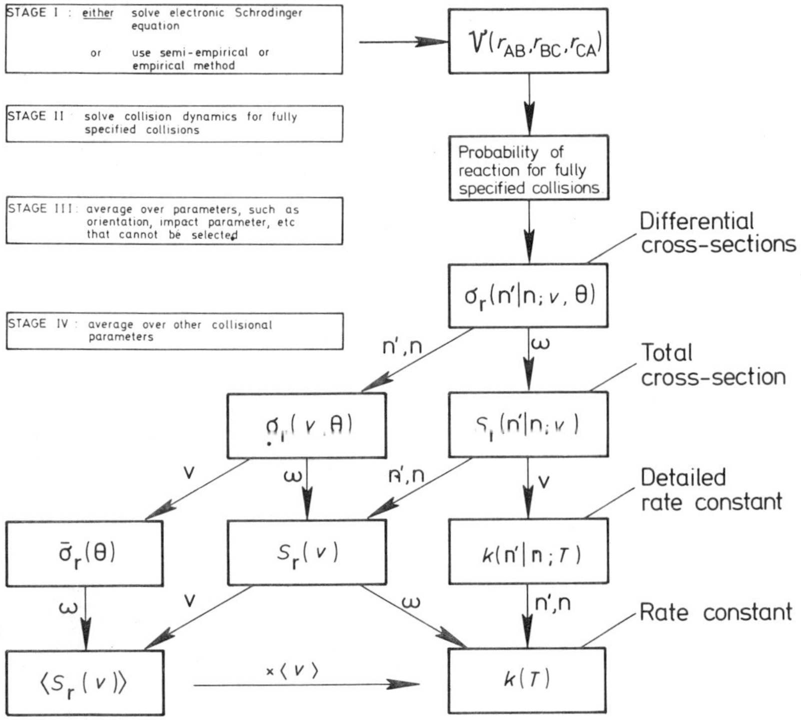

Figure 1.4. Flow diagram showing the steps in a fully detailed theory of collisional reactions and the relationship between various quantities introduced in Chapter 1. In the lower half of the diagram the presence of a symbol close to one of the lines connecting two boxes indicates averaging or integration over that parameter

of the theory has already been alluded to. In addition, it is important to appreciate the difficulty of trying to 'swim against the current' of the flow diagram. This can be forcefully illustrated by considering the connection between the state-averaged excitation function and the total rate constant. If $S_r(v)$ is known, the determination of $k(T)$ is trivial, since it simply requires a numerical integration over the thermal distribution of relative velocities as is shown in equation (1.43). On the other hand, to derive $S_r(v)$ from measurements of $k(T)$ over a reasonable range of temperatures for a typical reaction would require $k(T)$ to be measured with unattainable accuracy. This difficulty arises because of the dominance of the Boltzmann distribution term in the equation relating $S_r(v)$ to $k(T)$. In general, to test the first stages of detailed calculations or to discover something about the potential hypersurface for the reaction, it is important that the experiments provide values for quantities that appear high up on *Figure 1.4*[5]. Otherwise, information at the *molecular* level is hidden under too many layers of averaging.

References

The scientific literature can be divided into three broad categories. Firstly, there are journals, such as the primary publications of the national chemical and physical societies, which are devoted to papers reporting new theoretical or experimental research. Secondly, there are collections of review articles, usually written for the specialist, which attempt to distil and review the latest work in a particular, often rather narrow, area. Finally, there are books whose aim is generally to present a personal and coherent view of a subject at either an undergraduate or graduate level. In preparing the present volume, I have relied heavily on all three kinds of source. Three earlier books deserve a special mention:

BUNKER, D. L., *Theory of Elementary Gas Reaction Rates*, Pergamon Press, Oxford (1966)
JOHNSTON, H. S., *Gas Phase Reaction Rate Theory*, Ronald, New York (1966)
LEVINE, R. D., and BERNSTEIN, R. B., *Molecular Reaction Dynamics*, Oxford University Press, New York (1974)

My book is close in spirit, if not in competence, to these three excellent texts.

References which are provided with the headings of individual sections or sub-sections are usually to review articles. These normally supply more detailed coverage of a topic than is possible here and they should serve as suggestions for further reading. The number of references in the body of the text has been minimized, especially in the first four, 'foundation', chapters. In particular, only the latest of any series of papers on a given topic is quoted.

Sections 1.1 and 1.2

1 PRATT, G. L., *Gas Kinetics*, John Wiley, London (1969)

Sections 1.3 and 1.4

2 GREENE, E. F., and KUPPERMANN, A., *Journal of Chemical Education*, **45**, 361 (1968)
3 MENZINGER, M., and WOLFGANG, R., *Angewandte Chemie, International Edition*, **8**, 438 (1969)
4 POLANYI, J. C., and SCHREIBER, J. L., in *Physical Chemistry: An Advanced Treatise*, Eds. Eyring, H., Henderson, D. and Jost, W., vol VIA, chap. 6, Academic Press, New York (1974)
5 TOENNIES, J. P., in ref. 4, vol. VIA, chap. 5

Chapter 2

The electronic potential energy in molecular systems

2.1 General introduction

If the diverse subject of modern chemistry has a single unifying theme it must be the attempt to explain observed behaviour in terms of the forces or potentials[a] between atoms, ions and molecules. The electronic potential in an isolated polyatomic system determines the energy of the states that are associated with the relative motion of the nuclei. Consequently, where the system is stable, accurate information about the potential can be obtained by observing the spacing of its vibrational–rotational states spectroscopically.

The all-important role of the potential in molecular collisions, and therefore in ultimately determining the rates of bimolecular processes, has already been emphasized. Unfortunately, there is rarely any *direct* way to obtain information about those unstable regions of the potential which are then most important, since the instantaneous concentration of species actually in the process of undergoing reaction is much too small to observe spectroscopically. A similar situation arises, for example, for pairs of noble-gas atoms, since they form sufficient molecular species to be observed spectroscopically only under extreme conditions. Because of this difficulty, empirical information about the potential in both these cases can only be obtained by interpreting the magnitude of properties that depend on the final outcome of many collisions. When these are the results of bulk experiments—for example, virial or viscosity coefficients in non-reactive systems, or rate constants for reactions—only limited information about the potential can be extracted from the data. More sophisticated experiments—particularly those employing molecular beams—yield more detailed information. Some purists go so far as to say that the whole purpose of experiments of this kind should be to discover the form of the potential, on the grounds that, once this is known, calculating the results of molecular collisions is only a (complicated) exercise in applied mathematics.

The previous two paragraphs have indicated how information about intermolecular potentials can be obtained empirically. Of course, the chemical

[a] The relationship between *force* and *potential* is simply given by $\boldsymbol{F}(q) = -\partial \mathscr{V}/\partial q$, where $\boldsymbol{F}(q)$ is the force acting along the direction defined by the coordinate q. Although $\boldsymbol{F}(q)$ may be easier to conceptualize, it is usually more convenient to work with $\mathscr{V}$.

theoretician (an even purer purist than the one already introduced) might be dissatisfied with this approach, but then he must set to and calculate the potentials quantum mechanically. Despite the difficulty of this task, the advent of the high speed computer—together with much ingenuity on the part of theoreticians—has made such calculations possible. Indeed, the energies determined for some two and three atom systems with relatively few electrons—for example, H_2, HF, H_3 and FH_2—are now approximately as accurate as measured dissociation and activation energies.

Anyone now carrying out research on elementary gas-phase reactions should be aware of a variety of theoretical and semi-theoretical methods for estimating kinetic data. These range from sophisticated *a priori* calculations of potential energy hypersurfaces to totally empirical procedures for estimating an activation energy by interpolating between the known values for closely related reactions. Most methods are based either on theoretical treatments originally applied to systems of two atoms or on properties, such as dissociation energies, of diatomic molecules. Therefore, before considering reactive systems comprised of three atoms and more, we consider the form of the potentials that operate between two atoms and how these can be estimated and represented.

2.2 Diatomic systems[1–5]

2.2.1 Bound and repulsive states

Because the potential energy between two atoms is simply a function of r, the distance between the nuclei, each electronic state can be represented by a single *potential energy curve* of the type shown later in *Figure 2.1*. The complete electronic energy level diagram consists of a family of such curves with the symmetry of the spin and orbital wavefunctions for each denoted by a *term symbol*[3]. Except where both atoms are in $^1S_0$ states, i.e., have 'closed shell' configurations, each state-specified pair of atoms transforms smoothly into more than one molecular state. The number of these states must correspond to the total degeneracy associated with the separated atoms[a], and their term symbols can be related to those of the atoms by well-established *correlation rules* (Section 2.4.1).

Broadly speaking, it is possible to distinguish two types of potential curve. In the first, $\mathscr{V}(r)$ never falls below its value at $r = \infty$ by more than about 10 kJ mol^{-1} and the state is chiefly characterized by steep repulsive forces. Such states arise, for example, when two noble-gas atoms in their ground electronic states interact. The second type of interaction can arise when one or

[a] A good example is provided by the interaction of two oxygen atoms in their ^3P electronic ground states. Neglecting the splitting caused by spin–orbit coupling each is nine-fold degenerate, and consequently no less than 81 molecular states correlate with the separated atoms.

both of the atoms possess unpaired electrons. Now the possibility exists of a much stronger *attractive* interaction, i.e., chemical bonding, and this leads to a potential curve with a deep minimum, usually > 100 kJ mol^{-1} below the energy of the separated atoms, at relatively small internuclear separations (about 0.1 to 0.2 nm).

Strong interactions arise when there is appreciable overlap of the electron clouds originally associated with the individual atoms. To estimate $\mathscr{V}$ realistically in this region, quantum mechanical calculations are needed. At larger nuclear separations, simpler calculations of $\mathscr{V}$ frequently suffice. A number of examples of these 'long-range' potentials are considered next.

2.2.2 Long-range potentials

Probably the most familiar example of a fundamental 'long-range' interaction is the Coulombic potential between two electrically charged particles, which is given by

$$\mathscr{V}(r) = z_A z_B e^2/(4\pi\varepsilon_0)r \tag{2.1}$$

where the individual charges, $z_A e$ and $z_B e$, are expressed in SI units as multiples of e, the charge carried by a single electron, and ε_0 is the permittivity of a vacuum. If the species bear opposite charges $\mathscr{V}(r)$ is negative and the force between the particles is attractive, whereas if z_A and z_B have the same sign the particles repel one another.

It is useful to consider next the interaction between a point charge (ze) and an electrical dipole ($\boldsymbol{\mu}_D$) that are well separated. Instantaneously, the potential

$$\mathscr{V}(r, \theta) = -ze\mu_D \cos\theta/(4\pi\varepsilon_0)r^2 \tag{2.2}$$

where $\cos\theta$ is the cosine of the angle between $\boldsymbol{\mu}_D$ and $\boldsymbol{r}$, which connects the charge to the electrical centre of the dipole.

As well as any permanent dipole which a molecule may possess, a dipole can be *induced* by an electrical field. The induced dipole is related to the field (E) by

$$\boldsymbol{\mu}_{D,\text{ind}} = \alpha \boldsymbol{E} \tag{2.3}$$

where α is the polarizability and is, in general, a tensor. In the simplest case of an atom or a centrosymmetric molecule, α is a constant and the induced dipole is directed along the field. The electric field due to an ion, or strictly a point charge, at a distance r is given by

$$E = -\frac{\mathrm{d}}{\mathrm{d}r}(ze/(4\pi\varepsilon_0)r) = ze/(4\pi\varepsilon_0)r^2 \tag{2.4}$$

So the dipole induced in a centrosymmetric species is

$$\mu_D = \alpha ze/(4\pi\varepsilon_0)r^2 \tag{2.5}$$

To obtain an expression for the nett potential energy, it is necessary to take the difference between the attraction between the induced dipole and the field

$(=\alpha E^2)$ and the work expended in creating the induced dipole $(=\frac{1}{2}\alpha E^2)$. Hence

$$\mathscr{V}(r) = -\alpha(ze)^2/(4\pi\varepsilon_0)^2\,2r^4 \tag{2.6}$$

and, because $\mathscr{V}(r)$ is proportional to *ze squared*, the interaction is attractive for both negative and positive ions.

A dipole, as well as a charge, can induce a dipole in a molecule nearby, but then the resultant attractive potential is proportional to r^{-6}. Interactions of this kind form the basis of the *dispersion* or *London* forces that are the names usually applied to the long-range attractive interaction between neutral species. The forces arise because the electron distribution in an atom or molecule is subject to very high frequency fluctuations with the result that, although the distribution averaged over time is centrosymmetric, at any moment this is not so, and the atom possesses an instantaneous dipole moment. This dipole then induces a dipole in the second molecule and an attractive interaction, with an r^{-6} dependence, results. Likewise an instantaneous dipole in the second species can induce a dipole in the first, so that the magnitude of the potential is proportional to the product of the polarizabilities of the two species as well as depending on r^{-6}.

2.2.3 *A priori* calculations of short-range potentials[4, 5]

In this section we consider, from a quantum mechanical standpoint, the interaction between a pair of atoms that have approached one another closely enough for their initially independent electron clouds to overlap to a significant extent. The objective is to describe, though only briefly, the hierarchy of methods for estimating the electronic energy of molecules, and to indicate the level of accuracy that can be obtained with each of these procedures. All calculations of this kind are based on the Born–Oppenheimer principle introduced in Section 1.4. In order to determine the full potential curve for a diatomic state, a number of individual calculations must be performed, each yielding a value of $\mathscr{V}$ at a particular internuclear separation. Unfortunately, these calculations can only be carried out exactly for one-electron diatomic molecules such as H_2^+. Usually, energies can only be determined by approximate methods: in particular, the *variation method*.

The basis of the variation method is best explained by first summarizing the condition that any exact solution to the time-indepedent Schrödinger equation must satisfy. If this equation is written in its shortened form

$$\mathscr{H}\psi = E\psi \tag{2.7}$$

where $\mathscr{H}$ is the Hamiltonian operator for the system, for ϕ to be a solution of this equation, it must satisfy the eigenvalue equation

$$E = (1/\phi)\mathscr{H}\phi \tag{2.8}$$

This means that (2.8) must yield a value of E that is a constant and independent of the values of the coordinates in which ϕ is expressed.

In the variation method one evaluates the mean energy, $\overline{E}$, that is associated with a selected 'trial' wavefunction χ. If χ is not an exact solution to (2.7), the equation analogous to (2.8) will yield an expression for E involving the coordinates of the system. However, remembering that χ^2 (or $\chi\chi^*$ if χ is complex) corresponds to a continuously varying distribution function, a mean value of $\overline{E}$ can be determined by making use of the equation for the average of a quantity that is continuously distributed (Appendix 3). Then

$$\overline{E} = \int (\chi\chi^*)(1/\chi)\mathscr{H}\chi \, d\tau \Big/ \int (\chi\chi^*)\, d\tau$$

$$= \int \chi^* \mathscr{H} \chi \, d\tau \Big/ \int (\chi^* \chi)\, d\tau \qquad (2.9)$$

where the integrations are performed over all possible values of the coordinates of all the electrons in the system.

The success of the variation method rests upon the *variation theorem.* This states that the energy calculated in equation (2.9) is always greater than the true energy, and that the more closely χ corresponds to the true wavefunction the smaller the difference will become between $\overline{E}$ and the actual energy of the system. The theorem transforms what, at first sight, might appear to be a hit-or-miss method to a powerful procedure for systematically improving χ and with it the estimate of the energy.

To illustrate how the variation method is applied in practice, calculations on H_2 will now be considered in some detail. To begin with a description is given of two simplified but celebrated approaches to chemical bonding: the *valence bond* (VB) and *molecular orbital* (MO) treatments.

The VB method has as its starting point two hydrogen atoms infinitely far apart. For this system there are two exact solutions to the Schrödinger equation corresponding to the two lowest lying states of equal energy (i.e., they are degenerate). Allowing for the indistinguishability both of the electrons and of the nuclei, the most suitable form for these wavefunctions are

$$\psi_{\pm} = \frac{1}{\sqrt{2}}\{1s_A(1)1s_B(2) \pm 1s_A(2)1s_B(1)\} \qquad (2.10)$$

In this expression $1s_A(1)$ denotes a $1s$ atomic orbital in terms of the coordinates of electron 1 with reference to nucleus A. The factor $1/\sqrt{2}$ ensures normalization.

The simplest VB approach is to assume that this wavefunction does not change when the atoms are brought closer together. So the functions in equation (2.10) are substituted into equation (2.9), yielding two expressions that were originally derived by Heitler and London:

$$\overline{E}_+ = (Q + J)/(1 + S^2) \qquad (2.11a)$$

and

$$\overline{E}_- = (Q - J)/(1 - S^2) \qquad (2.11b)$$

where
$$\begin{aligned} Q &= \int 1s_A(1)1s_B(2)\mathscr{H}1s_A(1)1s_B(2)\mathrm{d}\tau \\ &= \int 1s_A(2)1s_B(1)\mathscr{H}1s_A(2)1s_B(1)\mathrm{d}\tau \\ J &= \int 1s_A(1)1s_B(2)\mathscr{H}1s_A(2)1s_B(1)\mathrm{d}\tau \\ &= \int 1s_A(2)1s_B(1)\mathscr{H}1s_A(1)1s_B(2)\mathrm{d}\tau \\ S &= \int 1s_A(1)1s_B(2)\mathrm{d}\tau = \int 1s_A(2)1s_B(1)\mathrm{d}\tau \end{aligned}$$

and Q, J and S are generally referred to as the Coulomb, exchange and overlap integrals. At $r = \infty$, both J and S are zero and $\overline{E}_+$ and $\overline{E}_-$ are both equal to twice the ionization potential of a single hydrogen atom. As r is reduced, the value of Q falls slightly at first but then at $r \approx 0.07$ nm it rises steeply. The degeneracy of the two states is split by J (and S) increasing in magnitude as the orbitals $1s_A$ and $1s_B$ begin to overlap. Because J is negative, evaluation of $\overline{E}_+$ leads to a potential curve with a minimum in it, whereas $\overline{E}_-$ rises monotonically at short range as r is lowered. Therefore, in spite of retaining wavefunctions that only accurately describe the system when $r = \infty$, the method correctly predicts the existence of a bound and an unbound state correlating with $H(1^2S_{\frac{1}{2}}) + H(1^2S_{\frac{1}{2}})$.

There is one fairly obvious way in which the trial wavefunction, ψ_+, for the H_2 electronic ground state can be improved, and this provides a simple illustration of the use of the variation theorem. It consists of using $1s$ orbital wavefunctions that include an adjustable parameter ζ in the exponent, i.e.,

$$1s' = N_0 \exp(-\zeta\rho) \tag{2.12}$$

where ρ is directly proportional to the distance between the electron and nucleus in question. The optimum value of ζ will be that for which $\overline{E}$ is a minimum, i.e., for which

$$\partial\overline{E}/\partial\zeta = 0 \tag{2.13}$$

This procedure for *scaling* the atomic wavefunctions in variational calculations is frequently used. The data in *Table 2.1* show how estimates of D_e and r_e, the depth and internuclear distance at the minimum of $\mathscr{V}(r)$, are improved by scaling the simple VB function for H_2.

The basis of the *MO method* is different. Molecular orbitals are formed by taking linear combinations of the atomic orbitals (LCAO) for the isolated atoms, i.e.,

$$\Phi = \sum_i a_i \phi_i \tag{2.14}$$

Table 2.1 Calculated and experimental values of D_e and r_e for the ground state of H_2

Source	D_e/kJ mol^{-1}	r_e/nm
Simple VB[a]	303	0.0869
VB; single scaling factor	363	0.076
Simple MO	259	0.085
MO; single scaling factor	335	
Weinbaum's function (equation 2.18)	386	0.0751
CI calculation, 10 orbital basis, 16 configurations (SSMK[b])	445.2	0.07418
CI calculation, 46 orbital basis, 126 configurations (Liu[c])	456.0	0.07416
Experimental[d]	458.0	0.07413

[a] Data in the first five rows of the table are taken from LINNETT, J. W., *Wave Mechanics and Valency*, Methuen, London (1960)
[b] SHAVITT, I., STEVENS, R. M., MINN, F. L., and KARPLUS, M., *Journal of Chemical Physics*, **48**, 2700 (1968)
[c] LIU, B., *Journal of Chemical Physics*, **58**, 1925 (1973); SIEGBAHN, P., and LIU, B., *Journal of Chemical Physics*, **68**, 2457 (1978)
[d] STOICHEFF, B. P., *Canadian Journal of Physics*, 35, 70 (1957); HERZBERG, G., *Journal of Molecular Spectroscopy*, 33, 147 (1970)

and these are then filled according to the Aufbau principle. The coefficients in equation (2.14) are determined according to the variational principle, i.e.,

$$\partial \bar{E}/\partial a_i = 0 \tag{2.15}$$

and the magnitude of each coefficient determines the contribution each atomic orbital makes to that particular molecular orbital.

In the simplest MO treatment the two lowest lying molecular orbitals are

$$1\sigma_g = (1/\sqrt{2})(1s_A + 1s_B) \tag{2.16a}$$

and

$$1\sigma_u^* = (1/\sqrt{2})(1s_A - 1s_B) \tag{2.16b}$$

where the g and u subscripts indicate that the functions are respectively symmetric (*gerade*) and unsymmetric (*ungerade*) on inversion of the wavefunction through the centre of symmetry, and the asterisk denotes that for the second function the energy rises as r decreases, so that this corresponds to an 'antibonding' orbital. By analogy with atomic systems, the wavefunction for the ground electronic state of H_2 is therefore

$$\chi = 1\sigma_g(1)1\sigma_g(2) \tag{2.17}$$

The values of D_e and r_e that are obtained from calculations which use this expression in equation (2.9) are listed in *Table 2.1*. They are somewhat poorer than their VB counterparts.

This is a suitable point at which to draw attention to an unfortunate feature of *a priori* calculations of electronic energies. These calculations provide a direct estimate of the *total* energy of the system relative to completely separated nuclei and electrons. An error in this quantity—in the present case, D_e *plus* twice the ionization potential of H—always leads to a much larger percentage error in the quantity of interest, which here is D_e.

The next step in improving the 'trial' wavefunction merges the MO and VB treatments. In the MO equations for χ, i.e., (2.17) combined with (2.16a), the terms $1s_A(1)1s_A(2)$ and $1s_B(1)1s_B(2)$ are accorded equal weight with $1s_A(1)1s_B(2)$ and $1s_A(2)1s_B(1)$, but they do not appear at all in the VB equation (2.10). It therefore seems reasonable to try a function of the form:

$$\chi = \{1s_A'(1)1s_B'(2) + 1s_A'(2)1s_B'(1)\} + \lambda\{1s_A'(1)1s_A'(2) + 1s_B'(1)1s_B'(2)\} \qquad (2.18)$$

where the 'primes' denote scaled wavefunctions and λ as well as ζ, the scaling factor, is optimized according to the variational principle. As the data in *Table 2.1* show, this procedure leads to a further improvement in the estimates of D_e and r_e.

This improvement has a straightforward physical explanation. Since the electrons repel one another they will tend not to be found close together (for the moment, we speak as if the electrons are identifiable particles). The terms $1s_A(1)1s_A(2)$ and $1s_B(1)1s_B(2)$ are given too much 'weight' in the simple MO expression since they tend to overemphasize 'structures' where the electrons are close together near one or other of the nuclei. The modification represented by equation (2.18) is a simple example of how χ can be improved by allowing for interelectronic repulsion or *electron correlation.*

From the viewpoint of MO theory, this procedure for improving χ can be looked upon as an example of a very important general method, in which electron correlation is allowed for by forming a judicious mixture of the first approximation to the wavefunction for the electronic ground state with one or more excited state wavefunctions of suitable symmetry. This is termed *configuration interaction* (CI), and the coefficients expressing the degree of mixing are determined by the usual variation technique. For H_2, the next two excited states above the ground state have wavefunctions

$$\chi = \frac{1}{\sqrt{2}}\{1\sigma_g(1)1\sigma_u^*(2) - 1\sigma_g(2)1\sigma_u^*(1)\} \qquad (2.19a)$$

and

$$\chi = \frac{1}{\sqrt{2}}\{1\sigma_g(1)1\sigma_u^*(2) + 1\sigma_g(2)1\sigma_u^*(1)\} \qquad (2.19b)$$

However, neither of these states has the same parity as the ground state, since one electron is in a *gerade* orbital, the other in an *ungerade* orbital. Consequently, they cannot mix with the ground state and the simplest MO–CI function for H_2 is

$$\chi = 1\sigma_g(1)1\sigma_g(2) + b1\sigma_u^*(1)1\sigma_u^*(2) \tag{2.20}$$

which corresponds to equation (2.18), if $\lambda = (1-b)/(1+b)$.

Thus far, the effects of *electron spin* have been virtually ignored, and the wavefunctions that have been considered depend only on the spatial coordinates of the electrons. With light atoms, to a good approximation, the orbital and spin motions of the electrons are uncoupled, so that a total electronic wavefunction can be written as a product of spin and spatial functions. Furthermore, electrons can take only one of two spins characterized by the quantum numbers $+\frac{1}{2}$ or $-\frac{1}{2}$. If the corresponding spin wavefunctions are written as α and β, for a two electron system the possible total spin wavefunctions are $\alpha(1)\alpha(2)$, $\beta(1)\beta(2)$ and $\alpha(1)\beta(2) \pm \alpha(2)\beta(1)$.

We are now in a position to consider some of the effects of the *Pauli exclusion principle*, which is of immense significance in chemistry. The familiar rule that no two electrons in an atom or molecule can have the same four quantum numbers arises from a more fundamental requirement that the *total* electronic wavefunction must change sign when any pair of electrons are interchanged. When both electrons occupy the same orbital, as in the H_2 ground state, the spatial wavefunction is symmetrical with respect to interchange of electrons so that this requirement can only be satisfied by combining it with the single unsymmetric spin wavefunction $\alpha(1)\beta(2) - \alpha(2)\beta(1)$. For the excited state wavefunctions given earlier in equations (2.19a) and (2.19b), only the following combinations are allowed:

$$\chi = \left\{ \begin{matrix} \alpha(1)\alpha(2) \\ \alpha(1)\,\beta(2) + \alpha(2)\beta(1) \\ \beta(1)\beta(2) \end{matrix} \right\} \times \{1\sigma_g(1)1\sigma_u^*(2) - 1\sigma_g(2)1\sigma_u^*(1)\} \tag{2.21a}$$

and

$$\chi = \{\alpha(1)\beta(2) - \alpha(2)\beta(1)\} \times \{1\sigma_g(1)1\sigma_u^*(2) + 1\sigma_g(2)1\sigma_u^*(1)\} \tag{2.21b}$$

Equation (2.21a) represents a *triplet* state, (2.21b) a *singlet* state. The fact that, the triplet state lies lower in energy is a manifestation of the general rule of Hund, which states that, for states arising from the same electronic configuration, the state of maximum multiplicity (*see* Section 2.4.1) lies lowest.

The requirements of the Pauli exclusion principle can be incorporated into the theory most easily by making use of *Slater determinants*. These have the form:

$$\chi = \begin{vmatrix} \phi_1(1)\phi_1(2) & \cdots\cdots & \phi_1(N) \\ \phi_2(1) & & \\ \vdots & & \\ \phi_N(1) & & \phi_N(N) \end{vmatrix} \tag{2.22}$$

where each element is the product of a spin and spatial wave function. The numbers in parentheses denote the electrons assigned to each of the *spin-orbitals* distinguished by the subscripts. The Pauli exclusion principle emerges as a direct consequence of the determinant terminology, since assigning two electrons to the same spin-orbital would result in the determinant having two identical columns so that it would equal zero.

Of course, in using a Slater determinant in a variational calculation of electronic energy, only spatial wavefunctions need be included, since it is only the electrostatic interactions, which are functions of only the spatial coordinates, that determine the energy. In general, the elements of the determinant are written as linear combinations of the atomic orbitals (LCAO) which constitute the *basis set* for the calculation. Frequently, the basis set includes more than one orbital of a particular type—for example, 1*s* orbitals with different scaling coefficients in the exponent—and in some calculations the orbitals are also allowed to 'float', that is to have their origin at a position removed from the nucleus that they are nominally associated with.

It should now be obvious that the computational labour in a full variational treatment of a molecule is very considerable. One important way in which it can be reduced is to use the *self-consistent field* (SCF) technique which provides a method for systematically improving the one-electron wavefunctions in the Slater determinant. In this procedure one sets up a one-electron Schrödinger equation with an electrostatic potential energy due to all the nuclei and the average effect of all but the selected electron, this average being calculated using one's first estimate of ϕ_2, ϕ_3 . . . . . . etc. This equation is solved by the variation method to give the best values of the coefficients in the function ϕ_1. The procedure is repeated successively for all the ϕ to complete one cycle of the calculations. More cycles are carried out until there is no further improvement in the wavefunctions. These are then used to construct the determinant representing the total wavefunction for the system. For closed-shell configurations, a single Slater determinant with LCAO–SCF wavefunctions can provide a moderately good estimate of the electronic energy. Where this is not the case—and this includes most reactive systems—or to improve the calculations on stable molecules, a linear combination of determinants must be taken. This corresponds to the inclusion of configuration interaction of which a simple example was given earlier.

Several large calculations on the electronic ground state of H_2 have been carried out in recent years. In *Table 2.1* results are given from two of these, which each formed part of a larger effort to calculate the energy surface for the reactive H_3 system. A still more massive calculation[6] on H_2 has given results in essentially exact agreement with experimental data. In this work, the potential curve of the repulsive triplet state on H_2 was also calculated accurately; variational techniques can be applied to excited states so long as one is careful to include only wavefunctions of the appropriate symmetry. In *Figure 2.1*, curves for these two states are presented; the expanded curves at long range show how the states are identical until a point is reached where 'chemical forces' take over.

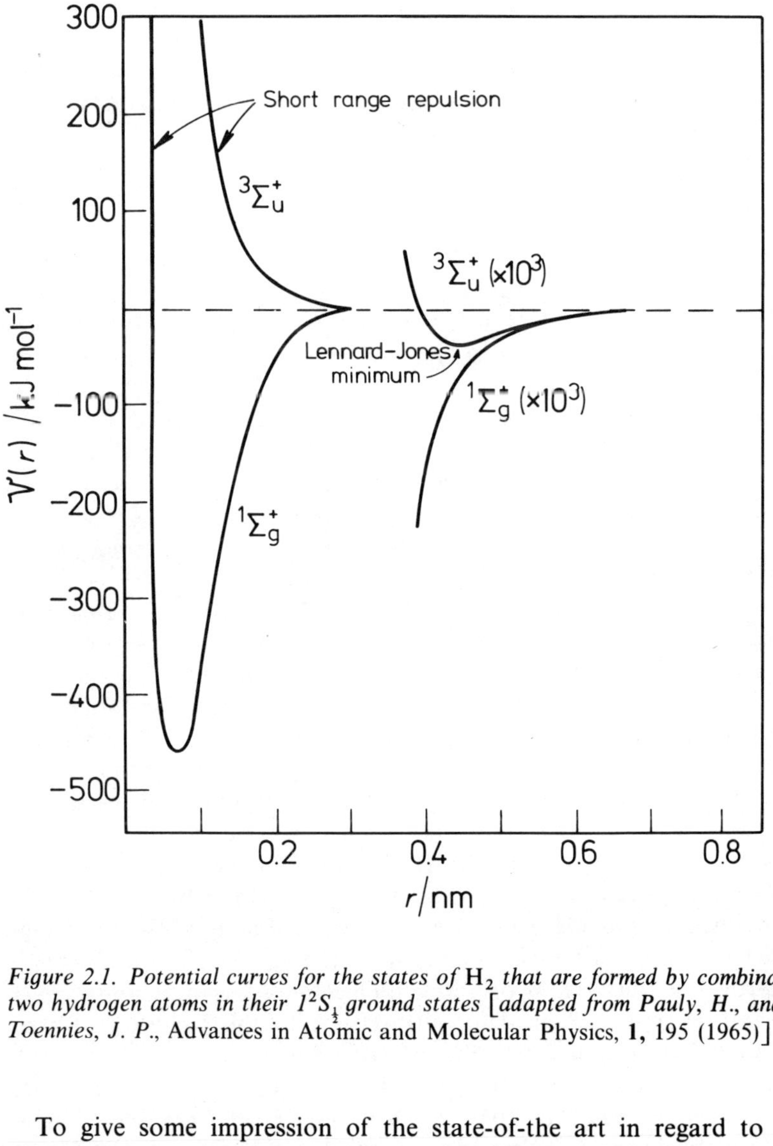

Figure 2.1. Potential curves for the states of H_2 *that are formed by combination of two hydrogen atoms in their* $1^2S_{\frac{1}{2}}$ *ground states [adapted from Pauly, H., and Toennies, J. P.,* Advances in Atomic and Molecular Physics, **1**, 195 (1965)]

To give some impression of the state-of-the art in regard to *a priori* calculations, some results for diatomic molecules other than H_2 are compared with experimental data in *Table 2.2.* The case of F_2 is particularly interesting. Here, the value calculated for D_e is sufficiently accurate for it to have played a part in deciding the correct value of the dissociation energy of this molecule.

Table 2.2 Comparison of experimental and theoretical values of D_e and r_e for some diatomic molecules other than H_2

	D_e/kJ mol^{-1} (calc.)	(expt.)	r_e/nm (calc.)	(expt.)
OH[a]	$445._8$	$445._9$($\pm$1)	0.09726	0.09706
HF[b]	567	589	0.0920	0.0917
F_2[c]	161	162 ($\pm$10)	0.141	0.142
O_2[d]	455	503	0.122	0.1207

[a] STEVENS, W. J., DAS, G., WAHL, A. C., KRAUSS, M., and NEUMANN, D., *Journal of Chemical Physics*, **61**, 3686 (1974); ARNOLD, J. O., WHITING, E. E., and SHARBAUGH, L. F., *Journal of Chemical Physics*, **64**, 3251 (1976)
[b] BONDEBEY, V., PEARSON, P. K., and SCHAEFFER III, H. F., *Journal of Chemical Physics*, **57**, 1123 (1972); DUNNING, T. H., Jr., *Journal of Chemical Physics*, **65**, 3854 (1976)
[c] DAS, G., and WAHL, A. C., *Physical Review Letters*, **24**, 440 (1970); *Journal of Chemical Physics*, **56**, 3532 (1972)
[d] SCHAEFFER III, H. F., *Journal of Chemical Physics*, **54**, 2207 (1971)

2.2.4 Empirical potentials

It has already been implied that potential curves for the stable states of diatomic molecules are usually constructed from spectroscopic data. Where the spectroscopic measurements are extremely accurate, 'true' potential curves can be determined using the numerical RKR inversion technique[7]. This method, like *a priori* calculations, does not rely on or supply an expression in which $\mathscr{V}$ is equated to an algebraic function of r.

For many purposes, the two-parameter Morse potential is adequate. If energies are related to a zero where the atoms are completely separated, it takes the form:

$$\mathscr{V}(r) = D_e\{\exp(-2\beta\Delta r) - 2\exp(-\beta\Delta r)\} \qquad (2.23)$$

where $\Delta r = r - r_e$. A particular advantage of this form is that, when it is substituted into the Schrödinger equation for relative motion of the nuclei, it yields analytical wavefunctions for the vibrational states. The corresponding energy levels are simply related to the vibrational quantum number, v, by

$$\varepsilon_v = hc\tilde{\nu}_e(v + \tfrac{1}{2}) - hc\tilde{\nu}_e x_e(v + \tfrac{1}{2})^2 \qquad (2.24)$$

where $\tilde{\nu}_e$ is the frequency (in wavenumbers, i.e., cm^{-1}) of the hypothetical classical oscillator making infinitesimally small vibrations about r_e. Clearly, the two parameters, $\tilde{\nu}_e$ and $\tilde{\nu}_e x_e$, must be related to D_e and β. The relationships are

$$\beta = \pi c\tilde{\nu}_e(2\mu/D_e)^{\frac{1}{2}} \qquad (2.25)$$

and

$$D_e = hc\tilde{\nu}_e(\tilde{\nu}_e/\tilde{\nu}_e x_e)/4 \qquad (2.26)$$

where μ is the reduced mass of the diatomic oscillator.

The energies of the lower vibrational levels of many diatomic bound states can be matched to equation (2.24) with rather high accuracy and values of $\tilde{\nu}_e$ and $\tilde{\nu}_e x_e$ obtained. They can be used to calculate a value of D_e from equation (2.26), which can then be compared with the actual value of D_e to give some indication of how closely the true potential follows the Morse curve. Morse parameters are listed and this comparison is made for a number of diatomic molecules in *Table 2.3*.

Table 2.3 Morse parameters for the ground electronic states of some diatomic molecules[a]

	D_e / kJ mol^{-1}	$\tilde{\nu}_e$ / cm^{-1}	$\tilde{\nu}_e x_e$ / cm^{-1}	β / nm^{-1}	D_e / $N_A hc\tilde{\nu}_e^2/4\tilde{\nu}_e x_e$
H_2	457.9	4401.2	121.3	19.4	0.96
HF	589	4138.7	90.05	22.1	1.04
HCl	445.2	2990.95	52.82	18.6	0.88
N_2	955.7	2358.0	14.13	26.9	0.81
Cl_2	242.6	559.7	2.70	20.0	0.70
NaBr	372	315	1.15	9.15	1.44

[a] Data from *Spectroscopic Data relative to Diatomic Molecules*, Ed. ROSEN, B. Pergamon, Oxford (1970), and HERZBERG, G., *Molecular Spectra and Molecular Structure. I Spectra of Diatomic Molecules*, 2nd edn., Van Nostrand, New York (1950)

Only limited information about non-bound molecular states can be obtained from studying the continuous spectra that arise as a result of transitions to or from these states. However, molecular beam measurements on the elastic scattering of pairs of noble-gas atoms have now been made[8, 9] with sufficient resolution, for the potential curves controlling the collisions to be determined accurately, by a numerical inversion technique which is the analogue of the RKR method that is used to treat spectroscopic data.

Less sophisticated experiments are frequently interpreted in terms of a two-parameter potential (in this way, similar to the Morse potential) in which a term representing the steep, short-range, repulsive forces is added to that for the appropriate long-range attraction. The most familiar example of such a function is the Lennard–Jones (6–12) potential:

$$\mathscr{V}(r) = 4\varepsilon_{LJ}\left[\left(\frac{\sigma_{LJ}}{r}\right)^{12} - \left(\frac{\sigma_{LJ}}{r}\right)^{6}\right] \tag{2.27a}$$

or

$$\varepsilon_{LJ}\left[\left(\frac{r_m}{r}\right)^{12} - 2\left(\frac{r_m}{r}\right)^{6}\right] \tag{2.27b}$$

Table 2.4 Parameters of the potentials acting between pairs of noble-gas atoms

	He—He	Ne—Ne	Ar—Ar	Kr—Kr	Xe—Xe	He—Ne	He—Ar	He—Kr	He—Xe
$(\varepsilon_{LJ}/k_B)/K$[a]	10.2	35.6	120.	171.	221.	19.1	35.0	41.8	47.5
$(\varepsilon/k_B)/K$[b]	10.57	42.7	143.	201.9	281.0	14.3	24.2	24.7	25.2
r_m/nm[a]	0.287	0.319	0.383	0.361	0.460	0.298	0.335	0.346	0.374
r_m/nm[b]	0.297	0.309	0.376	0.4007	0.4362	0.321	0.354	0.375	0.415

[a] Data on homonuclear pairs taken from tabulation of Lennard-Jones parameters in HIRSCHFELDER, J. O., CURTISS, C. F., and BIRD, R. B., *Molecular Theory of Gases and Liquids*, John Wiley, New York (1964). Parameters for mixed pairs calculated using the combining rules, equations (2.40a) and (2.40b)

[b] Depth and position of potential well determined from elastic scattering measurements by Y. T. Lee and his coworkers; CHEN, C. H. SISKA, P. E., and LEE, Y. T., *Journal of Chemical Physics*, **59**, 601 (1973); BARKER, J. A., WATTS, R. O., LEE, J. K., SCHAFER, T. P., and LEE, Y. T., *Journal of Chemical Physics*, **61**, 3081 (1974), and references therein.

where ε_{LJ}, σ_{LJ} and r_m are defined in *Figure 2.2.* Values of ε_{LJ} and r_m for pairs of noble-gas atoms are listed in *Table 2.4.*

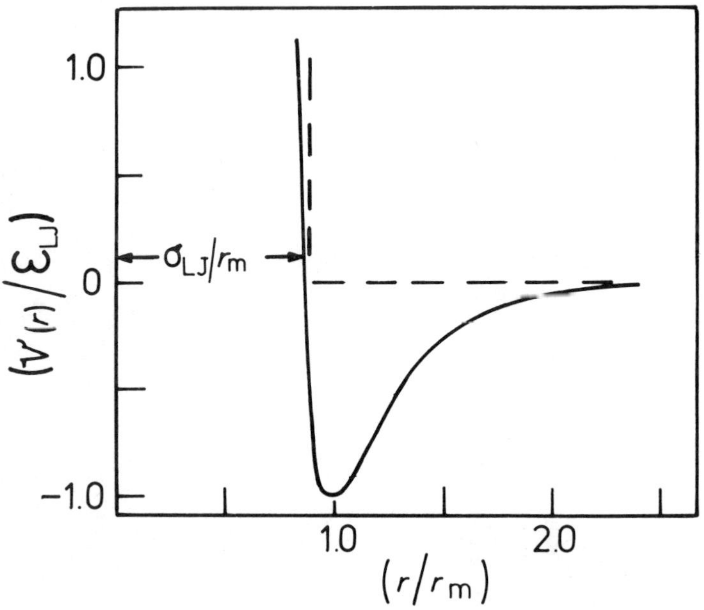

Figure 2.2. Potential curve for the Lennard-Jones function [equation (2.27)], in terms of the reduced coordinates, $\mathcal{V}(r)/\varepsilon_{LJ}$ *and* r/r_m. *The dashed curve shows the corresponding 'hard-sphere' potential*

For some (very simple) purposes it is useful to think of atoms (and molecules) as 'hard-spheres' for which $\mathcal{V}$ is zero until some value of r is reached at which $\mathcal{V}$ rises infinitely steeply. In making, for example, calculations of the frequency of collisions within this approximation (Section 3.3.1), it is usual to take the Lennard–Jones parameter σ_{LJ} as the hard-sphere diameter. The potential corresponding to this assumption is also shown in *Figure 2.2.*

2.3 Triatomic systems[10—12]

2.3.1 Types of potential

The potential energy for a system of three atoms depends on the values of three coordinates which specify the relative position of the nuclei. They are frequently chosen as r_{AB}, r_{BC} and r_{CA}, the distances between the pairs of atoms

specified as subscripts. An alternative is to use $r_{A,BC}$, r_{BC} and ϕ, where $r_{A,BC}$ is the distance from A to the centre-of-mass of BC and ϕ is the angle between this line and the BC axis. We have already seen that it is impossible to represent the variation of $\mathscr{V}$ diagrammatically, since this would require four orthogonal axes. A second problem is that, whereas about 10 to 20 calculations of $\mathscr{V}(r)$ suffice to define the potential curve of a diatomic state, many more individual calculations are required if an adequate 'map' of a potential hypersurface for three atoms is to be obtained. In view of this, it is scarcely surprising that potential functions containing a high degree of empiricism are frequently used to describe reactive systems (Section 2.3.3.), but it will increasingly be possible to test the validity of these against *a priori* calculations, as well as against experimental data.

For triatomic systems, four broad categories of intermolecular potential can be distinguished. The first (of no importance here) is that between, say, three noble-gas atoms. The second arises where two of the atoms form a chemical bond but the third is inert. A potential of this type controls the dynamics of collisions between atoms and diatomic molecules, e.g., $Ar + N_2$, where vibrational and rotational energy transfer, as well as elastic scattering, can take place. For some purposes, the A + BC interaction may be approximated by a Lennard–Jones potential, but this clearly ignores the anisotropy of the intermolecular forces. This type of potential is also important when considering atomic recombination, in which A can act as a 'moderator' removing energy, so that BC is stabilized in the 'valley' of the potential (Section 5.8).

The second and third surfaces shown in *Figure 2.3* represent cases where AB as well as BC is a bound diatomic molecule. In (b), if reaction is to occur, the (classical) system must pass over a barrier of potential energy separating the reactants, A + BC, from the products, AB + C. In contrast, on potential (c), there is a deep 'well' or 'basin', corresponding to the possibility of forming a bound triatomic molecule, ABC. There is, of course, no reason to doubt that triatomic potentials exist ranging all the way from those with high barriers to atom-transfer to those with deep wells.

Atom-transfer reactions can be expected to occur across potentials of both type (b) and type (c), but different dynamical behaviour should be expected. Collisions on potentials of type (b) with barriers to atom-transfer are likely to proceed *directly*. On the other hand, trajectories on potentials of type (c) may get 'trapped' in the potential well, resulting in the formation of a *collision complex*[a], which may survive for at least several vibrations before the system finds its way down the 'exit' valley on the surface leading to products, or back down the 'entry' valley, thus reforming separated reactants. At high total

[a] The concept of a *collision complex* must be distinguished from that of the *activated complex* of transition state theory (Chapter 4). The former implies the existence of a potential well on the reaction hypersurface and the survival of a single 'supermolecule' for a time comparable with a rotational period, the latter does not.

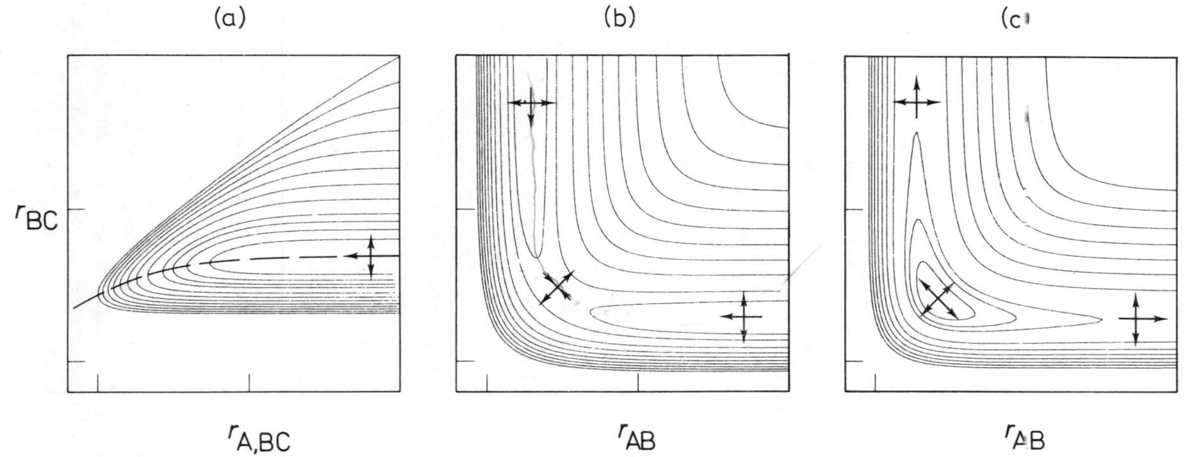

Figure 2.3. Surfaces representing three different types of potential in triatomic systems: (a) *the potential between a stable molecule* **BC** *and an inert-gas atom plotted as a function of* r_{BC} *and* $r_{A,BC}$, *the distance from* A *to the centre-of-mass of* **BC**; *the dashed line represents the path of minimum energy, and shows how the repulsive forces tend to compress the* **BC** *molecule;* (b) *here* **AB** *and* **BC** *are (identical) stable molecules but there is a potential-energy barrier to the reaction* $A + BC \rightarrow AB + C$; (c) *here* **ABC**, *as well as* **AB** *and* **BC**, *is stable so there is a deep 'well' at the symmetrical configuration. The arrows indicate the directions in which potential energy increases*

pressure, there is the additional possibility of ABC being formed by stabilization of the collision complex, before it redissociates, through a collision with some other species that removes some of the collision complex's high internal energy. Of course, if ABC is sufficiently stable, it will be possible to obtain information about the potential through spectroscopic measurements on the triatomic molecule. However, such experiments do not provide information about the potential in those regions, far from the equilibrium configuration, that are likely to determine whether a collision complex forms in A + BC collisions.

The methods which are used to calculate electronic potential energies for three or more atoms are essentially the same as for diatomic molecules, and the calculations are beset by similar difficulties. In particular, the energetic aspects of a potential are the most difficult to calculate accurately, and the quantity of greatest interest, the classical barrier height, is usually a very small difference between the large total energies (*a*) for the separated reactants and (*b*) at the maximum on the reaction path of minimum energy that leads from reactants to products. The calculations rapidly become larger, and therefore more difficult to perform accurately, as the number of electrons and nuclei in the system increase.

Because of their electronic simplicity the most accurate quantum mechanical calculations on reactive systems have been on H_3^+ and H_3. These systems can, in fact, serve as the prototypes of surfaces (b) and (c) in *Figure 2.3*, since the H_3^+ potential[13] has a deep well about 440 kJ mol^{-1} below the energy of $H_2 + H^+$ at $r_{AB} = r_{BC} = r_{CA} = 0.087$ nm, and the H_3 potential is of the kind generally associated with atom-transfer reactions between neutral species, having a barrier of about 40 kJ mol^{-1}. Although it is only recently that calculations on these systems have begun to yield results of real *quantitative* significance, already calculations have been performed[14] on the electronically much more complex reaction, $F + H_2 \rightarrow HF + H$, which appear to be in good agreement with the experimental values for the activation energy and exoergicity for the reaction.

2.3.2 *A priori* calculations

The starting point for our earlier discussion of quantum mechanical calculations on H_2 was the Heitler–London equation (2.11). The equivalent equation for three atoms, originally proposed by London for the lowest state of H_3, is

$$\mathscr{V}(r_{AB}, r_{BC}, r_{CA}) = Q_{AB} + Q_{BC} + Q_{CA} - \{\tfrac{1}{2}[(J_{AB} - J_{BC})^2 + (J_{BC} - J_{CA})^2 + (J_{CA} - J_{AB})^2]\}^{\frac{1}{2}} \quad (2.28)$$

where, as before, the Q and J are pairwise Coulomb and exchange integrals defined by the relationships below equation (2.11). However, it is important to recognize a basic difference between this equation and that of Heitler and London for H_2. In the latter, the expression for the mean electronic energy, $\bar{E}$,

is completely evaluated using the simple VB expression as an approximation to the wavefunction. The London equation for H_3, on the other hand, can only be derived if certain integrals in the full VB expression for the energy are neglected. This is typified by the absence of overlap integrals in equation (2.28) so that, as one atom is removed from the other two, the equation reduces to

$$\mathscr{V}(r_{AB}) = Q_{AB} + J_{AB} \tag{2.29}$$

rather than equation (2.11a)[a]. The fact that omitting S^2 increases the calculated value of D_e for H_2 to 450 kJ mol^{-1}, bringing it much closer to its true value, is no justification for this action. It is necessary to emphasize this point as the potential barrier which is estimated for H_3 from the London equation is quite close to the true value. However, when some of the assumptions that are required to derive equation (2.28) are removed, the barrier height rises considerably, demonstrating that the value originally calculated from the London equation is close to the experimental value for the activation energy of the $H + H_2$ reaction only as the result of a fortuitous cancellation of errors.

Two other properties of the London potential for H_3 are worthy of note. First, there is the prediction that there is a potential 'basin' with a minimum at $r_{AB} = r_{BC} = 0.09$ nm, indicating the possible formation of an H_3 collision complex. Relative to the calculated energy of $H + H_2$, the energy at this point is $+13.8$ kJ mol^{-1}, but this minimum is 23.0 kJ mol^{-1} below maxima which appear on either side of it along the reaction path of minimum energy. This feature does not appear in the best quantum mechanical calculations, nor is there any experimental evidence for its existence.

A second prediction, which is supported by more complete calculations, is that the energy barrier separating reactants from products is lowest when the three atoms are in a straight line. Therefore, at least for $H + H_2$, and possibly for other reactions involving species in similar electronic states, the transition state is linear. This result, that the lowest energy of H_3 in C_{2v} symmetry is associated with the collinear arrangement (whereas the three nuclei in H_3^+ form an equilateral triangle), could not be predicted on the basis of simple molecular orbital theory. Furthermore, one should resist any tendency to assume that the transition state for an atom-transfer reaction always has the three atoms participating most directly in the process in a straight line. In practice, increasing use is being made of qualitative theoretical arguments, overlaid with some empiricism, to predict the geometry of transition states. The arguments that are used follow the lead given by Walsh's famous studies[15] on small, stable, polyatomic molecules; examples are given in later chapters.

The London equation provided a springboard for early theoretical calculations of reactive potential surfaces, and it showed, within the framework of quantum mechanics, how an elementary chemical reaction can occur with an activation energy much lower than the energy of the bond that needs to be broken. Nearly all modern calculations on reactive systems are, however,

[a] It should be remembered that J_{AB} is *negative*. This explains the apparent difference in sign in equations (2.28) and (2.29).

founded on the variation method. The earliest of these were on H_3, and were carried out with a basis set comprising a single 1*s* orbital on each atom. The final calculation of this type was equivalent to Weinbaum's calculation on H_2 since, in a similar way, the trial wavefunction for H_3 can be considered as arising either from a proper VB treatment with the inclusion of ionic structures, or from a MO approach with configuration interaction. The barrier energy and the internuclear distances at the barrier predicted by these calculations are given in *Table 2.5*; they are—as, of course, they should be—the same within computational error.

Further improvement of the calculations requires that the variational calculations are carried out using a larger basis set of atomic orbitals and with the inclusion of more configurations. At the time of Laidler and Polanyi's review[16], written at the end of the 'early computer age', the most extensive calculations[17] had used a basis set of two 1*s* functions on each atom. The exponent of each of these pairs were kept in a fixed ratio but those on atom B could be varied independently of those on A and C. The results were rather disappointing; the barrier energy remained 122 kJ mol^{-1} above the true energy of $H + H_2$.

In recent years the greater speed and capacity of computers have made much larger calculations on H_3 possible. The two latest variational calculations are those of Shavitt, Stevens, Minn and Karplus (SSMK)[18] and of Liu[19]. Both predict that the minimum barrier separating reactants from products lies on the surface representing energies for collinear configurations. Furthermore, the barrier is at $r_{AB} = r_{BC}$; there is no sign of the well indicated by the London equation. Both these features of the H_3 potential must now be regarded as firmly established. The energies and internuclear separations at the barrier predicted by these two calculations are given in *Table 2.5*. Liu's calculations

Table 2.5 Calculated values of the energy[a] and the internuclear separations at the barrier on the H_3 potential

	$r^*_{AB} = r^*_{BC}$ / nm	$\mathscr{V}^*$/kJ mol^{-1} (i)	$\mathscr{V}^*$/kJ mol^{-1} (ii)
VB plus polar structures, single scaling factor[b]	0.0974	177.2	105.2
MO plus CI, single scaling factor[c]	0.1019	174.6	105.1
MO plus CI (SSMK)[d]	0.0915	58.5	46
MO plus CI (Liu)[e]	0.0930	43.0	41.0

[a] Values of the barrier energy are given (i) relative to the energy of $H + H_2$ determined experimentally and (ii) relative to the energy of $H + H_2$ found in the same calculation
[b] HIRSCHFELDER, J. O., EYRING, H., and ROSEN, B., *Journal of Chemical Physics*, **4**, 121 (1936)
[c] BOWEN, H. C., and LINNETT, J. W., *Transactions of the Faraday Society*, **60**, 1185 (1964)
[d] SHAVITT, I., STEVENS, R. M., MINN, F. L., and KARPLUS, M., *Journal of Chemical Physics*, **48**, 2700; **49**, 4048
[e] LIU, B., *Journal of Chemical Physics*, **58**, 1925 (1973); SIEGBAHN, P., and LIU, B., *Journal of Chemical Physics*, **68**, 2457 (1978)

show that the energy barrier must certainly lie below 43.0 kJ mol^{-1} (by the variation theorem) and he estimates that it is no lower than 39.7 kJ mol^{-1}. Other properties of these potentials for H_3 will be considered later.

The figures just quoted show conclusively that *a priori* calculations are now capable of 'chemical accuracy'[a], at least for H_3. This accuracy is already being approached for more complex triatomic systems, in particular FH_2 and HF_2. The calculations on the potential for the reaction

$$F + H_2 \rightarrow HF + H \tag{2.30}$$

are particularly interesting, since this is one of the reactions whose dynamical aspects have been most extensively studied. The collinear surface calculated by Bender *et al.*[14] is shown in *Figure 2.4*. The reaction path for this exoergic

[a] Traditionally, 'chemical accuracy' has been set at 1 kcal/mole. Presumably this transforms to 4 kJ mol^{-1} in SI units!

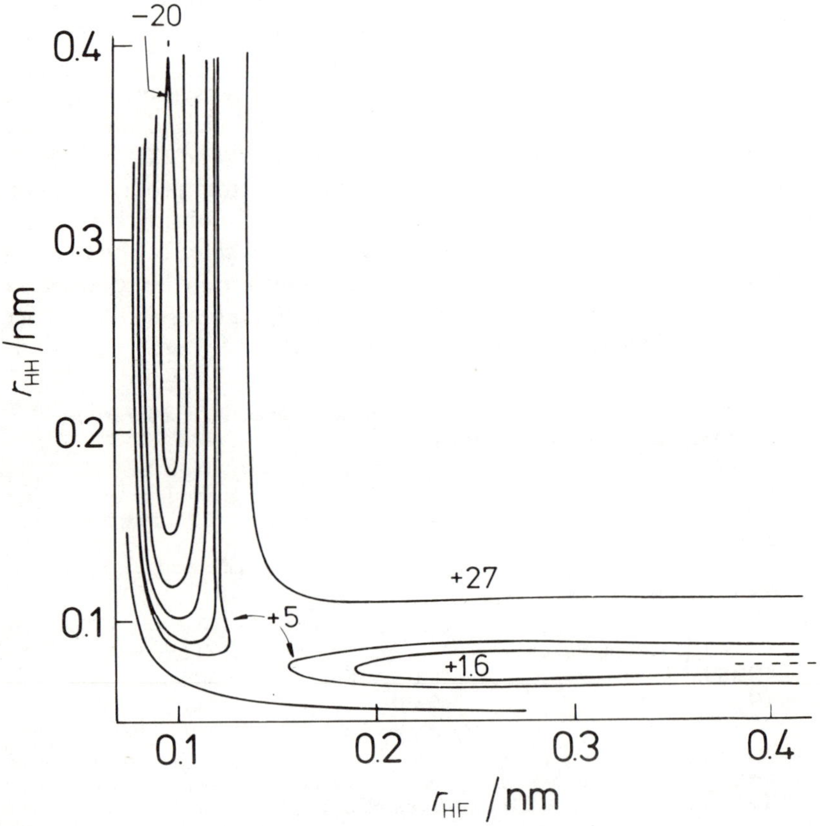

Figure 2.4. Ab initio *potential energy surface for collinear configurations of* FHH [*adapted from Bender, C. F., Pearson, P. K., O'Neill, S. V., and Schaefer III, H. F.,* Journal of Chemical Physics, **56,** 4626 (1972)]. *Contour energies are expressed in kcal/mole [1 kcal/mole = 4.184 kJ mol^{-1}]*

reaction is characterized by a low barrier which appears at a point where r_{HF} is still much greater than $r_{e,HF}$ but $r_{HH} \approx r_{e,H_2}$. The significance of this feature and its implication for the molecular reaction dynamics will be discussed later.

The FH_2 system can also be used to illustrate—to any experimentalist feeling sceptical—the practical value of *a priori* calculations. They predict[20a] (experimental evidence is now available[20b]) a high energy barrier for the transfer of the F atom between the two H atoms, i.e.,

$$H_A + F_B H_C \rightarrow H_A F_B + H_C \tag{2.31}$$

Although this process does not bring about any chemical change, it could provide a mechanism for the rapid relaxation of vibrationally excited HF in the medium of the important chemical laser fuelled by the H_2/F_2 chain reaction. Several dynamical calculations on the rate of relaxation by this reactive pathway have been reported. Unfortunately, the *a priori* calculations indicate that the majority of these have been carried out on potentials that have much too low a barrier to reaction (2.31).

2.3.3 Empirical potentials: the LEPS function

In view of the difficulty of rigorous quantum mechanical calculations, it is scarcely surprising that attempts have been made to represent the potentials for reactive systems by functions incorporating some degree of empiricism. They fall into two broad categories:

(1) those where the empirical information is not drawn from kinetic data, so that the estimated energy barrier can be compared with the experimental activation energy, and
(2) those in which a parameter is adjusted until the calculated and observed activation energies coincide, and then the accuracy of the hypersurface is evaluated either by comparison with the best *a priori* estimate or by calculating other attributes of the reaction, for example, detailed rate constants or the excitation function, and seeing how they compare with the observed quantities.

The two types of empirical surface are exemplified by the BEBO (bond-energy–bond-order) and LEPS (London–Eyring–Polanyi–Sato) potentials.

The LEPS potential is based on the London equation. The contribution of Eyring and Polanyi was to suggest that $Q+J$ could be equated for each diatomic pair to the appropriate Morse potential, and that independent expressions for Q and J could then be obtained by assuming that $Q/(Q+J) = 0.14$, a value based on the Coulombic contribution to the binding energy of H_2 at $r \approx 0.08$ nm. Unfortunately, the barrier height is sensitive to the choice of this parameter and, unless it is reduced to a value (0.07) which gives far too high a barrier (~ 84 kJ mol^{-1}), a well is found at the linear symmetrical configuration just as with the London equation.

An empirical modification which avoids this defect was proposed by Sato[21],

who suggested that a factor $1/(1+\Delta)$ be inserted on the right-hand side of the London equation, yielding

$$\mathscr{V}(r_{AB}, r_{BC}, r_{CA}) = [1/(1+\Delta)]\Big\{Q_{AB}+Q_{BC}+Q_{CA}-\{\tfrac{1}{2}[(J_{AB}-J_{BC})^2 + (J_{BC}-J_{CA})^2+(J_{CA}-J_{AB})^2]\}^{\frac{1}{2}}\Big\} \tag{2.32}$$

Although in this equation Δ clearly bears a *formal* resemblance to S^2, the square of the overlap integral in the Heitler–London equation (2.11), it should properly be viewed as no more than an adjustable parameter. Sato further proposed that expressions for the Q and J should be derived by equating $(Q+J)/(1+\Delta)$ to the appropriate Morse function [equation (2.23)] and $(Q-J)/(1-\Delta)$ to an 'anti-Morse' function:

$$\mathscr{V}(r) = (D_e/2)\{\exp(-2\beta\Delta r)+2\exp(-\beta\Delta r)\} \tag{2.33}$$

This was suggested because the curve calculated for the first excited $^3\Sigma_u^+$ state of H_2 using this expression could be fitted reasonably well to the best theoretical curve for this state, through the range of internuclear separations that are important from the point of view of the H_3 potential.

The LEPS procedure is to be preferred to that of Eyring and Polanyi simply on the grounds that it provides a potential free from spurious basins. There is no more theoretical justification for introducing a constant $1/(1+\Delta)$ factor than there is for assuming $Q/(Q+J)$ to be constant.

There have been more sophisticated attempts to improve the potential based on the London equation. Porter and Karplus's (PK) potential[22] is particularly worthy of note since it has been used in extensive investigations of the collision dynamics of the $H+H_2$ reaction (Section 3.5.3). Porter and Karplus's calculation of the potential was based on a non-ionic VB treatment but several of the approximations that are involved in deriving the London equation were not made. Thus, the three atomic orbitals were not assumed to be orthogonal, but overlap integrals and three-centre exchange integrals were evaluated. The values of the integrals which made the largest contributions to the total energy were determined from accurate $^1\Sigma_g^+$ and $^3\Sigma_u^+$ potential curves, using an

Table 2.6 Comparison of properties at the barrier of the H_3 potential as predicted by semi-empirical and theoretical calculations

Method	$\mathscr{V}^*$ / kJ mol^{-1}	$r^*_{AB} = r^*_{BC}$ / nm	$\tilde{\nu}^*$ / cm^{-1}
BEBO with $p = 1.041$	41.2	0.0922	2330, 695, 1700*i*
LEPS with $\Delta = 0.134$	41.2	0.0930	1858, 883, 2042*i*
Porter-Karplus[a]	38.2	0.0900	2190, 980, 2220*i*
SSMK[b]	46.0	0.0915	2024, 972, 1550*i*
Liu[b]	41.0	0.0930	2047, 913, 1513*i*

[a] PORTER, R. N., and KARPLUS, M., *Journal of Chemical Physics*, **40**, 1105 (1964)
[b] References given under *Table 2.5*.

improved version of Sato's method, and a final adjustment to the barrier energy was made to bring it into line with the experimental activation energy. The properties of the predicted transition state are compared with the best quantum mechanical calculations in *Table 2.6*. The surface corresponding to collinear configurations of the three atoms was shown in *Figure 1.2*.

2.3.4 Empirical potentials: the BEBO function[23, 24]

The BEBO method for constructing a potential differs from the LEPS method in three main ways. First, although the energy is considered, as in the LEPS potential, to arise only from pairwise interactions, the method of estimating the interactions is not the same. Secondly, the expression for the potential is in no way related to any theoretical expression, such as the London equation. Finally, the basic data are drawn entirely from non-kinetic information, so there are no adjustable parameters and predicted and observed expressions for the rate constant can be compared.

The BEBO model is of a reaction occurring via configurations in which there are partial or fractional bonds between A and B and between B and C. The method is confined to reactions where B is an H (or D) atom, although A and C may be polyatomic. At the outset of a reactive collision there is a single bond between B and C and zero bonding between A and B, whilst if reaction occurs the AB bond order eventually becomes unity and that of BC zero. The basic postulate is that all positions along the reaction path of minimum energy (defined by the coordinate x) correspond to a total bond order of unity, this being the sum of the orders of the fractional A–B and B–C bonds. In addition, it is argued that the electronic structure can be simplistically represented by the two resonance forms

$$\begin{array}{ccc} \uparrow & \downarrow & \uparrow \\ \mathrm{A} & \mathrm{B}\!-\!\!\!- & \mathrm{C} \end{array}$$

and

$$\begin{array}{ccc} \uparrow & \downarrow & \uparrow \\ \mathrm{A}\!-\!\!\!- & \mathrm{B} & \mathrm{C} \end{array}$$

where the directions of the arrows denote electrons of different spin. Since the electrons on A and C have the same spin, their interaction is supposed to have an antibonding effect, giving rise to the third term in the BEBO expression for the potential energy along x:

$$\mathscr{V}(x) = -E_{AB}(r_{AB}) - E_{BC}(r_{BC}) + \mathscr{V}_{rep}(r_{CA}) \tag{2.34}$$

E_{AB} and E_{BC} are the energies associated with the partial bonds between A + B and B + C and $\mathscr{V}_{rep}(r_{CA})$ is the repulsive energy between C + A.

For $\mathscr{V}_{rep}(r_{CA})$, Johnston multiplied Sato's anti-Morse function (2.33) by 0.5, arguing that this led to a closer fit to the best calculation on $H_2(^3\Sigma_u^+)$ in the

important range of internuclear separation between 0.1 and 0.2 nm. Estimates of the partial bond energies are based on empirical information about the relationships between bond energy, bond length, and bond order. These relationships and their validity are considered in the next few paragraphs.

Broadly speaking, it is reasonable to suppose that a number of molecular parameters alter in a predictable way when the strength of a chemical bond is changed. Thus, if a molecule is promoted to an electronic state with a lower dissociation energy, it is natural to expect that the force constant, and therefore $\tilde{\nu}_e$, will also decrease, but that r_e will increase. A large number of empirical relationships connecting these various quantities for different species, as well as for different states of the same molecule, have been proposed. One of the most successful has been Badger's rule, which relates force constant (f_e) to equilibrium internuclear separation (r_e) by the equation

$$r_e = a_{ij} - b_{ij} \log f_e \tag{2.35}$$

where a_{ij} and b_{ij} are constants for molecules formed by combining atoms from ith and jth rows of the Periodic Table.

To use the BEBO equation (2.34), a correlation is required between partial bond energy and internuclear separation. Johnston 'derived' such a relationship through the 'bond order'. The concept of a bond order is difficult to define rigorously, but is one that chemists learn to grasp almost intuitively. In its simplest terms, it corresponds to half the sum of the electrons held in common by the two species, those in bonding orbitals counting +1, those in non-bonding orbitals 0, and those in anti-bonding orbitals −1.

For bonds of order between 1 and 3 (for C bound to C, C to O, and O to O), there is quite a good log–log correlation between the bond energy (D_e) and bond order (n), i.e.,

$$D_e = D_{e,s} n^p \tag{2.36}$$

where $D_{e,s}$ is the energy of the single bond between a particular pair of atoms. In addition, Johnston made use of an equation originally suggested by Pauling relating n to bond length:

$$r = r_s - 0.26 \ln n \tag{2.37}$$

so that combining these two equations gave

$$\ln (D_e/D_{e,s}) = p(r_s - r)/0.26 \tag{2.38}$$

In order to use this equation to evaluate *partial* bond energies where $r > r_s$, the additional assumption was made that this relationship is not only valid for $n \geqslant 1$ but also holds as n approaches zero. As a consequence, the parameter p was evaluated, by substituting for D_e and r in equation (2.38) the Lennard–Jones parameters ε_{LJ} and r_m, for the noble-gas pair belonging to the same horizontal rows of the Periodic Table as the partially bound species in question; so

$$p = 0.26 \ln (D_{e,s}/\varepsilon_{LJ})/(r_m - r_s) \tag{2.39}$$

The use of the Lennard–Jones parameters was justified on the grounds that the force constants calculated from these potentials appeared to obey Badger's rule quite well, and the values of ε_{LJ} and r_m also fitted a smooth (but not linear) extrapolation of bond energy versus bond length for C–C and N–N bonds. As no information on the Lennard–Jones potentials for *mixed* pairs of noble-gas atoms, for example He + Ne, was available at the time the **BEBO** method was proposed, ε_{LJ} and r_m were evaluated, for these cases, using the combining rules:

$$r_m(\text{He–Ne}) = 0.5\{r_m(\text{He–He}) + r_m(\text{Ne–Ne})\} \qquad (2.40a)$$

and

$$\varepsilon_{LJ}(\text{He–Ne}) = \{\varepsilon_{LJ}(\text{He–He})\varepsilon_{LJ}(\text{Ne–Ne})\}^{\frac{1}{2}} \qquad (2.40b)$$

Combination of equations (2.34) and (2.36) and addition of $D_{e,BC}$, so that the potential energy is referred to separated A + BC rather than A + B + C, yields

$$\begin{aligned}\mathscr{V}(x) &= D_{e,BC} - D_{e,BC}n^{p_{BC}} - D_{e,AB}(1-n)^{p_{AB}} + \mathscr{V}_{rep}(r_{CA}) \\ &= D_{e,BC}(1-n^{p_{BC}}) - D_{e,AB}(1-n)^{p_{AB}} + \mathscr{V}_{rep}(r_{CA}) \qquad (2.41)\end{aligned}$$

where n and $(1-n)$ are the BC and AB bond orders. Using Pauling's relationship (2.37) and the fact that $r_{CA} = r_{AB} + r_{BC}$ for collinear configurations, r_{AB}, r_{BC} and $\mathscr{V}(x)$ can be evaluated at values of n from 1 to 0. The maximum value of $\mathscr{V}(x)$ corresponds to the predicted *energy barrier* for the reaction.

To determine *activation energies* for comparison with experimental results, use should be made of transition state theory (Chapter 4). At the time the **BEBO** method was proposed, it was claimed to predict activation energies within about 2 kcal/mole ($\sim$ 8 kJ mol^{-1}) for most H atom-transfer reactions, but since then more instances have been discovered where the agreement is poorer. Partly because of this, a number of modifications and re-evaluations of the **BEBO** method have been carried out since it was first proposed. Quite recently, attention was drawn[25] to the fact that equations (2.40a) and (2.40b) predict values of r_m and ε_{LJ} for mixed noble-gas atom pairs, differing significantly from those which have been accurately determined from measurements of elastic scattering in molecular beam experiments (*see Table 2.4*). If the latter data are used in equation (2.38) to find values of the parameter p, the predicted activated energies are changed considerably for the worse. For example, for the $H + H_2$ reaction, the calculated energy barrier ($\mathscr{V}^*$) is lowered from 41.4 to 27.6 kJ mol^{-1}. Even more recently, however, this gloomy prognosis has been disputed[26], it being pointed out that other small modifications of the **BEBO** technique, especially adjusting the numerical factor in equation (2.37) to 0.28, restores and even improves the success of the method.

What the various evaluations clearly, if inadvertently, demonstrate is the sensitivity of the **BEBO** predictions to the values chosen for the p parameters, since relatively large changes in the predicted barrier heights are brought about

by an alteration of only a few percent in these factors. Since it is extremely doubtful whether the numerical factor in Pauling's relationship remains constant for a variety of partially bound pairs, the earlier quantitative agreement of the BEBO predictions with experimental data must be seen as slightly fortuitous. The BEBO method provides a fine example of soundly based empiricism, and it has already assisted a generation of kineticists to acquire a clear understanding of the nature of atom-transfer reactions. However, it now seems unlikely that additional insight is to be gained by further tinkering with the basic equations. Furthermore, it appears doubtful if a straightforward BEBO calculation can provide a more reliable prediction of an unknown activation energy than the educated guess of an experienced kineticist who has access to data on related reactions. An alternative use of the BEBO expression is to have a single value of p in equation (2.41) which can be adjusted, like Δ in the LEPS equation, to bring the predicted and observed activation energies into coincidence. However, the use that can be made of this potential is limited, since the BEBO procedure does not generate even an energy surface, let alone a full hypersurface.

2.3.5 Comparison of potentials for H_3

Table 2.6 compares the properties of various H_3 surfaces at the energy barrier; an extensive summary of calculations up to 1966 can be found in Johnston's book[24]. The agreement between the first three values of $\mathscr{V}^*$ that are listed is scarcely surprising since semi-empirical potentials can be adjusted to give this coincidence (Porter and Karplus's value for $\mathscr{V}^*$ is low because they fitted their potential to an earlier estimate of the 'true' barrier height). Comparison of the other quantities in the table indicates how alike the potentials are in the region of the energy barrier.

All the methods predict very similar internuclear distances in the transition state—and agree that it is linear. A further test that can be made on the surfaces at this point is to compare their curvature, i.e., the second derivatives of $\mathscr{V}$ with respect to the normal mode coordinates. A normal mode analysis yields three frequencies. One of these is imaginary, reflecting the fact that the system is unstable with respect to the linear asymmetric displacement, i.e., there is a *negative* restoring force. The real frequencies are associated with motions that in a stable molecule would correspond to the symmetric stretching and doubly degenerate bending vibrations. In the standard form of transition state theory these motions are assumed to be separable, those leading to real frequencies are quantized, and the zero-point energies associated with them are important in determining the activation energy of the reaction. Furthermore, within this separable approximation, the potential energy profile along the reaction coordinate, the reduced mass associated with motion along this direction, and, therefore, the imaginary frequency are related to the extent of quantum mechanical tunnelling through the barrier (Sections 3.6 and 4.4.2).

The frequencies predicted by the five potentials referred to in *Table 2.5* are in quite good agreement. The most important difference would appear to be

between values of the imaginary frequency, which the semi-empirical potentials predict to be rather high. This corresponds to a narrower barrier, and could lead to the effects of tunnelling being overestimated. More detailed comparisons of the last three of these potentials have been carried out by Shavitt, Stevens, Minn and Karplus[18] and by Liu[19].

2.3.6 General properties of potentials for reactive systems

Attempts to relate the activation energies, or energy barriers, for a family of similar elementary reactions to their heats of reaction were first made in the 1930s by M. Polanyi and his associates, and they have continued since that time. One of the best known correlations was made by Fettis and Knox[27], who showed that the activation energies of the *endoergic* reactions of Cl, Br and I with H_2 and its isotopic analogues were given by the expression

$$E_{act} \approx \alpha(-\Delta H_0^\circ) + c \tag{2.42}$$

where ΔH_0° is the enthalpy change for reaction in the exoergic direction, $\alpha = 0.91$, and $c = 18$ kJ mol^{-1}. They proposed that this correlation should hold for H atom transfer reactions from alkanes (RH) to halogen atoms (X):

$$\mathrm{R + HX \leftarrow RH + X} \tag{2.43}$$

Then, using the measured activation energies for these reactions, they deduced thermodynamic data which could not be obtained by other methods at that time. Because $\Delta H_0^\circ \approx (E_{act,\,f} - E_{act,\,r})$, the relationship

$$E_{act} = (1-\alpha)\Delta H_0^\circ + c \tag{2.44}$$

should hold for the *exoergic* reactions occurring in the opposite direction to those considered by Fettis and Knox. In practice, the real test of any relationship of this type is its ability to predict activation energies for exoergic reactions, since, for endoergic reactions, $-\Delta H_0^\circ$ nearly always comprises a large fraction of E_{act} and the likelihood of a sizeable fractional error is reduced.

Mok and Polanyi[28], after a thorough investigation of LEPS potentials, suggested that a suitable relationship between the height of the energy barrier and the difference between the electronic potential energy of product and reactants ($\Delta\mathscr{V}$) might be

$$\log \mathscr{V}^* = \alpha' \Delta\mathscr{V} + \log c' \tag{2.45}$$

which has the merit that $\mathscr{V}^*$ tends to zero as $\Delta\mathscr{V} \to -\infty$. They also explored how other features of the potential hypersurface alter as parameters in the LEPS expression are systematically changed. As well as the energy barrier decreasing as the exoergicity of the reaction was increased, the barrier moved to a progressively 'earlier' position along the reaction path. This finding can be explained most easily by reference to simple diagrams showing the variation in potential energy along the reaction path of minimum energy. At some point on this path $(r_{AB}/r_{BC}) = (r_{e,AB}/r_{e,BC})$, and we can consider that this separates a region where the system is best described as 'approaching reactants' from one

where the 'products separate'. For exoergic reactions, Mok and Polanyi found that the barrier lies on the reactants side of the $(r_{AB}/r_{BC}) = (r_{e,AB}/r_{e,BC})$ line, that is at the head of the 'entry valley' on a surface such as that shown in *Figure 2.4*. As the exoergicity is increased, other parameters staying the same, the barrier moves to a position further from the separating line. This behaviour is illustrated in *Figure 2.5*. The general features for endoergic reactions can, of course, be obtained by suitably reversing what has been said for exoergic reactions. The diagrams in *Figure 2.5* are then 'read' from right to left.

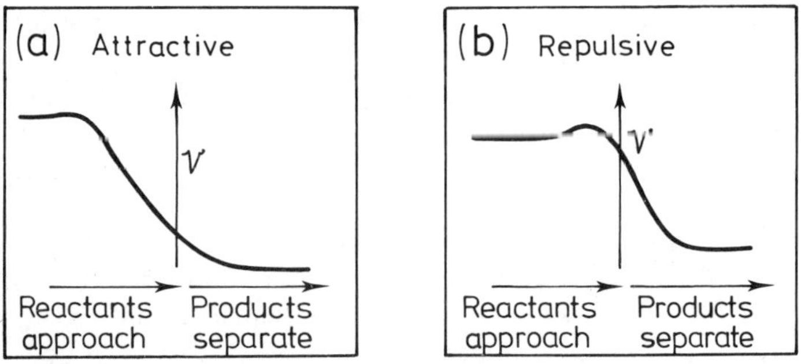

Figure 2.5. Diagram illustrating the release of energy during exoergic chemical reactions [adapted from Polanyi, J. C., Accounts of Chemical Research, **5**, 161 (1972)]. *The bold lines represent the change in potential along the reaction path of minimum energy. On the attractive surface, most of the energy is released as the reactants approach and before the point is reached where $r_{AB}/r_{e,AB} = r_{BC}/r_{e,BC}$. The less exoergic reaction has a repulsive surface with most of the energy being released as the products separate*

The simple diagrams in *Figure 2.5* can also be used to illustrate a distinction between different types of potential surface that is important in discussions of reactive collision dynamics that will come later[29]. In case (b), where the barrier is close to the separating line, the potential energy has fallen little from its initial (reactant) value at the time this line is crossed. Energy is released largely as the products of reaction separate and this surface is broadly classed as being *repulsive*. On the other hand, in case (a), a large fraction of the potential energy is released before the separating line is crossed, i.e., as the reagents approach, and this surface is termed *attractive*. We shall see that these different kinds of potential can lead to quite different modes of energy disposal between the products of simple bimolecular reactions.

2.4 Electronic states of molecules and the adiabatic correlation rules[3, 4, 30—32]

So far in this chapter emphasis has been placed on the theoretical and empirical determination of how the electronic potential energy of a *single*

molecular state varies with the nuclear dispositions. In this section, we consider how one can predict what electronic states arise when molecular fragments of known electronic structure are brought together, and how one can use this knowledge in interpreting the results of molecular collisions. The fact that the interaction between two atoms with non-closed electronic shells gives rise to more than one molecular electronic state has already been mentioned and it emerged clearly when calculations on H_2 were being discussed. These showed that two potential curves correlates with $H(1^2S_{\frac{1}{2}}) + H(1^2S_{\frac{1}{2}})$. It was demonstrated that the ground bound state corresponded to a single spin–orbital molecular wavefunction which was assigned the term symbol $^1\Sigma_g^+$, whereas the upper unbound state was shown to be triply degenerate, since three spin functions could combine with the orbital wavefunction with the products satisfying the requirements of the Pauli exclusion principle [equation (2.21)].

The *correlation rules* for determining the molecular states that arise when two atoms interact were derived almost 50 years ago by Wigner and Witmer and they continue to be of immense value. Extension of these rules to polyatomic systems and the systematic application of the results to collisional phenomena are more recent developments[30–32]. As our concern is largely with reactions involving light atoms, only the correlation rules that apply to systems where Russell–Saunders coupling is a good approximation will be considered here. The validity of Russell–Saunders coupling implies that the total electronic wavefunction can be written as a product of spin and spatial parts without significant loss of accuracy, so that S and L, which specify the electronic spin and orbital angular momenta, are 'good' quantum numbers. First a brief resumé is given of the Wigner–Witmer rules for diatomic systems.

2.4.1 The Wigner–Witmer rules for diatomic systems[3,4]

In the line joining the nuclei, diatomic molecules possess a unique internal axis with respect to which the symmetry of internal electric fields can be defined. As a consequence of this, it is the *component* of the total electronic orbital angular momentum, rather than the overall value, that is important in defining the electronic state of a diatomic system. This component, $\mathbf{\Lambda}$, has magnitude Λ, and the value of Λ is indicated in the term symbol by the use of Σ if $\Lambda = 0$, $\Pi(\Lambda = 1)$, $\Delta(\Lambda = 2)$, etc. The first part of *Table* 2.7 lists the most important cases derived from the correlation rules for calculating Λ.

In contrast to orbital momentum, electron spin in an atom is unaffected by the axially symmetric electric field that is generated by the approach of a second atom. As a result the correlation rules for spin are particularly straightforward and hold in the same form for polyatomic systems. The molecular spin, S, can take all values that can be obtained by combining the individual spins, S_1 and S_2, i.e.

$$S = |S_1 + S_2|, |S_1 + S_2 - 1| \ldots\ldots |S_1 - S_2| \tag{2.46}$$

Examples of the *multiplicities* ($= 2S + 1$) that arise on combining fragments

Table 2.7 Correlation table for (i) atom + atom, and (ii) atom + linear molecule to give species of C_s symmetry, in the approximation that L and S are good quantum numbers

	S_g^a / S_u	P_g / P_u	D_g / D_u
(i)			
S_g	Σ^+	Σ^-, Π	Σ^+, Π, Δ
	Σ^-	Σ^+, Π	Σ^-, Π, Δ
P_g	Σ^-, Π	$\Sigma^+(2), \Sigma^-, \Pi(2), \Delta$	$\Sigma^+, \Sigma^-(2), \Pi(3), \Delta(2), \Phi$
	Σ^+, Π	$\Sigma^+, \Sigma^-(2), \Pi(2), \Delta$	$\Sigma^+(2), \Sigma^-, \Pi(3), \Delta(2), \Phi$
D_g	Σ^+, Π, Δ	$\Sigma^+, \Sigma^-(2), \Pi(3), \Delta(2), \Phi$	$\Sigma^+(3), \Sigma^-(2), \Pi(4), \Delta(3), \Phi(2), \Gamma$
	Σ^-, Π, Δ	$\Sigma^-(2), \Sigma^+, \Pi(3), \Delta(2), \Phi$	$\Sigma^-(2), \Sigma^+(3), \Pi(4), \Delta(3), \Phi(2), \Gamma$
(ii)			
Σ^+	A'	$A' + 2A''$	$3A' + 2A''$
	A''	$2A' + A''$	$2A' + 3A''$
Σ^-	A''	$2A' + A''$	$2A' + 3A''$
	A'	$A' + A''$	$3A' + 2A''$
Π, Δ, etc.	$A' + A''$	$3A' + 3A''$	$5A' + 5A''$
	$A' + A''$	$3A' + 3A''$	$5A' + 5A''$

[a] Combination of two ungerade atoms gives the same as that of two gerade atoms

with various individual multiplicities are given in *Table 2.8*. Term symbols for all the molecular states that correlate with state-specified atoms can be obtained by taking all permissible combinations from the two tables.

Table 2.8 Correlation of atomic and molecular multiplicities

Separated species	*Multiplicities of resultant molecule*
singlet + singlet	singlet
singlet + doublet	doublet
singlet + triplet	triplet
doublet + doublet	singlet, triplet
doublet + triplet	doublet, quartet
doublet + quartet	triplet, quintet
triplet + triplet	singlet, triplet, quintet
triplet + quartet	doublet, quartet, sextet
quartet + quartet	singlet, triplet, quintet, septet

Whilst considering the electronic states of diatomic molecules, it is important to consider to what extent the 'zero approximation' wavefunctions provide an adequate description of the electronic states, particularly in the regions where the potentials are close together. An important part of the answer to this problem is supplied by the *non-crossing rule*, which states that potential curves of the same species, i.e., identical symmetry type, never cross. The physical explanation for this rule is quite straightforward. We have already seen that one way to improve the approximate wavefunctions for a molecular state is to allow for configuration interaction, that is to take mixtures of the zero approximation wavefunctions for different states. The requirement that it should not be possible to distinguish the electrons in the combined wavefunction means that the states that are mixed must be of the same symmetry species. It is this requirement which prevents the wavefunction for the $^3\Sigma_u^+$ state of H_2 being included in a CI calculation on the ground electronic state. The lowest excited state wavefunction which possesses the correct symmetry is that for the $(1\sigma_u^*)^2\ ^1\Sigma_g^+$ state, which correlates with separated ions. Because of the high ionization potential and low electron affinity of the H atom, this ionic state lies far above the ground state of the molecule. Consequently, the extent of the mixing is slight, the energies of the two states are only perturbed by a relatively small amount compared with their 'initial' separation, and the essential characteristics of the two states as covalent and ionic are retained.

In the case of NaCl, which is illustrated in *Figure 2.6*, the situation is quite different. The energy of $Na^+ + Cl^-$ is only 137 kJ mol^{-1} above that of the separated atoms. Both give rise to $^1\Sigma^+$ states, but the long-range Coulombic attraction causes the ionic curve to fall much more rapidly at large r. In zero approximation these curves would cross at $r \approx 1$ nm. In a higher order of

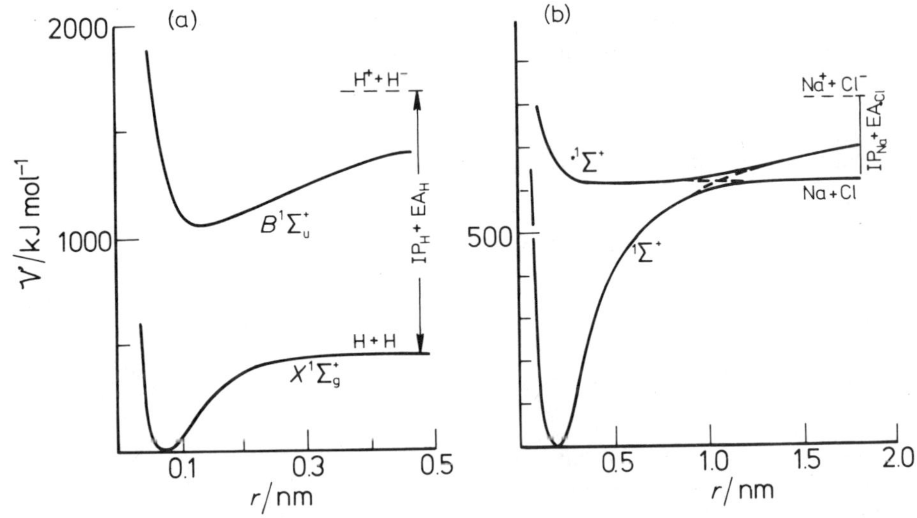

Figure 2.6. Potential curves of the bound states resulting from ground state atoms and ions in (a) H_2 *and* (b) NaCl [*adapted from Figure* 170 *in Herzberg, G.,* Molecular Spectra and Molecular Structure. I. Spectra of Diatomic Molecules, *Van Nostrand, Princeton* (1950)]. *In* (b) *the adiabatic states are represented by full curves and the magnitude of the interaction is slightly exaggerated for clarity. The energy difference represented by the vertical arrows is that between the ionization potential* (IP) *and electron affinity* (EA) *of the atoms.* [*The original from which Figure 2(b) is adapted is copyright by Litton Educational Publishing Inc., and is reprinted by permission of Van Nostrand Reinhold Company*]

approximation, however, the electronic wavefunctions can 'mix' strongly in this region. The result of such a perturbation is *always* to push the levels apart and this mutual interaction will be strongest at the avoided crossing point. Consequently, the curves avoid one another in the way shown in *Figure 2.6*, where the adiabatic potentials are represented by the solid lines.

Potential curves of different electronic type can cross. Perturbations between states associated with these curves may still occur, but only to the extent that the Born–Oppenheimer or the Russell–Saunders approximations break down, so that perturbations arise as a result of terms in the total Hamiltonian which are associated with coupling between electronic and nuclear motions or with spin–orbit coupling.

2.4.2 Adiabatic correlations for triatomic systems[30—32]

For three atoms, rules exist for correlating the electronic states of the whole system both with those of all three separated atoms, i.e., A + B + C, as well as with a smaller number of fragments, e.g., A + BC. The spin correlation rules, which are particularly useful and important, remain the same as for diatomic systems. In considering the adiabatic correlation of orbital symmetry, only that of the least symmetrical configuration is of interest, since this is the situation at nearly all points during a molecular collision. This least symmetrical configuration has only a single plane of symmetry and belongs to

the point group $\mathbf{C}_s$. All molecular wave functions must be either symmetric, i.e., remain the same, or antisymmetric, i.e., change sign, with respect to reflection in this plane. Thus, there are two, and only two, kinds of electronic state, given the symbols A′ and A″. The correlation of atom plus diatomic molecule states with these species are listed in *Table 2.7*.

For triatomic and larger systems, the non-crossing rule does not *strictly* apply. Nevertheless, the larger number of degrees of freedom gives rise to a situation such that the non-crossing of states of the same type does effectively hold. This is usually assumed in constructing correlation diagrams to establish whether collisions between species in selected electronic states can lead to products in specific states via an electronically adiabatic pathway, i.e., on a single hypersurface. Of course, as with diatomic systems, non-adiabatic transitions between electronic states can take place under the influence of coupling between electronic and nuclear motions or spin and orbital motions. A reduced probability should be expected for these non-adiabatic processes, but they may still provide the dominant channel for reaction if the energy barrier associated with the adiabatic route is high.

The benefits and limitations of applying the adiabatic correlation rules to collisional processes are best illustrated by some examples. Although the rules have been most useful in helping to interpret the quenching of electronically excited atoms[31], here we shall emphasize chemical reactions of ground state species.

2.4.3 Application of the adiabatic correlation rules

To find whether there is an adiabatic route between reactants and products in specified electronic states is perfectly straightforward. One first uses the tables to construct lists of all the compound states that correlate with (1) the reactants, and (2) the products. The states common to both lists constitute the adiabatic pathways.

As examples, let us take the reactions of $N(^4S)+NO(^2\Pi)$ and of $O(^3P)+CS(^1\Sigma^+)$. The following products are energetically possible:

$$N(^4S)+NO(^2\Pi) \rightarrow N_2(^1\Sigma_g^+)+O(^3P);\ \Delta H_0^\circ = -313.8\ \text{kJ mol}^{-1} \tag{2.47a}$$

$$\rightarrow N_2(^1\Sigma_g^+)+O(^1D);\ \Delta H_0^\circ = -124.0\ \text{kJ mol}^{-1} \tag{2.47b}$$

and

$$O(^3P)+CS(^1\Sigma^+) \rightarrow CO(^1\Sigma^+)+S(^3P);\ \Delta H_0^\circ = -356\ \text{kJ mol}^{-1} \tag{2.48a}$$

$$CO(^1\Sigma^+)+S(^1D);\ \Delta H_0^\circ = -245\ \text{kJ mol}^{-1} \tag{2.48b}$$

$$CO(^1\Sigma^+)+S(^1S);\ \Delta H_0^\circ = -90\ \text{kJ mol}^{-1} \tag{2.48c}$$

In both cases, application of the spin correlation rules alone is sufficient to suggest strongly that production of the excited atoms will not occur, and this confirms what is found in practice. The spin conservation rule also requires that the reactions yielding products in their ground electronic states must occur across *triplet* potential hypersurfaces. Consequently, neither reaction should occur via the ground states of N_2O and OCS, both of which are singlets and which would, in the context of these reactions, correspond to deeply bound collision complexes.

The ground electronic state of $N_2O(^1\Sigma^+)$ and $OCS(^1\Sigma^+)$ actually correlate with $N_2(^1\Sigma_g^+)+O(^1D)$ and $CO(^1\Sigma^+)+S(^1D)$. This situation has a considerable effect on the kinetics of the dissociation–association reactions. In fact, because the electronically adiabatic process, e.g.,

$$OCS(^1\Sigma^+) \rightarrow CO(^1\Sigma^+)+S(^1D); \quad \Delta H_0^\circ = +415.5 \text{ kJ mol}^{-1} \qquad (2.49a)$$

would have a much higher energy barrier than the non-adiabatic reaction,

$$OCS(^1\Sigma^+) \rightarrow CO(^1\Sigma^+)+S(^3P); \quad \Delta H_0^\circ = +305.0 \text{ kJ mol}^{-1} \qquad (2.49b)$$

the reactions do occur via the non-adiabatic channel. However, they are much slower than they would be if the triatomic molecules correlated directly with ground state fragments. This point will be considered further in Section 4.4.3.

As a final example of the application of the adiabatic correlation rules, we take the reactions of oxygen atoms with H_2 for which a partial correlation diagram is presented in *Figure 2.7.* In this system, the ground state reactants, $O(^3P)+H_2(^1\Sigma_g^+)$, correlate adiabatically with two $^3A'$ and one $^3A''$ state, and the ground state products with $^1A'$, $^1A''$, $^3A'$ and $^3A''$. Clearly two

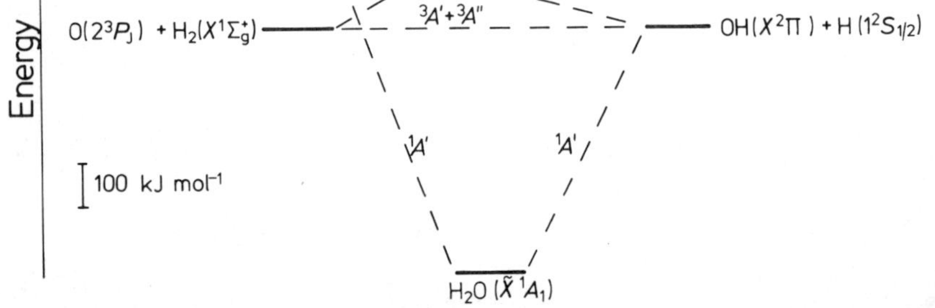

Figure 2.7. Correlation diagram connecting some of the states of $O+H_2$ *and* OH + H *[adapted from Donovan, R. J., and Husain, D.,* Chemical Reviews, **70**, 489 (1970)]. *The energy of the* $^4\Sigma^-$ *state of* OH *is uncertain*

adiabatic pathways ($^3A' + {}^3A''$) connect these species. In addition, there are singlet adiabatic routes from $O(^1D) + H_2(^1\Sigma_g^+)$ to $OH(^2\Pi) + H(^2S)$, one of which correlates with the ground state of H_2O. Consequently, the reactions of both $O(^3P)$ and $O(^1D)$ could be expected to yield $OH(^2\Pi)$ and $H(^2S)$. In practice, the reaction of the ground-state atoms is found to be rather slow with a high activation energy (37.2 kJ mol^{-1}), whereas that of $O(^1D)$ goes extremely rapidly and yields $OH(^2\Pi) + H(^2S)$, rather than $O(^3P) + H_2(^1\Sigma_g^+)$, with essentially unit probability.

Hopefully, these few examples have indicated what the adiabatic correlation rules *cannot* predict as well as what they can. Thus, they do not of themselves provide much guide concerning the *nature* of the adiabatic potentials. For example, they do not suggest why there are apparently large barriers on the $^3A'$ and $^3A''$ surfaces connecting $O(^3P) + H_2(^1\Sigma_g^+)$ and $OH(^2\Pi) + H(^2S)$. Furthermore, the approximations, such as the assumption of Russell–Saunders coupling, on which they are based must be borne in mind. The rules provide no quantitative guide as to how rigorously, for example, conservation of S will apply.

References

Section 2.2

1 PAULY, H., and TOENNIES, J. P., *Advances in Atomic and Molecular Physics*, **1**, 195 (1965)
2 *Intermolecular Forces*. Ed. Hirschfelder, J. O., John Wiley, New York (1967), vol. XII in the series *Advances in Chemical Physics*
3 HERZBERG, G., *Molecular Spectra and Molecular Structure. I. Spectra of Diatomic Molecules*, 2nd edn., Van Nostrand, New York (1950)
4 MURRELL, J. N., KETTLE, S. F. A., and TEDDER, J. M., *Valence Theory*, John Wiley, London (1965)
5 MULLIKEN, R. S., and ERMLER, W. C., *Diatomic Molecules: Results of ab initio Calculations*, Academic Press, New York (1977)
6 KOLOS, W., and WOLNIEWICZ, L., *Journal of Chemical Physics*, **43**, 2429 (1965) and **49**, 404 (1968)
7 MASON, E. A., and MONCHICK, L., in ref. 2, p. 351
8 CHEN, C. H., SISKA, P. E., and LEE, Y. T., *Journal of Chemical Physics*, **59**, 601 (1973)
9 BARKER, J. A., WATTS, R. O., LEE, J. K., SCHAFER, T. P., and LEE, Y. T., *Journal of Chemical Physics*, **61**, 3081 (1974), and references therein

Section 2.3

10 BALINT-KURTI, G. G., in *Molecular Scattering: Physical and Chemical Applications*, Ed. Lawley, K. P., John Wiley, New York (1975), vol. XXX in the series *Advances in Chemical Physics*
11 EYRING, H., and LIN, S. H., in *Physical Chemistry: An Advanced Treatise*, Eds Eyring, H., Henderson, D, and Jost, W., vol. VI A, chap. 3, Academic Press, New York (1974)
12 KUNTZ, P. J., in *Dynamics of Molecular Collisions*, Part B, Ed. Miller, W. H., chap. 2, Plenum, New York (1976)
13 CSIZMADIA, I. G., KARI, R. E., POLANYI, J. C., ROACH, A. C., and ROBB, M. A., *Journal of Chemical Physics*, **52**, 6205 (1970)
14 BENDER, C. F., PEARSON, P. K., O'NEILL, S. V., and SCHAEFER III, H. F., *Journal of Chemical Physics*, **56**, 4626 (1972)
15 WALSH, A. D., *Journal of the Chemical Society*, 2260–2332 (1953)

16 LAIDLER, K. J., and POLANYI, J. C., *Progress in Reaction Kinetics*, **3,** 1 (1965)
17 BOYS, S. F., and SHAVITT, I., University of Wisconsin Naval Research Laboratory Technical Report WIS-AF-13, 1959
18 SHAVITT, I., STEVENS, R. M., MINN, F. L., and KARPLUS, M., *Journal of Chemical Physics*, **48,** 2700; **49,** 4048
19 LIU, B., *Journal of Chemical Physics*, **58,** 1925 (1973); SIEGBAHN, P., and LIU, B., *Journal of Chemical Physics*, **68,** 2457 (1978)
20 (*a*) BENDER, C. F., GARRISON, B. J., and SCHAEFER III, H. F., *Journal of Chemical Physics*, **62,** 1188 (1975); (*b*) (added in proof) BARTOSZEK, F. E., MANOS, D. M., and POLANYI, J. C., *Journal of Chemical Physics*, **69**, 933 (1978)
21 SATO, S., *Journal of Chemical Physics*, **23,** 592, 2465 (1955)
22 PORTER, R. N., and KARPLUS, M., *Journal of Chemical Physics*, **40,** 1105 (1964)
23 JOHNSTON, H. S., and PARR, C., *Journal of the American Chemical Society*, **85,** 2544 (1963)
24 JOHNSTON, H. S., *Gas Phase Reaction Rate Theory*, Ronald, New York (1966)
25 JORDAN, R. M., and KAUFMAN, F., *Journal of Chemical Physics*, **63,** 1691 (1975)
26 GILLIOM, R. D., *Journal of Chemical Physics*, **65,** 5027 (1976)
27 FETTIS, G. C., and KNOX, J. H., *Progress in Reaction Kinetics*, **2,** 1 (1964)
28 MOK, M. H., and POLANYI, J. C., *Journal of Chemical Physics*, **51,** 1451 (1969)
29 POLANYI, J. C., *Accounts of Chemical Research*, **5,** 161 (1972)

Section 2.4

30 SHULER, K. E., *Journal of Chemical Physics*, **21,** 624 (1953)
31 DONOVAN, R. J., and HUSAIN, D., *Chemical Reviews*, **70,** 489 (1970)
32 MAHAN, B. H., *Accounts of Chemical Research*, **8,** 55 (1975)

Chapter 3

Molecular collision dynamics

3.1 General introduction

In Chapter 2 we were concerned with the first stage in the fully detailed theory of collisional reactions that was outlined in *Figure 1.4*: that is, the calculation of potential energy hypersurfaces for chemically reactive systems. In this chapter, we turn our attention to the other stages in that theoretical approach: that is, how one can calculate the result of individual fully specified collisions, and then proceed from these results to estimates of the cross-sections and rate constants that describe the results of practical, less than fully specified, experiments.

Studies of molecular collision dynamics using realistic (if not exact) potential hypersurfaces and the *classical* laws of motion have now become almost routine. On the other hand, *semi-classical* and *quantum mechanical* scattering calculations are in an earlier stage of development and remain the prerogative of the specialist. Comparison of the results from these calculations with those where classical mechanics have been employed are particularly important in showing up the circumstances in which the latter are unlikely to provide a proper description of the dynamics. These situations will be discussed in Section 3.6. For the moment, we simply note that it appears that classical mechanics can usually provide an adequate description of reactive collisions, and that this is particularly so for the great majority of reactions for which detailed microscopic data are available. It is the classical approach that is considered first and emphasized in this chapter.

3.2 General principles of classical dynamics

To solve most problems in classical dynamics, it is convenient to use Hamilton's equations, which specify the state of a three-dimensional system comprised of N particles in terms of $3N$ positional coordinates $(q_1, q_2, \ldots\ldots q_{3N})$ and the same number of *conjugate momenta* $(p_1, p_2, \ldots\ldots p_{3N})$. Usually the potential energy of the system depends only on the q; for molecular systems this corresponds to making the Born–Oppenheimer assumption, whereas the kinetic energy, T, is generally a

function of both these coordinates and their time derivatives, i.e., $\dot{q}_1, \dot{q}_2, \ldots\ldots \dot{q}_{3N}$, the components of the particle velocities. Since the conjugate momenta are defined by

$$p_i = \partial T(q_1, q_2, \ldots\ldots q_{3N}; \dot{q}_1, \dot{q}_2, \ldots\ldots \dot{q}_{3N})/\partial \dot{q}_i \tag{3.1}$$

T can be expressed in terms of the p and q rather than the $\dot{q}$ and q. The sum of the kinetic and potential energies expressed in the form:

$$\begin{aligned} &H(q_1, q_2, \ldots\ldots q_{3N}; p_1, p_2, \ldots\ldots p_{3N}) \\ &\quad = T(q_1, q_2, \ldots\ldots q_{3N}; p_1, p_2, \ldots\ldots p_{3N}) + \mathscr{V}(q_1, q_2, \ldots\ldots q_{3N}) \end{aligned} \tag{3.2}$$

is called the *classical Hamiltonian* for the system. If the particles only interact with one another, H must remain constant and the system is said to be *conservative.*

The classical *equations of motion* in their Hamiltonian form are:

$$(\partial H/\partial p_i) = \dot{q}_i \tag{3.3a}$$

and

$$(\partial H/\partial q_i) = -\dot{p}_i \tag{3.3b}$$

For a given interaction potential and selected values of p_i and q_i at some initial time, the classical laws of motion completely define the subsequent dynamical behaviour of a system and the classical path or *trajectory* of the collision can be calculated by numerically integrating the coupled set of $6N$ equations represented by (3.3a) and (3.3b). On a modern high-speed computer this can be carried out in an acceptable time (about one second) for molecular collisions involving several particles.

Before computing a trajectory, the task is usually simplified by taking account of two general principles. First, if any of the coordinates, for example q_j, does not appear explicitly in the expression for H,

$$(\partial H/\partial q_i) = -\dot{p}_j = 0 \tag{3.4}$$

so p_j does not alter with time. The momentum p_j is therefore a *constant of the motion* and the coordinate q_j is *ignorable* Secondly, if the total Hamiltonian can be written as the sum of two sets of terms, i.e., $H = H_1 + H_2$, without any of the subscripts on the p and q appearing in both H_1 and H_2, H is said to be *separable* and H_1 and H_2 must individually remain constant.

A general and important example of separability is when the coordinates are chosen so that three of them specify the position of the centre-of-mass of the entire system. Because, if no external field of force is present, linear momentum must be conserved, the terms in H which involve the momentum components of the motion of the centre-of-mass can be separated, so that

$$H_{\text{total}} = H_{\text{CM}} + H_{\text{rel}} \tag{3.5}$$

and hence $(6N-6)$ equations describe the *relative* motion of the particles. In principle, this number can be reduced still further by separating terms in the

components of total angular momentum, since these must also be conserved, but generally the algebra that is required is formidable and this operation is rarely carried out.

3.3 Classical collisions between two particles[1—4]

Collisions between two particles are characterized by an interaction potential that depends only on their separation. At the molecular level, this is strictly true only for collisions between two atoms, and then, unless the system undergoes a radiative transition between electronic states (Section 5.8.1), the atoms will scatter elastically whatever the form of the potential. Nevertheless, it is useful to consider such collisions at some length, partly as an introduction to more complex systems, and also because some ultrasimplified theories of collisional reactions are based on the assumption of two-body behaviour.

The system under examination is one of two atoms of mass m_A and m_B. Their positions can be described by means of six Cartesian coordinates $q_1 \ldots\ldots q_6$ and their motion by either the velocities $\mathbf{v}_A$ and $\mathbf{v}_B$ or the momenta conjugate to the q, $p_1 \ldots\ldots p_6$. The classical Hamiltonian is

$$H = (1/2m_A) \sum_{i=1}^{3} p_i^2 + (1/2m_B) \sum_{i=1}^{3} p_{i+3}^2 + \mathscr{V}(q_1 \ldots\ldots q_6) \quad (3.6)$$

The choice of a more suitable set of coordinates is suggested (1) by the need to separate and then ignore the motion of the centre-of-mass, and (2) by noting that r, the separation of the two particles, is given by

$$r = \left\{ \sum_{i=1}^{3} (q_{i+3} - q_i)^2 \right\}^{\frac{1}{2}} \quad (3.7)$$

Accordingly, new coordinates are defined by the equations

$$Q_i = q_{i+3} - q_i \quad (3.8)$$

and

$$Q_{i+3} = (1/M)\,(m_A q_i + m_B q_{i+3}) \quad (3.9)$$

where $M = (m_A + m_B)$. Equation (3.9) gives the Cartesian coordinates of the centre-of-mass and (3.8) those of B relative to A as origin. $\mathscr{V}$ is a function only of the Q_i, so according to the earlier definition and in agreement with momentum conservation, P_{i+3}, the momentum components of the centre-of-mass, are constant. To relate the new momenta to the old, it is necessary to use the equation for a point-contact transformation

$$p_i = \sum_j P_j(\partial Q_j/\partial q_i) \quad (3.10)$$

This yields

$$p_i = -P_i + P_{i+3}\,(m_A/M) \quad (3.11)$$

and

$$p_{i+3} = P_i + P_{i+3}\,(m_B/M) \tag{3.12}$$

Substitution in equation (3.6) gives

$$H = (1/2\mu) \sum_{i=1}^{3} P_i^2 + (1/2M) \sum_{i=1}^{3} P_{i+3}^2 + \mathscr{V}\,(Q_1 Q_2 Q_3) \tag{3.13}$$

for the total Hamiltonian in terms of the new coordinates and momenta. Ignoring the term in P_{i+3}, the expression obtained for *relative* motion of A and B only is

$$H_{rel} = (1/2\mu) \sum_{i=1}^{3} P_i^2 + \mathscr{V}(Q_1 Q_2 Q_3) \tag{3.14}$$

This equation is identical to that for a single particle of mass μ and velocity $\boldsymbol{v} = \boldsymbol{v}_A - \boldsymbol{v}_B$ moving in a potential which is centrosymmetric, i.e., a function only of the distance of the particle from the origin of the coordinates.

Equation (3.14) can be simplified further by taking account of the need to conserve (orbital) angular momentum. The vector $\boldsymbol{l}$ representing this momentum is perpendicular to the plane containing $\boldsymbol{v}$ and the origin of the system, and since its direction, as well as its length, must not change during the course of a collision, it is clear that it must be possible to represent the whole of one trajectory on a single plane, as shown in *Figure 3.1*. If Cartesian coordinates

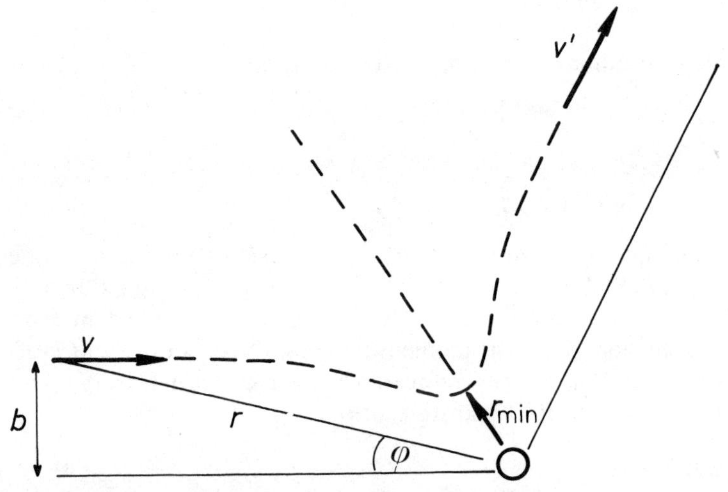

Figure 3.1. Reduction of a two-body elastic collision to the equivalent one-particle problem

are chosen, with the z-axis perpendicular to the plane of the trajectory, the kinetic energy is given by

$$T = \tfrac{1}{2}\mu(\dot{x}^2 + \dot{y}^2) \tag{3.15}$$

A trivial transformation to polar coordinates yields

$$T = \tfrac{1}{2}\mu(\dot{r}^2 + r^2\dot{\phi}^2) \tag{3.16}$$

and the Hamiltonian for relative motion can be written as

$$H_{\text{rel}} = \tfrac{1}{2}\mu(\dot{r}^2 + r^2\dot{\phi}^2) + \mathscr{V}(r) \tag{3.17}$$

The original problem involving two particles in three dimensions has now been reduced to an equivalent, one-body, two-dimensional problem. Two more things are worth noting about equations (3.16) and (3.17): first, the expression for T includes the position coordinate r as well as $\dot{r}$ and $\dot{\phi}$ [cf. equation (3.2)]; secondly, because the second term in the Hamiltonian involves r and $\dot{\phi}$, H_{rel} cannot be separated.

A collision between A and B can be characterized by two parameters: v, the initial magnitude of the relative velocity, and b, the *impact parameter*, which is the perpendicular distance from the extension of $\boldsymbol{v}$ to the origin, i.e., the distance of closest approach in the absence of any interaction between the particles. The collision energy, $\varepsilon_T = \tfrac{1}{2}\mu v^2$, is unchanged in an elastic collision, and the trajectory is symmetrical about a line joining the origin to r_{min}, the actual distance of closest approach during the collision, as is shown in *Figure 3.1*. For particular values of v and b, the deflection caused by the collision is (classically) uniquely defined.

It is possible to derive an equation from which the magnitude of the deflection can be calculated, starting from the laws of conservation of angular momentum and energy written in terms of the polar coordinates. At the outset of a collision (i.e., when $r = \infty$, $\phi = 0$), the orbital angular momentum is equal to μbv and the total energy is $\tfrac{1}{2}\mu v^2$, so

$$\mu bv = \mu r^2\dot{\phi} \tag{3.18}$$

and

$$\varepsilon_T = \tfrac{1}{2}\mu v^2 = \tfrac{1}{2}\mu(\dot{r}^2 + r^2\dot{\phi}^2) + \mathscr{V}(r) \tag{3.19}$$

Elimination of $\dot{\phi}$ yields

$$\varepsilon_T = \tfrac{1}{2}\mu\dot{r}^2 + \varepsilon_T b^2/r^2 + \mathscr{V}(r) \tag{3.20}$$

This equation represents the one-dimensional motion of a single particle of mass μ and total energy ε_T in an *effective potential*,

$$\mathscr{V}_{\text{eff}}(r) = \varepsilon_T b^2/r^2 + \mathscr{V}(r) \tag{3.21}$$

The term $\varepsilon_T b^2/r^2$ is frequently called the *centrifugal potential*, and the whole expression gives rise to a family of curves of the type shown in *Figure 3.2*.

When equations (3.18) and (3.20) are rearranged to provide expressions for $\dot{\phi}$ and $\dot{r}$ and these equations are divided into one another, the result is

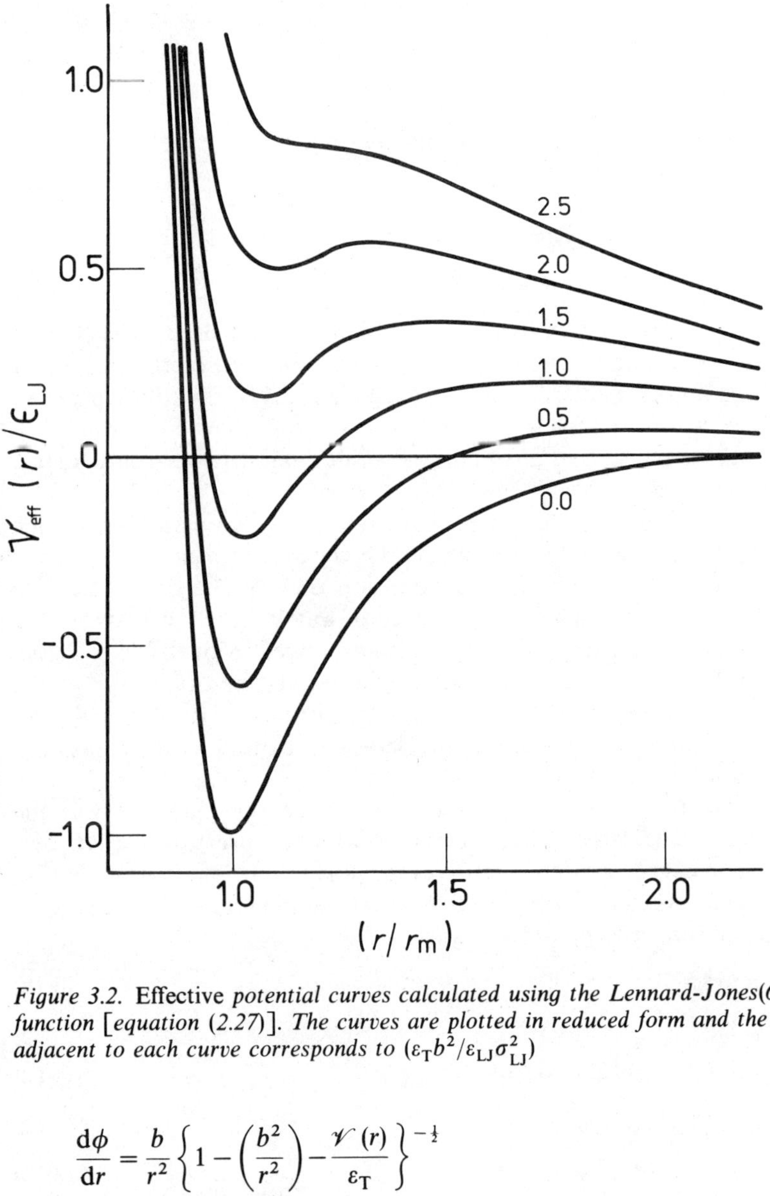

Figure 3.2. Effective *potential curves calculated using the Lennard-Jones(6–12) function [equation (2.27)]. The curves are plotted in reduced form and the number adjacent to each curve corresponds to* $(\varepsilon_T b^2/\varepsilon_{LJ}\sigma_{LJ}^2)$

$$\frac{d\phi}{dr} = \frac{b}{r^2}\left\{1 - \left(\frac{b^2}{r^2}\right) - \frac{\mathscr{V}(r)}{\varepsilon_T}\right\}^{-\frac{1}{2}} \tag{3.22}$$

Because of the symmetry of the trajectory about ϕ_{min}, the total angle of deflection χ is given by

$$\chi = \pi - 2\phi_{min} \tag{3.23}$$

and hence χ can be calculated by evaluating

$$\chi = \pi - 2b\int_{\infty}^{r_{min}} \frac{dr}{r^2\{1 - (b^2/r^2) - \mathscr{V}(r)/\varepsilon_T\}^{\frac{1}{2}}} \tag{3.24}$$

The upper limit to the integration in this equation is found by remembering that at the turning point of the collision where $r = r_{min}$ and $\phi = \phi_{min}$, the radial velocity $\dot{r}$ is zero. Therefore, from equation (3.20)

$$\varepsilon_T = \varepsilon_T b^2/r_{min}^2 + \mathscr{V}(r_{min}) \tag{3.25}$$

and so

$$r_{min} = b[1 - \mathscr{V}(r_{min})/\varepsilon_T]^{-\frac{1}{2}} \tag{3.26}$$

In order to determine χ as a function of b for collisions with a particular collision energy and interaction potential, the integration in equation (3.24) has usually to be carried out numerically.

Before discussing how χ varies with b, it is useful to consider how the angular scattering distributions derived from experiment are presented. This is done in terms of a differential scattering cross-section which is defined by the expression [cf. equation (1.26)]:

$$\sigma(\varepsilon_T, \theta) = dS/d\omega \tag{3.27}$$

where S is the total scattering cross-section (at the specified collision energy ε_T), $d\omega$ is an element of solid angle, and θ is the angle between the final and initial velocity vectors and is termed the *scattering angle.*

The difference between θ and χ is that, whereas for classical collisions at low energy, χ can range from π to $-\infty$ (corresponding to *orbiting* of the two particles), θ can only sensibly take values between π and zero. Since it is impossible in reality to restrict all collisions to a single plane, the azimuthal positions of the $\boldsymbol{v}$ vectors about the line $b = 0$ are distributed randomly, and

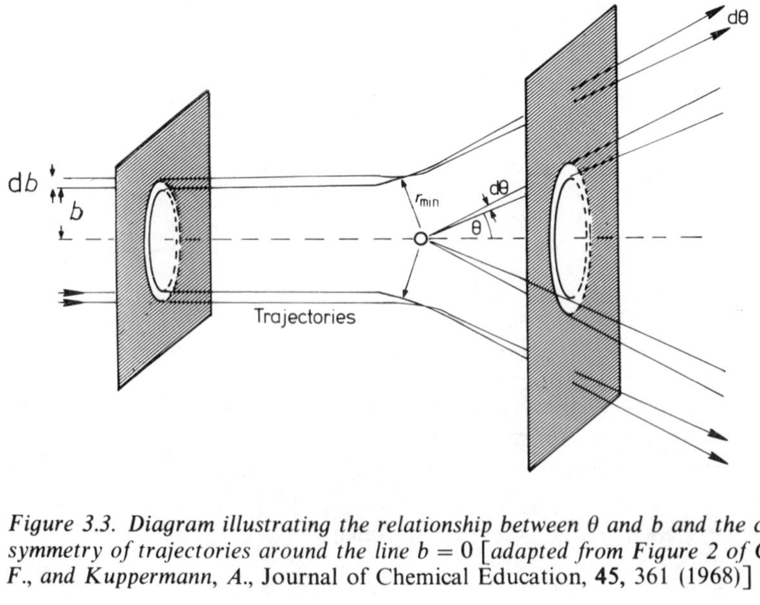

Figure 3.3. Diagram illustrating the relationship between θ and b and the cylindrical symmetry of trajectories around the line $b = 0$ [adapted from Figure 2 of Greene. E. F., and Kuppermann, A., Journal of Chemical Education, **45**, 361 (1968)]

the resultant scattering must be cylindrically symmetrical about this line—and therefore about the direction of the initial relative velocity vector. This is illustrated in *Figure 3.3*.

That different values of χ can correspond to the same value of θ is demonstrated in *Figure 3.4*. This shows a trajectory (A) in which the particles after appreciating long-range attractive forces rebound from one another as they run up against the repulsive wall of the potential. Trajectory (B), on the other hand, has a larger impact parameter and consequently only the influence of the attractive forces is felt. In practice, a detector placed at the scattering angle θ will be unable to distinguish between the results of collisions characterized by these different impact parameters. This effect will be referred to again when elastic scattering from realistic intermolecular potentials is considered in Sections 3.3.3 and 3.3.4.

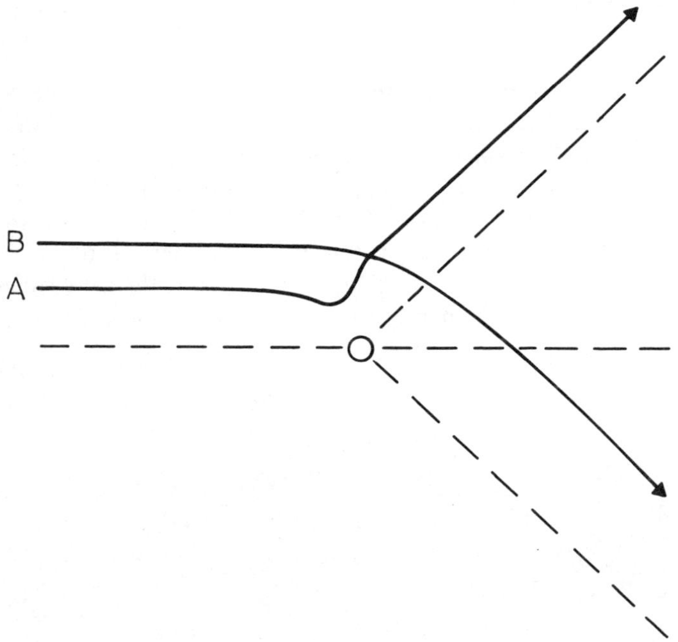

Figure 3.4. Diagram showing two elastic collisions with different impact parameters that have the same value of θ, the scattering angle

The relationship between $d\omega$ and $d\theta$, the elements of solid angle and of the scattering angle, can be established by simple geometrical arguments. The area of the elemental strip shown shaded on the spherical surface in *Figure 3.5* is $r^2\,d\omega$; alternatively, it can be calculated by taking the product of the width of the strip ($r\,d\theta$) and its length ($2\pi r \sin\theta$). Hence

$$d\omega = 2\pi \sin\theta\, d\theta \tag{3.28}$$

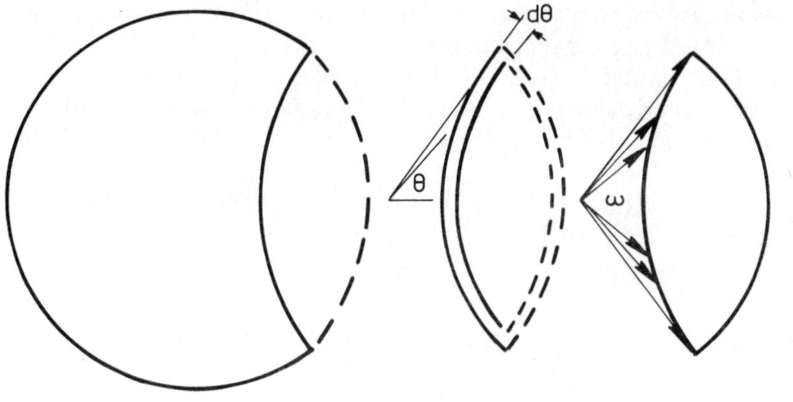

Figure 3.5. Diagram illustrating the relationship between the scattering angle θ and the solid angle ω

and so

$$\sigma(\varepsilon_T, \theta) = dS/d\omega$$
$$= dS/2\pi \sin\theta \, d\theta \qquad (3.29)$$

[cf. equation (1.26)]. Since $\sin\theta$ is zero when θ is equal to π or zero, it is obvious that $\sigma(\varepsilon_T, \theta)$ is infinite at these values of θ, unless $dS/d\theta$ converges to zero at least as rapidly as $\sin\theta$. Partly to offset this effect, scattering data are frequently presented by plotting $\sigma(\varepsilon_T, \theta)\sin\theta$, rather than $\sigma(\varepsilon_T\ \theta)$, against θ.

Just as S, the total scattering cross-section, can be thought of as a target area over which the representative mass point is deflected from its original path, so dS corresponds to a differential element of that area. Because of the cylindrical symmetry of the scattering, illustrated in *Figure 3.4*, the contribution to the total scattering resulting from collisions with impact parameters between b and $b + db$ is given by

$$dS = 2\pi b \, db \qquad (3.30)$$

The total cross-section is therefore given by

$$S = \int_0^{b_{max}} 2\pi b \, db = \pi b_{max}^2 \qquad (3.31)$$

This integration is performed between the limits $b = 0$, corresponding to 'head-on' collisions, and directly 'backward' scattering with $\theta = \pi$, and $b = b_{max}$, the value of the impact parameter at which the angular deflection becomes zero. Finally, in obtaining an expression for $\sigma(\varepsilon_T, \theta)$, it must be remembered that the scattering observed at some values of θ may arise from trajectories with different impact parameters, so that

$$\sigma(\varepsilon_T, \theta) = \sum_i b_i/[\sin\theta \, |(d\theta/db_i)|] \qquad (3.32)$$

where the summation is over the i different values of the impact parameter that

contribute to the observed scattering at the angle θ. The modulus sign arises because $(\mathrm{d}\theta/\mathrm{d}b)$ may be negative but the concept of a negative contribution to the classical differential cross-section is physically unreasonable.

The equations derived in this section permit one to calculate the variation of $\sigma(\varepsilon_T, \theta)$ with impact parameter for any choice of $\mathscr{V}(r)$. Then by suitable integrations, the cross-section and rate constant for the total rate of collisions can be obtained. Before describing the results of these calculations for a realistic potential, it is helpful to consider the nature of the classical scattering from some simplified potentials.

3.3.1 Collisions between 'hard spheres'

The 'hard-sphere' potential was illustrated in *Figure 2.2* and it can be defined by the relationships:

$$\mathscr{V}(r) = 0 \quad \text{for} \quad r > d$$

and

$$\mathscr{V}(r) = \infty \quad \text{for} \quad r \leqslant d \tag{3.33}$$

It is obvious that the total scattering cross-section for hard spheres must be equal to πd^2 at all collision energies. To calculate a rate constant, $Z_{hs}(T)$, associated with the rate of collisions, $S_r(n'|n; \varepsilon_T) = \pi d^2$ must be substituted into equation (1.48) yielding

$$\begin{aligned} Z_{hs}(T) &= (1/\pi\mu)^{\frac{1}{2}}(2/k_B T)^{\frac{3}{2}} \int_0^\infty \pi d^2 \varepsilon_T \exp(-\varepsilon_T/k_B T)\, \mathrm{d}\varepsilon_T \\ &= \pi d^2 (8k_B T/\pi\mu)^{\frac{1}{2}} \\ &= \pi d^2 \langle v_R \rangle \end{aligned} \tag{3.34}$$

where $\langle v \rangle$ is the mean relative velocity of A and B.

The angular distribution of the scattering is also easy to calculate. As there is no attractive component to the interaction potential, χ only takes values between π for $b = 0$ and zero for $b = d$, and χ and θ are identical. Furthermore, as *Figure 3.6* shows, the scattering corresponds to specular reflection, and so there are simple relationships between b, d, and the sine and cosine of α, the angle between the $b = 0$ line and the line joining the origin of the system to the point at which the two particles make contact. Thus

$$\begin{aligned} \chi = \theta &= \pi - 2\alpha \\ &= \pi - 2\sin^{-1}(b/d) \end{aligned} \tag{3.35}$$

To determine the differential cross-section, $\cos\theta$ is expressed in terms of b and d:

$$\cos\theta = -\cos 2\alpha = (2b^2/d^2) - 1 \tag{3.36}$$

which on differentiation yields

$$(\mathrm{d}\theta/\mathrm{d}b) = -(4b/d^2)/\sin\theta \tag{3.37}$$

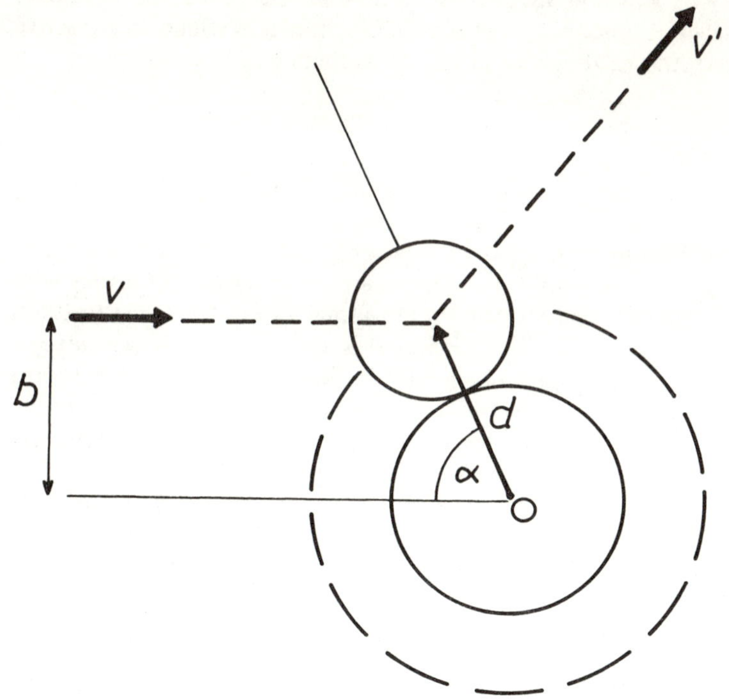

Figure 3.6. Collisions between hard spheres showing specular reflection

Substitution into equation (3.32), with $i = 1$ for all values of θ for collisions between hard spheres, gives

$$\sigma(\varepsilon_T, \theta) = d^2/4 \quad (3.38)$$

This confirms that the differential cross-section is, in this special case, independent of both ε_T and θ. Integration of $\sigma(\varepsilon_T, \theta)$ over the full range of solid angles yields the total cross-section:

$$S(\varepsilon_T) = (d^2/4)\int_0^{4\pi} d\omega = \pi d^2 \quad (3.39)$$

3.3.2 Hard spheres with attraction

The hard-sphere model provides a poor representation of real intermolecular forces ignoring as it does both the long-range attractive interactions and the

compressibility of atoms at short range. For some purposes, the second of these features can be ignored and use made of a model in which an attractive term is added to the hard-sphere potential so that

$$\mathscr{V}(r) = -Cr^{-n} \quad \text{for} \quad r > d$$

and

$$\mathscr{V}(r) = \infty \qquad \text{for} \quad r \leqslant d \tag{3.40}$$

In *classical* collisions controlled by a potential of this type, in which the attractive component only approaches zero as $r \to \infty$, even trajectories with very large impact parameters are deflected to a small extent and hence the total scattering cross-section is infinite. However, it is useful to determine a value of the cross-section that is associated with only those collisions in which the 'hard cores' of the particles come into contact, i.e., r becomes equal to d. From the viewpoint of molecular collisions, the important cases are when n in equation (3.40) is equal to 1, 4, and 6, corresponding to the potentials (*see* Section 2.2.2) between: (1) two oppositely charged ions, (2) an ion and a polarizable molecule, and (3) two neutral molecules. In each instance, it is assumed that the potential depends only on r.

For potentials of the type being considered, the effective potential at distances $r > d$ is given by

$$\mathscr{V}_{\text{eff}}(r) = (\varepsilon_T b^2/r^2) - Cr^{-n} \tag{3.41}$$

The procedure for calculating the 'core scattering cross-section' (S_{core}) depends on whether the effective potential curves possess a centrifugal maximum, as for $n \geqslant 3$, or not, as for $n \leqslant 2$.

For attraction between two oppositely charged ions

$$-Cr^{-n} = z_A z_B e^2/(4\pi\varepsilon_0)r \tag{3.42}$$

At the hard-sphere diameter d, the magnitude of the attractive potential, ε_d, is $-z_A z_B e^2/(4\pi\varepsilon_0)d$, and the total effective potential is

$$\mathscr{V}_{\text{eff}}(d) = (\varepsilon_T b^2/d^2) - \varepsilon_d \tag{3.43}$$

Equating this expression to ε_T defines a maximum value of b for which ions colliding with this energy come into 'hard' contact. Hence

$$\begin{aligned} S_{\text{core}}(\varepsilon_T) &= \pi b^2_{\text{max}} \\ &= \pi d^2 (\varepsilon_T + \varepsilon_d)/\varepsilon_T \end{aligned} \tag{3.44}$$

In contrast to the total scattering cross-section, S_{core} is finite. As ε_T is increased, S_{core} decreases monotonically to a value corresponding to the hard-sphere cross-section.

The expression for $Z(T)$, the rate constant for core collisions, is

$$\begin{aligned} Z(T) &= (1/\pi\mu)^{\frac{1}{2}}(2/k_B T)^{\frac{3}{2}} \int_0^\infty \pi d^2 (\varepsilon_T + \varepsilon_d) \exp(-\varepsilon_T/k_B T) \mathrm{d}\varepsilon_T \\ &= \pi d^2 (8k_B T/\pi\mu)^{\frac{1}{2}} (1 + \varepsilon_d/k_B T) \end{aligned} \tag{3.45}$$

The potential between two unit *point* charges of opposite sign at a separation of 0.3 nm is 7.7×10^{-19} J, i.e., $\varepsilon_d/k_B = 5.6 \times 10^4$ K. Consequently, the rate of core collisions will be two to three orders of magnitude greater than the hard-sphere collision rate as a result of the Coulombic attraction.

When $n \geqslant 3$ the procedure for calculating S_{core} and $Z(T)$ for core collisions is somewhat different. Now the effective potential curves pass through a 'centrifugal maximum', except at very high collision energies, and the cores of the colliding species come into contact whenever this centrifugal barrier is surmounted. If ε_d is again used to denote the magnitude of the attractive potential at $r = d$, then

$$\mathscr{V}_{eff}(r) = (\varepsilon_T b^2/r^2) - \varepsilon_d(d/r)^n \tag{3.46}$$

The calculation of S_{core} then proceeds in three stages. First, r_{max}, the value of r at the centrifugal maximum, is determined by setting $\partial \mathscr{V}_{eff}(r)/\partial r$ equal to zero:

$$0 = -2(\varepsilon_T b^2/r^3_{max}) + (n\varepsilon_d d^n/r^{n+1}_{max}) \tag{3.47}$$

so

$$r_{max} = (n\varepsilon_d d^n/2\varepsilon_T b^2)^{1/(n-2)} \tag{3.48}$$

Secondly, the value of $\mathscr{V}_{eff}$ at r_{max} is found:

$$\mathscr{V}_{eff}(r_{max}) = \varepsilon_T b^2(2\varepsilon_T b^2/n\varepsilon_d d^n)^{2/(n-2)}(1 - 2/n) \tag{3.49}$$

Finally, this expression for $\mathscr{V}_{eff}(r_{max})$ is equated to ε_T to discover the maximum impact parameter for which collisions of this energy will surmount the centrifugal barrier. This yields

$$b^2_{max} = d^2(n\varepsilon_d/2\varepsilon_T)^{2/n}.(n/[n-2])^{(n-2)/n} \tag{3.50}$$

$$S_{core} = \pi d^2(n\varepsilon_d/2\varepsilon_T)^{2/n}.(n/[n-2])^{(n-2)/n} \tag{3.51}$$

Substitution of particular values of n into this equation reduces its complicated appearance. Thus, for $n = 4$,

$$S_{core}(\varepsilon_T) = \pi d^2(4\varepsilon_d/\varepsilon_T)^{\frac{1}{2}} \tag{3.52a}$$

and for $n = 6$

$$S_{core}(\varepsilon_T) = \pi d^2(6.75\varepsilon_d/\varepsilon_T)^{\frac{1}{3}} \tag{3.52b}$$

As before, including the attractive component to the potential enhances the core cross-section over the hard-sphere value and again the core cross-section *decreases* with increasing collision energy. This behaviour is quite general for collisions dominated by long-range attractive interactions. Of course, S_{core} should not decline below πd^2; equations (3.51) and (3.52) apparently predict this, only because the hard-sphere repulsive term was omitted from equation (3.46) for the effective potential. The maximum value of the collision energy for which equations (3.52a) and (3.52b) hold are clearly $\varepsilon_T = 4\varepsilon_d$ and $\varepsilon_T = 6.75\varepsilon_d$ respectively. For higher collision energies, $S_{core}(\varepsilon_T) = \pi d^2$.

For typical ion–induced dipole interactions, ε_d is sufficiently large that, in calculating the rate constant for core collisions, one can use equation (3.52a) and carry out the integration over collision energies from zero to infinity without much loss of accuracy. This yields

$$Z(T) = (1/\pi\mu)^{\frac{1}{2}}(2/k_{\mathrm{B}}T)^{\frac{3}{2}} \int_0^\infty \pi d^2 (4\varepsilon_d/\varepsilon_{\mathrm{T}})^{\frac{1}{2}}\, \varepsilon_{\mathrm{T}} \exp(-\varepsilon_{\mathrm{T}}/k_{\mathrm{B}}T)\, \mathrm{d}\varepsilon_{\mathrm{T}}$$

$$= \pi d^2 (8k_{\mathrm{B}}T/\pi\mu)^{\frac{1}{2}} \int_0^\infty (4\varepsilon_d/k_{\mathrm{B}}T)^{\frac{1}{2}} x^{\frac{1}{2}} \exp(-x)\mathrm{d}x \qquad (3.53a)$$

where

$$x = (\varepsilon_{\mathrm{T}}/k_{\mathrm{B}}T), \text{ so}$$

$$Z(T) = \pi d^2 (8k_{\mathrm{B}}T/\pi\mu)^{\frac{1}{2}}(\varepsilon_d/k_{\mathrm{B}}T)^{\frac{1}{2}}\Gamma(1/2) \qquad (3.53b)$$

For a typical ion–induced dipole interaction, $(\varepsilon_d/k_{\mathrm{B}})$ is about 2.5×10^3 K and, at room temperature, $Z(T)$ is about 5 times larger than $Z_{\mathrm{hs}}(T)$ as a result of the r^{-4} attractive potential. Because ε_d is proportional to d^{-4}, $Z(T)$ does not depend on any assumption as to the value of d; furthermore, as equation (3.53b) clearly indicates, $Z(T)$ is actually independent of temperature.

A similar procedure can be adopted for the case where $n = 6$. The attractive potential then has the same form as the second term in the Lennard-Jones function given in equation (2.27). If d is equated to σ_{LJ} and ε_d to $4\varepsilon_{\mathrm{LJ}}$, the two potentials, i.e., the hard-sphere plus attraction potential and the Lennard-Jones potential, correspond in the limit of large r. Typically $(\varepsilon_{\mathrm{LJ}}/k_{\mathrm{B}}) \sim 120$ K so $(\varepsilon_d/k_{\mathrm{B}}) \sim 5 \times 10^2$ K. Once again a reasonably good approximation to $Z(T)$ for this model potential can be obtained by integrating the expression in equation (3.52b) for the collision cross-section over the full range of collision energies, i.e.,

$$Z(T) = \pi d^2 (8k_{\mathrm{B}}T/\pi\mu)^{\frac{1}{2}} \int_0^\infty (6.75\varepsilon_d/k_{\mathrm{B}}T)^{\frac{1}{3}}\, x^{\frac{2}{3}} \exp(-x)\, \mathrm{d}x$$

$$= \pi d^2 (8k_{\mathrm{B}}T/\pi\mu)^{\frac{1}{2}}(2\varepsilon_d/k_{\mathrm{B}}T)^{\frac{1}{3}}\Gamma(2/3) \qquad (3.54)$$

For $(\varepsilon_d/k_{\mathrm{B}}T) = 5 \times 10^2$, the effect of the attractive potential is to double the collision rate. However, it should be noted that this overestimates the effect of attractive forces in real intermolecular interactions between neutral species because, at finite values of $r > d$, $\mathscr{V}(r)$ for the Lennard-Jones function is greater than that for the hard-sphere plus attraction model, on account of the significant contribution of the first (repulsive) term in the Lennard-Jones expression.

The above calculations have emphasized how the presence of long-range attractive forces can enhance the rate of 'close' collisions. Where this occurs, the cross-section associated with these close collisions decreases with increasing collision energy and the rate constant may decrease as the temperature is

Table 3.1 Sample rate constants for bimolecular collisions controlled by simple potentials

Potential[a]	ε_d/k_B / K	10^{10} Z/cm^3 molecule^{-1} s^{-1} $T = 298$ K	$T = 1000$ K
Hard-spheres:			
$r \leqslant d, \mathscr{V}(r) = \infty$; $r > d, \mathscr{V}(r) = 0$		1.8	3.4
Hard-spheres with attraction:			
(i) $r \leqslant d, \mathscr{V}(r) = \infty$; $r > d, \mathscr{V}(r) = -e^2/(4\pi\varepsilon_0)r$	5.6×10^4	346	191
(ii)[b] $r \leqslant d, \mathscr{V}(r) = \infty$; $r > d, \mathscr{V}(r) = -\alpha e^2/(4\pi\varepsilon_0)^2 2r^4$	2.5×10^3	9.6	9.6
(iii) $r \leqslant d, \mathscr{V}(r) = \infty$; $r > d, \mathscr{V}(r) = -\varepsilon_d d^6/r^6$	4.8×10^2	3.6	4.4

[a] The hard-sphere collision diameter, d, is given the value 0.3 nm
[b] $(\alpha/4\pi\varepsilon_0)$ is assumed to equal 2.5×10^{-3} nm^3.

raised. To indicate the magnitude of the effects some typical data are listed in *Table 3.1.*

3.3.3 Scattering from realistic potentials

The calculations carried out in the previous section are essentially independent of the form assumed for the short-range repulsive part of the potential. Consequently, the results that were obtained there are valid for more realistic forms of the potential, only as long as the behaviour of the system in regard to the quantity being calculated is dominated by the long-range attractive part of the intermolecular potential. However, to calculate the classical deflection function, $\chi(b)$, and hence the differential cross-section as a function of θ, equation (3.24) must be solved. The resultant form of $\chi(b)$ depends on the full form of $\mathscr{V}(r)$, not just the long-range part of the potential.

For any realistic potential function, equation (3.24) can only be solved numerically. To illustrate the form of the solutions, two curves of $\chi(b)$, calculated for a Lennard-Jones potential and two different collision energies are shown in *Figure 3.7.* In both, $\chi(b)$ approaches zero asymptotically as $b \to \infty$, again demonstrating that the *classical* total cross-section for scattering (and the differential cross-section at $\chi = \theta = 0$) is infinite, if the potential contains an attractive component that only becomes zero as $r \to \infty$.

For case A in *Figure 3.7.* χ is greater than $-\pi$ for all values of b and the curve for $\theta(b)$ can be constructed simply by reflecting that part of the curve for which $\chi < 0$ in the $\chi = 0$ axis. At the maximum in this curve $d\theta/db = 0$, so according to equation (3.32), $\sigma(\varepsilon_T, \theta)$ becomes infinite at this scattering angle which is usually referred to as the *rainbow angle.* At values of θ below the rainbow angle, contributions to the differential cross-section come from

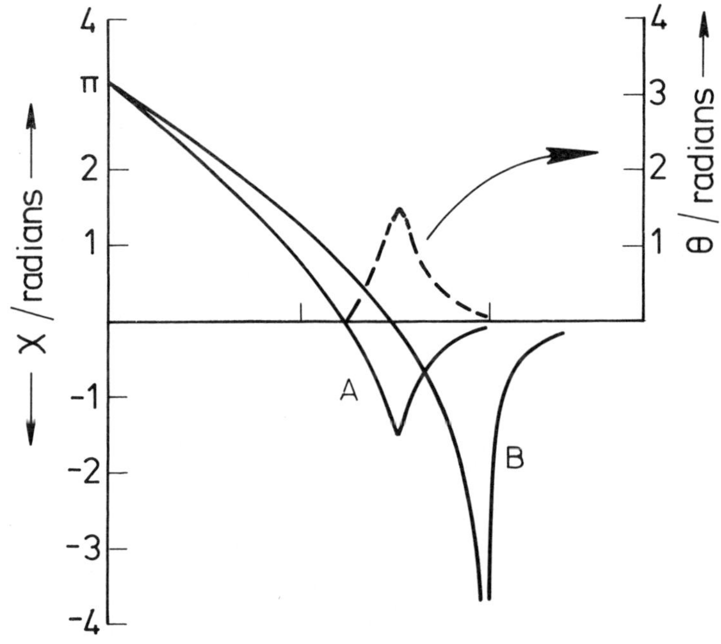

Figure 3.7. The classical deflection χ as a function of (b/σ_{LJ}) for a Lennard-Jones (6–12) *potential* A *with* $(\varepsilon_T/\varepsilon_{LJ}) = 1.6$, B *with* $(\varepsilon_T/\varepsilon_{LJ}) = 0.4$ [*calculated from data in Table I–R of Hirschfelder, J. O., Curtiss, C. F., and Bird, R. B.,* Molecular Theory of Gases and Liquids, *Wiley, New York* (1954)]

collisions with three distinctly different impact parameters. The variation of $\sigma(\varepsilon_T, \theta)$ with θ for this case, as calculated *classically*, is shown as the upper curve in *Figure 3.8*.

Case B represents collisions at a lower collision energy. The major difference is that now, at a certain impact parameter, the collision energy can correspond exactly to the effective potential energy at the centrifugal maximum. *Orbiting* occurs and $\chi \rightarrow -\infty$. It has been suggested that these collisions which classically give rise to an orbiting complex could play an important role in atomic recombination (Section 5.8). At *large* collision energies orbiting does not occur, because ε_T exceeds the height of the centrifugal barrier for all values of b for which a *maximum* appears in the effective potential.

3.3.4 Quantum effects in elastic scattering[3—6]

The analysis of two-body collisions that has been presented up to this point has ignored the fact that the motions of atomic nuclei are only approximately

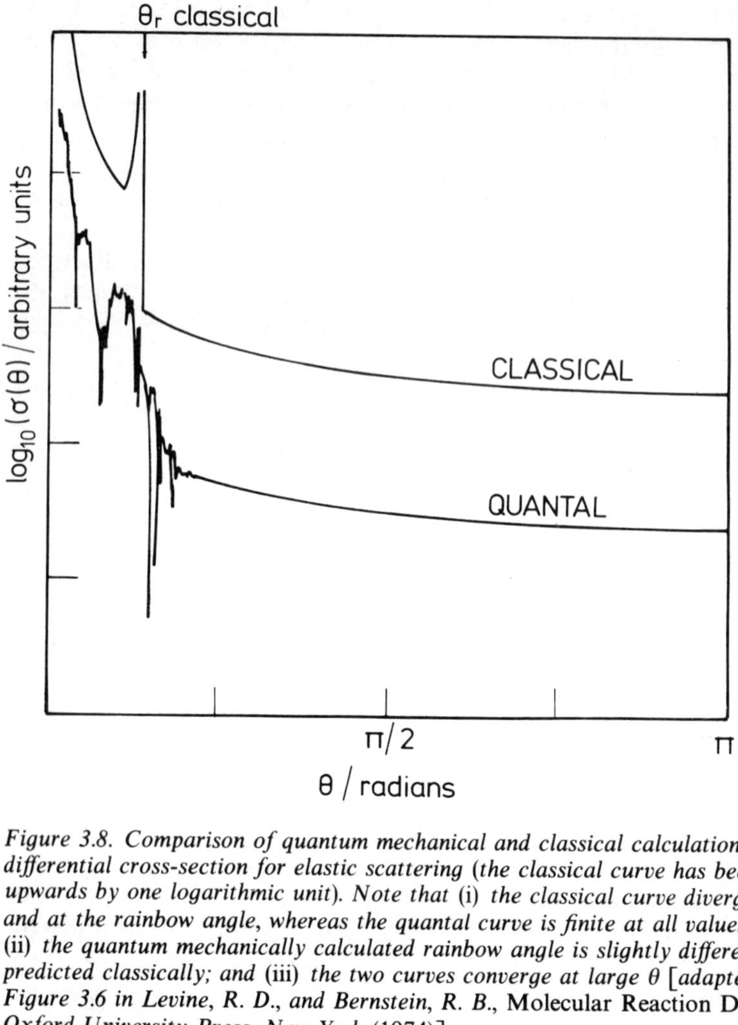

Figure 3.8. Comparison of quantum mechanical and classical calculations of the differential cross-section for elastic scattering (the classical curve has been displaced upwards by one logarithmic unit). Note that (i) *the classical curve diverges at* $\theta = 0$ *and at the rainbow angle, whereas the quantal curve is finite at all values of* θ; (ii) *the quantum mechanically calculated rainbow angle is slightly different from that predicted classically; and* (iii) *the two curves converge at large* θ [*adapted from Figure 3.6 in Levine, R. D., and Bernstein, R. B.,* Molecular Reaction Dynamics, *Oxford University Press, New York* (1974)]

described by the classical equations of motion. In comparison with a quantal analysis of *reactive* collisions, that for elastic scattering is straightforward, but a full description of these calculations would be out of place here. Consequently, all that is given in this section is a brief qualitative description of how and why the predictions of quantum mechanical and classical treatments of elastic scattering differ.

Reference was made in the previous section to some rather 'troublesome' features that arise from a classical analysis of two-body scattering. Among these were the infinite values obtained for the total scattering cross-section and for the differential scattering cross-section at certain values of θ (or b). They disappear when the scattering is treated quantum mechanically. Thus, both the

total cross-section and the differential cross-section at $b = 0$ become finite[a]. This is associated with the fact that, at sufficiently large impact parameters, the classical deflection of the particles would become so small, that the simultaneous observation of the small values of θ and of the relative velocity would conflict with the requirements of the uncertainty principle.

In addition, a quantum mechanical analysis predicts that the behaviour of the scattering in the region of the rainbow angle will be modified from the classical prediction. The classical dynamics of collisions at energies close to the centrifugal maximum in the effective potential curve can be described qualitatively in the following terms. At collision energies just below the potential maximum the two atoms will approach to a minimum separation that is greater than r_{max} given by equation (3.48). The particles feel the influence of the long-range attractive forces and deviate from their original paths in a trajectory of the type illustrated in *Figure 3.4*(B), but their cores do not come into contact. By contrast, in collisions at energies greater than $\mathscr{V}_{eff}(r_{max})$, the two particles surmount the centrifugal barrier and approach much more closely, their cores coming into 'hard' contact. Finally, at ε_T exactly equal to $\mathscr{V}_{eff}(\max)$, an orbiting complex of infinite lifetime is formed. These three contrasting types of classical behaviour are illustrated in *Figure 3.9*.

The application of quantum mechanics modifies the above description of the collision dynamics. It is recognized that a stream of particles colliding at a given energy can be represented on a diagram such as *Figure 3.9* by a plane wave whose wavelength at $r = \infty$ is related to ε_T through the de Broglie relationship, $\lambda = h/p = h/(2\mu\varepsilon_T)^{\frac{1}{2}}$. On solving the wave equation for the radial motion using $\mathscr{V}_{eff}(r)$ for the potential energy, one finds that there is a finite probability of collisions with $\varepsilon_T < \mathscr{V}_{eff}(r_{max})$ penetrating the centrifugal barrier and reaching quasi-bound states of the kind represented by the dashed line in *Figure 3.9*. In addition, at collision energies greater than $\mathscr{V}_{eff}(r_{max})$ there is a finite probability of the particles being 'reflected'. Because the probability of tunnelling through the barrier in either direction is not high, systems that do reach the quasi-bound states will be unusually long-lived relative to those undergoing 'direct' collisions. However, the formation of orbiting complexes of *infinite* lifetime does not occur.

The quantum mechanical treatment of one-dimensional tunnelling is considered later in Section 4.4. The quasi-bound states referred to here are analogous to predissociating states mentioned in Section 5.8.1. Finally, it should be pointed out that a quantum analysis both removes the classically predicted singularity at the rainbow angle and also slightly changes the calculated value of θ at the rainbow maximum.

As well as modifying the classical predictions, quantum mechanical treatments of elastic scattering also predict behaviour of an entirely non-classical kind. Because of the wave nature of molecular motion, molecular

[a] An exception to this general rule is provided by the case of interaction between oppositely charged ions. The Coulombic potential approaches zero so slowly that the quantum mechanical total cross-section is also infinite.

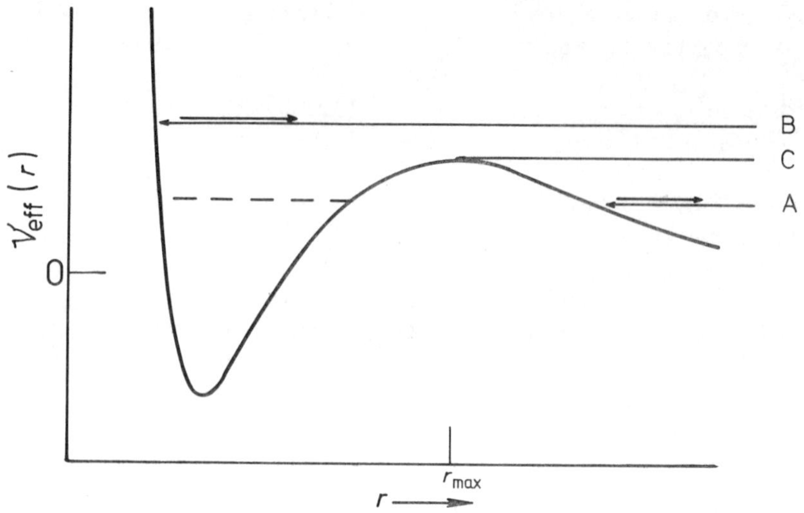

Figure 3.9. Classical two-body collisions at energies close to the maximum on the effective potential curve. In A $\varepsilon_T < \mathscr{V}_{eff}(r_{max})$ *and the cores of the species do not come into contact. By contrast, in* B *the barrier is surmounted and the particles scatter from the repulsive wall of the intermolecular potential. In* C, $\varepsilon_T = \mathscr{V}_{eff}(r_{max})$ *and the particles go into orbit around one another. According to quantum mechanics the system can tunnel through the centrifugal barrier and reach quasi-bound states of the type represented by the dashed line*

collisions are analogous to light scattering, and similar phenomena can be observed. Classically, the observed intensity of the particle flux reaching a detector at some scattering angle θ, is obtained by adding up the *probabilities* associated with different classical trajectories that lead to scattering at the angle θ. This is exemplified by equation (3.32). However, the correct quantum mechanical approach is to determine the *amplitudes* associated with the scattered waves reaching the detector and originating from different initial conditions. To find the resultant intensity, and the value of the differential cross-section, the amplitudes are summed and the square of this sum is proportional to $\sigma(\varepsilon_T, \theta)$. This quantum mechanical superposition of waves of differing phase gives rise to interference effects; two sorts can be distinguished.

First, interference occurs between the results of collisions with similar values of b or l, where l is the quantum number associated with the orbital angular momentum of the two particles and is related to b by $\mu v b = \sqrt{[l(l+1)\hbar]}$. This gives rise to a number of oscillations on the plot of $\sigma(\varepsilon_T, \theta)$ against θ. These are usually referred to as 'supernumerary rainbows' and are most pronounced at angles below the rainbow angle. In addition, there is the interference between waves scattered from different branches of the deflection function that was shown in *Figure 3.7*. As a result further high-frequency oscillations are superimposed on the pattern of supernumerary and main rainbow maxima as shown in *Figure 3.8*. Finally, a quantum mechanical analysis of elastic

scattering predicts that the total scattering cross-section varies with collision energy in a non-monotonic way, the oscillations being strongest at low collision energies.

The detailed 'structure' of these quantum effects depends sensitively on the intermolecular potential, and their accurate observation allows the potential to be constructed without assuming that it follows any particular analytical form. Experimental data can be obtained in experiments utilizing crossed molecular beams. If all the structure in the scattering curve is to be observed, it is essential that the relative velocity of the collisions is highly defined and that the observations are made with good angular resolution, i.e., that the detector subtends only a very small angle at the scattering zone. Such experiments have now been carried out and have provided the most accurate data available on the intermolecular potentials of non-bound, diatomic systems, such as noble-gas pairs (e.g., the data in *Table 2.4*).

3.4 Simple models of reactive collisions

In the crudest models of bimolecular reactions, the reagents are treated as structureless spheres so that the intermolecular potential is assumed to depend only on their separation. Whether reaction occurs in a close collision between the reagents depends on whether the collision energy or some component of it exceeds a critical value. Although these treatments are extremely approximate, they provide a useful introduction to more detailed collisional calculations and, in broad terms, they also correctly predict the different kinds of behaviour that can be expected, on the one hand, for reactions with an energy barrier separating reactants and products and, on the other, for reactions which proceed without an energy barrier.

3.4.1 Simple collision theory

The title 'simple collision theory' (SCT) is generally given to a particular hard-sphere model of reaction between two neutral species. The colliding species are viewed as structureless spheres which do not interact with one another until some critical separation, $r = d$, is reached, at which point the intermolecular potential rises as a step-function. It is usually assumed that reaction occurs at this point if the component of the collision energy directed *along the line-of-centres* of the two colliding particles, ε_{lc}, exceeds a critical value which, according to the model, corresponds to the threshold energy, ε_T^0, for the reaction. An alternative assumption would be that all of the collision energy, ε_T, could be utilized in overcoming the potential barrier, but this would fail to differentiate between collisions which vary in their 'violence of impact'. Thus it seems intrinsically reasonable that reaction is more probable in head-on ($b = 0$) collisions than in 'glancing' collisions with $b \approx d$.

The component of the relative velocity directed towards the origin of the system, v_{lc}, can be deduced easily from *Figure 3.6*:

$$v_{lc} = v\cos\alpha = v\{(d^2 - b^2)/d^2\}^{\frac{1}{2}} \tag{3.55a}$$

so

$$\varepsilon_{lc} = \varepsilon_T(1 - b^2/d^2) \tag{3.55b}$$

For a particular value of ε_T above ε_T^0, as b is increased ε_{lc} decreases, until a value of b is reached at which $\varepsilon_{lc} = \varepsilon_T^0$. This impact parameter corresponds to a maximum value at which reaction can occur at this collision energy; for $b \leqslant b_{max}$ the probability of reaction is unity, whilst for $b > b_{max}$, $P_r(\varepsilon_T, b) = 0$. Equation (3.55b) with the substitution $\varepsilon_T^0 = \varepsilon_{lc}$ can then be used to calculate b_{max}^2, i.e.,

$$b_{max}^2 = d^2(1 - \varepsilon_T^0/\varepsilon_T) \tag{3.56}$$

and hence the reaction cross-section is given by

$$S_r(\varepsilon_T) = \pi b_{max}^2 = \pi d^2(1 - \varepsilon_T^0/\varepsilon_T) \tag{3.57}$$

This expression for the excitation function, which is, of course, only valid for $\varepsilon_T \geqslant \varepsilon_T^0$, rises monotonically from zero at ε_T^0 to approach πd^2 asymptotically as $\varepsilon_T \rightarrow \infty$. Finally, an expression for the thermal rate constant can be obtained:

$$k_{SCT}(T) = (1/\pi\mu)^{\frac{1}{2}}(2/k_BT)^{\frac{3}{2}} \int_{\varepsilon_T^0}^{\infty} \pi d^2.\ (1 - \varepsilon_T^0/\varepsilon_T)\varepsilon_T \exp(-\varepsilon_T/k_BT)\mathrm{d}\varepsilon_T$$

$$= \pi d^2(8k_BT/\pi\mu)^{\frac{1}{2}}\exp(-\varepsilon_T^0/k_BT)$$

$$= Z_{hs}(T)\exp(-\varepsilon_T^0/k_BT) \tag{3.58}$$

where $Z_{hs}(T)$ is the rate constant describing the total rate of collisions, unreactive as well as reactive.

This simplistic treatment of collisional reactions has predicted the form of three quantities, $P_r(\varepsilon_T, b)$, $S_r(\varepsilon_T)$ and $k_{SCT}(T)$. The functional form of the first two of these are shown in *Figure 3.10*, where they are compared with curves calculated according to a modified version of SCT. Equation (3.58) for $k_{SCT}(T)$ shows clearly that, even in this ultrasimple theory, there is no correspondence between ε_T^0 and ε_{act} because the pre-exponential term in (3.58) has a functional dependence on T. Taking $-\mathrm{d}\ln k_{SCT}(T)/\mathrm{d}(1/RT)$ yields

$$E_{act} = E_T^0 + \tfrac{1}{2}RT \tag{3.59}$$

and hence

$$k_{SCT}(T) = Z_{hs}(T)e^{\frac{1}{2}}\exp(-E_{act}/RT) \tag{3.60}$$

Strictly 'tests' of SCT should compare the experimentally determined A factor for a reaction with $Z_{hs}(T)e^{\frac{1}{2}}$.

The model of reaction that is adopted for SCT is so crude that frequently it is cast aside without any pause to reflect on all the assumptions that are made in its derivation. The most glaring of these assumptions—and the one that is often allowed to obscure the others—is that the reactants are structureless spheres so that their orientation relative to one another when they collide does

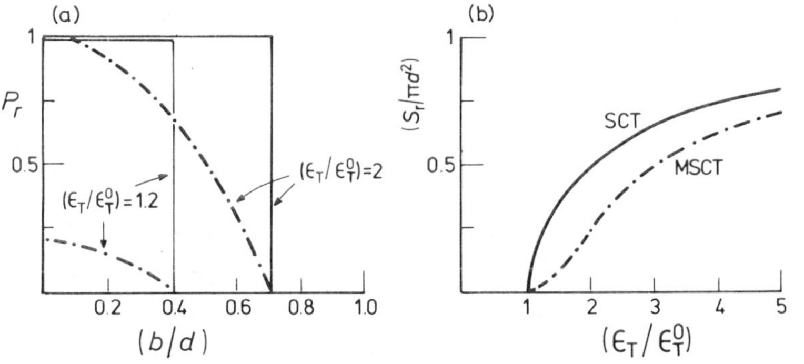

Figure 3.10. Predictions of simple collision theory (——) and of modified simple collision theory (—·—) for (a) *the opacity function, i.e., the dependence of reaction probability on b/d, and* (b) *the reduced excitation function,* i.e., $(S_r/\pi d^2)$ *against* $\varepsilon_T/\varepsilon_T^0$. *For this diagram it is assumed that d = D*

not influence the probability of reaction taking place. To allow for, or at least recognize, this effect, it is common practice to multiply the SCT expression (3.58) for the rate constant by a *steric factor*, *P*, so that

$$k_{SCT}(T) = PZ_{hs}(T)\exp(-\varepsilon_T^0/k_BT) \tag{3.61}$$

After rate data have been obtained, *P* can be calculated from the equation

$$P = A_{exp}(T)/Z_{hs}(T)e^{\frac{1}{2}} \tag{3.62}$$

Alternatively, it might be possible to *estimate* *P* empirically, if results are available on related reactions. However, as Bunker[7] has pointed out in his book, introduction of the steric factor does little more than absorb our ignorance and because of this it might more properly be termed a probability factor. On closer examination, it is clear that a number of effects can help to determine the magnitude of *P*.

What one does in SCT, as in all other less than fully dynamical theories, is to choose a surface in configuration space that one supposes divides reactants from products and then calculate the overall rate at which systems cross this surface. SCT is inadequate chiefly because of the way in which the 'dividing surface' is selected. Its representation in terms of coordinates that describe the separation of the centre-of-masses of the two reactants is quite reasonable. However, the assumption that the potential at this surface depends only on r, the radial coordinate, and not at all on ϕ, the angle defining the orientation of the molecular axis to $\mathbf{r}$, (we assume here that we are dealing with an A + BC → AB + C three-atom reaction) is not. A second unsatisfactory element in SCT is that the dividing surface is usually defined by $r = d$, with d taken to be a collision diameter derived from Lennard-Jones potentials for non-reacting species. However, as was seen in Chapter 2, at the energy barrier on a reactive hypersurface, the nuclei are generally much closer together than in collisions where chemical effects are absent. For example, at the energy barrier on the H_3

surface, $r_{A,BC} = 0.14$ nm, whereas an estimate of the hard-sphere collision diameter for $H + H_2$ would yield ~ 0.35 nm[a].

It is quite straightforward to remove both these 'unreasonable features' from the simple collision theory of reaction rates and to improve it as a result. This modified version of SCT is described before consideration is given to the approximations that remain in both versions of the theory.

3.4.2 Modified simple collision theory

The basic model for the modified version of simple collision theory (MSCT) is illustrated in *Figure 3.11*. It is once again assumed that the relative motion of the reactants up to the critical surface can be represented by motion of a single representative mass point moving in a plane, and that the component of the collision energy along the line of centres of the two species must exceed a critical value for reaction to occur. Now, however, the potential energy at the surface separating reactants from products, and defined by $r = D$, is assumed to depend on the angle ϕ between the line OP and an axis drawn in the sphere $r = D$ and passing through O according to the equation[b]:

$$\mathscr{V}(D) = \varepsilon_T^0 + 2\varepsilon'(1 - \cos\phi) \tag{3.63}$$

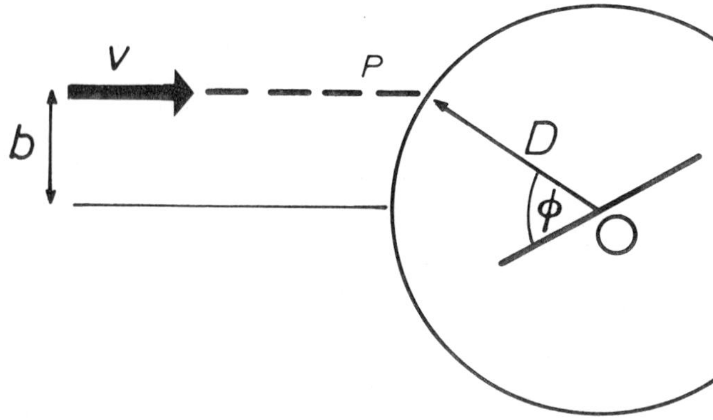

Figure 3.11. Parameters in modified simple collision theory

The condition for reaction to occur in collisions of specified ε_T and b is then

$$\varepsilon_{lc} = \varepsilon_T(1 - b^2/D^2) \geq \varepsilon_T^0 + 2\varepsilon'(1 - \cos\phi) \tag{3.64}$$

[a] The hard-sphere collision diameter for an H atom is estimated from the $^3\Sigma_u^+$ curve for H_2.

[b] $2(1 - \cos\phi)$ is a convenient mathematical form; as $\phi \to 0$, $2(1 - \cos\phi) \to \phi^2$.

This defines a maximum value that ϕ can have if reaction is to take place in these collisions, i.e.,

$$2(1-\cos\phi_{max}) = \{\varepsilon_T(1-b^2/D^2)-\varepsilon_T^0\}/\varepsilon' \tag{3.65}$$

To calculate the probability of reaction in these collisions with ε_T and b both specified, all possible orientations of the sphere $r = D$ are considered. $P_r(\varepsilon_T, b)$ is simply the fraction of the surface area of this sphere over which equation (3.64) is satisfied. If it is assumed that the axis drawn in the sphere represents a homonuclear diatomic molecule, so that reaction is possible at both ends of the molecule, this procedure yields:

$$P_r(\varepsilon_T, b) = 0 \quad \text{for} \quad \varepsilon_T < \varepsilon_T^0 \tag{3.66a}$$

$$\begin{aligned} P_r(\varepsilon_T, b \leqslant b_{max}) &= 2\int_0^{\phi_{max}} 2\pi D^2 \sin\phi \, d\phi/4\pi D^2 \\ &= (1-\cos\phi_{max}) \\ &= \{\varepsilon_T(1-b^2/D^2)-\varepsilon_T^0\}/2\varepsilon', \\ &\quad \text{for } \varepsilon_T^0 \leqslant \varepsilon_T(1-b^2/D^2) \leqslant (2\varepsilon'+\varepsilon_T^0) \end{aligned} \tag{3.66b}$$

$$P_r(\varepsilon_T, b \leqslant b_{max}) = 1, \text{ for } \varepsilon_T(1-b^2/D^2) > (2\varepsilon'+\varepsilon_T^0) \tag{3.66c}$$

Where reaction is possible with only one end of the molecule, either all these formulae could be divided by two, corresponding to an assumption that $\mathscr{V}(D)$ is infinite for $\pi/2 \leqslant \phi \leqslant \pi$, or it could be assumed that equations (3.63) to (3.65) hold for $0 \leqslant \phi \leqslant \pi$. In that case, $P(\varepsilon_T, b \leqslant b_{max})$ given by equation (3.66b) has to be divided by two and the upper limit on its use raised to $\varepsilon_T(1-b^2/D^2) \leqslant (4\varepsilon'+\varepsilon_T^0)$, and $P(\varepsilon_T, b \leqslant b_{max})$ becomes one for values of $\varepsilon_T(1-b^2/D^2) > (4\varepsilon'+\varepsilon_T^0)$.

Expressions for the reaction cross-sections and the rate constant can be derived easily from equations (3.66a) to (3.66c). Thus

$$S_r(\varepsilon_T) = \int_0^{b_{max}} P_r(\varepsilon_T, b) 2\pi b \, db \tag{3.67}$$

with b_{max} given by equation (3.56). Where $\varepsilon_T \leqslant (2\varepsilon'+\varepsilon_T^0)$, the integration is straightforward and yields

$$S_r(\varepsilon_T) = \pi D^2(\varepsilon_T-\varepsilon_T^0)^2/4\varepsilon_T\varepsilon' \tag{3.68}$$

Hence

$$\begin{aligned} k(T) &= \pi D^2(8k_BT/\pi\mu)^{\frac{1}{2}}(k_BT/2\varepsilon')\exp(-\varepsilon_T^0/k_BT) \\ &= k_{SCT}(T)(k_BT/2\varepsilon')(D^2/d^2) \end{aligned} \tag{3.69}$$

The factor $(k_BT/2\varepsilon')$ arises because the potential energy at the critical dividing surface is permitted to vary. The ratio (D^2/d^2) allows for the fact that the separation at this surface should be less than a mean hard-sphere collision

diameter (d). If reasonable values of ε' and D are inserted into equation (3.69), 'probability factors' of between 0.1 and 0.01 are predicted.

The curves in *Figure 3.10* show the functional dependence of (a) $P_r(\varepsilon_T, b)$, frequently referred to as the *opacity function*, and (b) $S_r(\varepsilon_T)$, the excitation function, for both the SCT and MSCT models. One result of the modification is to eliminate the intrinsically unreasonable step-function form for $P_r(\varepsilon_T, b)$. The curves for the opacity function suggest that for reactions with an energy barrier separating reactants from products, collisions with low impact parameters are the most likely to lead to reaction. In the absence of reaction these collisions would give rise to high-angle scattering (cf. *Figure 3.7*). Therefore, it is reasonable to expect that the angular distribution of non-reactive scattering of species that can undergo a reaction of this type will chiefly be altered as a consequence of reactive loss in the region of high θ. Furthermore, if one were to extend the theory and suppose that reaction occurs instantaneously, with energy released along the original line-of-centres direction, the chemical reaction products will be predominantly scattered at angles close to $\theta = \pi$. This is the characteristic feature of a *rebound* reaction.

The excitation function derived from the modified model rises much less steeply at values of ε_T at, and just above, threshold than is predicted by SCT. To infer this difference from thermal rate measurements, however, one would only have the temperature dependence of $k(T)$ as a guide. Despite the quite drastic change in the form of $S_r(\varepsilon_T)$, particularly near threshold, the temperature dependence of the rate constant is altered only slightly, which again illustrates how difficult it always is to infer microscopic kinetic information from thermal, or macroscopic, rate data.

It is clear that as a *simple* model for bimolecular reactions that proceed across a reactive potential hypersurface with an energy barrier separating reactants and products, the modified version of SCT is a significant improvement over SCT itself. Of course, a number of assumptions and approximations remain. This is a convenient point at which to list these; other rate theories can be checked against this list.

(1) The potential interaction is still grossly oversimplified. Thus, the interaction between the reactants prior to their reaching the dividing surface, $r = D$, is ignored and $\mathscr{V}$ is also assumed to be independent of coordinates describing the relative positions of atoms in the reactants.

(2) It is assumed that the dynamics of collisions are controlled by a single potential. No allowance is made for the possibility that the reactants may correlate with several electronic states of the complete system. This could, of course, be allowed for by introducing a simple numerical factor allowing for the statistical probability of the reactants colliding on the potential that is assumed to lead to products.

(3) All collisions which satisfy the criterion expressed in equation (3.64) are assumed to yield products. No allowance is made for the dividing surface being recrossed before the products separate, or for quantum effects which will 'soften' the rigidity of the criterion for reaction.

(4) It is assumed that only the energy of relative translational motion 'promotes' reaction. No mention is made of vibrational or rotational energies of the reactants.

(5) Finally, it should be noted that the model leads to no predictions about how energy released in the reaction is disposed of among the degrees of freedom or states of the products.

3.4.3 The Langevin model for ion–molecule reactions

In the last two sections it was assumed that there was an energy barrier on the potential hypersurface which had to be surmounted if reactants were to be transformed to products during a collision. Although this is the usual situation where reaction involves the transfer of an atom between an atomic or polyatomic free radical and a neutral molecule with a closed electronic shell, there are other types of reaction where there is essentially no electronic potential energy barrier preventing reaction. To illustrate some general characteristics for this kind of reaction, we now consider the Langevin model for reactions between an ion and a neutral molecule. In its simplicity, this treatment parallels simple collision theory.

The Langevin model is based on the supposition that, in ion–molecule encounters, the attractive ion–induced dipole forces (Section 2.2.2) can be considered independently of the 'chemical forces' and that they are sufficiently strong to counterbalance the chemical activation barrier associated with the breaking of one bond and the formation of another. That this is plausible can be demonstrated by a simple calculation: the ion–induced dipole potential energy between a singly charged ion and a molecule for which $(\alpha/4\pi\varepsilon_0) = 2.5 \times 10^{-3}$ nm^3 is 106 kJ mol^{-1} when the species are 0.2 nm apart. This is much larger than the activation energy of many elementary atom-transfer reactions. Of course, the accuracy of this calculation should not be taken too seriously; the use of an r^{-4} potential with an isotropic polarizability (α), obtained from measurements where the mean separation of molecular species is much greater than that used in the calculation, is of dubious validity.

The Langevin theory assumes that at long-range only the ion–induced dipole term in the potential is important and that all collisions which surmount the centrifugal barrier are reactive. The algebraic development then simply follows that set out in Section 3.3.2 for calculating the cross-section for 'core' collisions.

For an ion–induced dipole interaction, the attractive potential is proportional to r^{-4}. Substituting $n = 4$ into equation (3.50) yields an expression defining a maximum value of the impact parameter

$$b_{max} = d(4\varepsilon_d/\varepsilon_T)^{\frac{1}{4}} \tag{3.70}$$

such that for collisions at an energy ε_T, the probability of reaction is unity for $b \leqslant b_{max}$ and zero for $b > b_{max}$. The value of ε_d, the ion–induced dipole potential at $r = d$, is found by substituting $r = d$ in equation (2.6). For a singly charged ion this yields

$$-\mathscr{V}(d) = \varepsilon_d = \alpha e^2/(4\pi\varepsilon_0)^2 2d^4 \qquad (3.71)$$

Hence

$$b_{max} = \{2\alpha e^2/(4\pi\varepsilon_0)^2 \varepsilon_T\}^{\frac{1}{4}} \qquad (3.72)$$

The opacity function, $P_r(\varepsilon_T, b)$, predicted by the Langevin model, is a step-function, like that predicted by SCT. Now, however, b_{max} increases as ε_T *falls*, indicating the increasing ability of the attractive potential to pull the two particles into close contact as their relative velocity is lowered. The reaction cross-section is given by πb^2_{max}, i.e.,

$$S_r(\varepsilon_T) = \pi\{2\alpha e^2/(4\pi\varepsilon_0)^2 \varepsilon_T\}^{\frac{1}{2}} \qquad (3.73)$$

as long as $\{2\alpha e^2/(4\pi\varepsilon_0)^2\varepsilon_T\}^{\frac{1}{2}} \geqslant d^2$. However, because this equation holds for values of ε_T that make a significant contribution to the rate constant, $k(T)$ can be calculated in the usual way yielding

$$k(T) = \pi d^2 (8k_B T/\pi\mu)^{\frac{1}{2}} (\alpha e^2/(4\pi\varepsilon_0)^2 2d^4 k_B T)^{\frac{1}{2}} \Gamma(1/2) \qquad (3.74)$$

Equations (3.73) and (3.74) should be compared with their SCT and MSCT counterparts. In the model system treated by Langevin, $S_r(\varepsilon_T)$ decreases monotonically from an infinite value at $\varepsilon_T = 0$ to a value of πd^2, the hard-sphere value, at collision energies equal to and greater than $2\alpha e^2/(4\pi\varepsilon_0)^2 d^4$. The rate constant is independent of temperature. This behaviour contrasts strikingly with that predicted from the SCT and MSCT treatments. Although all these models of reaction are extremely simple, they correctly suggest appreciable differences in kinetic behaviour between reactions that occur over potentials with significant energy barriers and those where the interaction is dominated by long-range attractive forces.

3.4.4 Relationships between $S_r(\varepsilon_T)$ and $k(T)$

Discussing simple models of reaction in some detail has provided the opportunity to illustrate how the excitation function and the thermal rate constant are related. In particular, it has been shown how the temperature dependence of the rate constant depends on the form of the excitation function and that the variation of $k(T)$ with temperature can usually not be exactly interpreted in terms of a temperature-independent activation energy.

To emphasize this point, the averaging that is implied by the equation

$$k(T) = (1/\pi\mu)^{\frac{1}{2}} (2/k_B T)^{\frac{3}{2}} \int_0^\infty S_r(\varepsilon_T)\varepsilon_T \exp(-\varepsilon_T/k_B T)\, d\varepsilon_T \qquad (3.75)$$

is represented diagrammatically in *Figure 3.12*. Curve A corresponds to the integrand in equation (3.75) and can be identified with the distribution of collision energies in those collisions that lead to reaction; this function, $S_r(\varepsilon_T)\varepsilon_T \exp(-\varepsilon_T/k_B T)$, is frequently termed the *reaction function*. The shaded areas correspond, with proper normalization, to the rate constant for

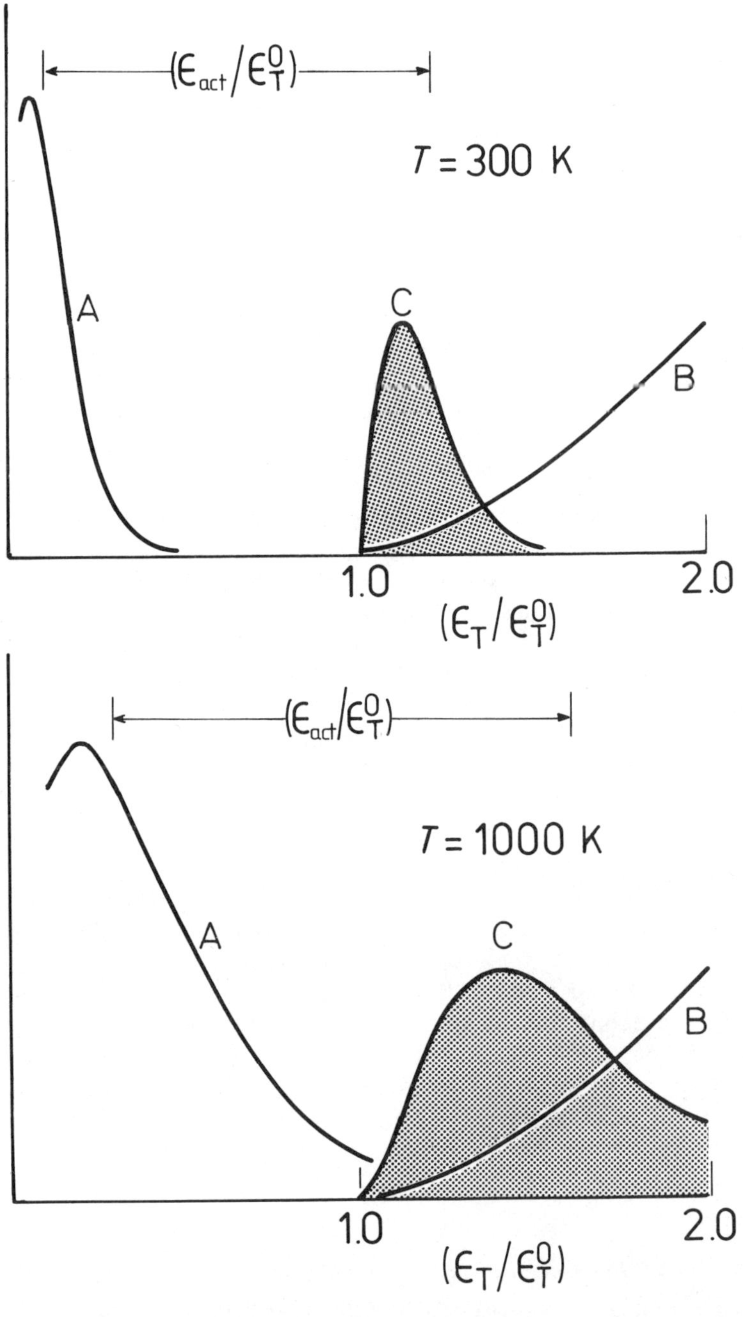

Figure 3.12. Curves showing the variation of A $\varepsilon_T[\exp(-\varepsilon_T/k_B T)]$, B *the reaction cross-section from modified simple collision theory, and* C *their product, i.e., the reaction function, at two different temperatures. The horizontal arrows show the*

the reaction. Recalling from Section 1.4.2 that the activation energy is equal to the difference between the mean energy in collisions that lead to reaction and the mean energy in all collisions, it is clear that, as the temperature is raised and the spread of collision energies widens, the activation energy will increase—at least, for the MSCT form of the excitation function that is given in *Figure 3.12*.

The form of the reaction function in these diagrams also brings out the difficulty of obtaining information about the excitation function from measurements of the rate constant. At low temperatures, the spread of collision energies leading to reaction is relatively narrow but confined to values just above threshold. At higher temperatures, a broader range of the excitation function is probed but the spread of the reaction function is greater.

The relationships between the excitation function and the rate constant and between the threshold and activation energies have been explored by Menzinger and Wolfgang[8] and by Le Roy[9] using various general—and intrinsically reasonable—forms for $S_r(\varepsilon_T)$. Their results are summarized in Table 3.2. The cases of SCT, modified SCT, and the Langevin model of ion–molecule reactions constitute special cases of these general relationships and the results obtained for these models are also listed in this table.

3.5 Classical reaction dynamics[10—12]

The modification to SCT that was described in Section 3.2.2 represents a sensible limit to extensions of the simplified theory of collisional reactions. Attempts to improve the model further soon founder on unmanageable algebra. In any case, such efforts would be essentially worthless, since it is now possible to study three-dimensional collision dynamics almost routinely by numerically solving the classical laws of motion—that is by calculating *classical trajectories.* This method allows reaction dynamics to be investigated in detail at the microscopic level. Of course, such calculations rest on an assumption of classical behaviour that is not strictly valid. However, there is now a good deal of evidence, that will be reviewed later, both from quantum mechanical calculations on collision dynamics and from experimental studies, that classical calculations can provide an adequate description of the collision dynamics for many purposes. In this section, the method itself is summarized briefly, and the general conclusions that have been obtained from trajectory calculations are reviewed.

The first stage in any classical trajectory study is to select a function to

---

difference between the mean energy of reactive collisions and the mean energy of all collisions, and corresponds to the (reduced) activation energy [equation 1.50]. The threshold energy ε_T^0 is chosen so that $(\varepsilon_T^0/k_B) = 5 \times 10^3$ K; $(\varepsilon_{act}/\varepsilon_T^0) = 1.1$ at 300 K and 1.3 at 1000 K

Table 3.2 Relationships between $S_r(\varepsilon_T)$, $k(T)$, ε_T^0 and E_{act}

	$S_r(\varepsilon_T)^a$	$k(T)^a$	$E_{act}{}^b$
I	$C(\varepsilon_T-\varepsilon_T^0)^n \exp[-m(\varepsilon_T-\varepsilon_T^0)]$	$\frac{C'(k_BT)^n \exp(-\varepsilon_T^0/k_BT)}{(1+mk_BT)^{n+2}} \times \{\Gamma(n+2)+\Gamma(n+1)(1+mk_BT)\varepsilon_T^0/k_BT\}$	$E^\circ + RT(n+\frac{1}{2}) - \frac{m(n+2)(RT)^2}{1+mRT} - \frac{E^\circ}{(n+1)+(1+mRT)E^\circ/RT}$
II	$C[(\varepsilon_T-\varepsilon_T^0)^n/\varepsilon_T] \exp[-m(\varepsilon_T-\varepsilon_T^0)]$	$\frac{C'(k_BT)^{n-1}\Gamma(n+1)\exp(-\varepsilon_T^0/k_BT)}{(1+mk_BT)^{n+1}}$	$E^\circ + RT(n-\frac{1}{2}) - \frac{m(n+1)(RT)^2}{1+mRT}$
SCT $m=0,\ n=1$	$C(\varepsilon_T-\varepsilon_T^0)/\varepsilon_T$	$C'\exp(-\varepsilon_T^0/k_BT)$	$E^\circ+\frac{1}{2}RT$
MSCT $m=0,\ n=2$	$C(\varepsilon_T-\varepsilon_T^0)^2/\varepsilon_T$	$2C'(k_BT)\exp(-\varepsilon_T^0/k_BT)$	$E^\circ+\frac{3}{2}RT$
III[c]	$C\varepsilon_T^n$	$C'(k_BT)^n[\Gamma(n+2)-\gamma(n+2,\varepsilon^0/k_BT)]$	$RT(n+\frac{1}{2}) + \frac{E^\circ(E^\circ/RT)^{n+1}\exp[-E^\circ/RT]}{\Gamma(n+2)-\gamma(n+2,E^\circ/RT)}$
Langevin $n=-\frac{1}{2},\ \varepsilon^0=0$	$C\varepsilon_T^{-\frac{1}{2}}$	$C'\Gamma(\frac{1}{2})/2(k_BT)^{\frac{1}{2}}$	0

[a] C is merely a proportionality constant; $C'=C(8k_BT/\pi\mu)^{\frac{1}{2}}$
[b] $(E^\circ/R)=(\varepsilon_T^0/k_B)$
[c] γ denotes an incomplete gamma function. Equations for E_{act} for this case have been tabulated by Le Roy[9]

represent the potential. Secondly, there must be methods for selecting the initial positions and momenta of the atoms for each individual trajectory, for integrating the equations of motion, and for calculating the final properties of the products of each collision. Finally, some averaging may have to be carried out to yield the values of quantities that might, for example, be compared with experiment. In contrast to the limited predictions of simple theories, the classical trajectory method yields values for a complete range of kinetic quantities, differential as well as total cross-sections, and detailed as well as total rate constants.

3.5.1 Selection of the potential function

Classical trajectory calculations are usually performed with one of two aims. In the first type of study, the objective is to identify what general features on the potential hypersurface, or in the kinematics, lead to particular results; for example, to energy being released preferentially into a particular degree of freedom in the products. The second and more ambitious aim is to discover the form of the potential for a particular reaction by successively adjusting the potential until the calculated results coincide with those observed experimentally. In this way, one hopes to move counter to the flow in *Figure 1.4* and deduce the potential from experiment through the medium of classical trajectories. This approach is only worthwhile when a good deal of microscopic data has been obtained by experiment.

In both kinds of investigation it is useful to have a flexible potential, that is one whose characteristics can be altered by the adjustment of one or more parameters. In order to solve the equations of motion at all, it is necessary that the potential is expressed as an algebraic function of coordinates that describe the relative positions of the three (or more) atoms in the system. The LEPS function and its modifications meet both these requirements and have been widely used.

For studies on a particular reaction, one should ideally like a potential based on *a priori* calculations of the hypersurface. However, even when calculations of this type have been carried out, the results cannot be used directly in a dynamical calculation, because $\mathscr{V}$ will not have been obtained as a function of the position coordinates. Consequently, where such a potential has been determined, the points from the quantum mechanical calculation are usually fitted to an analytical function containing several adjustable parameters, and this is then employed in the trajectory calculations. Alternatively, an empirical approach can be adopted. For example, the LEPS function can be adjusted until its barrier height is consistent with the measured value of the activation energy and then trajectories are run on this hypersurface to predict other attributes of the reaction. In assessing the results of these calculations it is important to examine critically the evidence for the potential that has been chosen. In certain cases, even the most rudimentary features of the potential, such as the height of the energy barrier, are not known with any certainty. As the potential is the single most important factor in determining the results of

molecular collisions, the 'predictions' from such calculations should be treated with caution.

3.5.2 Monte Carlo classical trajectories

Once the potential has been chosen, the computation of a single classical trajectory requires the solution of the equations of motion in the form that they were given in equation (3.3). For three atoms in three dimensions, with the motion of the centre-of-mass of the system separated, there are 12 equations given by:

$$(\partial H_{rel}/\partial p_i) = (\partial T_{rel}/\partial p_i) = \dot{q}_i \tag{3.76a}$$

and

$$-(\partial H_{rel}/\partial q_i) = -(\partial \mathscr{V}/\partial q_i) = \dot{p}_i \tag{3.76b}$$

which must be solved by numerical integration. At the start of an individual trajectory, a choice has to be made for the six coordinates that describe the relative position of the atoms and the six conjugate momenta. The calculation starts with the atomic reagent A sufficiently far from BC for their interaction to be negligibly small, and it ends when once again one atom is far away from the other pair and the collision can be said to be over.

The three-particle system is usually described in terms of three coordinates that specify the position of B relative to C and three that give the position of A relative to the centre-of-mass of BC. The selection of initial values of the p and q is not made directly, but other variables are chosen which can easily be related to the p and q and whose distribution of values in a practical situation is well known. Among others, these variables include the relative translational velocity of A and BC, the impact parameter, and the separation of A from the centre-of-mass of BC. These parameters either have their initial value assigned, or they are chosen by some pseudo-random, 'Monte Carlo', procedure. The purpose of Monte Carlo selection is to ensure that the distribution of each initial parameter within a sample of trajectories approaches the true statistical distribution as the size of the sample grows. The use of this technique normally allows one to obtain results of acceptable accuracy from a relatively small set of trajectories and means that 'after-the-event' averaging is not required over the initial parameters that are chosen in this way. There are three, somewhat different, pseudo-random procedures that can be followed, depending on the nature of the distribution function for the variable in question.

The most straightforward case is where the probability of a variable taking a particular value is constant over its entire range. For example, initial values of the angle between the BC axis and the line joining A to the centre-of-mass of BC and in the plane containing all three atoms are distributed in this way. This angle can be selected simply by multiplying π by a randomly chosen number between -1 and $+1$.

The second procedure can be illustrated by referring to selection of the impact parameter. The probability, $P(b)$, of the impact parameter having a

value between b and $b + \mathrm{d}b$ is proportional to $2\pi b\,\mathrm{d}b$, so that a technique is required that will generate a distribution appropriate to this weighting factor. To do this, a function $F(b)$ must be found such that b is uniformly distributed in F space, i.e., such that $\mathrm{d}P(b)/\mathrm{d}F$ is a constant $(1/\alpha)$. Since $\mathrm{d}P(b)/\mathrm{d}F = (\mathrm{d}P(b)/\mathrm{d}b)/(\mathrm{d}F/\mathrm{d}b)$,

$$F = \int_0^b \mathrm{d}F = \alpha \int_0^b 2\pi b\,\mathrm{d}b \tag{3.77}$$

Consequently F is proportional to b^2, and impact parameters are chosen using the equation

$$b = S(\xi b_{\mathrm{max}}^2)^{\frac{1}{2}} \tag{3.78}$$

where ξ is a freshly generated random number between 0 and 1, S is randomly -1 or $+1$, and b_{max} is chosen so that reaction is very improbable for $b > b_{\mathrm{max}}$.

Finally, there is a Monte Carlo method for selecting variables where no closed function such as F can be found. Here, another new random number is used to provide a value of the variable, for example the relative velocity, but a decision as to whether to accept this value is made on the outcome of a 'game of chance' against a second random number. The probability of a particular value being accepted is proportional to the probability density in the statistical distribution at that value. The procedure is repeated until a value of v passes the 'filter', and this value is then incorporated into the starting parameters.

Usually, the internal momenta of BC are chosen to correspond to the energies of the vibrational–rotational quantum states of the molecule referred to the minimum in its potential curve. Such calculations are referred to as *quasiclassical*. Of course, once a calculation is under way, no quantum mechanical restriction can be included and the internal energy of the final product molecule, whether it be AB or AC, reaction having taken place, or BC, will not correspond to that of a quantum state. In order, therefore, to obtain detailed cross-sections or detailed rate constants for production of state-specified products, some way must be found of slicing up the continuous distribution of final vibrational and rotational energies into quantum levels. This is usually done by first dividing the final internal energy of the molecular product between vibrational and rotational motions, and then assigning this molecule to the quantum states closest in energy to the classical rotational and vibrational energies that have been calculated. Thus, if the final vibrational energy of the product molecule is between

$$\{(\tfrac{1}{2}+\tfrac{1}{2})hc\tilde{v}_e - (\tfrac{1}{2}+\tfrac{1}{2})^2 hc\tilde{v}_e x_e\} \quad \text{and} \quad \{(1\tfrac{1}{2}+\tfrac{1}{2})hc\tilde{v}_e - (1\tfrac{1}{2}+\tfrac{1}{2})^2 hc\tilde{v}_e x_e\}$$

the collision is assumed to have produced the molecule in $\mathrm{v} = 1$ [cf equation (2.24)]. The indications are that this procedure is satisfactory where the distribution of final energies is wide relative to the spacing of the quantum states. Where this is not so, this process of assigning the results to quantum state 'boxes' can lead to large errors.

Within the limitations imposed by the use of classical mechanics, and

possibly by uncertainty concerning the potential, a Monte Carlo trajectory study of a reaction such as $A + BC \rightarrow AB + C$ can provide a complete inventory of the microscopic and macroscopic quantities referred to in *Figure 1.4*, in particular:

(1) state specified differential reaction cross-sections, $\sigma_r(n' | n; v, \theta)$;
(2) state specified total reaction cross-sections, $S_r(n' | n; v)$;
(3) detailed rate constants, $k(n' | n; T)$ or $k(n' |; T)$
(4) thermal rate constants, $k(T)$, and hence Arrhenius parameters, A and E_{act}.

However, the provision of *all* these quantities requires that only partial use is made of Monte Carlo techniques in selecting the initial parameters for each trajectory. This is because, if values of $\sigma_r(n' | n; v, \theta)$ and $S_r(n' | n; v)$ are to be calculated, initial values of n, representing the state of BC, and v cannot be selected by Monte Carlo procedures, but must be assigned particular values in each trajectory of a complete set. To derive detailed and overall rate constants, many sets of trajectories with different combinations of n and v must be calculated and 'after-the-event' averaging carried out. If one only requires values for the rate constants, much time (and expense!) are saved by selecting v and some, or all, of the initial internal energies (for example, the rotational energies but not the vibrational energies) by Monte Carlo methods.

A number of important general conclusions have emerged from the large number of classical trajectory studies that have been carried out. These are considered next. Results on particular reactions will be considered alongside experimental data in later chapters. Before concluding this section, however, it may be useful to see how far the classical trajectory method goes in meeting the criticisms of the simpler collision theories that were listed at the end of Section 3.4.2.

It can readily be seen that the uncertainties that remain centre on the use of classical mechanics to describe the collision dynamics and of potential functions that may not be accurate. Since there is usually only a single adiabatic hypersurface across which reaction will proceed with significant probability, allowance can be made for the existence of multiple hypersurfaces correlating with the reactants by dividing the results from a calculation of classical trajectories by a numerical factor. This is simply the degeneracy of the electronic state assumed to connect reactants to products divided by the product of the electronic partition functions (*see* Appendix 3) of the separated reactants. When the absolute values of calculated quantities are compared with those determined experimentally, this factor should be included.

3.5.3 Results: cross-sections and rate constants

As an example of an investigation of the classical collision dynamics of a simple atom-transfer reaction, the early study of Karplus, Porter and Sharma (KPS)[13] is outstanding. Although the potential that was used was the

Porter–Karplus (PK) one for H_3 and the calculations were carried out with the reaction

$$H + H_2 \rightarrow H_2 + H \tag{3.79}$$

in mind, several of their results can (with care) be carried over to other reactions of a similar type, particularly those with symmetrically placed barriers in which the three atoms involved have similar masses.

Karplus, Porter and Sharma found that the vast majority of their trajectories, whether or not reaction occurred, could be classified as *direct*, since the collision time was comparable with that required for an H atom to travel the same distance unimpeded by contact with an H_2 molecule. The variation in the internuclear distances during a typical reactive trajectory are shown in *Figure 3.13*. A particularly important feature to note is that at only one point is $r_{AB} = r_{BC}$. In only a very few trajectories, and these mostly at energies well above threshold, was the surface defined by $r_{AB} = r_{BC}$ crossed more than once. Where the reactive potential is free from hollows, direct collision dynamics are generally found. Only as the vibrational or translational energy of the reactants is increased is there much tendency for the $r_{AB} = r_{BC}$ surface to be crossed more than once. This feature is important in fundamental considerations of transition state theory (Section 4.2.2).

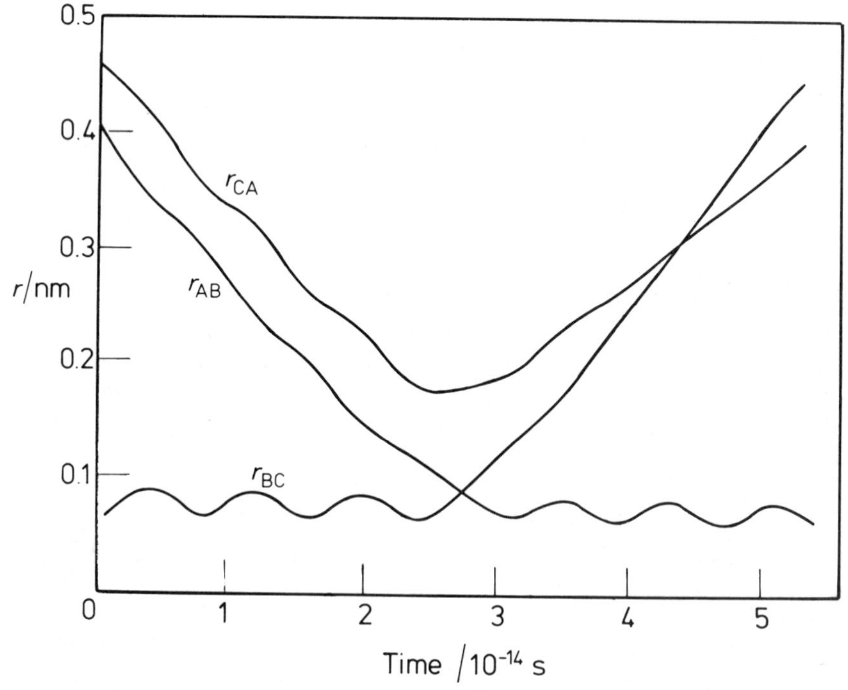

Figure 3.13. Typical reactive classical trajectory for an $H + H_2$ *collision [from Karplus, M., Porter, R. N., and Sharma, R. D.,* Journal of Chemical Physics, **43**, 3259 (1965)]. *Note that the* r_{AB} *and* r_{BC} *curves cross only once*

Because the rovibrational states (v, j) of H_2 are widely spaced, Karplus, Porter and Sharma were able to calculate reaction cross-sections, $S_r(\mathrm{n}; \varepsilon_T)$, for each of the levels that is significantly populated at readily accessible temperatures. Increasing the rotational energy of the molecular reactant changed ε_T^0, the translational energy threshold, and the shape of the (state-selected) excitation function only slightly. Excitation to v = 1, on the other hand, enhanced the reaction rate by about three orders-of-magnitude when the initial relative translational and rotational energies were chosen from a 300 K thermal distribution. However, the equilibrium population in v = 1 is too low for reactions of H_2(v = 1) to make a significant contribution to $k(T)$. Since the calculations were quasiclassical and some, though not all, of the H_2 zero-point vibrational energy could be utilized in surmounting the potential barrier, the results of KPS cannot be directly compared with predictions from simpler theories. However, the form of the excitation and opacity functions which they obtained are qualitatively very similar to those that were derived from the modified form of SCT and shown in *Figure 3.10*.

Finally, Karplus, Porter and Sharma compared the overall rate constants, $k(T)$, derived from their calculations, with (*a*) the Arrhenius equation, and (*b*) the rate expression derived from the 'usual' form of transition state theory (Section 4.2) and the properties of the PK potential at the energy barrier (*Table 2.6*). Their calculated rate constants obeyed the Arrhenius rate law quite well, yielding the expression

$$k(T) = 0.72 \times 10^{-10} \exp(-31.1 \text{ kJ mol}^{-1}/RT) \text{ cm}^3 \text{ molecule}^{-1} \text{ s}^{-1} \quad (3.80)$$

However, the Arrhenius plot [log $k(T)$ against $1/T$] showed slight positive curvature indicating some increase of the activation energy with temperature. Over the same range of temperature, evaluation of the standard transition state rate expression (equation 4.28) gave

$$k(T) = 1.23 \times 10^{-10} \exp(-36.9 \text{ kJ mol}^{-1}/RT) \text{ cm}^3 \text{ molecule}^{-1} \text{ s}^{-1} \quad (3.81)$$

Clearly, the transition-state values of $k(T)$ are smaller than those evaluated from equation (3.80) at low temperatures. However, when it is recalled that all trajectories commence with the H_2 molecule possessing 25.9 kJ mol^{-1} 'zero-point' vibrational energy and there is no quantum restriction at subsequent points in the trajectory, in particular at the energy barrier, it is rather remarkable that the difference between the two results is not greater. In fact, it appears to be generally true that kinematic features can limit the extent to which reactant vibrational energy can be utilized in overcoming energy barriers to reaction and that this goes a long way towards compensating for the lack of zero-point requirements in the transition state. Thus, even for *collinear* collisions, Karplus, Porter and Sharma found a translational energy threshold of 22.0 kJ mol^{-1} which is considerably greater than the value, 12.3 kJ mol^{-1}, required on the grounds of energy conservation, i.e., by subtracting the initial H_2 vibrational energy (25.9 kJ mol^{-1}) from the potential energy barrier (38.2 kJ mol^{-1}).

3.5.4 Results: detailed rate constants

A major use of classical trajectories has been to investigate the *specificity* of energy disposal and the *selectivity* of energy consumption in elementary reactions of the $A + BC \rightarrow AB + C$ type. In particular, Polanyi and his co-workers in connection with their experimental determinations of detailed rate constants by the technique of infrared chemiluminescence (Section 6.2.1), have studied how the preferential, i.e., non-statistical, release of energy depends on features of the potential hypersurface[10, 14–16].

The most important factor in determining the nature of the energy disposal is the position of the barrier and the character of the potential energy profile along the reaction path of minimum energy (*Figure 2.5*). Where the barrier is at a 'late' position along the reaction coordinate, so that a major portion of the energy is released after A and B have approached to a separation close to the value of r_e in the isolated AB molecule, energy is channelled chiefly into the relative translational motion of the separating products. When the barrier is positioned 'early' on the reaction path—which is its expected location for highly exoergic reactions—a higher fraction of the energy is released as the reactants approach and when r_{AB} is still much greater than $r_{e,AB}$. In this case, AB's vibration is preferentially excited.

The tendency for high vibrational excitation does not, however, increase indefinitely as the potential is made more attractive. First, there is an increasing probability of the products of reactive trajectories undergoing 'secondary encounters'. This happens in collisions where AB is formed, but so little energy is initially released into the relative translation of AB and C that AB vibrates with large amplitude before C has moved away. Consequently r_{BC}, having once increased, then decreases until the atom C once again 'comes into contact' with A or B. These secondary encounters tend to allow energy to flow from the molecular vibration into other degrees of freedom of the reaction products. Finally, if the potential is made sufficiently attractive, a hollow appears on the surface, genuine collision complexes can form, and the energy released in the reaction will be shared evenly among the various degrees of freedom of the reaction products.

In atom-transfer reactions involving only three atoms, the *rotation* of the molecular product absorbs only a limited amount of the energy that is released. This appears to be a consequence of the need to conserve total angular momentum during the collision. Initially, the total angular momentum can be represented by a vector $\boldsymbol{J}$, which is the vectorial sum of $\boldsymbol{l}$, representing the initial orbital momentum, and $\boldsymbol{j}$, the rotational angular momentum of BC, i.e.,

$$\boldsymbol{J} = \boldsymbol{j} + \boldsymbol{l} \tag{3.81}$$

In principle, $\boldsymbol{J}$ can be conserved and the molecular rotation highly excited if $\boldsymbol{l}'$ and $\boldsymbol{j}'$, the angular momentum vectors of the products, point in opposite directions and have roughly the same magnitude, but this does not appear to happen in a significant number of collisions.

Two sources of rotational excitation can be identified:

(1) on highly attractive surfaces, such as that for reactions of alkali-metal atoms with halogen molecules,

$$M + X_2 \rightarrow MX + X \qquad (3.83)$$

collisions with very large impact parameters, up to 0.8 nm, can lead to reaction, with most of the initial angular momentum being converted into rotational momentum of the product molecule, i.e., $l \sim j'$;

(2) repulsion between the products when they are in a bent configuration, i.e., A⁄B⁀C, can also cause rotational excitation.

The observation that very high rotational excitation is rare indicates that reaction generally occurs in nearly collinear configurations. Although systematic studies have not been made, it seems that angular momentum conservation may be less restrictive in reactions such as A + BCD → AB + CD, particularly if BCD is a bent molecule.

The 'shape' of the potential hypersurface is the major, but not the only, factor that determines the pattern of energy disposal in reactive collisions. The effect on the collision dynamics of changing the relative masses of the atoms A, B and C has also been investigated systematically by Polanyi's group. They found that the distribution of energy between product vibration and relative translation could be related more closely with the type of 'energy release' than with the attractive or repulsive nature of the potential[15]. They coined the terms *attractive* ($\mathscr{A}$), *mixed* ($\mathscr{M}$) and *repulsive* ($\mathscr{R}$) *energy release*, which were defined with reference to the dynamics of collinear trajectories at collision energies just above threshold and with no initial vibrational energy in BC. For each surface and mass combination, the path of a single trajectory is divided into three portions. The $\mathscr{A}$ energy is that released before r_{BC} increases beyond its maximum value whilst undergoing classical vibrations with the energy corresponding to the v = 0 level. The $\mathscr{R}$ energy is that released after r_{AB} first equals $r_{e,AB}$, and the $\mathscr{M}$ energy is the remainder, released as the trajector 'cuts the corner' on the potential surface.

That this classification distinguishes between systems on account of their relative masses as well as their surface characteristics is demonstrated diagrammatically in *Figure 3.14*. The pictures depict collinear collisions at the same energy on the same repulsive surface. In case (a), the collisional reduced mass is fairly large but the transferred atom is light. Consequently, product repulsion acts before the AB distance closes to the equilibrium bond length and there is a high degree of mixed energy release. In case (b), however, atom A is much lighter than B and C so the relative velocity of the reactants is greater. As a result the products only start to repel one another appreciably, after AB is essentially formed, and there is little mixed energy release.

Full three-dimensional trajectory studies indicate that there is a fairly good quantitative correlation between the amount of energy released as ($\mathscr{A} + \mathscr{M}$) and that which, on average, enters the vibration of the product molecule in

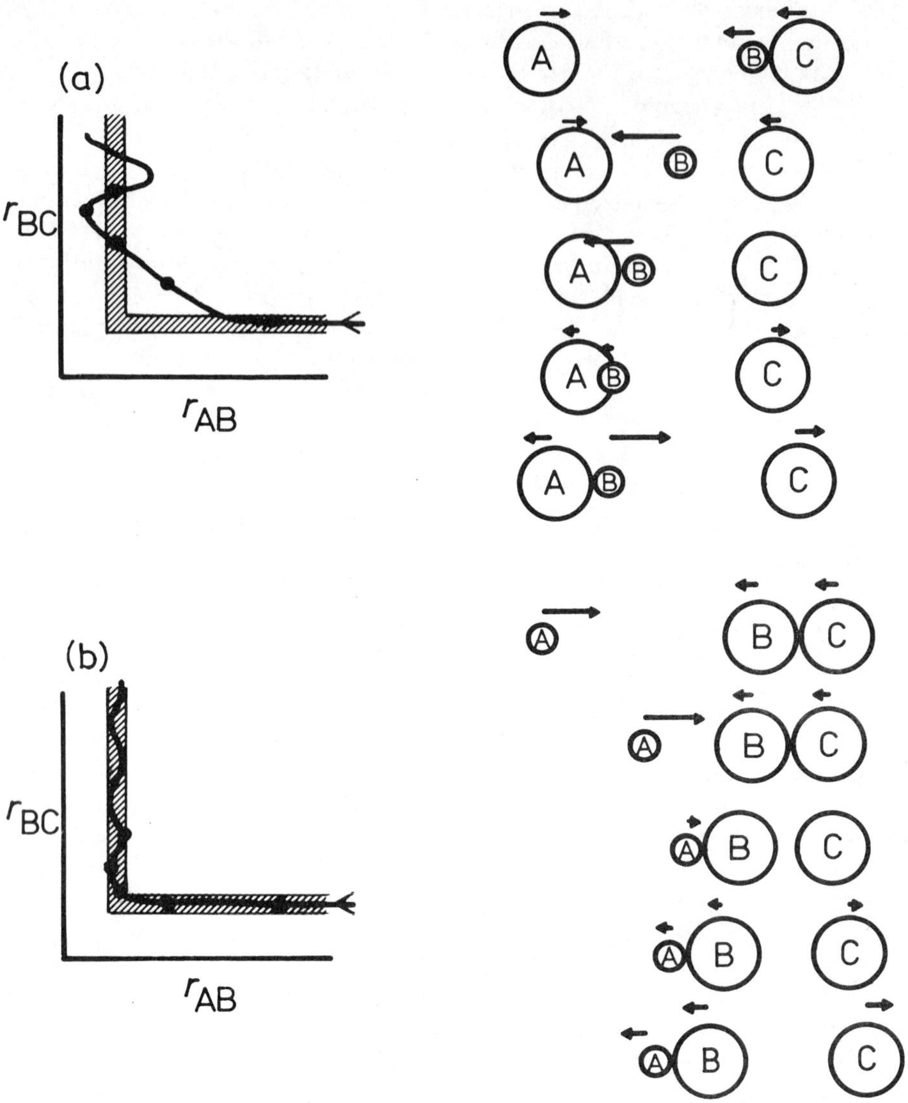

Figure 3.14. Pictorial representations of collinear collisions on a repulsive surface involving different mass combinations. In (a) *where a light atom is transferred there is considerable mixed energy release leading to appreciable vibrational excitation of* AB; *in* (b) *where the attacking atom is light, the trajectory follows a nearly rectilinear path across the surface and energy is released predominantly into the relative translational motion of the products*

direct encounters. One cautionary note should, however, be added. Mixed energy release and the results of corner cutting trajectories should not be confused with changes to the *potential* which cause the reaction path of minimum energy on the collinear surface to deviate further and further from

the rectilinear. Such a change causes the forces acting on the atoms to become less impulsive, trajectories tend to cling more closely to the reaction path of minimum energy, and the degree of internal excitation is lowered.

When fully detailed rate constants are available for a reaction in one direction, the principle of detailed balancing, as expressed in equation (1.39), can be used to calculate the detailed rate constants for the reverse reaction. In practice, most experiments and trajectory calculations on exoergic reactions yield only partially detailed rate constants, that is $k(\mathrm{n}'|; T)$ rather than $k(\mathrm{n}'|\mathrm{n}; T)$, as the reactants are thermally distributed among the rovibrational states represented by n. Nevertheless, because the spread in energy of the species that undergo reaction is fairly small, it is possible to derive partially detailed rate constants for reaction in the reverse endoergic direction, that is reaction *from* the states represented by n′, using an approximate form of equation (1.39). An example of this procedure is provided in Section 6.3.5. From arguments of this kind, it can be shown that, when reaction in the exoergic direction specifically channels energy into the vibration of the newly formed molecule, the reverse endoergic reaction will be selectively promoted by exciting this molecule vibrationally. This conclusion has been confirmed by calculating classical trajectories for endoergic reactions, where the barrier to reaction is located late along the reaction coordinate.

The effect of selective reactant excitation on *thermoneutral* reactions has also been studied[16]. With an appreciable barrier (29.3 kJ mol^{-1}) displaced into the 'exit valley' of the potential, vibrational energy was again found to promote reaction much more effectively than relative translational energy. However, when the position of the barrier was switched into the entry valley, the energy requirements altered dramatically and relative translation became the effective motion. Although these conclusions are based on full, three-dimensional, Monte Carlo calculations, they can be rationalized in terms of the single collinear trajectories shown in *Figure 3.15*.

On surface +I, the direction of the reaction coordinate at the barrier is almost parallel to the potential minimum along the floor of the reactant valley. Relative translational motion moves the system along the valley and over the barrier into the product zone. Vibration of BC, on the other hand, provides motion transverse to the reaction coordinate and does not help the system up to the head of the entry valley. On surface +II, however, the reaction coordinate in the region of the barrier is nearly perpendicular to that of the reactant valley. Consequently, vibrational motion, if it is of the right phase, provides momentum along the exit valley and carries the system over to products. Now relative translational motion is transverse to the reaction coordinate at the barrier and is ineffective.

For real thermoneutral reactions, the most likely position of the barrier is at or close to a symmetrical configuration defined by $(r^*_{AB}/r^*_{BC}) = (r_{e,AB}/r_{e,BC})$. There are then less dramatic differences between the effects of exciting the vibration of BC and of giving the same amount of energy to relative translation[17]. Even when BC's vibrational energy is well in excess of the barrier height, there remains a significant collision energy threshold, illustrating again

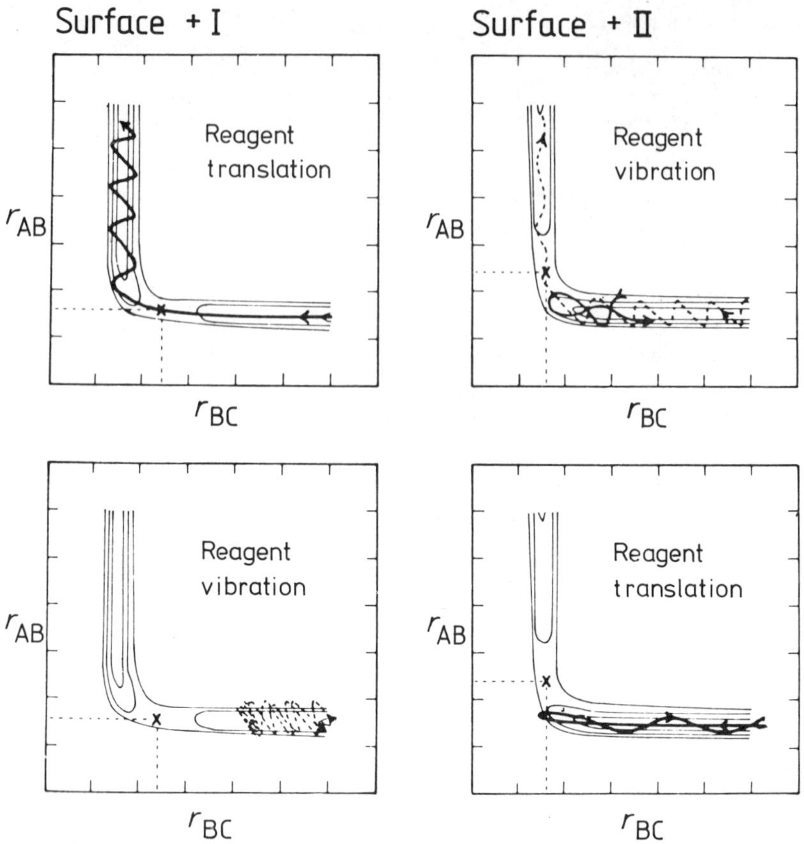

Figure 3.15. Representative classical trajectories showing how the effect of reagent translational energy and reagent vibrational energy in promoting reaction depends on the location of the energy barrier. On surface +I *with an 'early' barrier, relative translational energy is more effective. On surface* +II *with a 'late' barrier, vibrational excitation of* BC, *if the phase of the vibration is correct, brings about reaction* [*adapted from Figure 4 in Polanyi, J. C.,* Accounts of Chemical Research, **5**, 161 (1972)]

that the use of reactant energy in overcoming potential barriers is restricted by dynamic constraints. One important factor is the relative masses of the atoms.

In order to discuss differences which arise for different mass combinations, it is useful to consider the dynamics of collinear collisions in terms of the motion of a single sliding mass point on a properly modified potential surface. For this analogy to be correct, the kinetic energy must be diagonalized so that, for example, it takes the form:

$$T = \tfrac{1}{2}(\dot{X}^2 + \dot{Y}^2) \tag{3.83}$$

representing the motion of a point of unit mass in a new Cartesian frame. The

coordinates X and Y are related to r_{AB} and r_{BC} by the equations:

$$X = \mu^{\frac{1}{2}} r_{AB} + \mu'^{\frac{1}{2}} r_{BC} \cos\beta \tag{3.83a}$$

$$Y = \mu'^{\frac{1}{2}} r_{BC} \sin\beta \tag{3.83b}$$

The modified potential surface is constructed by *skewing* the r_{BC} axis towards the r_{AB} axis until they define the angle β where

$$\cos^2\beta = (m_A m_C / m_{AB} m_{BC}) \tag{3.83c}$$

and by *scaling* the r_{AB} and r_{BC} axis by multiplying them by μ and μ', respectively.

Figure 3.16 shows the effect of skewing the surface in two cases where $m_A = m_C \neq m_B$. When $m_A = m_C$, vibrational excitation of BC corresponds to motion of the mass point parallel to the Y-axis. In case (b), because B is light and the r_{AB} and r_{BC} axes are at an acute angle to one another, this motion is nearly perpendicular to the line $r_{AB} = r_{BC}$. Consequently, one would expect reactant vibrational excitation to be more effective in carrying the mass point over to the product zone—and, therefore, in promoting reaction—than in case (a). This expectation is borne out by three-dimensional quasiclassical trajectory calculations.

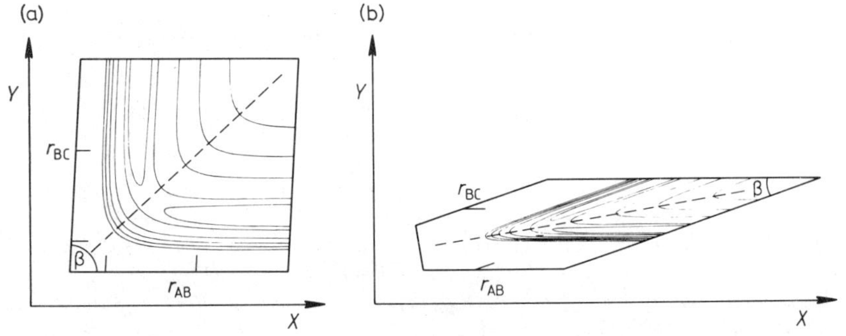

Figure 3.16. 'Skewed' potential-energy surfaces for collinear reactions where (a) $m_A = m_C = (m_B/17.5)$, *and* (b) $m_A = m_C = 17.5\ m_B$; $\mathcal{V}(r_{AB}, r_{BC}, r_{CA})$ *is the same in* (a) *and* (b). *In general the rectilinear coordinates are given by equation* (3.83). *In the special case where* $m_A = m_C$, $\cos\beta = (m_A/m_{BC})$ *and* $\mu = \mu'$

The trajectories plotted on the rectilinear surfaces in *Figure 3.15* do *not* correspond to the motion of a single mass point on these surfaces. Strictly, the surfaces should be properly skewed and scaled for each mass combination, and then the arguments relating the selective use of reactant energy to the position of the barrier on the potential surface need some rephrasing. Nevertheless, the correlation between barrier location and selective energy consumption appears to hold qualitatively for all mass combinations.

3.5.5 Results: differential cross-sections

For reactive systems, the initial direction of the A beam in the centre-of-mass frame of reference is taken to define $\theta = 0$ degrees, and 'forward' product scattering is the term used when AB is scattered preferentially into the hemisphere defined by $\theta \leqslant 90$ degrees. Where the reaction dynamics are determined by the formation of a *collision complex* capable of surviving several rotational periods, the system 'forgets' the details of its mode of formation, and the scattering of products is constrained only by the need to conserve energy and linear and angular momentum. This leads to an angular distribution of reaction products that is symmetrical about $\theta = 90$ degrees.

The angular scattering resulting from *direct* reactions has been thoroughly investigated by Polanyi's group. On repulsive hypersurfaces, AB is found to scatter backwards, irrespective of the relative atomic masses. This is reasonable, since such reactions occur predominantly in low impact parameter collisions and near linear configurations, so that the energy which is released as

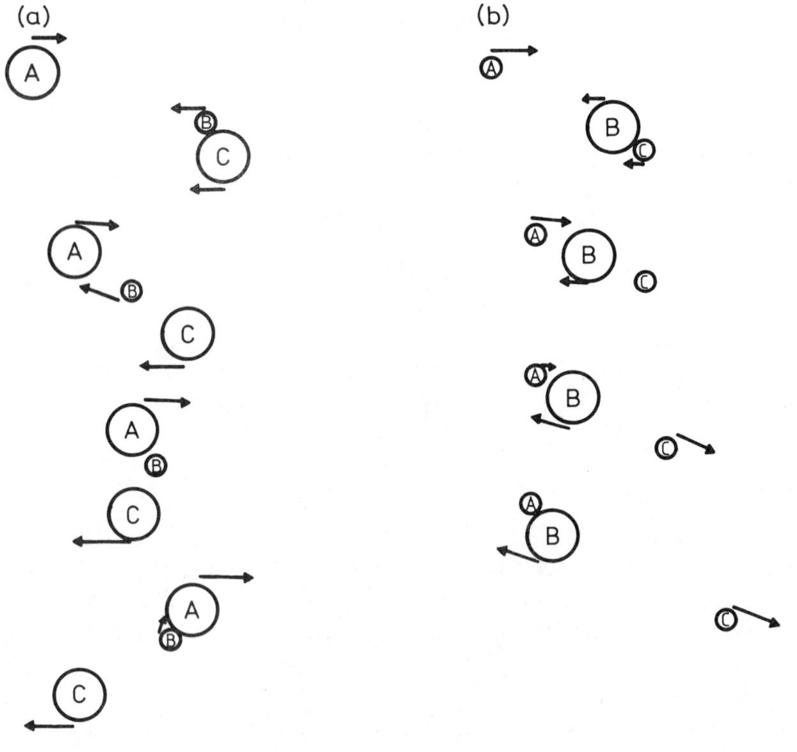

Figure 3.17. Effect of the relative atomic masses on reactive scattering on a moderately attractive hypersurface. In (a) *the atom transferred is light; in large impact collisions, the direction in which* A *and* C *move is scarcely disturbed and* AB *is scattered forward. In* (b), *the atom transferred is relatively heavy and the molecular product is scattered into the backward hemisphere*

repulsion between B and C tends to throw the molecule AB back along the direction from which A approached. As the potential is made more attractive, and reaction in collisions with larger impact parameters becomes more likely, there is a tendency towards more forward scattering of the molecular reaction product. However, the extent of this shift in the angular distribution depends quite strongly on the relative masses of A, B and C. *Figure 3.17* shows two quite different cases. In (a), with $m_B \ll m_A$ and m_C, on an attractive surface A can pick up the light B atom in large impact parameter collisions with relatively little disturbance of its initial momentum. Consequently, AB is highly excited internally and is scattered 'forward'. In (b), with $m_B \gg m_A$ and m_C, even on an attractive surface AB is scattered predominantly into the backward hemisphere. This is because as A approaches, roughly along the BC axis, what product repulsion there is is sufficient to expel the light C atom along the BC axis and therefore roughly along a continuation of the line of A's approach. Thus C is scattered forward and AB, by momentum conservation, rebounds backward in the centre-of-mass frame of reference.

3.6 Non-classical calculations on reaction dynamics[18, 19]

Up to the present time, conclusions regarding the relationship between the characteristics of reactive potentials and the results of collisions controlled by those potentials have been based almost entirely on the results of *classical* calculations. It is for this reason that these calculations were described in some detail in Section 3.5. Of course, a classical description of molecular dynamics is necessarily inexact, but non-classical scattering calculations are much more difficult to perform. Only one accurate three-dimensional quantal calculation on a reactive system has been completed[22]. In this section, an attempt is made to explain why, and what kind of, quantum effects arise, and to identify the circumstances under which they might be observed. The techniques for non-classical calculations on molecular collisions are not described in any detail.

3.6.1 Quantum description of molecular scattering

For a classical system selection of the momenta and position coordinates at any instant in time fully determines the state of that system, i.e., the momenta and coordinates, at any subsequent time. To find the (transition) probability of a particular result from many microscopically different collisions—for example, that the final vibrational energy of a molecule lies within a particular range—the number of trajectories (N^*) that lead to that particular result are counted and this number is divided by the total number of trajectories (N) in the sample. This amounts to assigning a probability of $1/N$ to each individual trajectory leading to the specified result and adding these to find the required overall probability,

$$P = \sum_{N^*} (1/N) \tag{3.84}$$

A quantum mechanical system in a selected state at $t = 0$ does not evolve with unit probability into a single state at some later time but rather there are finite probabilities of the system reaching all energetically accessible states. The fundamental quantity linking some initial state i to a final state f is the *S-matrix element*, $S_{i,f}$. This may be looked upon as a *transition amplitude*, which is related to the observable quantity, the *transition probability*, by the equation

$$P_{i,f} = |S_{i,f}|^2 \tag{3.85}$$

A basic difference between classical and quantum mechanical descriptions of scattering is that the former omit all reference to the transition amplitudes and simply sum probabilities directly, as in equation (3.84). The quantum principle of *superposition* requires, however, that one adds the probability amplitudes for indistinguishable processes and then squares the result to find the overall probability. Thus, if just two indistinguishable processes (I and II) give rise to the transition from state i to state f, classically one calculates the total probability as

$$P_{i,f} = P_I + P_{II} \tag{3.86}$$

whereas quantum mechanically

$$P_{i,f} = |S_I + S_{II}|^2$$

$$= P_I + P_{II} + \text{interference term} \tag{3.87}$$

The last term in this equation gives rise to 'quantum effects'. An example is the oscillatory structure on plots of differential cross-sections for elastic scattering that was referred to in Section 3.3.4 and shown in *Figure 3.8*.

3.6.2 Non-classical scattering calculations

S-Matrix elements are less difficult to calculate for inelastic collisions, e.g.,

$$A + BC(n) \rightarrow A + BC(n') \tag{3.88}$$

than for reactive collisions, where the atomic groupings as well as the quantum numbers change. Most fully quantal calculations are based on solutions of the *time-independent* Schrödinger equation[a]. For a non-reactive three-atom system, this is conveniently written in terms of three coordinates (r, θ, ϕ) that specify the position of A relative to the centre-of-mass of BC and three internal coordinates ($\mathbf{r}_{BC}$) of BC.

Solution of the wave equation is attempted by writing the total wave-

[a] A simple, one-dimensional example of the use of the time-independent Schrödinger equation to describe a dynamical problem is given in Section 4.4.1. The use of the time-independent equation implies that specified intermolecular collisions are occurring at a constant rate in the system.

function in terms of sets of basis functions, i.e.,

$$\Psi(r, \theta, \phi, \boldsymbol{r}_{BC}) = \sum_{nlm} r^{-1} u_{nlm}(r) Y_{lm}(\theta, \phi) \psi_n(\boldsymbol{r}_{BC}) \quad (3.89)$$

where $\psi_n(\boldsymbol{r}_{BC})$ are wavefunctions associated with internal states of BC and $Y_{lm}(\theta, \phi)$ are spherical harmonics depending on the orbital quantum number l and its projection m. Introducing this expression into the Schrodinger equation yields a series of coupled equations for the functions $u_{nlm}(r)$. These equations must be solved, subject to the boundary condition that as $r \to \infty$ the radial wave equation reduces to

$$d^2 u_{nlm}(r)/dr^2 = -k_n^2 u_{nlm}(r) \quad (3.90)$$

Consequently, at large r

$$u_{nlm}(r \to \infty) \to k_n^{-\frac{1}{2}}[A_{nlm} \exp(ik_n r + il\pi/2) - B_{nlm} \exp(ik_n r - il\pi/2)] \quad (3.91)$$

The two terms on the right-hand side of this equation represent incoming and outgoing spherical waves, and the S-matrix elements are related to the amplitudes of these waves:

$$B_{nlm} = \sum_{n'l'm'} S_{nlm,n'l'm'} A_{n'l'm'} \quad (3.92)$$

In practice, formal calculations of this kind are limited, because the expansion of Ψ must be truncated to a finite set of basis functions. For accurate results, all the internal states that are coupled during the strongest part of the collision should be included in the calculation. Unfortunately, the size of the computation increases steeply with the number of states in the close-coupled expansion[21], equation (3.89). Consequently, full quantal calculations on inelastic scattering have been restricted to cases where relatively few internal states need be included as, for example, in collisions involving molecules with widely spaced rovibrational levels. This situation is analogous to that in the field of *a priori* potential energy calculations, where for electronically more complex systems the basis set of atomic orbitals required for accurate calculation becomes too large to handle.

The study of *reactive* scattering by close-coupled techniques has additional difficulties. These stem from the fact that the eigenstates of A + BC are not orthogonal to those of AB + C. To overcome this problem, efforts have been made to devise and use systems of 'natural collision coordinates' that transform smoothly from describing the states of the reactants to describing the states of the products.

Because of the difficulty of carrying out exact quantum dynamical calculations, a good deal of interest has centred recently on *semiclassical* calculations. In these, S-matrix elements are calculated via *classical trajectories* and summed to yield transition probabilities, as in equation (3.87). The values of the transition amplitudes are determined via expressions that are derived as quantum mechanics is allowed to approach the classical limit. For example, Miller's formulation[20] proceeds via the Feynmann propagator, which de-

scribes the time evolution of a quantum mechanical system by relating its wavefunction at some time t_f to that at an earlier time t_i. In the classical limit, the matrix elements relating the probability of the system being formed at some position defined by a set of position coordinates q_f given that it was initially at q_i can be evaluated by identifying all the classical paths (differing in their initial momenta, p_i) that connect q_i to q_f. These matrix elements are directly related to the S-matrix elements.

Where it has been possible to compare the results of semiclassical and accurate quantal calculations, the agreement has been excellent. The semiclassical method has been particularly useful in defining those situations in which purely classical calculations are inadequate.

3.6.3 Comparison with results from classical trajectories

When the results of classical and non-classical scattering calculations are compared, two somewhat different quantum phenomena can be distinguished[a]. The first is the interference that arises as a result of quantum mechanical superposition. It causes the quantum mechanical transition probabilities to oscillate about the classical values as the sign of the interference term changes. This effect is observed, for example, in calculations on energy transfer in high energy *collinear* collisions between $H_2(v)$ and He. The quantum mechanical and semiclassical values of $P(v'|v; \varepsilon_T)$ oscillate around the classical values. The latter varies more smoothly with collision energy and provides a kind of average result.

Because of the difficulty of identifying the larger number of classical trajectories connecting initial and final points in three-dimensional space, even semiclassical three-dimensional calculations are difficult to perform. However, because of the much greater number of detailed pathways leading to any particular observable result, there is no doubt that interferences will essentially 'wash out' in real, i.e., three-dimensional, systems involving more than two atoms. This constitutes an important conclusion from comparison of classical and non-classical predictions: where the classical probabilities of a particular event are appreciable, i.e., the region being sampled is 'classically allowed', the classically calculated results are likely to be close to the true values and a classical analysis of the collision dynamics is valid. This applies, for example, to the calculation of reaction cross-sections at energies well in excess of threshold. It also means that the procedure of slicing up a final continuous distribution of vibrational energies to provide a quantum state distribution does not introduce large errors as long as the classical distribution is wide compared with the separation between states.

The second quantum effect is more important. It can be described generally as the 'penetration of classically forbidden regions'. The best publicized example of this phenomenon is that of reactive quantum mechanical

[a] Here, the formation of quasi-bound states is ignored. These are the quantum analogue of orbiting pairs and were considered briefly in Section 3.3.4.

tunnelling at energies below the classically determined threshold. This possibility has been recognized for many years. In the usual form of transition state theory, tunnelling is allowed for by first assuming that motion 'along the reaction coordinate' is separable, and then estimating a 'quantum correction' to the rate expression derived by assuming this motion to be classical. This estimate is usually made by carrying out single-particle, one-dimensional, wave mechanical calculations, examples of which are discussed in Section 4.4. These calculations indicate correctly that tunnelling will have its largest effect on reactions such as that between H and H_2, where the potential barrier is high and narrow, and the reduced mass associated with motion along the reaction coordinate in the transition state is small[a]. However, the numbers that these calculations yield should be treated with caution, since the assumption that motions in the transition state can be treated separately is a poor one, and also because there is disagreement about what potential should be used in the one-particle tunnelling calculations (Section 4.4.2).

More recently, the effects of quantum mechanical tunnelling in the $H + H_2$ reaction have been investigated more properly by means of non-classical calculations on collinear, and even three-dimensional, collisions. Using the semiclassical method, this is done by computing complex valued classical trajectories[20]. In discussing reactive tunnelling, it is usual to compare either the curves of reaction probability versus collision energy derived from collinear calculations or those of the excitation function from three-dimensional calculations. In comparing the results from classical and non-classical calculations, one should carefully distinguish just what kind of calculations are being compared. Thus, for a *truly* classical system, the threshold energy is set entirely by energetic consideration[b]. On the other hand, in *quasi*-classical calculations, kinematic factors limit the extent to which the zero-point vibrational energy of the molecular reactant can be employed in carrying the system over the energy barrier (*see* Section 3.5.3). Consequently, the threshold energy is well above that expected solely on energetic grounds. *Figure 3.18* shows that the collinear $H + H_2$ reaction occurs to an appreciable extent below the threshold energy established in quasiclassical calculations but not significantly below the energetic threshold.

Finally, *Figure 3.19* shows the results of a three-dimensional quantum mechanical treatment of the reaction,

$$H + H_2(v = 0, j = 0) \rightarrow H_2(v = 0, j = 1) + H \qquad (3.93)$$

[a] These requirements can be combined by saying that tunnelling is likely to be important when the complex frequency, derived from a normal mode analysis of the system at the barrier, is large (cf. *Table* 2.6).

[b] To see that there must be a classical trajectory that carries the system from reactants to products with total initial energy $\varepsilon = \mathscr{V}^* + d\varepsilon$, consider starting with a system 'balanced' on the top of the barrier. An infinitesimal displacement in one direction along the reaction coordinate carries this system to products, a similar displacement in the reverse direction to reactants.

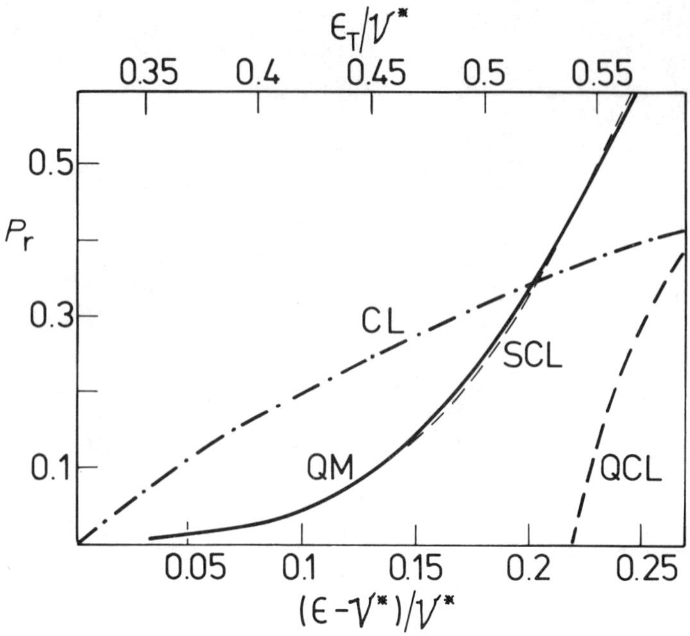

Figure 3.18. Reaction probability in collinear collisions of $H + H_2$ *plotted against* (a) *the reduced collision energy* ($\varepsilon_T/\mathscr{V}^*$) *as given by quantum mechanical* (QM), *semi-classical* (SCL) *and quasiclassical* (QCL) *calculations, and* (b) *the reduced total energy over the barrier height* ($(\varepsilon - \mathscr{V}^*)/\mathscr{V}^*$), *this also including the classical* (CL) *trajectory result* [*adapted from figures in Hornstein, S. M., and Miller, W. H.,* Journal of Chemical Physics, **61**, 745 (1974) *and Chapman, S. Hornstein, S. M., and Miller, W. H.,* Journal of the American Chemical Society, **97**, 893 (1975)]

in which a close coupling technique was used to solve the Schrödinger equation[22]. Except in the immediate threshold region, there is remarkably good agreement between these results and those from quasiclassical trajectory calculations on the same potential hypersurface. At the present time it is not clear whether this agreement is peculiar to the H_3 system. Of course, collisions with energies close to threshold play a dominant part in determining the value of the thermal rate constant, especially at low temperatures. The higher values of the reaction cross-sections derived from Schatz and Kuppermann's quantum mechanical calculations cause the corresponding values of $k(T)$ at 600, 300 and 200 K to exceed the quasiclassical values by factors of 1.07, 3.3 and 18, respectively. The disagreement with values of $k(T)$ calculated from the standard expression of transition state theory—which was evaluated earlier in equation (3.81)—is much worse, rising to a factor of 427 at 200 K. These comparisons foreshadow a difficulty in the application of conventional transition state theory: namely, that the usual rate expression is derived *via* a

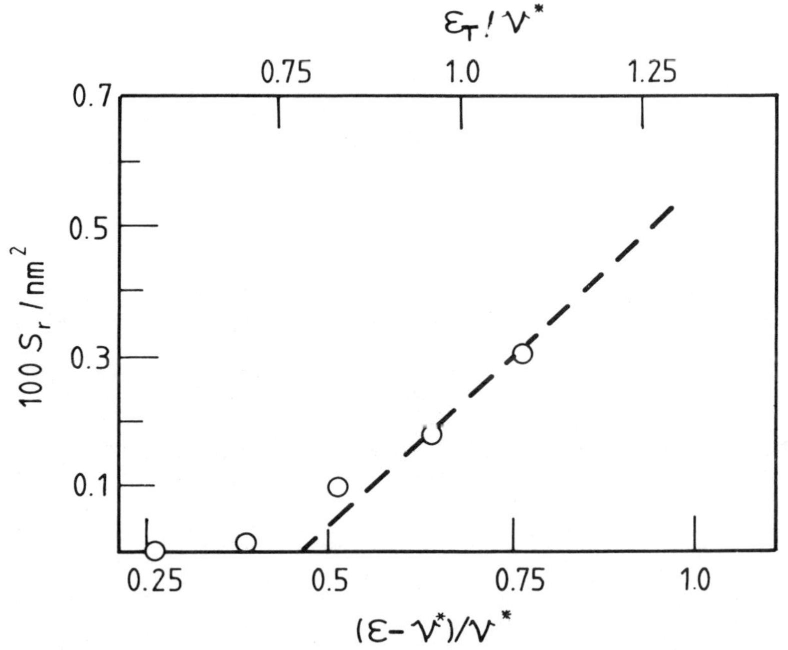

Figure 3.19. Comparison of the reaction cross-sections for H + H_2(v = 0, j = 1) → H_2 + H *from three-dimensional quantum mechanical scattering calculations* (○) *and three-dimensional quasiclassical trajectory calculations* (---) [*adapted from Schatz, G. C., and Kuppermann, A.*, Journal of Chemical Physics, **65**, 4668 (1976)]

procedure involving several assumptions. Consequently, it may be impossible to attribute any disagreement between the rate constant predicted by this expression and that observed experimentally to a single cause. This problem is considered in some detail in the next chapter.

3.6.4. A Postscript

The present situation, as revealed by comparisons of classical and non-classical scattering calculations, is not without irony. It appears that for reactive systems where the most detailed measurements are available from experiment, i.e., from reactive scattering measurements using crossed molecular beams, the results can be explained in terms of classical (i.e., approximate) dynamics. This is because the reactions that can be studied using beams must possess large reaction cross-sections. In spite of a tendency—perhaps evident in this book—to look down on 'simple' thermal rate measurements, it appears likely that measurements of rate constants at low temperature are likely to provide the sternest test of quantum mechanically based rate theories in the immediate future.

References

Section 3.3

1 HIRSCHFELDER, J. O., CURTISS, C. F., and BIRD, R. B., *Molecular Theory of Gases and Liquids,* John Wiley, New York (1954)
2 JOHNSTON, H. S., *Gas Phase Reaction Rate Theory,* chap. 7 and chap. 9, Ronald, New York (1966)
3 FLUENDY, M. A. D., and LAWLEY, K. P., *Chemical Applications of Molecular Beam Scattering*, Chapman and Hall, London (1973)
4 CHILD, M. S., *Molecular Collision Theory,* Academic Press, London (1974)
5 BERNSTEIN, R. B., in *Molecular Beams,* Ed. Ross J., chap. 3, John Wiley, New York (1966), vol. X in the series *Advances in Chemical Physics*
6 BUCK, U., in *Molecular Scattering: Physical and Chemical Applications,* Ed. Lawley, K. P., p. 313, John Wiley, New York (1975), vol. XXX in the series *Advances in Chemical Physics*

Section 3.4

7 BUNKER, D. L., *Theory of Elementary Gas Reaction Rates,* Pergamon Press, Oxford (1966)
8 MENZINGER, M., and WOLFGANG, R., *Angewandte Chemie, International Edition,* **8,** 438 (1969)
9 Le ROY, R. L., *Journal of Physical Chemistry,* **73,** 4338 (1969)

Section 3.5

10 POLANYI, J. C., and SCHREIBER, J. L., in *Physical Chemistry: An Advanced Treatise,* Eds. Eyring, H., Henderson, D. and Jost, W., vol. VI A, chap. 6, Academic Press, New York (1974)
11 PORTER, R. N., in *Dynamics of Molecular Collisions,* Part B, Ed. Miller, W. H., chap. 1, Plenum, New York (1976)
12 KUNTZ, P. J., chap. 2 in ref. 11
13 KARPLUS, M., PORTER, R. N., and SHARMA, R. D., *Journal of Chemical Physics,* **43,** 3259 (1965)
14 POLANYI, J. C., *Accounts of Chemical Research,* **5,** 161 (1972)
15 KUNTZ, P. J., NEMETH, E. M., POLANYI, J. C., ROSNER, S. D., and YOUNG, C. E., *Journal of Chemical Physics,* **44,** 1168 (1966)
16 POLANYI, J. C., and WONG, W. H., *Journal of Chemical Physics,* **51,** 1439 (1969)
17 SMITH, I. W. M., and WOOD, P. M., *Molecular Physics,* **25,** 441 (1973)

Section 3.6

18 GEORGE, T. F., and ROSS, J., *Annual Review of Physical Chemistry,* **24,** 263 (1973)
19 MICHA, D. A., in ref. 6, p. 7.
20 MILLER, W. H., *Accounts of Chemical Research,* **4,** 161 (1971) and in ref. 6, p. 77
21 GORDON, R. G., *Faraday Discussions of the Chemical Society,* **55,** 22 (1973)
22 SCHATZ, G. C., and KUPPERMANN, A., *Journal of Chemical Physics,* **65,** 4642, 4668 (1976)

Chapter 4

Statistical theories of reaction rates

In view of the difficulty of determining complete potential hypersurfaces for reactive systems and then carrying out full-blown quantum scattering calculations using these potentials, it is scarcely surprising that considerable efforts have been invested in devising simpler theories of reaction rates. In this chapter a description is given of a number of theoretical approaches that have in common their basis in the laws of statistical mechanics and in requiring only limited information about the reactive potential. The first section reviews very briefly some relationships from statistical mechanics that will be required subsequently.

4.1 Relationships from statistical mechanics

Statistical mechanics considers the average behaviour of very large assemblies of atoms and molecules, and its fundamental tenet is that all states of a molecular system at the same energy are equally probable. For an assembly of molecules, with the total number of molecules, the volume, and the total energy fixed, this principle can be shown to lead to the Boltzmann laws that determine the distribution of molecules over states not of the same energy

$$\frac{n_i}{n} = \frac{g_i \exp(-\varepsilon_i/k_B T)}{\sum_i g_i \exp(-\varepsilon_i/k_B T)} \qquad (4.1)$$

where (n_i/n) is the fraction of the total number of molecules that are in the energy level i, which has a degeneracy of g_i and an energy ε_i above the lowest state of the system. The denominator on the right-hand side of this equation, $\sum_i g_i \exp(-\varepsilon_i/kT)$, is the *partition function* for the system. Clearly, the probability of any particular molecule having energy ε_i, i.e., $P(\varepsilon_i)$, is equal to (n_i/n).

Usually, ε_i can be written to a good approximation as the sum of energies associated with motions in different degrees of freedom, e.g., translational, rotational, vibrational, and electronic,

$$\varepsilon_i = \varepsilon_T + \varepsilon_R + \varepsilon_V + \varepsilon_{El} + \, \ldots\ldots\ldots \qquad (4.2)$$

so the total partition function is the product of partition functions associated with motions in these (approximately) separable modes:

$$\begin{aligned} Q_{\text{total}} &= \sum g_i \exp(-\varepsilon_i/k_B T) \\ &= \sum g_T \exp(-\varepsilon_T/k_B T) \sum g_R \exp(-\varepsilon_R/k_B T) \ldots\ldots \\ &= Q_T Q_R Q_V Q_{El} \ldots\ldots\ldots \end{aligned} \tag{4.3}$$

Expressions for 'separable partition functions' are given in Appendix 3. When the system is in 'internal thermodynamic equilibrium', a relationship such as equation (4.1) describes the distribution of molecules among the states associated with each of the individual degrees of freedom, and the temperature parameterizing these distributions will be the same for all modes.

A somewhat different description must be given of a system obeying classical, rather than quantum mechanical, laws. Whereas assigning a quantum number defines as closely as possible the state of a quantum mechanical system with respect to the motion in a particular degree of freedom, in classical mechanics both momentum (p) and position (q) coordinates must be specified. In order then to define *completely* the state of a classical system composed of N particles, it is necessary to specify $6N$ coordinates (cf. Section 3.2). Such a state could be represented 'graphically' by a point 'plotted' in $6N$-dimensional *phase space*, and an assembly of such systems would be represented by an unrestricted distribution of such points in phase space.

The classical version of the fundamental principle of statistical mechanics is that all elements of equal 'volume' in phase space at the same total energy have the same probability of containing a phase space point. The classical distribution law may be written in either of two useful ways. First, the probability, $d^{6N}P$, of finding a system within a $6N$-dimensional phase space element, $d\tau_{6N} = dq_1 \ldots\ldots\ dq_{3N}\, dp_1 \ldots\ldots\ dp_{3N}$, can be expressed as

$$d^{6N}P = \frac{\exp(-H/k_B T)\,d\tau_{6N}}{\int \exp(-H/k_B T)\,d\tau_{6N}} \tag{4.4}$$

where the integration sign represents $6N$ integrations over all possible values of the q and p, and H is the total energy written as the classical Hamiltonian [equation (3.2)]. Alternatively, the probability, $P(H)\,dH$, of finding a system with total energy between H and $H + dH$ can be expressed as

$$P(H)\,dH = \frac{g(H)\exp(-H/k_B T)\,dH}{\int g(H)\exp(-H/k_B T)\,dH} \tag{4.5}$$

where $g(H)\,dH$ is a multidimensional volume of phase space enclosed by the hypersurfaces H and $H + dH$. This formulation brings out clearly the analogy between g_i in equation (4.1) and $g(H)$ and $g(H)dH$.

Comparison of equation (4.1) with (4.4) or (4.5) also establishes that the *phase integral*, $\int \exp(-H/k_B T)\,d\tau_{6N}$ occupies the same position in classical statistical mechanics as the partition function in quantum statistical mechanics. It is clear, however, that these quantities cannot correspond exactly,

since their dimensionality is quite different: the partition function is purely a number, whereas the phase integral must have dimensions corresponding to those of (length × momentum)^{3N}. In practice, it can be shown that the ratio of the phase integral to the partition function is h^{3N}, i.e.,

$$\sum_i g_i \exp(-\varepsilon_i/k_B T) = (1/h^{3N}) \int \exp(-H/k_B T)\,d\tau_{6N} \tag{4.6}$$

The origin of this factor can be understood by considering the Heisenberg uncertainty principle. Because the product of the uncertainties in each conjugate pair of q and p must be at least equal to h, phase space cannot be truly continuous. In order to obtain correspondence between classical and quantal formulations, phase space can be thought of ('semi-classically') as being divided up into cells of volume $(1/h^{3N})$, each corresponding to a single quantum state. By dividing the phase integral by h^{3N} the number of phase space cells and hence the quantum states are correctly counted. This can be emphasized by writing equation (4.4) as

$$d^{6N}P = \frac{\exp(-H/k_B T)\,d\tau_{6N}/h^{3N}}{\int \exp(-H/k_B T)\,d\tau_{6N}/h^{3N}} \tag{4.7}$$

Any reactive chemical system can be considered as a single system of the appropriate total number of atoms, with the separated reactants and separated products merely being associated with quite different, and easily recognisable, regions of phase space—or sets of quantum states[a]. According to statistical mechanics, the relative probabilities of systems of a given total energy existing, at equilibrium, as reactants or as products is the ratio of the incremental volumes of phase space associated with the two forms at that energy, or in quantum terms the ratio of their density of states. To find the equilibrium constant, that is the ratio of product to reactant molecules where all species are distributed according to the Boltzmann laws, either the phase space expressions must be integrated over the regions corresponding to separated reactants and separated products, or the appropriate summation over energy states must be performed, leading to

$$\frac{\sum_i n_{i,\,\text{reactants}}}{\sum_j n_{j,\,\text{products}}} = \frac{\sum_i g_i \exp(-\varepsilon_i/k_B T)}{\sum_j g_j \exp(-\varepsilon_j/k_B T)} \tag{4.8}$$

In this equation, $\sum_i n_{i,\,\text{reactants}}$ and $\sum_j n_{j,\,\text{products}}$ correspond to the total number of molecular systems identifiable as reactants or products. To obtain an expression in terms of concentrations of the individual species, one must make the substitutions:

[a] The *proportion* of systems that are not clearly identifiable as reactants or products at any instant, i.e., those in the actual process of reacting, is extremely small.

$$\sum_i n_{i,\text{reactants}} = \prod_i [\text{Reactant}]_i V$$

$$\sum_j n_{j,\text{products}} = \prod_j [\text{Product}]_j V \tag{4.9}$$

where V is the total volume. This yields the following expression for the equilibrium constant in terms of concentrations:

$$K_c(T) = \frac{\prod_j [\text{Product}]_j}{\prod_i [\text{Reactant}]_i}$$

$$\frac{\prod_j (Q_j/V)}{\prod_i (Q_i/V)} \exp(-\Delta\varepsilon_0/k_B T) \tag{4.10}$$

where the (Q/V) are partition functions per unit volume, the subscripts i and j indicating reactants and products respectively, and $\Delta\varepsilon_0$ is the difference between their zero-point levels.

4.2 Transition state theory[1,2]

The 'grandfather' of all statistically based theories of chemical kinetics is *transition state theory* (TST), also referred to as *activated complex theory*. Since its formulation in the 1930s the reputation of TST has suffered a number of ups and downs. At its inception it appeared so superior to the simple collisional theories that were then available that it created great enthusiasm. At the peak of this enthusiasm it even became quite common to refer to the theory as 'absolute rate theory', a terminology that ignored both its inherent assumptions and also the fact that to evaluate the TST expression for the rate constant it is necessary to have data available which can only be provided by accurate *a priori* calculations on the properties of a region of the potential hypersurface. Recognition of this led to a state of affairs in which many kineticists felt that the theory was sufficiently 'elastic' that it could explain everything but predict nothing.

The lowest ebb in the fortunes of TST probably came in the 1960s as the application of new experimental techniques began to provide data outside the realms touched by the theory, and as the advent of large, high-speed computers made quantum scattering calculations feasible and classical scattering calculations fairly straightforward. More recently still, however, there has been an upsurge of interest in TST as attention has been refocused on the fundamental assumptions of the theory and as ways have been sought to circumvent the secondary assumptions[3,4]. In what follows, an attempt is made to identify the primary assumptions of TST as clearly as possible and to differentiate these from the ancillary postulates which are added to obtain an easily evaluated expression for the rate constant.

4.2.1 The usual derivation of the TST expression for $k(T)$

The idea has already been introduced that the reactants and products of a chemical reaction can be thought of as the same molecular system occupying different regions of phase space, and it was also indicated earlier that in a large assembly of such systems the number that cannot at any instant be clearly identified as reactants or products is proportionally very small. The basis of TST is to define a surface in coordinate space and to identify the flux of trajectories which pass through this surface *without turning back* as reactive flux. Because this definition is so wide, the dividing surface can be placed anywhere, but evaluation of the reactive flux, and hence the reaction rate, is facilitated if the surface is chosen so that the ratio of reactive to total flux in either direction (i.e.,

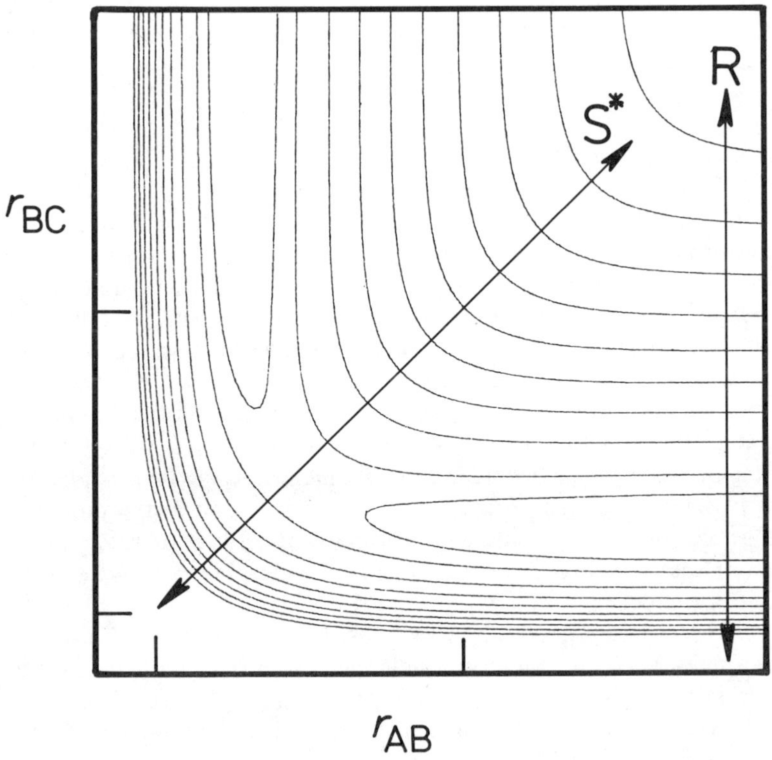

Figure 4.1. Potential energy surface for a symmetric (i.e., A ≡ C*) atom-transfer reaction showing two possible choices for the critical 'dividing line' across which the reactive flux might be calculated.* S* *is a better choice than* **R** *because the total and reactive fluxes will nearly coincide at all collision energies. The line* S* *on this surface corresponds to a 'cut' through the surface defined by* $r_{AB} = r_{BC}$ *on the full hypersurface*

reactants → products, products → reactants) is maximized. This point is quite general but can be illustrated most easily by referring to a symmetrical (A ≡ C), collinear, atom-transfer reaction with an appreciable energy barrier. The energy surface for such a reaction is drawn in *Figure 4.1*. In passing, it should be noted that, in common with nearly all reaction rate theories, TST has as its first assumption the validity of the Born–Oppenheimer separation of electronic and nuclear motions.

Because, for the moment, only collinear collisions are being considered, the dividing surface that has been referred to previously reduces to a dividing line. The line R, drawn perpendicular to the floor of the entry valley well into the reactant zone on the potential surface, could serve as the dividing line for our purposes. However, although it would be quite straightforward to calculate the *total* flux across this line, it would be difficult to evaluate the reactive flux, since only a small proportion of the trajectories passing through this line would continue on over the energy barrier and into the regions of phase space associated with products. The ratio of reactive to total flux could only be established by classical or quantal scattering calculations, which is scarcely what we want!

Obviously a more sensible choice for the dividing line in our example is the line S* defined by $r_{AB} = r_{BC}$. In general, when a reaction has an appreciable energy barrier, we should expect the dividing surface that best meets the previously stated criterion to pass through the configuration corresponding to the energy barrier and to be orthogonal to the reaction path of minimum energy at this point.

The usual procedure for deriving the TST rate expression is based on the supposition that motion along the reaction coordinate in the region of the dividing surface is *separable*. The profile of potential energy along the reaction path of minimum energy for our representative potential surface is shown in *Figure 4.2*. Any molecular system whose position on the potential hypersurface can be described in terms of a value of the reaction coordinate within a small length δ around the barrier crest, and $(3N-1)$ other position coordinates whose values are unimportant, is said to be in the transition state or to be an activated complex. The aim of TST is to calculate the flux of systems through this small region[a].

Before proceeding to derive the rate expression, it is useful to consider the situation at *chemical* equilibrium when the rates of the forward and reverse reactions are equal, i.e.,

$$k_f[A][BC] = k_r[AB][C] \tag{4.11a}$$

If, under these conditions, the *total* concentration of transition-state species, irrespective of their direction of motion through the transition state, is

[a] The number of systems actually in the transition state *at any instant* may seem to be too small for the application of statistical laws, i.e., the fluctuations in the instantaneous flux may be considerable. However, such fluctuations are unobservable so what is calculated is a 'time-averaged' flux. This is a common procedure in statistical mechanics.

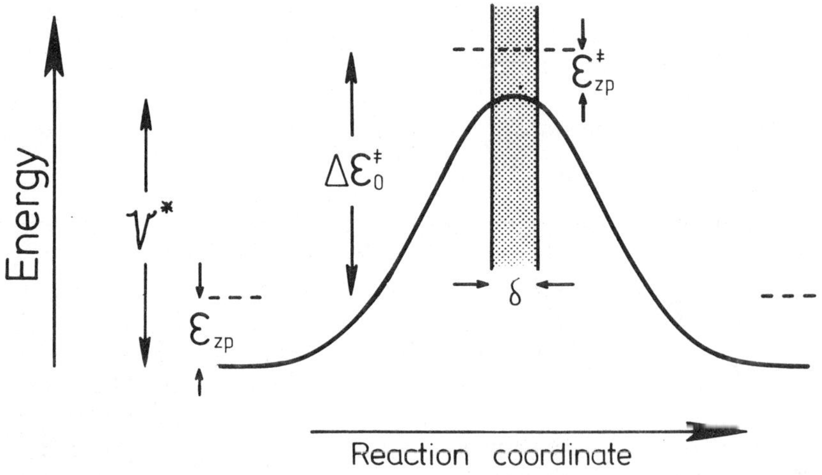

Figure 4.2. Diagram illustrating the relationship between $\mathcal{V}^*$ and $\Delta\varepsilon_0^{\ddagger}$; *transition state species are generally defined as those systems instantaneously 'contained' within a length* δ *along the reaction coordinate and centred at the maximum of the potential energy profile*

$[ABC^+]$, the rate of reaction in the forward direction (and, of course, the equal rate in the reverse direction) is equal to $[ABC^+]\ \langle \overrightarrow{v}_x \rangle/2\delta$, *where* $\langle \overrightarrow{v}_x \rangle$ is the mean velocity along x averaged over the 50 per cent of the complexes moving in the forward direction. The existence of a *dynamic* equilibrium between reactants, products *and* activated complexes can be expressed by the equation:

$$k_f[A][BC] = [ABC^+]\langle \overrightarrow{v}_x \rangle/2\delta \tag{4.11b}$$

$$= [ABC^+]\langle \overleftarrow{v}_x \rangle 2\delta = k_r[AB][C] \tag{4.11c}$$

The term 'dynamic equilibrium' is used to describe the relationships between the concentrations of reactants and 'forward-moving' complexes, i.e., $k_f[A][BC] = [ABC^+]\langle \overrightarrow{v}_x \rangle/2\delta$, and of products and 'reverse-moving' complexes, i.e., $k_r[AB][C] = [ABC^+]\ \langle \overleftarrow{v}_x \rangle/2\delta$. In deriving these relationships, it is implicitly assumed that no systems passing through the transition state region 'turn back': systems starting as A + BC go directly through to AB + C and vice versa. Consequently, equations (4.11b) and (4.11c) describe a quite different situation from that expressed by (4.11a), since chemical equilibrium is established as a result of processes occurring at equal rates in both directions.

Next we should consider the question of whether the reactants, products and activated complexes are in *internal thermodynamic equilibrium*, that is, whether the states or phase space cells associated with these different regions of phase space are populated according to the Boltzmann laws. At the position of chemical equilibrium, there is no doubt that the reactants and products must be individually in internal thermodynamic equilibrium. This is a consequence

of the principle of microscopic reversibility. Each unit phase space cell associated with the reactants will (classically) be connected to a unit phase space cell at the same total energy in the product region. At chemical equilibrium, the rate of the forward and reverse reactions connecting these states must be equal so the occurrence of reaction does nothing to disturb the state of internal thermodynamic equilibrium.

However, when the system is far from equilibrium, for example when the reaction in the forward direction is being studied, there has to be a mechanism to ensure that the equilibrium distribution among reactant states is maintained and those that undergo reaction do not become significantly underpopulated. In unimolecular reactions, where reaction occurs out of internal molecular states whose energy exceeds some critical value, disequilibrium effects can frequently be investigated by observing how the reaction rate varies with the total pressure in the reaction mixture. This is because equilibrium over internal molecular states of the reactant can only be maintained by intermolecular collisions. Consequently, the characteristic 'relaxation time' depends on the total pressure. Only when this relaxation time is much shorter than that characterizing the unimolecular decay of those molecules with sufficient energy to react will the reactant state distribution be effectively undisturbed by the reaction and transition state theory should then apply. With bimolecular reactions, if the relative translational energy is crucial in determining whether reaction takes place, similar effects are much less likely, since the characteristic relaxation time for translational energy (which is exchanged at essentially every collision) is much shorter than the time characteristic of reaction. This may not be the case where a bimolecular reaction proceeds preferentially out of vibrationally excited states under conditions where these states are populated according to the Boltzmann laws, but no instance of this effect has been definitely identified for reactions of this type.

In any case, the assumption that the *reactants* are in thermal equilibrium is often stated as the second fundamental assumption of transition state theory. Alternatively, one can regard this as a requirement on the experimentalist to ensure that non-equilibrium effects are absent by, for example, confirming that the rate of reaction is unaffected by the addition of different concentrations of inert diluent gas. However, as long as the reactants are in internal thermodynamic equilibrium, then the actual rate of reaction in the forward direction, i.e., $k_f[A][BC] = (\langle \vec{v}_x \rangle / 2\delta)[ABC^+]$, is unaffected by the extent to which the reverse reaction is occurring in the system.

It is more difficult to establish that the activated complexes are in internal thermodynamic equilibrium, since their lifetime will be far too short ($\leqslant 10^{-14}$ s) for energy-transferring collisions to play any role. When this difficulty is not simply glossed over, it often appears as if the assumption of internal thermodynamic equilibrium in the transition state is one of the fundamental assumptions of transition state theory. That this is not so will be demonstrated in the next section. For the moment we shall proceed on the basis that the transition state species are distributed among their phase space cells or states according to the Boltzmann laws.

Equation (4.11b) can now be rearranged to yield

$$\begin{aligned} k_f(T) &= (\langle \vec{v}_x \rangle /2\delta)([\mathrm{ABC}^+]/[\mathrm{A}][\mathrm{BC}] \\ &= (\langle \vec{v}_x \rangle /2\delta)K^+(T) \end{aligned} \tag{4.12}$$

and in the light of the discussion in the previous paragraph, $K^+(T)$ can be evaluated by analogy with equation (4.10) for a chemical equilibrium constant, i.e.,

$$K^+(T) = \{(Q^+/V)/\prod_i (Q_i/V)\} \exp(-\Delta\varepsilon_0^+/k_B T) \tag{4.13}$$

Here, Q^+ is the total partition function for the activated complexes and $\Delta\varepsilon_0^+$ is the difference between its zero point energy level and that of the separated reactants. If motion along the reaction coordinate is then assumed to be separable,

$$Q^+ = Q_x Q^{\ddagger} \tag{4.14}$$

where Q_x is the partition function associated with this single mode of motion, and $Q^{\ddagger}$ is the partition function associated with all the other degrees of freedom in the transition state.

All that now remains to be done is to adopt suitable expressions from statistical mechanics for Q_x and $(\langle \vec{v}_x \rangle /2\delta)$. The expression for one-dimensional motion in a restricted length δ is

$$Q_x = (2\pi\mu_x k_B T)^{\frac{1}{2}}\delta/h \tag{4.15}$$

μ_x being the appropriate reduced mass. For $\langle \vec{v}_x \rangle$, the equation

$$\langle \vec{v}_x \rangle = \frac{\int_0^\infty v_x \exp(-p_x^2/2\mu_x k_B T)\,dp_x}{\int_0^\infty \exp(-p_x^2/2\mu_x k_B T)\,dp_x} = (2k_B T/\pi\mu_x)^{1/2} \tag{4.16}$$

gives the average velocity of all forward-moving complexes, i.e., those with $v_x > 0$. The product $(\langle \vec{v}_x \rangle/2\delta)Q_x$ conveniently simplifies to $(k_B T/h)$, *apparently* ridding us of any difficulty in choosing δ or finding μ_x, and combining equations (4.13) to (4.16) yields

$$k_f(T) = (k_B T/h)\{(Q^{\ddagger}/V)/\prod_i (Q_i/V)\} \exp(-\Delta\varepsilon_0^+/k_B T) \tag{4.17}$$

This is the commonest form of the TST expression for the thermal rate constant of a reaction.

What has just been given is essentially the standard derivation of the TST rate expression. Two assumptions, that the Born–Oppenheimer separation is valid and that the reagents are distributed among their states according to the Boltzmann laws, have been identified as fundamental assumptions of the theory. However, a number of additional assumptions were made in arriving at equation (4.17). These were:

(1) that species in the transition state could clearly be identified as 'forward-moving' or 'reverse-moving' and that there was no need to allow for the possibility that systems might cross the dividing surface more than once in their passage from reactants to products or from products to reactants;
(2) that the activated complexes are distributed among their states according to the Boltzmann laws;
(3) that, in the transition state, motion along the reaction coordinate is separable and can be treated classically.

Finally, it is usually assumed that the 'internal' motions in the transition state can be separated so that $Q^{\ddagger}$can be evaluated as the product of partition functions associated with different modes of motion according to equation (4.3).

In the next section an alternative derivation of the TST rate expression is presented and the connection between the assumptions that have just been listed is closely examined.

4.2.2 An alternative derivation and the fundamental assumptions of TST[3,4]

In this derivation, attention is once again focused on the dividing surface—or, for the case of collinear collisions, dividing line—denoted by S*. Initially the assumption is made that the probability of finding a molecular system in a volume element of phase space lying on S* is given by the standard expression from classical statistical mechanics, i.e.,

$$d^{6N}P^* = \frac{\exp(-H/k_B T)\,d\tau_{6N}/h^{3N}}{\int^{S^*} \exp(-H/k_B T)\,d\tau_{6N}/h^{3N}} \tag{4.18}$$

In order to examine this assumption more closely, the following 'thought experiment' can be carried out.

First, q_1 is defined as the coordinate orthogonal to S*, i.e., the reaction coordinate, so that positive values of p_1 contribute to the flux through S* in the direction reactants → products. Now let us suppose that, for a system in any unit phase space cell lying on S*, if 'time runs forward' the system proceeds smoothly to products, and if 'time runs backward' it proceeds smoothly to reactants, with no trajectories in either set recrossing S*. The former condition implies that all the forward flux across S* is reactive flux. The latter condition can be interpreted as follows. Each unit phase space cell at the surface S* has a direct and exact correspondence with a unit phase space cell at the same total energy and located on any surface displaced, as far as we choose, into the reactant region of phase space. For collinear reaction over the surface in *Figure 4.1* this cell could be positioned, for example, at the line R. Furthermore, when any unit phase space cell in the reactant zone which is connected by a (classical) trajectory to a similar cell at S* is occupied, it must inevitably lead some time

later to occupation of the phase space cell at S* and then to products being formed.

Now the probability of any volume element of *reactant* phase space being filled can be written as

$$d^{6N}P = \frac{\exp(-H^\circ/k_B T)\, d\tau^\circ_{6N}/h^{3N}}{\int^{R} \exp(-H^\circ/k_B T)\, d\tau^\circ_{6N}/h^{3N}} \qquad (4.19)$$

where the addition of superscripts and change of integration symbol is made to emphasize that this relationship is written for a region of phase space that can be clearly identified as lying in the reactant zone. However, because of the 'one-to-one' relationship between phase space cells that lie along a reactive trajectory, and because the total energy along such a trajectory must be conserved (i.e., $H = H^\circ$)[a], equations (4.18) and (4.19) can be combined yielding

$$d^{6N}P^* = \frac{n^+}{\sum_i n_{i,\,\text{reactants}}} = \frac{\exp(-H/k_B T)\, d\tau_{6N}/h^{3N}}{\int^{R} \exp(-H^\circ/k_B T)\, d\tau^\circ_{6N}/h^{3N}} \qquad (4.20)$$

Here $\sum_i n_{i,\,\text{reactants}}$ has the same meaning as it had in equations (4.8) and (4.9) and the denominator on the right-hand side of the equation is clearly the product of the partition functions for the reactants. Consequently,

$$n^+ = \prod_i [\text{Reactants}]_i \frac{\exp(-H/k_B T)\, d\tau_{6N}/h^{3N}}{\prod_i (Q_i/V)} \qquad (4.21)$$

All that has been invoked at this point is that the reactants are in internal thermodynamic equilibrium and that all flux across the chosen dividing surface in one direction can be identified as reactive flux in that direction. This is the third, and last, *fundamental* assumption of transition state theory. Its validity will be considered in more detail later. Our justification of the assertion that, if this third fundamental assumption holds, it is unnecessary to make a separate assumption of internal thermodynamic equilibrium in the transition state has, of course, been based on a classical model of the dynamics. However, quantum mechanical[5], as well as more formal classical[4,6], treatments of this problem have been published. Furthermore, TST calculations have been carried out without assuming separability at S*[7]. The results of one such calculation will be referred to in Section 4.3.4.

For the moment, q_1 is identified as the reaction coordinate and motion along this direction is assumed to be separable. Then the incremental rate of reaction, i.e., the contribution to the total rate by passage through a particular element of phase space volume at S* is (dn^+/dt), i.e.,

[a] This is an informal statement of Louville's theorem from classical statistical mechanics.

$$(\mathrm{d}n^{+}/\mathrm{d}t) = \prod_i [\text{Reactant}]_i \frac{(\mathrm{d}q_1/\mathrm{d}t)\exp(-H/k_\mathrm{B}T)\,\mathrm{d}p_1\,\mathrm{d}\tau_{6N-2}}{h^{3N}\prod_i (Q_i/V)} \qquad (4.22)$$

To obtain the total rate of reaction *per unit volume*, integration must be carried out over all phase space at S* having positive values of p_1, and the result divided by V. To carry out the integration, the total Hamiltonian is separated,

$$H = (p_1^2/2\mu_1) + H'$$

and, because $p_1 = \mu_1(\mathrm{d}q_1/\mathrm{d}t)$,

$$\text{Rate} = \prod_i [\text{Reactant}]_i$$

$$\times \frac{(1/\mu_1 h)\int_0^{\infty} p_1 \exp(-p_1^2/2\mu_1 k_\mathrm{B}T)\,\mathrm{d}p_1 \int^{S^*} \exp(-H'/k_\mathrm{B}T)\,\mathrm{d}\tau_{6N-2}}{Vh^{3N-1}\prod_i (Q_i/V)} \qquad (4.23)$$

hence

$$k_\mathrm{f}(T) = (k_\mathrm{B}T/h)\frac{\left\{\int^{S^*} \exp(-H'/k_\mathrm{B}T)\,\mathrm{d}\tau_{6N-2}/Vh^{3N-1}\right\}}{\prod_i (Q_i/V)} \qquad (4.24)$$

The term in { } braces is clearly recognizable as $(Q^{\ddagger}/V)$, i.e., the partition function per unit volume for the transition state species less the contribution from motion along the reaction coordinate. Evaluating this quantity in the usual manner then yields equation (4.17):

$$k_\mathrm{f}(T) = (k_\mathrm{B}T/h)\{(Q^{\ddagger}/V)/\prod_i (Q_i/V)\}\exp(-\Delta\varepsilon_0^{\ddagger}/k_\mathrm{B}T)$$

In this section an attempt has been made to demonstrate that—at least, for systems obeying classical mechanics—the assumptions that

(1) all flux across the critical dividing surface is reactive flux, and
(2) phase space at this surface is occupied according to the Boltzmann laws,
are not independent assumptions.

Only (1) is a *fundamental* assumption of TST; if it is valid, internal thermodynamic equilibrium in the transition state is assured. In proceeding to an easily manageable expression for the rate constant, separability of the motions at S* was again invoked. However, it should be recognized that this assumption is convenient rather than essential.

To summarize, TST should provide an exact prediction of the rate of a reaction between thermally equilibrated reactants provided that

(1) the Born–Oppenheimer separation is valid and the potential energy across the dividing surface S* is known;
(2) the assumption of separability of the relative nuclear motions at S* is not made;
(3) there is some way of identifying the trajectories from phase space cells located at S* which do not lead smoothly to reactants in one direction and products in the other.

The last difficulty can be overcome by running trajectories in both directions starting from phase space cells located on S*. This method has been applied recently, using Monte Carlo methods to sample the phase space at S*, but this must be viewed as a statistical *plus* dynamical procedure. The possibility that all flux through S* may not be reactive flux can be recognized by including a factor κ into equation (4.21) for the incremental rate. For each phase space cell at S*, κ would have to be 0 or 1. Clearly, a maximum value of $k(T)$ or of the reaction cross-section (Section 4.3) will be calculated if $\kappa = 1$ for all these phase space cells. Therefore, the 'exact' TST rate expression calculated with conditions (1) and (2) satisfied, but all flux through S* assumed to be reactive flux, provides an upper limit to the actual rate.

Of course, when equation (4.17) is used to calculate $k(T)$, discrepancies may also arise from assuming that the motions can be separated at S* and that motion along the reaction coordinate can be treated classically. Attempts to correct the standard TST rate expression for quantum dynamical effects are described in Section 4.4.

4.2.3 Evaluation of the usual TST expression for $k(T)$

The 'usual' TST expression for $k(T)$ is equation (4.17). It can immediately be simplified by recognizing that, for both the reactant and transition state species, the overall translational motion is truly separable and the volume V only enters into the partition function associated with this motion (Q_T). Consequently,

$$(Q^{\ddagger}/V) = (Q_T^{\ddagger}/V)Q_{int}^{\ddagger}$$

and

$$\prod_{\text{reactants}} (Q/V) = \prod_{\text{reactants}} (Q_T/V) \prod_{\text{reactants}} (Q_{int}) \tag{4.25}$$

where the Q_{int} represent partition functions associated with all motions other than that of relative translation.

For a *unimolecular* reaction, $(Q_T^{\ddagger}/V)$ and $\prod_i (Q_T/V)$ clearly cancel leaving

$$k(T) = (k_B T/h)(Q_{int}^{\ddagger}/Q_{int})\exp(-\Delta\varepsilon_0^{\ddagger}/k_B T) \tag{4.26}$$

and it is useful to note that $(k_B T/h)$ has the dimensions of $(\text{time})^{-1}$ and has a value of $10^{13}\ \text{s}^{-1}$ at 480 K. For a *bimolecular* reaction,

$$(Q_T^{\ddagger}/V)/\prod_i (Q_T/V)_i = 1/(Q_T/V) \qquad (4.27)$$

where $(Q_T/V) = (2\pi\mu k_B T)^{3/2}/h^3$ is now the partition function *per unit volume* associated with *relative* motion of the reagents, these having a reduced mass of μ. Consequently, for a bimolecular reaction

$$k(T) = (k_B T/h)\{Q_{int}^{\ddagger}/(Q_T/V)\prod_i Q_{int}\} \exp(-\Delta\varepsilon_0^{\ddagger}/k_B T) \qquad (4.28)$$

This procedure corresponds, in dynamical terms, to locating the origin of the system at its centre-of-mass.

Evaluation of Q_{int}, that is the partition function associated with the internal motions for each of the reagents, poses few problems. For these species, the separation of internal motions into overall rotations and normal vibrational modes is usually a good approximation, and Q_{int} can be written as $Q_R Q_V Q_{El} \ldots$, the product of partition functions associated with independent rotational, vibrational and electronic motions. To calculate the individual partition functions, one needs to know (*see* Appendix 3) moments of inertia, vibrational frequencies and electronic states, but such information is readily available from spectroscopic measurements.

Rotational partition functions can nearly always be calculated using the appropriate classical limit formulae (Appendix 3). A slight problem does arise when symmetrical polyatomic reagents allow more than a single detailed reaction path to proceed. At first sight, it might appear that this will be adequately allowed for if symmetry numbers are included in the rotational partition functions of the transition state species and the reactants, but this is not always the case. For example, in the unimolecular isomerization of cyclopropane to propylene, the symmetry numbers of the reactant and activated complex are respectively 6 and 1, but each of the six hydrogen atoms can migrate to either of two neighbouring carbon atoms, so there are a total of 12 equivalent pathways. This factor is correctly treated by omitting symmetry numbers from all the rotational partition functions but multiplying the rate expression by the number of equivalent reaction paths.

Electronic partition functions must be evaluated using the quantum formulation of equation (4.1). In many cases, Q_{El} is simply the electronic degeneracy of the ground state of the species involved, as excited electronic states lie at too high an energy to be populated to any significant extent. However, for atoms with $L > 0$ or molecules with $\Lambda > 0$, spin–orbit coupling causes the ground states to split and proper account should be taken of this via equation (4.1). In accurate calculations on a few molecules, such as NO and OH, which have Π electronic ground states, allowance should also be made for the interaction between rotational and electronic orbital motions, i.e., Q_{El} and Q_R should not be evaluated separately.

Vibrational partition functions are usually calculated assuming separable, harmonic, normal modes. Within this approximation, with $u_s = hc\tilde{\nu}_s/k_B T$,

$$Q_V = \prod_s (1 - \exp[-u_s])^{-1} \qquad (4.29)$$

In most systems, the use of the harmonic approximation in treating the vibrations of the reagents introduces little error. One exception is in the high-temperature unimolecular dissociation of small molecules (*see* Section 4.5). Here, where molecules with sufficient energy to react possess several quanta per vibrational mode, a correction for anharmonicity should be applied.

If we are to use equations (4.26) and (4.28) to determine rate constants, we are left with the problem of evaluating $Q^{\ddagger}_{int}$. It is the impossibility of carrying out this part of the calculation satisfactorily which precludes a proper test of TST. There are three identifiable problems. First, there is our ignorance of the form of the potential in the region of the transition state in all but a handful of reactions. Spectroscopy cannot help, so we must rely on estimates of the potential, preferably from *a priori* quantum mechanical calculations. Of course, one can usually choose a model for the transition state so that, when the rest of the calculation is carried through, the calculated rate constants match those determined experimentally over a limited range of temperature. However, this procedure is only useful if other results, such as the form of the excitation function for a bimolecular reaction, are available, so that these can be cross-checked against the TST predictions without further adjustment in the model chosen for the transition state. Certainly, matching observed and calculated rates neither constitutes a rigorous test of TST nor characterizes the transition state geometry properly.

The remaining problems are closely interrelated and are concerned with how $Q^{\ddagger}_{int}$ should be evaluated, even if the geometry of the transition state species and the potential in this region is known. Usually, the methods which are correctly used to calculate Q_{int} are also employed to find $Q^{\ddagger}_{int}$; that is, motions are separated and treated in the same way as one does when dealing with stable molecules occupying stationary states. This procedure can only be justified on the (dubious) grounds that it makes the subsequent calculation straightforward. It is customary—and we shall later follow this custom—to speak of the separated motions as 'vibrations' and 'rotations'. This terminology is convenient, but it ignores the fact that the 'lifetime' of the transition state is much too short for any such periodic motions to occur. This brings us to a fundamental problem which has received very little attention. It is whether the quantum states associated with the passage of reactive systems through the critical dividing surface are identical, or even similar, to the *stationary* states associated with the periodic motions of some hypothetical stable species which has the same geometry and force constants as the transition state species.

The final problem is whether $Q^{\ddagger}_{int}$ can reasonably be calculated within the assumption of separable motions and small amplitude harmonic 'vibrations', even if the quantum states in the transition state do correspond to those in its hypothetical stable analogue. The validity of this approximation has been considered in several places and is discussed later. *Within the normal mode harmonic approximations*, the 'vibrational' partition function of the transition state is given by an expression like (4.29). Using this equation for Q_V and $Q^{\ddagger}_V$, the zero-point energies of the activated complex and reactants must be

incorporated into the exponential term, i.e.,

$$\Delta\varepsilon_0^{\ddagger}/k_B T = \mathscr{V}^*/k_B T - \sum_s (u_s/2) - \sum_s (u_{s,\,\text{reactants}}/2) \tag{4.30}$$

An alternative formulation of the TST rate expression is

$$k(T) = (k_B T/h)\{(Q^{\ddagger}/V)/\prod_i (Q_i/V)\}\exp(-\mathscr{V}^*/k_B T) \tag{4.31}$$

where now the vibrational partition functions are evaluated with respect to the classical minimum of energy, i.e.,

$$Q_V = \prod_s \exp(-u_s/2)(1 - \exp[-u_s])^{-1} \tag{4.32}$$

Equation (4.31) is useful in treating kinetic isotope effects (Section 5.6) since $\mathscr{V}^*$ does not change on isotopic substitution.

The rotational partition function, $Q_R^{\ddagger}$, is calculated according to the classical limit formulae mentioned earlier. The electronic partition function is usually assumed to be the degeneracy of the lowest electronic hypersurface that leads adiabatically from reactants to products. The adiabatic correlation rules help identify the possible electronic states, although it may be difficult to be certain which of several potentials lies lowest in energy.

4.2.4 The Arrhenius parameters

Transition state theory improves on simple collision theory by incorporating into the basic rate expression quantities, i.e., the partition functions, which reflect the molecular complexity of the reactants and of the single transitory 'supermolecule' through which the system must pass to become products. This improvement is concentrated entirely in the pre-exponential part of the basic rate expression. It is not part of either theory to predict $\mathscr{V}^*$, the height of the energy barrier separating reactants from products.

To compare the TST and SCT rate expressions directly, it is instructive to consider how equation (4.31) is evaluated for the SCT model of reaction between two structureless spheres with the reaction coordinate being the line joining their centres. The only remaining 'internal' modes in the transition state are the two mutually orthogonal rotations of the system around axes that are perpendicular to the line joining the centre of the spheres when they are in contact. Hence,

$$k(T) = (k_B T/h)\frac{(8\pi^2 I k_B T/h^2)}{(2\pi\mu k_B T)^{3/2}/h^3}\exp(-\mathscr{V}^*/k_B T) \tag{4.33}$$

Since I, the moment of inertia of the 'activated' complex, is equal to μd^2, this equation reduces to

$$k(T) = \pi d^2 (8k_B T/\pi\mu)^{1/2} \exp(-\mathscr{V}^*/k_B T) \tag{4.34}$$

This expression is clearly identical to equation (3.58) which was derived from a collisional treatment.

Next, we examine how the pre-exponential term in equation (4.28) might be evaluated in the case of reaction between two complex species, which are both non-linear and together contain N atoms. Assuming that rotational and vibrational modes can be separated, $\prod_i Q_{\text{int}}$ is the product of partition functions associated with six rotational degrees of freedom and $(3N-12)$ vibrations. Bearing in mind that Q_x, the partition function associated with motion along the reation coordinate, has been factorized out, $Q^{\ddagger}_{\text{int}}$ is the product of partition functions associated with three rotations and $(3N-7)$ vibrations or internal rotations. This can be represented by the equation

$$Q^{\ddagger}_{\text{int}}/(Q_{\text{T}}/V)\prod_i Q_{\text{int}} = (Q^3_{\text{R}}Q^{3N-7}_{\text{V}})^{\ddagger}/(Q^3_{\text{T}}/V)Q^6_{\text{R}}Q^{3N-12}_{\text{V}} \tag{4.35}$$

where the indices represent the number of degrees of freedom of the type identified by the subscript.

For the SCT model $N = 2$ and the ratio of partition functions in equation (4.35) reduces to $Q^2_{\text{R}}/(Q^3_{\text{T}}/V)$. In a real reaction, as well as in the SCT model, the three-dimensional relative translational motion associated with reactant collisions transforms (roughly) in the transition state to the motion along the reaction coordinate and two rotational motions around axes perpendicular to one another and to the 'reaction axis'. Consequently, the ratio of the pre-exponential part of the TST rate expression to the rate constant for hard-sphere collisions is given by:

$$A_{\text{TST}}/Z_{\text{hs}}(T) = (Q_{\text{R}}Q^{3N-7}_{\text{V}})^{\ddagger}/(Q^6_{\text{R}}Q^{3N-12}_{\text{V}}) \tag{4.36}$$

In some texts the ratio of partition functions in equation (4.36) is reduced to $\sim(Q^5_{\text{V}}/Q^5_{\text{R}})$ on the grounds that the variation between values of Q_{R} (and of Q_{V}) for different molecules is usually less than that between typical values of Q_{R} and of Q_{V} (*see* Appendix 3). Furthermore, because for stable molecules values of Q_{R} lie roughly in the range 10—100, whereas those of Q_{V} are usually close to unity, it is argued that pre-exponential factors for reactions between complex species may be several orders-of-magnitude below the hard-sphere collision number.

Johnston[8], in particular, has pointed out that the argument in the previous paragraph is dangerously oversimplified. The main reason for this is that most of the 'new' internal modes in the activated complex are ones that have no direct parallel in most stable molecules, as they generally consist of very low-frequency motions, such as essentially free internal rotations and unusually free bending vibrations. We follow Johnston in illustrating the correctness of this remark by considering the transition state for the reaction

$$CH_3 + DC(CH_3)_3 \rightarrow CH_3D + C(CH_3)_3 \tag{4.37}$$

In this system $N = 18$, but it is reasonable to suppose that the only force constants that are appreciably changed in passing from the separated reactants to the transition state are those directly affected by the D-atom transfer. Consequently, one need only consider 12 modes of motion in estimating the

ratio of partition functions in equation (4.35). In the reagents, these 12 degrees of freedom comprise three of relative translation, six of overall rotation, and the three normal vibrational modes (one stretching and two bending) that are closely associated with the C–D bond in $(CH_3)_3CD$. In the transition state these become: motion along the reaction coordinate, three overall rotations, one internal rotation, one stretching vibration, a pair of doubly degenerate bending motions and two pairs of doubly degenerate 'wagging' motions. In evaluating equation (4.36), the motion along the reaction coordinate and the two rotations about axes perpendicular to the reaction axis, together with the relative translational motions of the reactants, need not be considered. The remaining nine degrees of freedom in the transition state are illustrated in *Figure 4.3*. Clearly, nearly all of the 'vibrational' motions in the transition state are likely to yield partition functions appreciably greater than unity, and the kind of crude cancellation indicated in the previous paragraph would grossly underestimate the pre-exponential factor.

Johnston has also shown that the pre-exponential factors for bimolecular reactions involving species of different molecular complexity can be predicted much more satisfactorily if the partition functions are expressed in terms of 'local bond properties' rather than normal mode coordinates. This analysis indicates that pre-exponential factors should fall by only about three orders-of-magnitude in going from the SCT model to reactions involving polyatomic non-linear species. This result is confirmed by experimental measurements, although exceptions are occasionally found when special effects, such as electronically non-adiabatic transitions, occur. This qualitative analysis of the transition state for reaction (4.37) can also be used to point out that it is frequently a poor approximation to assume that rotations and vibrations can be separated in the transition state and that partition functions for the latter can be evaluated using formulae derived for small-amplitude harmonic oscillations.

Although it is no part of TST to predict values of $\mathscr{V}^*$, the rate expression can be used to explore how the Arrhenius activation energy might vary with temperature. Operating on equation (4.28) according to the equation

$$-\mathrm{d}\ln k(T)/\mathrm{d}(1/RT) = E_{\mathrm{act}} \qquad (4.38)$$

that defines the activation energy yields

$$E_{\mathrm{act}} = \Delta E_0^{\ddagger} - \tfrac{1}{2}RT - \frac{\mathrm{d}\ln Q_{\mathrm{int}}^{\ddagger}}{\mathrm{d}(1/RT)} + \sum_{\mathrm{reactants}} \frac{\mathrm{d}\ln Q_{\mathrm{int}}}{\mathrm{d}(1/RT)} \qquad (4.39)$$

Usually this equation can be simplified further by separating vibrational and rotational partition functions and using the classical limit formulae for the latter. Thus, for a reaction such as (4.37), between two non-linear reactants proceeding via a non-linear activation complex, all three rotational partition functions are proportional to $T^{\frac{3}{2}}$ and equation (4.39) becomes

$$E_{\mathrm{act}} = \Delta E_0^{\ddagger} - 2RT - \frac{\mathrm{d}\ln Q_{\mathrm{V}}^{\ddagger}}{\mathrm{d}(1/RT)} + \sum_{\mathrm{reactants}} \frac{\mathrm{d}\ln Q_{\mathrm{V}}}{\mathrm{d}(1/RT)} \qquad (4.40)$$

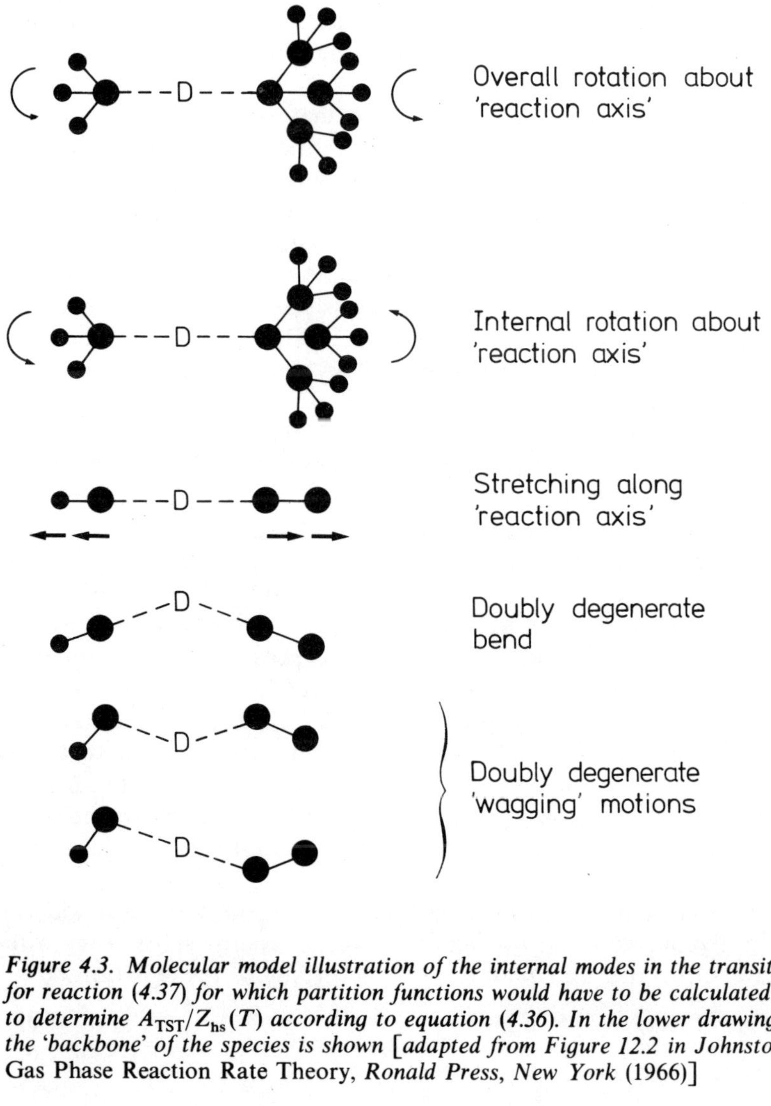

Figure 4.3. ***Molecular model illustration of the internal modes in the transition state for reaction (4.37) for which partition functions would have to be calculated in order to determine $A_{TST}/Z_{hs}(T)$ according to equation (4.36). In the lower drawings only the 'backbone' of the species is shown [adapted from Figure 12.2 in Johnston, H. S.,*** Gas Phase Reaction Rate Theory, *Ronald Press, New York* (1966)]

Generally, $hc\tilde{\nu} \gg k_B T$ for the vibrations of the reactants, and hence the contribution of the last term is small. However, this will not be so for the nine frequencies of the transition state identified in *Figure 4.3*. In practice, one should expect the positive contribution of the term $-\mathrm{d}\ln Q_V^{\ddagger}/\mathrm{d}(1/RT)$ to exceed $2RT$, even at quite low temperatures, and so the activation energy will generally increase with temperature.

4.3 Microcanonical transition state theory[4,6,9]

Up to this point we have been concerned with the development of a statistical theory of reaction rates for a *canonical* assembly of molecules; that is, one in which the distribution of reactant molecules among states is defined by the Boltzmann laws and a single temperature. The underlying basis of statistical mechanics, however, is that all states of a molecular system *at the same energy* are equally probable, so there is no reason why this principle cannot be applied to the derivation of expressions describing the rate of reaction in molecular events (collisions or unimolecular decay) at a specified energy. This leads to what is usually referred to as the *microcanonical* version of transition state theory.

4.3.1 The specific rate constant

The aim in microcanonical TST is to derive an expression for $k(\varepsilon)$, the specific rate constant, which is defined by

$$k(T) = \frac{\int_0^\infty k(\varepsilon) f(\varepsilon; T)\,\mathrm{d}\varepsilon}{\int_0^\infty f(\varepsilon; T)\,\mathrm{d}\varepsilon} \tag{4.41}$$

where $f(\varepsilon; T)$ is a function describing the distribution of reactant total energies at temperature T. The form of this relationship emphasizes again how the thermal rate constant describes the average result of microscopically different events. To calculate the specific rate, i.e., the rate of reaction between species whose total energy lies between ε and $\varepsilon + \mathrm{d}\varepsilon$, a similar procedure can be adopted to that which led to equations (4.20) to (4.24). Before doing this, however, it is useful to determine the rate of reactive flux through the critical surface S* of species with defined total energy.

It is once again assumed that at S* the total energy of transition state species can be expressed as the sum of ε_x, associated with motion along the reaction coordinate, and $\varepsilon_{\mathrm{int}}$, associated with all the other, 'internal', motions of the activated complex. To find the reactive flux through a single internal state, the number of translational states per unit length along the reaction coordinate must be multiplied by $(\mathrm{d}q_1/\mathrm{d}t)$, the velocity along this direction. Since $(\mathrm{d}q_1 \mathrm{d}p_1/h)$ is the number of states in the length $\mathrm{d}q_1$ with momentum directed along x in the range p_1 to $p_1 + \mathrm{d}p_1$, the number of states *per unit* length along x is $(\mathrm{d}p_1/h)$ and, if all these states were occupied, the required flux would be given by

$$\begin{aligned}(\mathrm{d}p_1/h)(\mathrm{d}q_1/\mathrm{d}t) &= (\mathrm{d}p_1/h)(p_1/\mu_1)\\ &= \mathrm{d}\varepsilon_x/h\end{aligned} \tag{4.42}$$

since $\varepsilon_x = p_1^2/2\mu_1$.

For a defined internal state of the activated complex, $\varepsilon_{\mathrm{int}}$ is fixed, so that $\mathrm{d}\varepsilon_x = \mathrm{d}\varepsilon$ and $(\mathrm{d}\varepsilon/h)$ corresponds to the contribution to the total flux from

passage through a single unit phase space cell located at S*. To find the total flux, $\boldsymbol{R}(\varepsilon)\mathrm{d}\varepsilon$, of systems with total energies between ε and $\varepsilon+\mathrm{d}\varepsilon$, allowance must be made for contributions to the flux arising from all possible partitionings of the energy $(\varepsilon-\mathscr{V}^*)$ between ε_x and ε_{int}, all of these being assumed equally probable. If there are N^* internal states in the range $\varepsilon_{\text{int}} \leqslant (\varepsilon-\mathscr{V}^*)$, there are N^* different reaction 'pathways', and

$$\boldsymbol{R}(\varepsilon)\,\mathrm{d}\varepsilon = \sum_{N^*} (\mathrm{d}\varepsilon_x/h) = \sum_{N^*} (\mathrm{d}\varepsilon/h) \tag{4.43}$$

so that

$$\boldsymbol{R}(\varepsilon) = \sum_{N^*} (1/h) = N^*(\varepsilon-\mathscr{V}^*)/h \tag{4.44}$$

where $N^*(\varepsilon-\mathscr{V}^*)$ is the number of internal states of the activated complex with energy range between $\mathscr{V}^*$ and ε. $\boldsymbol{R}(\varepsilon)$ has the dimensions of (energy time)$^{-1}$ and can be regarded as a differential total flux. Appropriate averaging over ε will yield the thermal reactive flux.

To find $k(\varepsilon)$, however, it is best to start with the microcanonical version of equation (4.20). This is

$$d^{6N}P^* = \frac{n^+}{\sum_i n_{i,\,\text{reactants}}} = \frac{\exp(-H/k_{\mathrm{B}}T)\,\mathrm{d}\tau_{6N}/h^{3N}}{g(H^\circ)\exp(-H^\circ/k_{\mathrm{B}}T)\mathrm{d}H^\circ/h^{3N}} \tag{4.45}$$

Since we are now concerned only with systems with energy between $\varepsilon = H^\circ$ and $\varepsilon + d\varepsilon = H^\circ + \mathrm{d}H^\circ$, the denominator in equation (4.20) is replaced by $g(H^\circ)\exp(-H^\circ/k_{\mathrm{B}}T)\,\mathrm{d}H^\circ/h^{3N}$, where $g(H^\circ)\,\mathrm{d}H^\circ/h^{3N}$ corresponds to the number of unit cells in the volume of phase space bounded by the surfaces H° and $H^\circ + \mathrm{d}H^\circ$ [cf. equation (4.5)]. Cancelling the exponential terms in equation (4.45) yields

$$d^{6N}P^* = \frac{n^+}{\sum_i n_{i,\,\text{reactants}}} = \frac{\mathrm{d}\tau_{6N}/h^{3N}}{g(H^\circ)\,\mathrm{d}H^\circ/h^{3N}} \tag{4.46}$$

If this equation is rearranged and the derivative is then taken with respect to time, an expression is obtained similar to equation (4.22), i.e.,

$$(\mathrm{d}n^+/\mathrm{d}t) = \sum_i n_{i,\,\text{reactants}} \frac{(\mathrm{d}p_1/h)\,(\mathrm{d}q_1/\mathrm{d}t)\,(\mathrm{d}\tau_{6N-2}/h^{3N-1})}{g(H^\circ)\,\mathrm{d}H^\circ/h^{3N}} \tag{4.47}$$

The numerator in the second term on the right-hand side of this equation is clearly what was identified earlier as the total reactive flux of systems possessing total energies between ε and $\varepsilon+\mathrm{d}\varepsilon$. It can, according to equations (4.43) and (4.44), be written as $N^*(\varepsilon-\mathscr{V}^*)\,\mathrm{d}\varepsilon/h$. The evaluation of other terms in equation (4.47) differs slightly depending on the molecularity of the reaction under consideration.

For a *unimolecular* reaction, overall translation can be ignored. To calculate

the rate of reaction per unit volume, each side of equation (4.47) is divided by V yielding

$$\text{Rate} = k(\varepsilon)[\text{Reactant}]$$

$$= [\text{Reactant}]\frac{N^*(\varepsilon - \mathscr{V}^*)\,\mathrm{d}\varepsilon}{g(H^\circ)\,\mathrm{d}H^\circ/h^{3N}} \qquad (4.48)$$

$g(H^\circ)\,\mathrm{d}H^\circ/h^{3N}$ is, of course, the number of *internal* reactant states in the specified energy range. In the classical limit, it can be written as $\rho(\varepsilon)\,\mathrm{d}\varepsilon$, where $\rho(\varepsilon)$ is the density of these states at energy ε, i.e., the number of states per unit increment of energy. The specific rate constant is then given

$$k(\varepsilon) = N^*(\varepsilon - \mathscr{V}^*)/h\rho(\varepsilon) \qquad (4.49)$$

Other methods of evaluating $g(H^\circ)\,\mathrm{d}H^\circ/h^{3N}$ are discussed in Section 4.5.

For *bimolecular* reactions, analogous relationships exist, but, in evaluating the appropriate volume of reactant phase space, states associated with *relative* translational motion of the reactants must also be taken into account. The expression for $k(\varepsilon)$ is the same, but $\rho(\varepsilon)$ is the density of internal and relative translational states of the reactants *per unit volume.*

4.3.2 Relationship of the differential reaction rate to the reaction cross-section[10, 11]

The results of encounters between reactants whose collision energy and internal states are selected is usually expressed in terms of the reaction cross-section, $S_r(|n; \varepsilon_T)$. To relate the cross-section to the differential reactive flux or rate constant is fairly straightforward.

In terms of l, the quantum number specifying the orbital angular momentum in a collision, the reactive cross-section is given by

$$S_r(|n; \varepsilon_T) = \frac{\sum_0^{l_{max}} P_l(2l+1)}{\sum_0^{l_{max}} (2l+1)} \pi b^2_{max}$$

$$= \pi(b_{max}/l_{max} + 1)^2 \sum_0^{l_{max}} P_l(2l+1) \qquad (4.50)$$

where P_l is the probability of reaction in a collision with orbital momentum defined by l. Now, because

$$\mu v = p = k\hbar \qquad (4.51a)$$

and

$$\mu v b \approx (l+1)\hbar \qquad (4.51b)$$

$$(b_{max}/l_{max} + 1)^2 \approx 1/k^2 \tag{4.52}$$

so that

$$S_r(|n; \varepsilon_T) = (\pi/k^2) \sum_{0}^{l_{max}} P_l(2l + 1) \tag{4.53}$$

Because of the 'conservation of reactive states' in going from the reactant zone of phase space to the transition state

$$\sum_{0}^{l_{max}} P_l(2l + 1) = N^*(\varepsilon - \mathscr{V}^*) \tag{4.54}$$

Consequently,

$$S_r(|n; \varepsilon_T) = (\pi/k_n^2) h \boldsymbol{R}(\varepsilon_T + \varepsilon_n) \tag{4.55}$$

where $\varepsilon = \varepsilon_T + \varepsilon_n$, and the subscript on k is introduced to emphasize that different partitionings of the initial energy between ε_T and ε_n are possible. This leads to the general expressions:

$$\boldsymbol{R}(\varepsilon) = (1/\pi h) \sum_{n} k_n^2 S_r(|n; \varepsilon) \tag{4.56}$$

and

$$k(T) = \frac{\displaystyle\int_0^\infty (1/\pi h) \sum_{n} k_n^2 S_r(|n; \varepsilon) \exp(-\varepsilon/k_B T)\, d\varepsilon}{\displaystyle\prod_i (Q_i/V)} \tag{4.57}$$

4.3.3 Applications of microcanonical TST

The equations from the microcanonical version of TST have found their widest applications in regard to unimolecular reactions and these will be considered in Section 4.5. Their use with bimolecular systems can be illustrated with the simple example provided by the SCT model of reaction.

In order to evaluate equation (4.49) for this simple model system, expressions are needed for the density of relative translational states and the number of internal states in the transition state lying in the range ε_T to $\mathscr{V}^*$. The number of relative translational states per unit volume is $4\pi p^2\, dp/h^3$ where $p = (2\mu\varepsilon_T)^{\frac{1}{2}}$ and $p\, dp = \mu\, d\varepsilon_T$. This leads to the following equation for the density of reactant states

$$\rho(\varepsilon_T) = (2^{\frac{1}{2}} \hbar^3 \pi^2)^{-1} \mu^{\frac{3}{2}} \varepsilon_T^{\frac{1}{2}} \tag{4.58}$$

In the transition state, only rotational states require counting. For a rigid linear rotor in the classical limit, the density of states is uniform and equal to $(2I/\hbar^2)$. Consequently

$$\mathscr{N}^*(\varepsilon_T - \mathscr{V}^*) = (2I/\hbar^2)(\varepsilon_T - \mathscr{V}^*) \tag{4.59}$$

with $I = \mu d^2$. When (4.58) and (4.59) are substituted into equation (4.49), one obtains

$$k(\varepsilon_T) = (2\varepsilon_T/\mu)^{\frac{1}{2}}\pi d^2(\varepsilon_T - \mathscr{V}^*)/\varepsilon_T$$
$$= v\pi d^2(\varepsilon_T - \mathscr{V}^*)/\varepsilon_T \qquad (4.60)$$

This expression is identical to that obtained by substituting the hard-sphere excitation function, equation (3.57), into equation (1.24) for the rate coefficient for reaction in collisions at a selected collision energy or relative velocity, v.

4.3.4 A test of transition state theory

Microcanonical transition state theory is based on exactly the same assumptions as the usual canonical form. In particular, the total flux of systems across the dividing surface S* is identified with the reactive flux. Recently, tests of this fundamental assumption have been made by comparing the results of

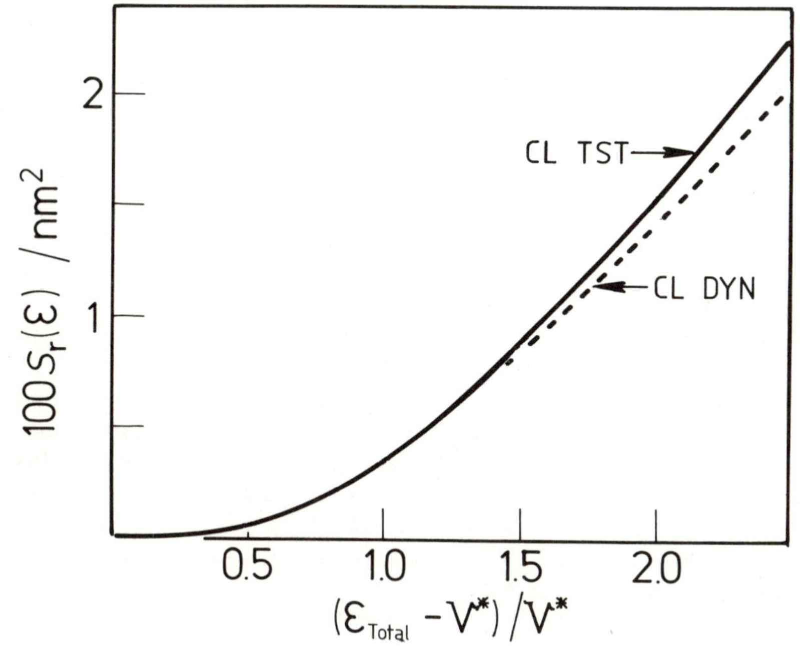

Figure 4.4. Comparison of excitation functions for the $H + H_2$ *reaction on the Porter–Karplus potential calculated from* classical TST *without any assumption of separability* (CL TST) *and by classical trajectories* (CL DYN) [*adapted from Chapman, S., Hornstein, S. M., and Miller, W. H.*, Journal of the American Chemical Society, **97**, 893 (1975)]

TST calculations, made *without* the secondary assumption of separability of the motions at S*, with the results of scattering calculations.

Some results from *classical* calculations of this kind are shown in *Figure 4.4*. The excitation function labelled (CL DYN) was obtained from three-dimensional classical trajectory calculations on the Porter–Karplus potential for H_3. These trajectories differed from those calculated by Karplus, Porter and Sharma, and referred to earlier, in that the initial internal energies of H_2 were selected by Monte Carlo procedures from a true classical distribution and not assigned values corresponding to quantum states. The curve labelled CL TST was determined by evaluating classical flux integrals without any assumption of separability of the various modes of motion.

The two curves deviate appreciably only at energies some way above threshold. This indicates that only at high energies do a significant proportion of classical trajectories turn back having passed through the transition state. Consequently, the third fundamental assumption of TST is valid for this classical system over an important range of energies, since the contribution to the thermal rate constant of collisions at total energies in the range where the two curves deviate will be extremely small.

4.4 'Special' quantum effects in transition state theory[12]

In deriving the usual TST rate expression, motions in the transition state were assumed, without justification, to be separable. In subsequently evaluating partition functions for these modes, it was further assumed that the motion along the reaction coordinate could be treated classically. On the other hand, it was suggested that the partition function associated with the other motions be determined in the quantized rigid rotor–harmonic oscillator approximation that is used for stable molecules, although it was pointed out that this may be a poor approximation in the transition state of many reactions.

The last of these assumptions leads, of course, to 'quantum effects' in transition state theory, since the rate expression would be quite different if the vibrational partition functions of the activated complex were evaluated classically. For example, zero-point energy in the transition state would disappear. However, nothing more will be said about this effect. Rather, in this section, attempts that have been made, within the framework of transition state theory, to allow for the non-classical nature of motion along the reaction coordinate will be examined. In addition, some remarks will be made about electronically non-adiabatic reactions.

Some simple considerations suggest strongly that the quantum nature of motion along the reaction coordinate is important for some reactions. For example, the threshold energy for

$$H + H_2(v = 0, j = 0) \rightarrow H_2 + H \qquad (4.61)$$

according to standard TST, corresponds to $\Delta\varepsilon_0^{\ddagger}$, the difference in zero-point energies in the transition state and in the reagents. For the Porter–Karplus

potential (*see* Table 2.6), $\Delta\varepsilon_0^{\ddagger}$ amounts to 36.7 kJ mol^{-1}, compared with a barrier height of $\mathscr{V}^* = 38.2$ kJ mol^{-1}. But Schatz and Kuppermann's quantum scattering results[13], displayed on *Figure 3.19*, demonstrate that there is appreciable reaction at much lower collision energies. This diagram also indicates, however, that the occurrence of reaction at energies $\varepsilon_T < \Delta\varepsilon_0^{\ddagger}$ may not entirely be the result of 'quantum penetration' of the potential barrier. Quasiclassical trajectories in this regime also lead to reaction. This means that the collision dynamics, *even when treated classically*, allow systems to pass through the critical dividing surface at energies less than the zero-point energy given by the simplistic, separable, description of transition state motions outlined in Section 4.2.3. For the present we shall set aside these complicated questions, and merely consider attempts that have been made to correct the usual TST rate expression for quantum features of the motion carrying the system along the reaction coordinate.

Next, it is useful to consider when the assumption of separable classical motion along the reaction coordinate is valid. A *first* condition is that the de Broglie wavelength associated with this motion is small relative to the 'width' of the potential barrier. For H_3, the appropriate reduced mass, μ_x, is 2/3 atomic mass units and for an energy of $\varepsilon_T \equiv 38.2$ kJ mol^{-1}, corresponding to the barrier height of the PK potential, the de Broglie wavelength is 0.056 nm[a].

Inspection of the PK surface in *Figure 1.2* shows clearly that the change in potential over this distance along the reaction path of minimum energy is considerable. Not only is the *first* requirement not satisfied for the H_3 system, but there is a *second*, more stringent condition. This is that the de Broglie wavelength should be short compared with the distance along the reaction coordinate for which separation of this motion is a reasonably good approximation.

For H_3, the reaction coordinate *at the transition state* can be defined as

$$x = (r_{AB} - r_{BC}) \qquad (4.62)$$

Separation of the motion along x from others in the transition state is only valid over the range of x in which $\mathscr{V}$ is proportional to x^2. Examination of the PK surface shows that the line $r_{AB} - r_{BC}$ is a tangent to the reaction path of minimum energy at the crest of the energy barrier. Furthermore, the variation of $\mathscr{V}$ along either of these paths is parabolic over a distance much shorter than the de Broglie wavelength.

Because the second of these conditions is more difficult to satisfy than the first, calculations that estimate a correction to the standard TST rate expression to allow for tunnelling in one-dimensional calculations are necessarily inexact. Recognition of this fact has led to some dispute as to what potential is best used in a calculation of this kind, if errors are to be kept to a

[a] This is the wavelength associated with the free, i.e., where $\mathscr{V} = 0$, motion of a particle of mass 2/3 amu and translational energy 38.2 kJ mol^{-1}. The wavelength of such a particle in the region of the barrier, where its kinetic energy must approach zero, will be appreciably longer.

minimum. Before considering this, the results of calculations on some idealized one-dimensional systems are described to illustrate the nature of the tunnelling phenomenon and to indicate the magnitude of the resultant quantum effects.

4.4.1 Quantum mechanical tunnelling in one dimension

Figure 4.5 shows three one-dimensional energy barriers of the same maximum height $\mathscr{V}^*$. According to classical mechanics, particles approaching any of these barriers from $x = -\infty$ would definitely continue to $x = \infty$ if their initial translational energy ε_x exceeds $\mathscr{V}^*$, but would equally surely be reflected if $\varepsilon_x < \mathscr{V}^*$. A plot of the probability of a particle passing the barrier would be a step function, with P_{cl} rising from 0 to 1 at $\varepsilon_x = \mathscr{V}^*$. In quantum mechanics, however, there is both a finite probability that a particle with $\varepsilon_x < \mathscr{V}^*$ will penetrate the barrier and a chance that a particle with $\varepsilon_x > \mathscr{V}^*$ will be reflected.

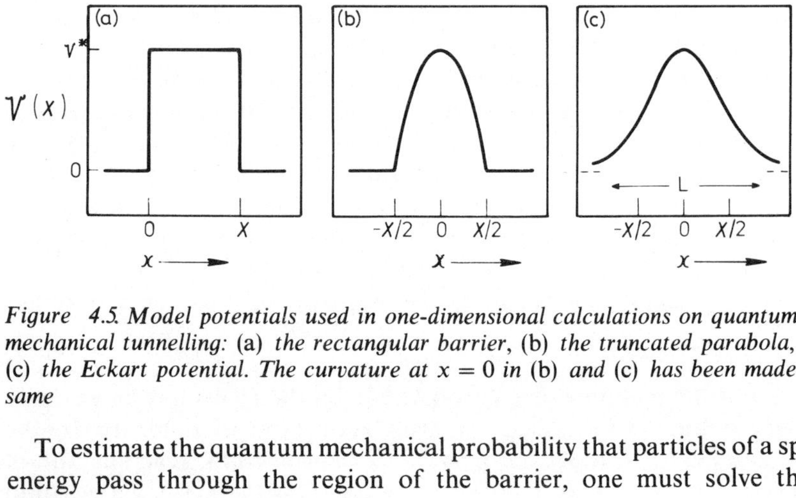

Figure 4.5 Model potentials used in one-dimensional calculations on quantum mechanical tunnelling: (a) *the rectangular barrier,* (b) *the truncated parabola, and* (c) *the Eckart potential. The curvature at x = 0 in* (b) *and* (c) *has been made the same*

To estimate the quantum mechanical probability that particles of a specified energy pass through the region of the barrier, one must solve the one-dimensional Schrödinger equation:

$$(-\hbar^2/2\mu_x)\mathrm{d}^2\psi/\mathrm{d}x^2 = [E - \mathscr{V}(x)]\psi \tag{4.63}$$

The use of the *time-independent* equation[a] reflects the fact that a situation is being considered where a constant flux of particles approach the barrier either from $x = -\infty$ or from both $x = -\infty$ and $+\infty$. The wave equation is easiest to solve for the rectangular energy barrier shown in *Figure 4.5*(a), and this case provides a simple illustration of the general procedure.

For the rectangular barrier, three regions of x can be distinguished:

[a] The use of the time-independent Schrödinger equation in multi-dimensional, quantal, scattering calculations was referred to briefly in Section 3.6.2.

(1) $x < 0$ where $\mathscr{V}(x) = 0$, (2) $0 < x < X$ where $\mathscr{V}(x) = \mathscr{V}^*$, and (3) $x > X$ where again $\mathscr{V}(x) = 0$. In each of these regions, $\mathscr{V}(x)$ is constant and the form of the wave function is well-known:

$$\psi_1 = A_1 \exp(i\alpha x) + B_1 \exp(-i\alpha x) \tag{4.64a}$$

$$\psi_2 = A_2 \exp(\beta x) + B_2 \exp(-\beta x) \tag{4.64b}$$

$$\psi_3 = A_3 \exp(i\alpha x) + B_3 \exp(-i\alpha x) \tag{4.64c}$$

where $\alpha = (2\mu_x \varepsilon_x)^{\frac{1}{2}}/\hbar$ and $\beta = (2\mu_x[\mathscr{V}^* - \varepsilon_x])^{\frac{1}{2}}/\hbar$.
If $\varepsilon_x < \mathscr{V}^*$, β is a positive real number, whereas if $\varepsilon_x > \mathscr{V}^*$, β is imaginary and ψ_2 has exactly the same form as ψ_1 and ψ_3. In each of these functions, the first term is associated with the motion of particles moving in the positive x direction, the second with those moving to negative x. Each of the A and B coefficients, i.e., the amplitudes of the waves, multiplied by its complex conjugate is proportional to the intensity of the stream of particles in the region and direction identified by the coefficient. To determine the quantum mechanical probability of particles passing through the region of the barrier, we consider the situation with particles approaching only from $x = -\infty$, so that $B_3 = 0$, and calculate

$$P_{qu}(\varepsilon_x) = |A_3 A_3^*| / |A_1 A_1^*| \tag{4.65}$$

To find this ratio, the properties of the wavefunctions at the discontinuities of the potential are examined. Here, as elsewhere, ψ must be single-valued and continuous. Applied at $x = 0$, this leads to the conditions

$$A_1 + B_1 = A_2 + B_2 \tag{4.66a}$$

and

$$i\alpha(A_1 - B_1) = \beta(A_2 - B_2) \tag{4.66b}$$

and applied at $x = X$,

$$A_2 \exp(\beta X) + B_2 \exp(-\beta X) = A_3 \exp(i\alpha X) \tag{4.67a}$$

and

$$\beta A_2 \exp(\beta X) - \beta B_2 \exp(-\beta X) = i\alpha A_3 \exp(i\alpha X) \tag{4.67b}$$

These four equations allow one to determine the ratio of any two of the coefficients. Manipulation of the equation yields:

$$1/P_{qu}(\varepsilon_x) = \cosh^2(\beta X) + \frac{(\alpha^2 - \beta^2)^2}{4\alpha^2\beta^2} \sinh^2(\beta X) \tag{4.68}$$

The form of this expression indicates that quantum effects will be prominent when X and β are both small. For particular values of $\mathscr{V}^*$ and ε_x, β will be smallest, and therefore tunnelling will be most pronounced, for light particles. Similarly, for particles of any specified mass, the difference between P_{cl} and P_{qu} will be greatest when $\varepsilon_x \approx \mathscr{V}^*$, so that again β is small. These conclusions are consistent with the de Broglie condition: classical behaviour is a good

approximation as long as the width of the barrier is appreciably greater than the de Broglie wavelength associated with the motion of the particles.

Figure 4.5(b) shows a truncated parabolic potential of the type first suggested by Bell, and *Figure 4.5*(c) a symmetric Eckart potential. The latter has the general form:

$$\mathscr{V}(y) = \frac{Ay}{(1+y)} + \frac{By}{(1+y)^2} \tag{4.69}$$

where $y = \exp(2\pi x/L)$. In this form, the Eckart function, with $A \neq 0$, can be used to represent systems where the potential energies at $x = +\infty$ and $-\infty$ are not the same. Furthermore, in contrast to the rectangular and truncated parabolic potentials, there are no discontinuities in the Eckart potential and $\mathscr{V}$ approaches zero asymptotically as x approaches $+\infty$ or $-\infty$.

The Schrödinger equation can be solved inexactly, but accurately, for the truncated parabola, and analytically for the Eckart potential, but details of these procedures will not be given here. However, the results of tunnelling calculations on the three potentials shown in *Figure 4.5* are compared in *Table 4.1*. Parameters for the potentials were chosen as follows. The value of $\mathscr{V}^*$ and the curvature at the maximum of potentials (b) and (c) were chosen to match those from the calculations of Shavitt *et al.* on H_3 (*see* line 4 of *Table 2.6*). These properties determine values of 0.127 nm for L, the length parameter in the Eckart potential, and 0.081 nm for X, the width across the truncated parabolic potential. The width across the rectangular potential was also taken to be 0.081 nm and the mass of the particles was 2/3 atomic mass unit, corresponding to the reduced mass of motion along $(r_{AB} - r_{BC})$ in the H_3 system.

The transmission probabilities in *Table 4.1* emphasize the strong influence

Table 4.1 Tunnelling factors, P_{qu}, as a function of reduced energy, $(\varepsilon_x/\mathscr{V}^*)$, for the simple one-dimensional barriers shown in Figure 4.5

$(\varepsilon_x/\mathscr{V}^*)$	*Rectangular* $(\mu_x = 2/3)^a$	*Parabolic* $(\mu_x = 2/3)$	*Eckart* $(\mu_x = 2/3)$	*Eckart* $(\mu_x = 32/3)$
0.2	5.0 (−8)[b]	3.8 (−6)	3.8 (−8)	
0.4	8.1 (−7)	8.7 (−5)	1.2 (−5)	
0.6	1.3 (−6)	2.0 (−3)	1.0 (−3)	
0.8	3.6 (−4)	4.2 (−2)	4.2 (−1)	2.0 (−6)
0.9	2.7 (−3)	0.17	0.19	1.7 (−3)
0.95	9.1 (−3)	0.31	0.35	4.2 (−2)
1.0	3.9 (−2)	0.5	0.54	0.5
1.05	0.99	0.69	0.72	0.96
1.1	1.0	0.83	0.84	1.0
1.2	1.0	0.96	0.96	1.0

[a] Masses in atomic mass units
[b] $5.8(-8) \equiv 5.0 \times 10^{-8}$.

of the barrier width in determining the values of $P_{qu}(\varepsilon_x)$. When $\varepsilon_x/\mathscr{V}^*$ is just less than unity, probabilities for potentials (b) and (c) are similar and are both appreciably larger than those for the wider rectangular barrier. However, as $\varepsilon_x/\mathscr{V}^*$ decreases the Eckart potential becomes widest, and at low energies the transmission probabilities are smallest for this potential. The figures in the last column indicate that quantum effects are very much reduced when μ_x is increased, and indeed the variation of $P_{qu}(\varepsilon_x)$ is, in this case, quite close to the classical step-function.

In order to calculate the ratio of quantum to classical rates of passage through the region of the barrier, it is necessary first to consider regions of one-dimensional space far from the barrier where $\mathscr{V}(x) = 0$. If the concentration of particles per unit length of x is n_x and if these particles all have the same momentum $p_x = (2\mu_x\varepsilon_x)^{\frac{1}{2}}$ in the positive x direction, the flux of particles through any point in this region is $n_x p_x/\mu_x$. If $\varepsilon_x > \mathscr{V}^*$ and the particles behave classically, they will all cross the barrier. In order then to calculate the classical rate where the momenta of the particles are distributed thermally, it is necessary to carry out the normal Boltzmann averaging and evaluate

$$\boldsymbol{R}_{cl}(T) = \frac{\displaystyle\int_{\varepsilon_x=\mathscr{V}^*}^{\infty} (n_x p_x/\mu_x)\exp(-p_x^2/2\mu_x k_B T)\,dp_x/h}{\displaystyle\int_0^{\infty} \exp(-p_x^2/2\mu_x k_B T)\,dp_x/h} \tag{4.70}$$

The corresponding equation for particles obeying quantum mechanics has the same denominator, but the integrand in the numerator must be multiplied by $P_{qu}(\varepsilon_x)$ and the integration performed from zero to infinity. Hence

$$\frac{\boldsymbol{R}_{qu}(T)}{\boldsymbol{R}_{cl}(T)} = \frac{\displaystyle\int_0^{\infty} P_{qu}(\varepsilon_x)\exp(-\varepsilon_x/k_B T)\,d\varepsilon_x}{\displaystyle\int_{\mathscr{V}^*}^{\infty} \exp(-\varepsilon_x/k_B T)\,d\varepsilon_x}$$

$$= \frac{\exp(\mathscr{V}^*/k_B T)}{k_B T}\int_0^{\infty} P_{qu}(\varepsilon_x)\exp(-\varepsilon_x/k_B T)\,d\varepsilon_x \tag{4.71}$$

This equation is the one on which corrections to the standard TST rate expression for quantum tunnelling effects are usually based. Even where $P_{qu}(\varepsilon_x)$ is an analytical function, the integral has to be evaluated numerically. However, the form of the equation indicates clearly that $\boldsymbol{R}_{qu}(T)/\boldsymbol{R}_{cl}(T)$ will always be greater than unity, since those trajectories that penetrate the barrier at energies $\varepsilon_x < \mathscr{V}^*$ are weighted more strongly by the Boltzmann term than those which have $\varepsilon_x > \mathscr{V}^*$ but are reflected from the barrier. This difference, and therefore the effect of tunnelling, will be most marked at low temperatures. In canonical transition state theory, this increase in rate is generally allowed for by multiplying the standard rate expression in equation (4.17) by a

transmission coefficient, so that

$$k(T) = \kappa_{qu}(T)(k_B T/h)\{(Q^{\ddagger}/V)/\prod_i (Q_i/V)\}\exp(-\Delta\varepsilon_0^{\ddagger}/k_B T) \qquad (4.72)$$

where

$$\kappa_{qu}(T) = \boldsymbol{R}_{qu}(T)/\boldsymbol{R}_{cl}(T)$$

Because $\kappa_{qu}(T)$ increases as the temperature is lowered, it is frequently assumed that if quantum mechanical tunnelling is important ln $k(T)$ will not vary linearly with $1/RT$. Conversely, if the experimental Arrhenius plot shows curvature, it is assumed that tunnelling is the cause. One should beware of accepting either of these conclusions too easily. Thus, significant tunnelling can occur without its leading to observable curvature in the Arrhenius plot but simply to a lowering of the activation energy relative to its value in the absence of tunnelling. Secondly, $\kappa_{qu}(T)$ is not the only temperature-dependent factor in the pre-exponential part of the rate expression in equation (4.72). It may be that variation of the partition functions with temperature is the cause of observed curvature in a plot of ln $k(T)$ against $1/RT$ (Section 5.5.3).

4.4.2 Tunnelling corrections to the TST rate expression

So far only the effects of tunnelling through one-dimensional barriers of simplified form have been considered. However, numerical techniques enable one to solve the Schrödinger equation in one dimension for any choice of $\mathscr{V}(x)$ and several calculations of this kind have been performed, particularly with the H_3 system in mind.

The simplest approach to this problem is to determine the potential energy profile for the path of minimum reaction energy on the 'skewed' potential surface and to solve the one-dimensional Schrödinger equation for the motion of particles of mass 2/3 amu under the influence of this potential. Values of $\kappa_{qu}(T)$ are then calculated according to equation (4.71). This approach certainly suffers from two defects. First, it ignores the fact that the variation of potential energy in the region of the barrier is not accurately represented by the sum of terms involving the squares of the symmetric and antisymmetric stretching coordinates over distances that are large compared with the de Broglie wavelength. Consequently, motion along the reaction coordinate is not properly separable. Secondly, account is not correctly taken of zero-point effects since the energy in motion along the reaction coordinate is assumed not to vary. This amounts to an assumption of conservation of vibrational energy (CVE). Examination of the data in *Table 2.6* demonstrates that this is not justified: the zero-point energies in the transition state and reactants are not the same.

In an attempt to overcome these difficulties, Johnston has carried out a number of calculations on H_3 and D_3. In his book[12], values of $\kappa_{qu}(T)$ are tabulated for different temperatures and a variety of potential functions and procedures. Just two of his results will be compared here. With a CVE Eckart

potential function chosen to match the profile of a LEPS potential along the reaction path of minimum energy, $\kappa_{qu}(T)$ was found to be equal to 23.0, 3.9 and 1.4 at temperatures of 333.3, 500 and 1000 K respectively. However, arguing that this would incorrectly estimate the tunnelling correction by not allowing for the energy possessed by the reactants as zero-point energy of H_2, Johnston recalculated $\kappa_{qu}(T)$ using an Eckart potential which was again fitted to the curvature of the LEPS potential at the maximum but with the energy maximum reduced by the zero-point energy of H_2 (thus making *no* allowance for zero-point energy in the transition state). This gave values of $\kappa_{qu}(T)$ of 6.7, 3.0 and 1.5 at 333.3, 500 and 1000 K[a], the most noticeable change being the much reduced value at 333.3 K.

More recently, Truhlar and Kupperman[14] have extended Johnston's ideas for incorporating zero-point effects into simple tunnelling calculations. Following Marcus[15], they argue that the most satisfactory way to derive the STST rate expression is to assume that each degree of freedom in the separated reactants transforms smoothly into a degree of freedom in the transition state and then into one in the separated reactants. Thus, in collinear collisions, the initial relative translation becomes the asymmetric (reaction coordinate) mode in the transition state and then the final relative translation of the products, and the initial reactant (BC) vibration becomes the symmetric stretching mode of the transition state and finally the vibration of the AB product. The quantum number associated with the latter mode (and other modes in a multidimensional system) should be conserved and this model is described as vibrationally adiabatic (VA). According to this description, the conserved total energy is the sum of energies in the reactant and other modes and both of these are a function of the distance of the system along the reaction path as the zero-point energy changes but the total energy must remain the same. To allow for this effect, Truhlar and Kuppermann argue convincingly that the potential energy which should be used in tunnelling calculations is not the classical potential energy but the zero-point potential energy calculated all the way along the reaction path.

According to this model, $\kappa_{qu}(T)$ is only 1.19 at 300 K. However, it must be emphasized that only linear configurations were considered so zero-point effects arising from the transformation of H_2 rotations into bending vibrations in the transition state were not included. Consequently, the maximum of the VA potential by Truhlar and Kuppermann was much less than $\Delta\varepsilon_0^{\ddagger}$. Therefore, although the case for using a VA potential is rather convincing, it seems that inclusion of adiabatic changes in the non-linear degrees of freedom will cause the *full* VA potential to resemble the CVE potential more closely than the one used in Truhlar and Kuppermann's calculations. Nevertheless, it seems likely that the full VA potential will be broader than the CVE profile leading to somewhat smaller value of $\kappa_{qu}(T)$.

[a] It is worth noting that these 'quantum corrections' would change the activation energy derived from the adjusted TST rate expression, equation (4.72), but not lead to appreciable curvature in a plot of $\ln k(T)$ *versus* $1/RT$.

Not only do the treatments of tunnelling which have been considered up to this point analyse only collinear collisions, they also implicitly assume that the most important reaction path is that leading directly over the saddlepoint on the potential surface. That this need not be so was first recognized by Johnston and Rapp[16]. Their method, which involved the calculation of tunnelling corrections for a series of parallel cuts through the barrier region, is described briefly in Section 5.6.3. More recently, Marcus and Coltrin[17] have described calculations which indicate that the most important tunnelling pathways for $H + H_2$ are those cutting the corner on the skewed surface. This route is much shorter than the standard path along the reaction coordinate and yields tunnelling corrections which bring the transition state predictions into line with the exact quantum scattering results.

In summary, it has to be said that there is no simple established procedure for accurately estimating the value of the quantum correction factor for use with the standard TST rate expression. The reasons for this should, by now, be clear. If the de Broglie wavelength associated with motion along the reaction coordinate in the transition state region is sufficiently large for appreciable tunnelling, then separation of this motion is a poor approximation. Conversely, where separation of this motion is a valid approximation, tunnelling effects are negligible. As a result it appears that quantum dynamical effects can only be accounted for accurately by comparing the results of multidimensional classical and non-classical scattering calculations (*see* Section 3.6) or by comparing the results of 'exact' classical and non-classical transition state theory calculations in which the secondary assumption of separable motions in the transition state is not made.

What the results of simple calculations can do is indicate when quantum effects may reasonably be neglected. For this purpose, it is simplest, and sufficient, to use the Eckart potential and to assume separation of motion along the reaction coordinate. If the Eckart function, equation (4.69), is matched to the profile of potential energy along the reaction path of minimum energy, the value of $\kappa_{qu}(T)$ is likely to be overestimated. Consequently, if this value is close to unity, the quantum correction to the transition state rate can safely be ignored.

The use of the Eckart potential is convenient because $\kappa_{qu}(T)$ can be tabulated in terms of a limited number of parameters. These are $(\mathscr{V}_f^*/-i\hbar c\tilde{\nu}_x)$, $(\mathscr{V}_r^*/-i\hbar c\tilde{\nu}_x)$ and $(-i\hbar c\tilde{\nu}_x/k_B T)$; $\mathscr{V}_f^*$ and $\mathscr{V}_r^*$ are the potential energy barriers to reaction in the forward and reverse directions and in terms of the parameters in equation (4.69) are given by

$$\mathscr{V}_f^* = (A+B)^2/4B \qquad (4.73a)$$

$$\mathscr{V}_r^* = (A-B)^2/4B \qquad (4.73b)$$

$\tilde{\nu}_x$ is the imaginary frequency (in cm^{-1}) obtained from $d^2\mathscr{V}/dx^2$ at the maximum of the potential profile.

Johnston has determined values of $\kappa_{qu}(T)$ for a large range of combinations of these three parameters. The tunnelling corrections are rather insensitive to

Table 4.2 Quantum mechanical transmission coefficients, $\kappa_{qu}(T)$, for symmetric Eckart barriers as a function of $(\mathscr{V}^*/-i\hbar c\tilde{\nu}_x)$ and $(-i\hbar c\tilde{\nu}_x/k_B T)$[a]

	$-i\hbar c\tilde{\nu}_x/k_B T$			
$\mathscr{V}^*/-i\hbar c\tilde{\nu}_x$	2	4	8	16
1	1.27	1.62	2.72	8.19
2	1.32	1.91	4.55	34.0
4	1.30	2.02	7.60	304
8	1.24	2.04	13.8	–
16	1.2	2.1	32	

[a] Taken from a fuller compilation in *Table* 2.2 in JOHNSTON, H. S., *Gas Phase Reaction Rate Theory*, Ronald Press, New York (1966)

$\mathscr{V}_f^* - \mathscr{V}_r^*$. Accordingly, in *Table 4.2*, we present a shortened version of Johnston's tabulation including data only on reactions with $\mathscr{V}_f^* = \mathscr{V}_r^*$. For the $H + H_2$ reaction with the Eckart parameters chosen to match those of the Liu potential, $(\mathscr{V}^*/-i\hbar c\tilde{\nu}_x) = 14.2$ and $(-i\hbar c\tilde{\nu}_x/k_B T) = 5.5$ at 400 K. These values determine $\kappa_{qu}(T = 400\text{ K}) \sim 4$ and clearly indicate that quantum dynamical effects cannot be ignored. On the other hand, for a reaction involving three atoms each of 16 amu on the same potential, $\tilde{\nu}_x$ would decrease by a factor of 4 and $\kappa_{qu}(T) < 1.2$. Thus, for most purposes, quantum tunnelling could be ignored.

4.4.3 Electronically non-adiabatic reactions[18]

A second specifically quantum mechanical effect, which occasionally has to be allowed for, arises when the electronic states of the reactants and of the products do not correlate with any state of the same symmetry species (Section 2.4). For reaction to occur, an electronically non-adiabatic transition must occur. In estimating a reaction rate, the partially forbidden nature of the process can be allowed for by introducing a transmission coefficient into the rate expression with a value less than unity.

The best known examples of electronically non-adiabatic reaction are ones where the total electron spin must change in going from reactants to products. In particular, the dissociation of N_2O and the reverse reaction, i.e.,

$$N_2O(^1\Sigma) \rightleftarrows N_1(^1\Sigma_g^+) + O(^3P) \tag{4.74}$$

and the analogous reactions of the isoelectronic molecules CO_2, OCS and CS_2 have been much studied. The products of these reactions correlate with $^3\Sigma$ and $^3\Pi$ states for collinear confirmations and these transform to $^2A'$ and $^3A + {^3A}$ in C_s symmetry, whereas the reactant correlates directly with the diatomic molecule in its $^1\Sigma_g^+$ ground state but the O or S atom in an excited ^1D state.

In the case of N_2O, neither the $^3\Sigma$ nor $^3\Pi$ state is bound and consequently

they cross the ground-state potential some way above the energy of $N_2(^1\Sigma_g^+)$ $+O(^3P)$, as is shown in a very simplified manner by the potential curves in *Figure 4.6*. This gives rise to an energy barrier for dissociation which is appreciably in excess of the endothermicity of the reaction, but the electronically non-adiabatic route still remains energetically favoured over the adiabatic production of $N_2(^1\Sigma_g^+)+O(^1D)$ by a considerable margin. Consequently, the preferred reaction pathway is via 'spin-forbidden' transitions from the discrete states associated with the $^1\Sigma$ state to the continuum states of $^3\Pi$ near the lowest-energy crossing between these states. This mechanism is clearly very closely analogous to that of 'spin-forbidden' predissociation. The analogy would suggest a first-order rate constant for molecules with energy above the crossing of $\sim 10^{11}\ s^{-1}$, corresponding to a transmission coefficient of $\sim 10^{-2}$.

Any full theoretical treatment requires calculations in multidimensional space and an accurate knowledge of the potential hypersurfaces in the region

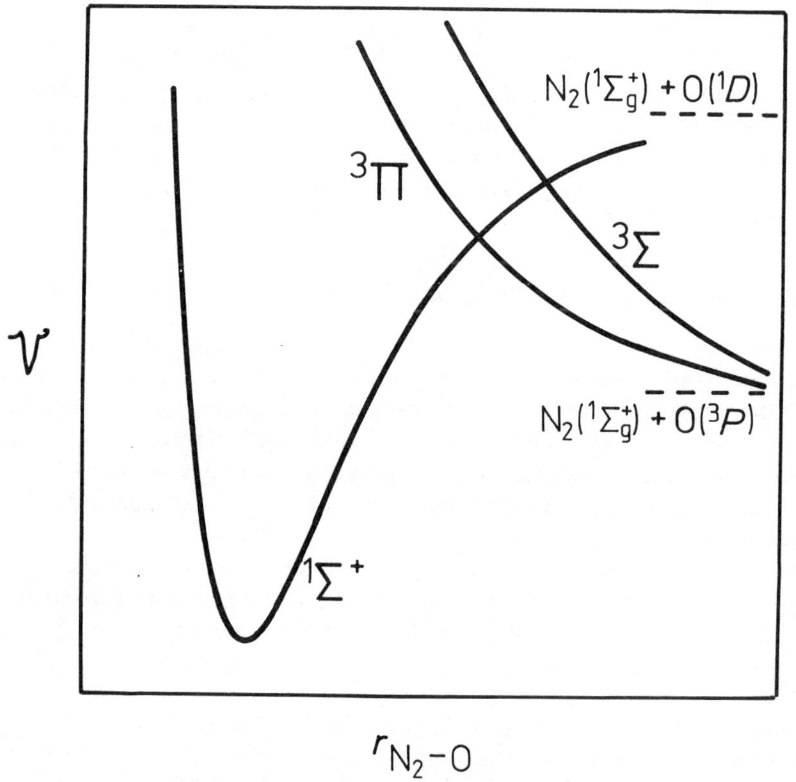

Figure 4.6. Schematic potentials for the N_2–O *system showing that the adiabatic potential for the* $^1\Sigma^+$ *ground state of* N_2O *crosses the triplet potentials correlating with* $N_2(^1\Sigma_g^+)+O(^3P)$ *some way above the energy of the separated species*

where they intersect. Not surprisingly, only very simplified—although not simple—calculations have been carried out. The transition probability is fundamentally linked to the magnitude of the spin–orbit operator which is itself related to the mass of the atoms in the system. For molecules containing atoms from the second and third rows of the Periodic Table, Nikitin[18] has shown that the probability of a non-adiabatic transition can be $\sim 10^{-2}$ when the wavefunctions of the states involved are quite different, because the symmetry of the atomic orbitals used in their construction changes. This is the case in the reactions considered here which occur via $^1\Sigma \rightarrow {}^3\Pi$ transitions. Much smaller probabilities are found where the molecular wavefunctions only change slightly, as in intersystem crossings among excited states of large aromatic molecules.

4.5 Unimolecular reactions[18—22]

Up to this point, transition state theory has been developed in a quite general way; thus equation (4.72) and (4.49) for the overall and specific rate constants can be applied to both bimolecular and unimolecular processes. However, in illustrating how these expressions might be evaluated, bimolecular atom-transfer reactions that occur over a potential with an appreciable energy barrier to the reaction have been taken as examples. This choice brings about two important simplifications in the TST treatment. First, the distribution of reactants over the energy states above threshold is very unlikely to be non-Boltzmann on account of removal of these species by the reaction. This is because the relaxation times characteristic of translational and rotational energy-transferring collisions is much shorter than the average lifetime of reactants before they undergo reaction, and excited vibrational states are usually too thinly populated to make an important contribution to the overall reaction rate. Secondly, the choice of the transition state is straightforward. To accord with the third fundamental assumption of the theory, this should be made so that the total flux orthogonal to the surface S^* coincides as closely as possible with direct reactive flux in one or other direction and multiple crossings are minimized. Where there is a well-defined maximum on the path of minimum energy leading from reactants to products, and the total energy is not too high, S^* passes through this configuration. There is, however, growing evidence that, as the total energy is increased, the best location for S^* may change (*see* note on p. 170).

In this section we shall primarily be concerned with unimolecular reactions. Although the theories of such reactions are almost all statistically based, one or both of the simplifying features that were mentioned in the previous paragraph are frequently absent. As a result, features arise in the treatment of unimolecular reactions which need not be considered in bimolecular reactions with appreciable energy barriers. The treatment described is that developed by Marcus and Rice[23] and by Marcus[24] from the earlier theory which was independently proposed by Rice and Ramsperger and by Kassel[25].

4.5.1 Categories of reaction and types of potential

The major difference between unimolecular and bimolecular reactions is that the individual reactive event in the former involves a solitary species which undergoes an *intramolecular* rearrangement that leads to the molecule either dissociating or isomerizing. The general supposition of unimolecular rate theories is that the rearrangement of atoms corresponding to the chemical reaction is immediately preceded by a rearrangement of *internal energy* which allows the reaction to occur. Molecules that contain internal energy, in this section denoted by ε, in excess of a critical or threshold value, ε^0, *can* undergo reaction. Reaction *will* occur when this energy is distributed in such a way as to carry the point representing the dynamical behaviour of the molecule on the multidimensional potential hypersurface along the reaction coordinate leading from reactants to products.

As well as the *intramolecular* energy transfer process that transforms a molecule with sufficient energy to react, i.e., with $\varepsilon > \varepsilon^0$, into an activated molecule and thence to products, a full mechanism for unimolecular reactions includes bimolecular processes in which *intermolecular* energy transfer takes place. However, in the limit of high total pressure or number density[a], [M], these collisions become so frequent that the distribution of reactant molecules over internal states is effectively undisturbed by reaction. Then the thermal rate constant is obtained from the specific rate constant by evaluating the right-hand side of equation (4.41)[b], or

$$k_{[M]=\infty}(T) = k^{\infty}(T) = \int_{\varepsilon_0}^{\infty} k(\varepsilon)P(\varepsilon)\,d\varepsilon \qquad (4.75)$$

where $P(\varepsilon)$ is the probability of finding a reactant molecule in an internal state with energy between ε and $\varepsilon + d\varepsilon$, when the distribution accords with the Boltzmann laws; $k(\varepsilon)$ is the first-order specific rate constant describing the rate of reaction for reactant molecules within this specified range of internal energy. Further theoretical analysis of equation (4.75), which will be considered later, shows that the activation energy derived from rate-constant measurements in the limit of high total gas density corresponds quite closely to the threshold or critical energy, and is related, but not equal, to the electronic potential energy barrier to the reaction.

In *Table 4.3*,the limiting high-pressure activation energy, E^{∞}_{act}, is compared with ΔE, the change in thermodynamic internal energy, for a number of unimolecular reactions of different types. For the *isomerization* reactions, one of the two main categories of unimolecular reaction that were identified at the start of Chapter 1, $E^{\infty}_{act} \gg \Delta E$. This indicates that there is an appreciable activation energy for reaction in the reverse, as well as the forward direction.

[a] It is usual, in papers on unimolecular reactions, to read of how the observed rate constant varies with total *pressure*. It seems preferable, however, to refer to the total gas concentration in the same units as are used in expressing the rate constant.

[b] For the moment, it is assumed that the distribution of internal energy and variation of $k(\varepsilon)$ are, to a good approximation, continuous. This will be justified later.

Table 4.3 Energies of activation for some unimolecular reactions

	Reaction	ΔE_{298} / kJ mol^{-1}	E_{act} / kJ mol^{-1}	*Temp. range* / K	*Ref.*
(i)	isomerizations				
	$CH_3NC \rightarrow CH_3CN$	−72	160.5	472.5–532.9	*a*
	cyclo-$C_3H_6 \rightarrow CH_3.CH{=}CH_2$	−33	274	693.6–808.3	*b*
(ii)	dissociation to stable molecules				
	$C_2H_5Cl \rightarrow C_2H_4 + HCl$	+70	254	671–766	*c*
	cyclo-$C_4H_8 \rightarrow 2C_2H_4$	+73	262	693–741	*d*
(iii)	dissociation to free radicals				
	$C_2H_6 \rightarrow 2CH_3$	+366	360	823–999	*e*
	$C_2H_6 \leftarrow 2CH_3$		~0	823–450	*f*
	$HNO_3 \rightarrow OH + NO_2$	+205	205	890–1200	*g*
	$HNO_3 \leftarrow OH + NO_2$		~ −5	220–550	*h*

[a] SCHNEIDER, F. W., and RABINOVITCH, B. S., *Journal of the American Chemical Society*, **84**, 4215 (1962)
[b] FALCONER, W. E., HUNTER, T. F., and TROTMAN-DICKINSON, A. F., *Journal of the Chemical Society*, 609 (1961)
[c] HOWLETT, K. E., *Journal of the Chemical Society*, 3695, 4487 (1952)
[d] GENAUX, C. T., KERN, F., and WALTERS, W. D., *Journal of the American Chemical Society*, **75**, 6196 (1953)
[e] WAAGE, E. V., and RABINOVITCH, B. S., *International Journal of Chemical Kinetics*, **3**, 105 (1971)
[f] PARKES, D. A., PAUL, D. M., and QUINN, C. P., *Journal of the Chemical Society, Faraday Transactions* I, **75**, 1935 (1976)
[g] GLÄNZER, K., and TROE, J., *Berichte der Bunsengesllschaft für Physikalische Chemie*, **78**, 71 (1974)
[h] ANASTASI, C., and SMITH, I. W. M., *Journal of the Chemical Society, Faraday Transactions* II, **72**, 1459 (1976)

This characteristic implies the existence of a high potential barrier along the reaction coordinate and can be represented in a simplified manner by the potential energy profile in *Figure 4.7*(a). One consequence of the large energy barrier is that, *in principle*, the choice of transition state poses no problems.

Dissociation reactions, the second major category of unimolecular reaction that was distinguished earlier, differ from isomerizations in that the reverse reactions are bimolecular. Now the variation of electronic potential energy along the reaction coordinate may fall anywhere between the two limiting cases illustrated in *Figure 4.7*(b) and *4.7*(c). In *Figure 4.7*(b), a reaction is represented where there is a well-defined barrier; once again, the choice of the transition state is straightforward. Reactions yielding products which both have closed electronic configurations are of this type. Two examples are given in part (ii) of *Table 4.3*.

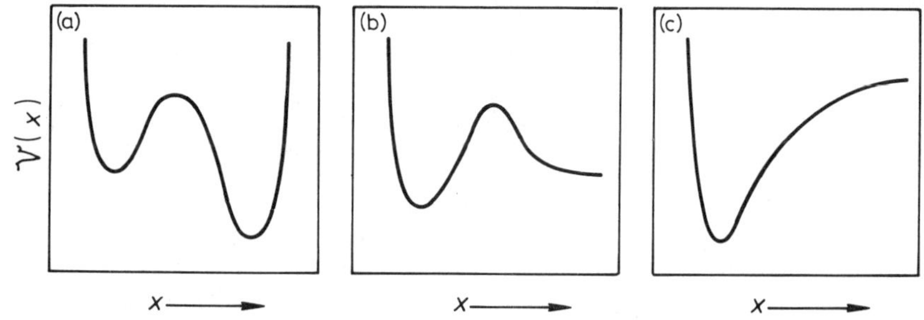

Figure 4.7. Schematic potential energy profiles for three types of unimolecular reaction: (a) *isomerization,* (b) *dissociation* $\rightleftarrows$ *association, where there is a high energy barrier and therefore a large activation energy for reaction in both directions,* (c) *dissociation* $\rightleftarrows$ *association, where there is no energy barrier to the association reaction*

However, when the products of dissociation are two free radicals, the energy of activation is usually found to be equal to or slightly less than the value of ΔE. This is true of the decompositions of C_2H_6 and HNO_3. In these cases, the reverse combination reactions have also been studied, although at lower temperatures, and the activation energies for these reactions are also given in *Table 4.3*. The experimental observations imply that there is little or no potential maximum along the reaction coordinate for these two reactions. This is the situation illustrated in *Figure 4.7*(c) and it presents some additional theoretical problems. In particular, it is no longer immediately obvious how one picks the transition state configuration, i.e., locates the dividing surface S*.

The problem identified in the previous paragraph is not peculiar to unimolecular dissociation reactions or to the bimolecular association reactions that are their reverse. It is also present when a bimolecular reaction occurs over a potential surface with no maximum on the reaction path of minimum energy. Examples are provided by many ion–molecule reactions, and by radical–radical reactions, such as

$$O + OH \rightarrow O_2 + H \tag{4.76}$$

$$D + OH \rightarrow OD + H \tag{4.77}$$

where reaction may proceed via initial association to form the energized polyatomic collision complexes, $HO_2^{\dagger}$ and $DOH^{\dagger}$.

It is important to appreciate that the *fundamental* criterion for locating the transition state is no different when there is no significant potential energy barrier. This criterion, which has already been stated several times, is that the dividing surface S*, and hence the transition state, should be positioned so that the reactive and total fluxes through S* coincide as closely as possible. An exactly equivalent statement of this requirement is that, for systems of specified energy ε, S* should be the surface which is both orthogonal to the reaction coordinate and contains the minimum number of states in the range ε

to $\varepsilon + d\varepsilon$, i.e., which has the minimum density of states at energy ε. That these two statements are equivalent can be illustrated in the following manner.

Let us consider that S* is located in keeping with the 'minimum density of states' criterion and, to simplify matters, that reactive and total flux through S* correspond exactly. Then each unit phase space cell located at S* and in the range ε to $\varepsilon + d\varepsilon$ has an equal probability of being populated. In the limit of classical mechanics, each of these phase space cells connect with a unit cell located over each of the surfaces S that are also orthogonal to the reaction coordinate but are displaced further towards the reactant zone of space. However, more states, i.e., more phase space cells, are located at these surfaces, since this was determined by our choice of S*. These additional phase space cells correspond to trajectories that cross S more than once; consequently, the ratio of reactive to total flux is lower here than at S* where it is a maximum.

Reactions where there is no maximum on the electronic potential energy profile along the reaction coordinate, x, have been treated by assuming that the transition state is located at the centrifugal barrier of the effective potential, $\mathscr{V}_{\text{eff}}(x)$. For present purposes, $\mathscr{V}_{\text{eff}}(x)$ can best be written in terms of J, the quantum number associated with the overall rotations of the reacting molecule about the two axes orthogonal to x:

$$\mathscr{V}_{\text{eff}}(x) = \mathrm{J}(\mathrm{J}+1)\hbar^2 2\mu x^2 - C/x^n \qquad (4.78)$$

[cf. equation (3.21)]. The second term represents the long-range attractive potential between the two separating products, it being assumed that the activated complex is so 'loose' that 'chemical forces' are negligible; the separating species are free to rotate independently at this point, so that this attraction depends only on the magnitude of x, their separation.

The value of x at the centrifugal maximum can be found by setting $\partial \mathscr{V}_{\text{eff}}/\partial x$ equal to zero, yielding

$$x_{\text{max}} = (\mu n C/\mathrm{J}(\mathrm{J}+1)\hbar^2)^{1/(n-2)} \qquad (4.79)$$

This equation demonstrates that x_{max}, and therefore the location of the transition state according to this model, depends on J. The components of the overall angular momentum should be conserved as the reagent separates to products (the inclusion of angular momentum effects is one way in which unimolecular reaction rate theories differ from 'conventional' transition state theory).

4.5.2 The specific rate constant for the unimolecular process

To determine the limiting high-pressure rate constant with the transition state located at the centrifugal maximum, a calculation in three stages must be performed. First, the specific rate constant $k_{\mathrm{J}}(\varepsilon)$ for a particular J state will have to be determined according to equation (4.49) but with the restrictions imposed by the requirement to conserve total angular momentum. In doing

this, it is convenient to identify the modes of motion in the reactant as either *active* or *adiabatic*. The overall rotations are classed as adiabatic (so, of course, is the overall translational motion) because the total angular momentum cannot change. On the other hand, energy can flow between active modes (vibrations and internal rotations) without restriction, and contribute to carrying the system to reaction. If ε and ε^+ are the energies of the *active* modes in the reactant molecules and in the transition state respectively, the correct expression for the specific rate constant is

$$k_J(\varepsilon) = \frac{N^*(\varepsilon^+)}{h\rho(\varepsilon)} \tag{4.80}$$

where $N^*(\varepsilon^+)$ is the total number of active mode states at energies $\leqslant \varepsilon^+$ and $\rho(\varepsilon)$ is the density of reactant states at the active internal energy ε. Centrifugal effects are correctly accounted for when

$$\varepsilon + \varepsilon_J = \varepsilon^+ + \varepsilon_J^+ + \Delta\varepsilon_0^{\ddagger} \tag{4.81}$$

where ε_J and ε_J^+ are the energies associated with the adiabatic motions. The relationship of these energy quantities are shown diagrammatically in *Figure*

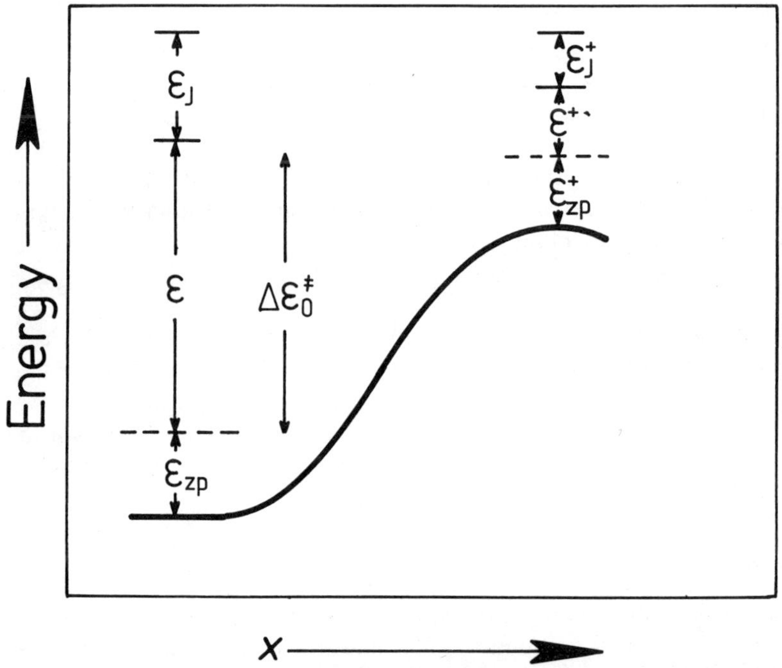

Figure 4.8. Diagram illustrating the relationship between the various energy quantities that enter into the formulation of RRKM *theory*

4.8. When J = 0, $\varepsilon_J^+ - \varepsilon_J$ is, of course, zero. In general,

$$\varepsilon_J^+ - \varepsilon_J = \frac{J(J+1)\hbar^2}{2\mu}\left[\frac{1}{x^{*2}} - \frac{1}{x_e^2}\right] \tag{4.82}$$

To obtain an expression for $k^\infty(T)$ from equation (4.80), it is necessary to integrate over $P(\varepsilon)$, the distribution over active internal states, and sum over J, the adiabatic state distribution.

In practice, this procedure can be replaced with little loss of accuracy[26] by evaluating the transition state expression:

$$k^\infty(T) = (k_B T/h)(Q_J^{\ddagger} Q_V^{\ddagger}/Q_J Q_V)\exp(-\Delta\varepsilon_0^{\ddagger}/k_B T) \tag{4.83}$$

where $Q_J^{\ddagger}$, in the case of a very 'loose' activated complex, is an effective rotational partition function

$$Q_J^{\ddagger} = \sum_{J=0}^{\infty} (2J+1)\exp(-\mathscr{V}_{\text{eff}}(x)/k_B T) \tag{4.84}$$

and Q_J is given by

$$Q_J = \sum_{J=0}^{\infty} (2J+1)\exp(-\varepsilon_J/k_B T) \tag{4.85}$$

The Q_V are partition functions associated with the other internal modes of motion.

In the model of a reaction passing through a 'loose' transition state that has just been described, the assumption that S* passes through a centrifugal maximum on an effective long-range potential was made without justification. In fact, this choice neglects the fact that, as the system moves along the reaction coordinate, there is a continuous and marked change in several modes of motion. Thus, bending vibrations and hindered rotations in the reactant molecule gradually become less restricted and finally transform into free rotations in the separated products. This brings about a continuous change in the density of states associated with these motions, and it is by no means certain that the centrifugal maximum coincides with the position along x that is associated with the minimum density of states. These ideas have been incorporated into the 'statistical adiabatic channel model' recently proposed by Quack and Troe[22, 27]. It will be described briefly in Section 4.5.6; now it is time to consider the role played in the mechanism of unimolecular reactions by intermolecular collisions and the resultant dependence of the observed rate on total gas density.

4.5.3 Dependence of the observed reaction rate on total gas concentration

Equation (4.80) clearly implies that the specific rate constants, $k_J(\varepsilon)$, depends on ε. An important assumption of statistically based theories of unimolecular reactions is that the lifetime of any molecule, with active internal energy between ε and $\varepsilon+\varepsilon$ and in an angular momentum state J, does not

depend on the initial state of the molecule, i.e., how the active internal energy is initially distributed among the active modes. This is the *random state assumption* and, if it is valid, the mean lifetime of the specified molecules with respect to reaction is $1/k_J(\varepsilon)$.

When the reactant molecule is small and its active internal energy is greater than the threshold energy, $1/k_J(\varepsilon)$ can be shorter than the mean lifetime between collisions. In these circumstances, the steady-state population in these excited states is reduced below the equilibrium thermodynamic value as a consequence of reactive loss. To allow for this effect, the overall mechanism for the reaction must include energy-transferring collisions. The mechanism can be written as

$$A + M \rightleftarrows A^{\dagger} + M \qquad (4.86a, -a)$$

$$A^{\dagger} \rightarrow \text{products} \qquad (4.86b)$$

where $A^{\dagger}$ is used to denote reactant molecules with active internal energy in the range ε to $\varepsilon + d\varepsilon$, all J states being included.

The bimolecular forward and reverse processes (4.86a and 4.86 − a) transfer molecules to and from levels within the specified energy range and their second-order specific rate constants will be denoted by $k_a(\varepsilon)$ and $k_{-a}(\varepsilon)$. In reality, equation (4.86) represents the nett effect of a multitude of processes that are microscopically different, because $k_a(\varepsilon)$ and $k_{-a}(\varepsilon)$ will depend on the energy of A as well as that of $A^{\dagger}$ and also possibly on the actual internal states of A and $A^{\dagger}$. In the limit $[M] \rightarrow \infty$, the forward and reverse rates of these energy-transferring collisions will be the same, hence $k_a(\varepsilon)/k_{-a}(\varepsilon)$ is equal to $P(\varepsilon)$, where $P(\varepsilon)\,d\varepsilon$ is the probability of finding a reactant molecule with internal energy between ε and $\varepsilon + d\varepsilon$. When the total gas density is lower, so that the Boltzmann distribution is not maintained, reactive loss will not only modify the steady-state population in the specified range. As a result, although $k_{-a}(\varepsilon)$ will be the same at finite gas densities as at $[M] = \infty$, since there is no way in which colliding $A^{\dagger} + M$ pairs can 'know' about the steady-state distribution in other states, the value of $k_a(\varepsilon)$ may change, as the distribution of A molecules over which averaging is carried out will depend on [M]. A crucial supposition in simpler theories of unimolecular reactions is that this change can be neglected. This is usually referred to as the *strong collision assumption*, since, for the assumption to be a good approximation, the collisions which receive greatest weight in the averaging procedure for $k_a(\varepsilon)$ should be those in which A molecules are transferred between states above threshold and states whose population is scarcely effected by any change in [M][a]. Clearly, this would require the average amount of energy transferred to be quite large.

Because the intermolecular processes of energy transfer are non-chemical,

[a] It is not only the states above threshold that are effected by reactive loss. This is because, at equilibrium, states below but close to threshold are populated from states above threshold by collisions in which energy is transferred. The rate of this process—and therefore the steady-state population of the lower states—falls when the pressure is lowered so that equilibrium is no longer maintained.

the rate constants $k_a(\varepsilon)$ and $k_{-a}(\varepsilon)$ do not generally depend *strongly* on the actual nature of M; this symbol is used to denote the dominant species in the reacting gas and may either be the reactant itself (i.e., $M \equiv A$) or a chemically inert diluent gas. $P(\varepsilon)$ cannot, of course, depend on the nature of M, so change in M must alter the limiting high pressure values of $k_a(\varepsilon)$ and $k_{-a}(\varepsilon)$ equally.

To find the rate of reaction of molecules possessing internal energy between ε and $\varepsilon + d\varepsilon$, an expression is first obtained for the steady-state concentration of $A^\dagger$:

$$k_a(\varepsilon)[A][M] = k_{-a}(\varepsilon)[A]_{ss}[M] + k(\varepsilon)[A^\dagger]_{ss} \tag{4.87}$$

$$[A^\dagger]_{ss} = \frac{k_a(\varepsilon)[A][M]}{k_{-a}(\varepsilon)[M] + k(\varepsilon)} \tag{4.88}$$

Hence, the rate of formation of products from reagents with specified internal energy is

$$\begin{aligned} d[\text{Products}]/dt &= k(\varepsilon)[A^\dagger]_{ss} \\ &= \frac{k(\varepsilon)k_a(\varepsilon)[A][M]}{k_{-a}(\varepsilon)[M] + k(\varepsilon)} \\ &= \frac{k(\varepsilon)P(\varepsilon)}{1 + k(\varepsilon)/k_{-a}(\varepsilon)[M]}[A] \end{aligned} \tag{4.89}$$

This equation is, of course, the microscopic equivalent of the famous Lindemann–Hinshelwood equation. It defines a phenomenological first-order rate constant:

$$k'(\varepsilon) = \frac{k(\varepsilon)P(\varepsilon)}{1 + k(\varepsilon)/k_{-a}(\varepsilon)[M]} \tag{4.90}$$

To derive the thermal rate constant[a] one must solve the equation

$$k(T) = \int_{\varepsilon^0}^{\infty} k'(\varepsilon)\, d\varepsilon \tag{4.91}$$

In the limit $[M] \rightarrow \infty$ it is clear that this yields

$$k^\infty(T) = \int_{\varepsilon^0}^{\infty} k(\varepsilon)P(\varepsilon)\, d\varepsilon \tag{4.92}$$

which is identical to equation (4.75). In the limit $[M] \rightarrow 0$,

$$k^0(T) = \int_{\varepsilon^0}^{\infty} k_{-a}(\varepsilon)[M]P(\varepsilon)\, d\varepsilon \tag{4.93a}$$

or

[a] Here, we follow the general convention of referring to the observed rate constant as a thermal rate constant at all [M], although only as $[M] \rightarrow \infty$ is the complete internal state distribution of the reagent in thermal equilibrium with other degrees of freedom in the gas.

$$\int_{\varepsilon^0}^{\infty} k_a(\varepsilon)[\mathrm{M}]\,\mathrm{d}\varepsilon \tag{4.93b}$$

and the rate constant is directly proportional to [M]. As part of the strong collision assumption, it is usual to suppose that the rate of de-excitation of molecules above threshold is independent of ε and is directly proportional to the rate of bimolecular collisions. Conventionally, this is expressed by

$$\begin{aligned} k_{-a}(\varepsilon)[\mathrm{M}] &= \beta Z[\mathrm{M}] \\ &= \beta\omega \end{aligned} \tag{4.94}$$

The coefficient β reflects the efficiency of a particular gas M in inducing energy transfer. Substitution of (4.94) into equation (4.93a) yields

$$k^0(T) = \beta\omega \int_{\varepsilon^0}^{\infty} P(\varepsilon)\,\mathrm{d}\varepsilon \tag{4.95}$$

so the relative values of β can be found by determining the rate constants for reaction in the limit of low pressure in the presence of different diluent gases. Of course, the implication, that in a fraction β of collisions $A^{\dagger}$ molecules lose sufficient energy for the strong collision assumption to hold, whereas in $(1-\beta)$ of the collisions no energy is transferred, cannot be justified. Other more sophisticated models have been treated[22], but these will not be considered here.

At intermediate [M], the right-hand side of equation (4.90) cannot be simplified and it forms the integrand in equation (4.91). This expression can be viewed as being the product of $k(\varepsilon)$ and $P'(\varepsilon)$, the latter being the distribution function over internal energies distorted by the occurrence of reaction, i.e.,

$$\begin{aligned} P'(\varepsilon) &= \frac{P(\varepsilon)}{1 + k(\varepsilon)/k_{-a}(\varepsilon)[\mathrm{M}]} \\ &= \frac{P(\varepsilon)}{1 + k(\varepsilon)/\beta\omega} \end{aligned} \tag{4.96}$$

In order to evaluate the various equations for the thermal rate constant, it is first necessary to obtain expressions for $P(\varepsilon)$ and $k(\varepsilon)$. It is these procedures that are considered next.

4.5.4 The classical RRK theory

Although any realistic calculation on unimolecular reaction rates should now employ the quantum formulation of Rice and Marcus, it is useful—not solely for historical reasons—to begin by considering the classical limit RRK (Rice–Rampsperger–Kassel) treatment. The most straightforward approach to RRK theory is through a model reactant whose internal modes are s identical harmonic oscillators, each of frequency ν. It is assumed that energy can flow between these modes—despite their harmonicity—but the levels associated with different modes do not perturb one another. The overall

rotational motions are not considered. Apart from the motion along the reaction coordinate, the $(s-1)$ frequencies in the transition state are also all assumed to equal ν.

To evaluate $k(\varepsilon)$ it is easiest to start with a quantum representation. The internal energy ε is taken to correspond to j vibrational quanta. The number of degenerate states at the energy $\varepsilon_j (= jh\nu)$ is then given by

$$g_j = \frac{(j+s-1)!}{j!(s-1)!} \tag{4.97}$$

The density of states in the classical limit, i.e., as $\nu \to 0$ and j becomes very large, is $g_j/h\nu$. Since $j \gg s$, $(j+s-1)/j$ can be replaced by j^{s-1} and one makes the substitution $j = \varepsilon/h\nu$. Hence

$$\rho(\varepsilon) = \left(\frac{1}{h\nu}\right)\frac{(\varepsilon/h\nu)^{s-1}}{(s-1)!} \tag{4.98}$$

In the transition state, the number of states associated with the $(s-1)$ internal motions must be counted. This can be done by integrating an expression similar to (4.98) from zero to ε^+, i.e.,

$$\mathcal{N}^*(\varepsilon^+) = \int_0^{\varepsilon^+} \left(\frac{1}{h\nu}\right)\frac{(\varepsilon^+/h\nu)^{s-2}}{(s-2)!}\,\mathrm{d}\varepsilon^+$$

$$= \frac{(\varepsilon^+/h\nu)^{s-1}}{(s-1)!} \tag{4.99}$$

Substituting (4.98) and (4.99) into equation (4.49) yields

$$k(\varepsilon) = \nu(\varepsilon^+/\varepsilon)^{s-1}$$

$$= \nu\left(\frac{\varepsilon-\varepsilon^0}{\varepsilon}\right)^{s-1} \tag{4.100}$$

which is the usual RRK equation.

The classical Boltzmann distribution function can be derived using equation (4.98) for the density of states. Thus

$$P(\varepsilon)\,\mathrm{d}\varepsilon = \frac{\left(\frac{1}{h\nu}\right)\frac{(\varepsilon/h\nu)^{s-1}}{(s-1)!}\exp(-\varepsilon/k_\mathrm{B}T)\,\mathrm{d}\varepsilon}{\int_0^\infty \left(\frac{1}{h\nu}\right)\frac{(\varepsilon/h\nu)^{s-1}}{(s-1)!}\exp(-\varepsilon/k_\mathrm{B}T)\,\mathrm{d}\varepsilon} \tag{4.101}$$

The denominator on the right-hand side of this equation is simply the total vibrational partition function for our model molecule in the limit $h\nu \ll k_\mathrm{B}T$ and is equal to $(k_\mathrm{B}T/h\nu)^s$. Hence

$$P(\varepsilon)\,\mathrm{d}\varepsilon = \frac{(\varepsilon/k_\mathrm{B}T)^{s-1}}{(s-1)!}\exp(-\varepsilon/k_\mathrm{B}T)(\mathrm{d}\varepsilon/k_\mathrm{B}T) \tag{4.102}$$

Equations (4.100) and (4.102) for $k(\varepsilon)$ and $P(\varepsilon)\,\mathrm{d}\varepsilon$ can be substituted into

(4.92) and (4.95) to find expressions for the thermal rate constant in the limits of high and low [M]. The mathematical manipulation is straightforward and yields the results:

$$k^{\infty}(T) = \nu \exp(-\varepsilon^0/k_B T) \tag{4.103}$$

and

$$k^0(T) = \beta\omega \frac{(\varepsilon^0/k_B T)^{s-1}}{(s-1)!} \exp(-\varepsilon^0/k_B T) \tag{4.104}$$

although the expression for $k^0(T)$ depends on ε^0 being very much greater than $k_B T$. This is usually the case (*see Table 4.3*). At intermediate [M], the following equation, with $x = \varepsilon/k_B T$ and $x^0 = \varepsilon^0/k_B T$, must be solved numerically.

$$k(T) = \int_{\varepsilon^0}^{\infty} \frac{[\nu(x-x^0)^{s-1}/(s-1)!]\exp(-x)}{1+\nu[(x-x^0)/x]^{s-1}/\beta\omega}\, dx \tag{4.105}$$

4.5.5 RRKM theory

The Rice–Marcus approach to unimolecular reactions, usually referred to as RRKM theory, improves on the RRK theory by treating properly three aspects of these processes that are neglected or not treated properly in the RRK model. These improvements are: first, account is taken of molecular rotation and related centrifugal effects; secondly, the active internal states are counted with due regard to the actual molecular frequencies; finally, and in connection with the counting of active states, the existence of zero-point energies is recognized and allowed for.

To explain how rotations are dealt with, it is necessary to recall that angular momentum should be conserved when an energized molecule passes through the transition state; this was recognized in equation (4.80) for $k_J(\varepsilon)$. In order to obtain an expression for the thermal rate constant, it is necessary to take a properly weighted average of $k_J(\varepsilon)$ over J and ε, i.e., one must evaluate

$$k(T) = \int_{\varepsilon^0}^{\infty} \sum_{J=0}^{\infty} \frac{k_J(\varepsilon)P(\varepsilon, J)}{1+k_J(\varepsilon)/\beta\omega}\, d\varepsilon \tag{4.106}$$

where $P(\varepsilon, J)\,d\varepsilon$ is the probability of finding a reactant molecule in the adiabatic state J and with active internal energy between ε and $\varepsilon + d\varepsilon$:

$$P(\varepsilon, J) = \frac{\rho(\varepsilon)\exp(-\varepsilon/k_B T)}{Q_V} \times \frac{(2J+1)\exp(-\varepsilon_J/k_B T)}{Q_J} \tag{4.107}$$

Using equation (4.80) for $k_J(\varepsilon)$ and equation (4.81) relating the energies in the transition state to those in the reagent, one obtains

$$k_J(\varepsilon)P(\varepsilon, J) = (1/h)\frac{\exp(-\Delta\varepsilon_0^{\ddagger}/k_B T)}{Q_J Q_V}(2J+1)\exp(-\varepsilon_J^{+}/k_B T)$$

$$\times N^*(\varepsilon^+)\exp(-\varepsilon^+/k_B T) \tag{4.108}$$

To simplify the denominator on the right-hand side of equation (4.106), equations (4.80) and (4.81) are first combined yielding

$$k_J(\varepsilon) = \frac{N^*(\varepsilon^+)}{h\rho(\varepsilon^+ + \Delta\varepsilon_0^{\ddagger} + \varepsilon_J^+ - \varepsilon_J)} \tag{4.109}$$

An approximate method is then used to average this expression over J, mean values of ε_J and ε_J^+ being defined by

$$\langle \varepsilon_J^+ \rangle = l k_B T/2 \tag{4.110a}$$

$$\langle \varepsilon_J \rangle = (I^+/I) l k_B T/2 \tag{4.110b}$$

where l is the number of adiabatic rotations and I^+ and I the moments of inertia associated with these rotations in the transition state and in the reactant respectively. Then $k(\varepsilon)$ is written as

$$k(\varepsilon) = \frac{N^*(\varepsilon^+)}{h\rho(\varepsilon^+ + \Delta\varepsilon_0^{\ddagger})F} \tag{4.111}$$

where F is a factor, generally close to unity, given by

$$F = \frac{\rho(\varepsilon^+ + \Delta\varepsilon_0^{\ddagger} + \langle \varepsilon_J^+ \rangle - \langle \varepsilon_J \rangle)}{\rho(\varepsilon^+ + \Delta\varepsilon_0^{\ddagger})} \tag{4.112}$$

Substitution of equations (4.108) and (4.109) into (4.106), followed by summation over J, yields the general expression for $k(T)$:

$$k(T) = (k_B T/h)\frac{Q_J^{\ddagger}}{Q_J Q_V}\exp(-\Delta\varepsilon_0^{\ddagger}/k_B T) \times \int_0^{\infty} \frac{N^*(\varepsilon^+)\exp(-\varepsilon^+/k_B T)}{1 + k(\varepsilon)/\beta\omega}\left(\frac{d\varepsilon^+}{k_B T}\right) \tag{4.113}$$

In the limit $[M] \to \infty$, this equation reduces, as it must since the internal energy distribution is then the thermodynamic one, to the TST expression

$$k^{\infty}(T) = (k_B T/h)\frac{Q_J^{\ddagger} Q_V^{\ddagger}}{Q_J Q_V}\exp(-\Delta\varepsilon_0^{\ddagger}/k_B T) \tag{4.114}$$

As $[M] \to 0$, one obtains

$$k^0(T) = \beta\omega F \frac{Q_J^{\ddagger}}{Q_J Q_V}\exp(-\Delta\varepsilon_0^{\ddagger}/k_B T) \times \int_0^{\infty} \rho(\varepsilon^+ + \Delta\varepsilon_0^{\ddagger})\exp(-\varepsilon^+/k_B T)\, d\varepsilon^+ \tag{4.115}$$

The final stage in the RRKM theory is the derivation of formulae for determining the total number and density of the active internal states required in evaluating equations (4.113) to (4.115). For the moment, the question of how

the transition state geometry and frequencies are chosen will be left on one side; and it will also be assumed that all active modes are vibrations (i.e., no internal rotations).

The suggestion of Marcus and Rice[23] was that the appropriate formulae for $N(\varepsilon)$ and $\rho(\varepsilon)$ are [cf. equations (4.98) and (4.99)]

$$N(\varepsilon) = (\varepsilon + \varepsilon_{zp})^s/s! \prod_i h\nu_i \qquad (4.116)$$

and

$$\rho(\varepsilon) = (\varepsilon + \varepsilon_{zp})^{s-1}/(s-1)! \prod_i h\nu_i \qquad (4.117)$$

where ε_{zp} is the zero-point energy and ε the active internal energy measured from this level (*see Figure 4.8*). At first sight, the use of a classically based expression for these functions may seem surprising. However, in molecules containing more than a few atoms, the density of vibrational states rises very quickly. Thus, equation (4.97) can be used to calculate the degeneracies of states for an idealized molecule with ten simple harmonic frequencies each with $\tilde{\nu} = 1000\ \text{cm}^{-1}$. One finds $g_s = 1, 10, 55, 220, 715 \ldots$ for $j = 0, 1, 2, 3, 4 \ldots$. On the other hand, such a molecule has a large zero-point energy ($5000\ \text{cm}^{-1}$) relative to the spacing between degenerate states. As a result, equations (4.116) and (4.117) do seriously overestimate $N(\varepsilon)$ and $\rho(\varepsilon)$ at low energies, as the comparison in *Figure 4.9* shows. Since $\varepsilon^+ + \Delta\varepsilon_0^\ddagger$ is usually appreciably greater than ε_{zp}, equation (4.117) can be used to determine $\rho(\varepsilon^+ + \Delta\varepsilon_0^\ddagger)$; the neglect of anharmonicity of the vibrations will almost certainly introduce larger errors. However, the corresponding equation to (4.116) is likely to provide a poor estimate for $N^*(\varepsilon^+)$.

An empirical modification of (4.116) has been proposed by Whitten and Rabinovitch[28] based on a comparison with a direct count of states in a range of molecules. They suggest the use of the equation (written here for the $s-1$ frequencies of the transition state):

$$N^*(\varepsilon^+) = (\varepsilon^+ + a\varepsilon_{zp}^+)^{s-1}/(s-1)! \prod_i h\nu_i \qquad (4.118)$$

where a is given by

$$a = 1 - w\{(s-2)\langle \tilde{\nu}^2 \rangle/(s-1)\langle \tilde{\nu} \rangle^2\} \qquad (4.119)$$

the averages being taken over all $s-1$ frequencies of the transition state. The parameter w is dependent on $\varepsilon^+/\varepsilon_{zp}^+$ (or $\varepsilon/\varepsilon_{zp}$) but does not vary much from molecule to molecule. Some values of w are listed in *Table 4.4*; a fuller tabulation is given in Whitten and Rabinovitch's paper[28].

Figure 4.9 shows that the Whitten–Rabinovitch method provides a better estimate of $N^*(\varepsilon)$ for our model molecule than the Rice–Marcus expression. For real molecules, the values of $N^*(\varepsilon)$ given by direct count vary more smoothly and the Whitten–Rabinovitch calculations provide a close approximation to this curve for a wide variety of molecules. However, it should be

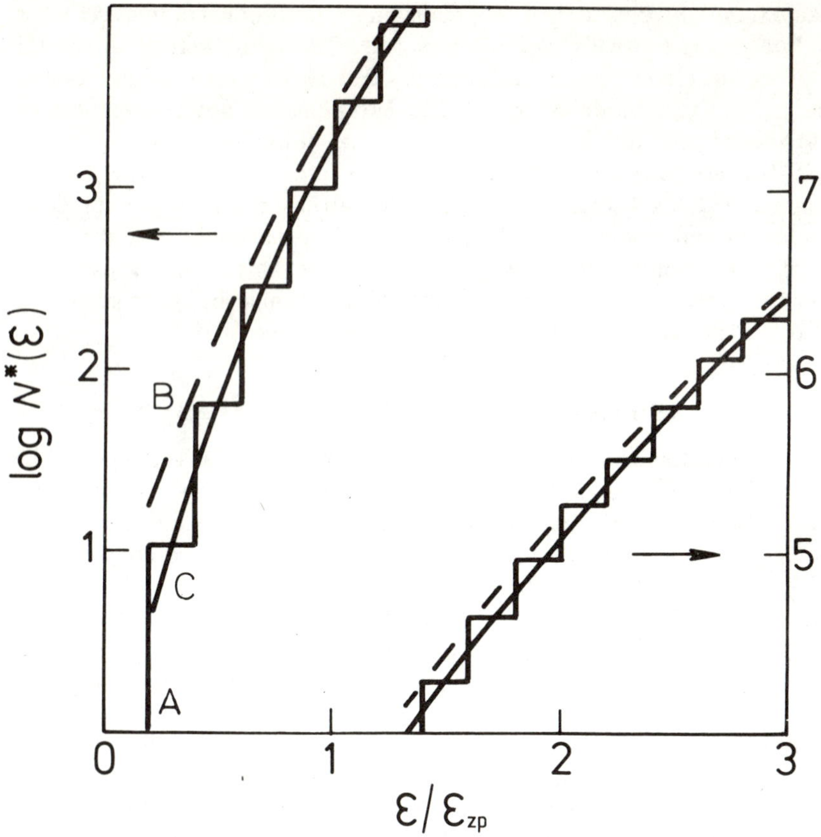

Figure 4.9. Comparison of the number of vibrational levels for a model molecule with ten identical harmonic oscillators obtained from A *direct count using equation* (4.97); B *the* RRKM *expression, equation* (4.116); *and* C *the Whitten–Rabinovitch modification of the* RRKM *formula*

Table 4.4 Recommended values of the Whitten–Rabinowitch parameter w^a

$\varepsilon/\varepsilon_{zp}$	w	$\varepsilon/\varepsilon_{zp}$	w
0.02	0.2437	0.42	0.1356
0.10	0.2055	0.50	0.1258
0.18	0.1791	1.0	0.0890
0.26	0.1608	2.0	0.0571
0.34	0.1470	5.0	0.0278

[a] Taken from a fuller compilation in Table 3 in WHITTEN, G. Z., and RABINOVITCH, B. S., *Journal of Chemical Physics*, **38**, 2466 (1963)

emphasized that this comparison was made on the basis of harmonic oscillator states. For processes involving large molecules, this is unlikely to introduce large errors, since even for critically energized species the average number of quanta per normal mode is small. The Whitten–Rabinovitch version of RRKM theory provides a good estimate of the unimolecular reaction rate constant because zero-point effects are quite accurately allowed for. On the other hand, in systems composed of only a few atoms, the average number of quanta per normal mode can be much larger, and then account should also be taken of anharmonic effects. A prescription for doing this, as well as introducing other correction factors to the basic formulae, has been given in a recent paper by Troe[29].

4.5.6 Choice of the transition state and extensions to RRKM theory

The major limitation in applying RRKM theory arises because the expressions for both the canonical and microcanonical rate constants can only be evaluated *theoretically* if information about the transition state, such as its electronic potential energy, nuclear geometry and bond force constants, is available from quantum mechanical calculations on the potential hypersurface[a]. This is, of course, a general defect of transition state theory of which RRKM theory is a special case. Furthermore, it is possible that, for reactions occurring on potentials without pronounced energy barriers, the transition state may not be uniquely defined but may be a function of the internal energy in the system. This may cause special difficulties in the important class of reactions in which a molecule dissociates to two free radicals.

The first stage in the usual procedure for interpreting experimental data on unimolecular reactions in terms of RRKM theory is to choose a model of the transition state *empirically* which satisfies two criteria. The first of these is that equation (4.114) should correctly predict the absolute values of $k^{\infty}(T)$ throughout the range of temperature in which measurements of the reaction rate have been made. The second requirement is that the nuclear geometry and vibrational frequencies selected for the transition state interpolate reasonably between the corresponding parameters in the reactants and products of the reaction. Having done this, the value of β for the particular diluent gas M is obtained. In principle, β can be determined by comparing the observed value of $k^0(T)$ with that given by equation (4.115). In practice, β is usually estimated empirically. This is done by comparing the relative values of $k^0(T)$ for different M. For large polyatomic gases, $\beta \sim 1$, i.e., the strong collision assumption appears to hold quite well. Hence, lower values of β for less complex M can be

[a] Most of the unimolecular reactions for which well-defined rate data exist are sufficiently complex to discourage full-scale quantum mechanical calculations on the potential hypersurface. An area which has been undeservedly neglected is the calculation of electronic potential energies where two radicals are just close enough to interact 'chemically'. The form of the potential in these regions must have an important bearing on the rates of unimolecular decomposition to radicals and of the reverse association reaction.

estimated. Finally, and without further adjustments to the transition state model, the value of $k(T)$ calculated from equation (4.113) is compared with the observed rate constants through the 'fall-off' region where they vary in a non-linear fashion with [M]. Application of this procedure to two reactions, the isomerization of CH_3NC and the dissociation of HNO_3, will be described in Chapter 5.

The partially empirical RRKM procedure described in the previous paragraph has its defects. In particular, the range of experimental data is frequently rather limited, especially in the case of reactions of types (i) and (ii) in *Table 4.3* which have large activation energies and can only be investigated experimentally over a limited temperature range. Because of this, the nature of the transition state may not be sharply characterized by the procedure of matching the experimental and calculated values of $k^\infty(T)$, nor is the fall-off behaviour very sensitive to the choice of transition-state parameters. This means that such comparisons do not provide a very searching test of the theory nor do they closely define the transition state structure. With reactions of type (iii), however, data can also be obtained on the reverse association of the radicals. Consequently, values of $k^\infty(T)$ are available over a wider range of temperatures and a more thorough test of the theory is provided.

Recently, some studies have been made of the special problems that arise when there is no electronic energy barrier in the reaction coordinate. Bunker and Pattengill[30] identified the transition state, in reactions of specified active internal energy, with the configuration along the reaction coordinate, x, at which the density of states reached a minimum. In a number of model triatomic systems they found the critical value of x, x^*, to be appreciably smaller than that at the centrifugal maximum but not apparently a strong function of the internal energy. Correcting their results for the effects of molecular anharmonicity, they obtained very good agreement between the results of classical RRKM calculations and those from a classical trajectory study of dissociation in the same triatomic model systems.

Troe, individually and with Quack, has recently published several papers on the treatment of unimolecular reactions. Pointing out that the form of 'fall-off' curves, i.e., the variation of $k(T)$ with [M], is not very sensitive to details of the theoretical description of the reaction, Troe[31] suggests an empirical method of estimating the two limiting rate constants, $k^0(T)$ and $k^\infty(T)$, when the available experimental data cover only a limited range of [M] in the fall-off region. This procedure is based upon a definition of the 'effective' values of two parameters s and $x^0 = \varepsilon^0/k_B T$. Subsequently the experimental data are matched to RRK curves that are obtained by substituting these parameters into equation (4.105) and evaluating the necessary integrals numerically for various values of $k^0(T)/k^\infty(T)$.

These calculations serve to emphasize that, in data from experimental studies of unimolecular reactions or their reverse, information about the nature of the transition state for the unimolecular chemical process is really only provided by the absolute value of $k^\infty(T)$ and how this rate constant at high [M] varies with temperature. Furthermore, as with bimolecular reactions,

the Boltzmann weighting factor, i.e., $P(\varepsilon)$ in equation (4.92), exerts such a dominating influence that the majority of reacting molecules have internal energy close to the threshold energy and little can be discovered about the form of $k(\varepsilon)$. In the limit of low [M], it is the intermolecular energy transfer process that is rate-determining, so values of $k^0(T)$ contain any information that can be deduced regarding the collisional efficiencies of different species M in inducing energy transfer[22,29].

In order to examine the variation of $k(\varepsilon)$ with ε more closely, *chemical* and *photochemical activation* studies have been carried out. In these experiments, reactant molecules are produced with a fairly closely specified internal energy above threshold by means of either a preceding chemical reaction or by absorption of radiation (*see* Section 6.4). The lifetime of these species relative to a subsequent unimolecular reaction can be estimated by observing how the yield of products of the unimolecular decay varies as the gas concentration, [M], and therefore the frequency of stabilizing collisions, is changed. The estimated lifetime can then be compared with the value of $1/k(\varepsilon)$ predicted by RRKM theory.

Compared with thermal experiments on unimolecular reactions, studies using chemical and photochemical activation allow reactant molecules to be prepared with energies well in excess of threshold; realistic interpretation of the results requires that the assumption of 'strong collisions', i.e., the removal of sufficient energy in a single collision to reduce ε to below threshold, is abandoned. Troe[22,32] has also applied more sophisticated models of the intermolecular energy transfer processes to the interpretation of kinetic data from thermal studies and shown that the 'collision efficiencies' β derived from a comparison of the values of $k^0(T)$ for different M can be related to the mean energy, $\langle \Delta\varepsilon \rangle$, that is removed from an energized reactant molecule per collision. In these calculations, an adjustable form was taken for the distribution of final internal energies in A following collisions between M and A molecules with specified initial energy. The derived relationship between $\langle \Delta\varepsilon \rangle$ and β was rather insensitive to the form of this distribution function. To a first approximation, for $\beta \leqslant 0.85$, these two quantities are related by the equation

$$\langle \Delta\varepsilon/k_B T \rangle = -\beta/(1-\beta^{\frac{1}{2}}) \qquad (4.120)$$

Finally, Quack and Troe[27] have suggested a novel approach to reactions where $\mathscr{V}(x)$ shows no maximum, which essentially dispenses altogether with the concept of a transition state. Their 'statistical adiabatic channel model' rests on the assumption that, during progress of a system through the critical or transition state region of phase space, it remains in a single 'channel'. This channel corresponds to a quantum state that correlates on one side of the transition region with a state of the reactant molecule and on the other side with a state of the products. For each channel an 'adiabatic potential' is constructed. It can be considered as the sum of three terms:

(1) the electronic potential energy along x, the reaction coordinate;

(2) energies associated with modes that change their character radically as dissociation occurs;
(3) a centrifugal term.

Examples of the motions considered under (2) are the stretching vibration that first becomes motion along the reaction coordinate and then relative translational motion of the separating products, and the bending vibrations in the molecule that eventually become rotations of the separated fragments. The energies associated with these modes and the electronic potential energy are made to vary continuously with x and to depend on a single empirical, but universal, parameter. The centrifugal term takes its usual form. To determine $k(\varepsilon)$, one must determine for each channel at the specified energy whether ε exceeds the maximum in the adiabatic potential for that channel. If so, it constitutes an 'open' channel and is counted in evaluating $k(\varepsilon)$ according to equation (4.49); if not, the channel is 'closed' and not counted.

The results of calculations with the statistically adiabatic channel model match experimental results quite well, when the same parameter is used to describe the variation of the adiabatic potential with x for all systems. Nevertheless, it is not yet entirely clear that this treatment provides a major improvement over RRKM calculations, given its greater computational complexity. Most recently, Quack and Troe[33] have suggested that their method of estimating how the adiabatic potential varies along the reaction coordinate may be used to establish the location of the transition state at the point of maximum free energy. For systems where there is no well defined maximum in $\mathscr{V}(x)$, the transition state occurs at progressively smaller values of x as the temperature is raised. Values of $k^{\infty}(T)$ can be calculated, at each temperature, using the equations from canonical TST. However, Quack and Troe recommend that the partition function associated with internal motions in the transition state is evaluated by an interpolation procedure, rather than by the unrealistic assumption that these motions are separable and the vibrations are small-amplitude harmonic oscillations. This treatment of unimolecular reactions in the high-pressure limit bears some resemblance to TST calculations on atom-transfer reactions with the BEBO method providing the potential.

4.6 Detailed reaction cross-sections and rate constants

4.6.1 Phase space theories[34]

Transition state and RRKM theories are concerned with predicting the total rates of reaction of microcanonical or canonical ensembles of reactants. The reactive flux through the critical dividing surface is only related to the properties and states of the system in reactant phase space through the density of state, or partition, functions. In order to determine the selectivity of energy consumption and the specificity of energy disposal in a reaction, it is necessary

to identify the transition probabilities connecting states of the activated complex with the actual states in reactant and product phase space. Although the connection between states (i.e., unit phase space cells) at the dividing surface S* and states at some 'earlier' surface R was considered in deriving classical TST in Section 4.2.2, it was neither necessary nor possible to deduce the nature of the reactant states that lead to activated states and then to products.

Statistically based predictions of kinetic quantities other than the thermal and specific rate constants are founded on the hope that the probabilities of different detailed results are determined only by the need to conserve total energy, linear momentum, total angular momentum, and its projection on one axis. Consequently, from collisions where these quantities are defined, all final states which accord with these restrictions will be equally populated. Where reaction is 'direct', dynamical factors will certainly distort the final state distribution from that expected purely on this statistical basis. The populations are most likely to be close to those calculated statistically, when reaction proceeds via the formation of a 'collision complex' and its subsequent unimolecular decay, and where there are no significant potential energy maxima in either the entry or exit channels for this process. Under these conditions, the detailed reaction cross-section can be written as

$$S_r(\mathrm{n}'|\mathrm{n};\varepsilon_T) = S_c(|\mathrm{n};\varepsilon_T)P_c(\mathrm{n}'|) \qquad (4.122)$$

where $S_c(|\mathrm{n};\varepsilon_T)$ is the cross-section for formation of the complex in collisions of energy ε_T of reactants in states defined by n, and $P_c(\mathrm{n}'|)$ is the probability that this complex once formed decays into the final product state denoted by n′.

In the 'best' situation for application of the 'phase space theory', formation of the strong coupled complex is limited only by centrifugal effects, so that $P_l = 1$ for $l \leqslant l_{max}$ and $P_l = 0$ *for* $l > l_{max}$, and

$$S_c(|\mathrm{n};\varepsilon_T) = (\pi/k^2)\sum_0^{l_{max}}(2l+1) \qquad (4.123)$$

[cf. equation (4.53)]; this cross-section is independent of the internal states of the reactant. In order to determine the final product states, the total cross-section defined by equation (4.123) must be divided into detailed cross-sections associated with formation of complexes in different states J of *total* angular momentum. It is assumed that in collisions of specified orbital and rotational momenta, all states of the complex within the limits defined by

$$|l+j| \geqslant J \geqslant |l-j| \qquad (4.124)$$

with $l \leqslant l_{max}$ have an equal probability of being formed. Therefore, to find the cross-section for formation of complexes of specified J, it is necessary to evaluate

$$S_{c,J}(|\mathrm{n};\varepsilon_T) = (\pi/k^2)\sum_l(2l+1)\frac{(2J+1)}{(2l+1)(2j+1)} \qquad (4.125)$$

where the last factor is the fraction of collisions with l, j that couple to yield a total angular momentum J, and the summation over l is restricted by (4.124).

Having obtained the cross-sections for formation of complexes with specified J, a similar procedure to that just described is applied to the calculation of the probabilities that a complex with specified energy and J breaks down to yield a particular specified pair of products. Thus, the probability of a particular outcome is assumed to be proportional to the ratio of the volumes of phase space associated with the state-specified products and with all the states that can be reached from the complexes with conservation of J and total energy. This procedure may also take account of centrifugal effects in the exit channel.

The foregoing discussion indicates that for products to be statistically distributed a strongly coupled complex must form which essentially 'forgets' the characteristics of its mode of formation. This is likely to happen only when there is a deep potential well between reactants and products. The production of a collision complex of this kind is a necessary but not sufficient condition for a final statistical distribution. If there is an appreciable potential-energy maximum between the regions on the surface associated with the complex and with the products, energy will be released non-statistically after the system has passed through the transition state and in a region of the potential hypersurface where all the motions are no longer strongly coupled. The proportion of the total energy that is released during this part of the trajectory is, broadly speaking, reflected in the deviation of the actual product distribution from that predicted statistically, and to treat this part of the collision additional calculations are necessary.

4.6.2 The information-theoretic approach[35–37]

Phase space calculations cannot be expected to yield accurate product-state distributions for 'direct' reactions. A particularly glaring defect is that they *always* predict a monotonically decreasing vibrational state population with increasing v′ (the product's vibrational quantum number). This is a consequence of the greater number of states or volumes of phase space associated with translational and rotational motion compared with vibrations. However, phase space concepts comprise one of the foundations of a valuable approach to detailed kinetic data which is based on concepts, borrowed by Bernstein and Levine, from information theory.

One principle of their method is that phase space calculations provide the 'least biased' estimate of the possible results of molecular events. As we have seen, owing to dynamic factors, this 'prior' expectation is rarely realized, that is the observed probability, $P(\mathrm{n}')$, of formation of a product in a particular state n′ nearly always differs from the probability, $P^0(\mathrm{n}')$, suggested by a phase space calculation. The deviation of $P(\mathrm{n}')$ from $P^0(\mathrm{n}')$ can be expressed in terms of the *surprisal*, which is defined by the relationship

$$I(\mathrm{n}') = -\ln[P(\mathrm{n}')/P^0(\mathrm{n}')] \qquad (4.126)$$

In calculating values of $P^0(n')$ it is usual to assume that *all* product states at a particular total energy have the same probability of being populated. Conservation of angular momentum therefore only enters into the picture as one of the dynamical factors causing the actual probability distribution to differ from the prior expectation. What is nearly always, but not invariably, found is that the surprisal associated with the distribution in a particular set of quantum states varies in a very straightforward fashion. As an example, we consider a distribution over product vibrational levels v′.

It is usual to express the probabilities of forming a product in a particular v′ as $P(f_{v'})$, where $f_{v'}$ is the energy of the v′ level expressed as a fraction of the total energy available for distribution in the reaction. The surprisal for a particular level is then defined as

$$I(f_{v'}) = -\ln[P(f_{v'})/P^0(f_{v'})] \qquad (4.127)$$

In most cases where data are available, $I(f_{v'})$ has been found to vary linearly, i.e.,

$$dI(f_{v'})/df_{v'} = \lambda_{v'} \qquad (4.128)$$

where $\lambda_{v'}$ is a constant, characteristic of the vibrational distribution from a particular reaction.

A more suggestive form of these relationships is obtained by substituting the integral form of equation (4.128), i.e.,

$$I(f_{v'}) = \lambda^0_{v'} + \lambda_{v'} f_{v'} \qquad (4.129)$$

into equation (4.127) yielding

$$P(f_{v'})\exp(\lambda^0_{v'}) = P^0(f_{v'})\exp(-\lambda_{v'} f_{v'}) \qquad (4.130)$$

Since $\sum_{v'} P(f_{v'}) = 1$,

$$\exp(\lambda^0_{v'}) = \sum_{v'} P^0(f_{v'})\exp(-\lambda_{v'} f_{v'}) \qquad (4.131)$$

and (4.130) may be written as

$$P(f_{v'}) = \frac{P^0(f_{v'})\exp(-\lambda_{v'} f_{v'})}{\sum_{v'} P^0(f_{v'})\exp(-\lambda_{v'} f_{v'})} \qquad (4.132)$$

Bearing in mind that $f_{v'} = (\varepsilon_{v'}/\varepsilon_{total})$, this equation clearly resembles a Boltzmann relationship with the equations

$$k_B T_{v'} = \varepsilon_{total}/\lambda_{v'} \qquad (4.133a)$$

and

$$Q_{v'} = \sum_{v'} P^0(f_{v'})\exp(-\lambda_{v'} f_{v'}) \qquad (4.133b)$$

defining parameters analogous to a temperature and a vibrational partition function. For most *direct* reactions, $\lambda_{v'}$ is negative and it follows that the temperature-like quantity $T_{v'}$ is also negative.

It has to be emphasized that the information-theoretic approach does not pretend to make *a priori* predictions of detailed kinetic data; the values of $P^0(\mathrm{n}')$ are *expected* to differ from the actual values of $P(\mathrm{n}')$. However, it does provide a means of compacting experimental data down to a relatively small number of parameters, such as $\lambda^0_{\mathrm{v}'}$ and $\lambda_{\mathrm{v}'}$, defining the slope and intercepts of surprisal plots of various product state distributions. Perhaps the most important aspect of this treatment is that it suggests a method for extrapolating from limited detailed information to provide data not obtained directly. For example, values of $P(\mathrm{v}')$ may only have been measured for some values of v′. The surprisal analysis then provides a means of estimating $P(\mathrm{v}')$ for levels other than those directly observed.

4.6.3 Product energy distributions from unimolecular decay processes[38, 39]

There have recently been a number of measurements of the product energy distributions from reactions that proceed via the bimolecular formation of a strongly coupled complex and its subsequent unimolecular decomposition. These experiments, which are discussed in Section 7.5.3, are the microscopic equivalent of the chemical activation studies that were mentioned in Section 4.5.6 and are considered in Section 6.4.3. Examples of reactions where distributions of the relative translational energy of the products have been determined in crossed molecular beams experiments[40] are provided by

$$F + C_2H_4 \rightarrow (C_2H_4F)^{\dagger} \rightarrow C_2H_3F + H \qquad (4.134)$$

and

$$F + C_2H_2Cl_2 \rightarrow (C_2H_2Cl_2F)^{\dagger} \rightarrow C_2H_2ClF + Cl \qquad (4.135)$$

Bulk experiments have also been performed on the related processes in which vibrationally excited molecules are deactivated in collisions with atoms with which they can form a transient complex[41], e.g.,

$$O_2(\mathrm{v}) + O \rightarrow (O_3)^{\dagger} \rightarrow O_2(\mathrm{v}' < \mathrm{v}) + O \qquad (4.136)$$

These types of experimental investigation have helped to stimulate theoretical work on the sharing of energy among the products of unimolecular decomposition.

As was pointed out in Section 4.6.1, phase space theory properly describes the sharing of energy in instances where the transition state is in the exit channel for the dissociative process and is extremely ‘loose’. However, it is essential that centrifugal effects are properly allowed for. Exactly equivalent results to those from a phase space calculation can be obtained from an approach based on RRKM theory. This considers the case where the transition state for decay of a complex of given total energy and angular momentum is located at the centrifugal maximum far out in the exit channel. The specific rate constant for dissociation of this specified complex is given by equation (4.80), i.e., $k_J(\varepsilon) = N^*(\varepsilon^+)/h\rho(\varepsilon)$. The probability of these complexes having internal energy equal to $(\varepsilon^+ - \varepsilon_x^+)$, where ε_x^+ is the energy in motion along the

reaction coordinate, is directly proportional to the density of internal states at that internal energy. Consequently, the distribution function for ε_x^+ is given by

$$P(\varepsilon_x^+) = \rho(\varepsilon^+ - \varepsilon_x^+). \tag{4.137}$$

which falls monotonically from $\varepsilon_x^+ = 0$.

However, equation (4.137) does not correspond to the distribution function for $\varepsilon_{T'}$, the final relative translational energy, since the energy of the centrifugal barrier, $l'(l'+1)\hbar^2/2I^+$, as well as ε_x^+, becomes final relative translational energy of the products. In considering the distribution of $\varepsilon_{T'}$ from collisions of a particular total energy but a distribution of angular momenta resulting from a mode of formation limited only by a centrifugal barrier in the entry channel, one can write

$$P(\varepsilon_T') = \rho(\varepsilon^+ - \varepsilon_x^+)A(\varepsilon_{T'}) \tag{4.138}$$

where the factor $A(\varepsilon_{T'})$ allows for centrifugal effects.

At $\varepsilon_{T'} = 0$, $A(\varepsilon_{T'})$ is also zero. Physically, this is because the probability of forming a collision complex which has zero angular momentum and which can therefore decompose without having to surmount a centrifugal barrier is vanishingly small. Consequently, the distribution function $P(\varepsilon_{T'})$ does not fall monotonically from $\varepsilon_{T'} = 0$ but passes through a maximum at some finite value of $\varepsilon_{T'}$.

The model that has just been described has been generalized by Marcus to cover reactions where there is a potential energy maximum in the exit channel from the collision complex. The existence of such a barrier can be inferred if the reverse association process is known to have a positive activation energy. For example, the addition of an H atom to a fluorinated olefin, i.e., the reverse of the second stage of reaction (4.134), has an activation energy of ~ 10–20 kJ mol^{-1}, whereas the activation energy for the analogous addition of a Cl atom is close to zero.

The presence of an energy barrier in the exit channel means that the transition state for decomposition of the energized $(C_2H_4F)^\dagger$ radical in reaction (4.134) will be 'tighter' than that for reaction (4.135). This implies that, whereas in a very 'loose' transition state the separating species are free to rotate, there are, in the 'tight' activated complex, bending vibrations or restricted rotations that only transform to *free* rotations after the system passes through the transition state. The basis of Marcus's model is that RRKM theory provides an appropriate description of the state distribution in the transition state; subsequently, the conversion of the bending motions to rotations is assumed to be 'statistically adiabatic' (i.e., adiabatic on average). This implies an 'on average' conservation of quantum numbers and, since the vibrational bending states in the activated complex will be more widely spaced than the final rotational states, extra energy will be supplied to the relative motion of the separating fragments. Consequently, decomposition of a strong collision complex through a 'tight' activated complex should lead to distributions of final translational energy shifted to higher $\varepsilon_{T'}$ relative to those from decays through a 'loose' transition state.

The statistical adiabatic channel model of Quack and Troe[27] that was referred to earlier in Section 4.5.6 is based on a somewhat different premise from that of Marcus, in that statistical adiabadicity is assumed right along the path from reactants to products and not just from transition state to products. Quack and Troe[27] have carried out calculations on inelastic–reactive processes such as that represented by equation (4.136) but not on larger systems such as those that have been studied in molecular beam experiments. It would appear that the assumption of adiabatic channels right through from reactants to products is likely to predict the release of more energy as relative translation of the products as the decrease in internal energy in any adiabatic channel going from reactants to products will be greater than that going only from the transition state to products. However, a direct comparison of the two methods for the same system would be valuable.

References

Section 4.2

1 GLASSTONE, S., LAIDLER, K. J., and EYRING, H., *Theory of Rate Processes*, McGraw-Hill, New York (1941)
2 LAIDLER, K. J., and POLANYI, J. C., *Progress in Reaction Kinetics*, **3**, 1 (1965)
3 MAHAN, B. H., *Journal of Chemical Education*, **51**, 709 (1974)
4 MILLER, W. H., *Accounts of Chemical Research*, **9**, 306 (1976)
5 MILLER, W. H., *Journal of Chemical Physics*, **61**, 1823 (1974)
6 PECHUKAS, P., and McLAFFERTY, F. J., *Journal of Chemical Physics*, **58**, 1622 (1973)
7 CHAPMAN, S., HORNSTEIN, S. M., and MILLER, W. H., *Journal of the American Chemical Society*, **97**, 893 (1975)
8 JOHNSTON, H. S., *Gas Phase Reaction Rate Theory*, chap. 12, Ronald, New York (1966)

Section 4.3

9 LEVINE, R. D., and BERNSTEIN, R. B., *Molecular Reaction Dynamics*, Sections 4.3 and 4.4, Oxford University Press, Oxford (1974)
10 MARCUS, R. A., *Journal of Chemical Physics*, **45**, 2138 and 2630 (1966)
11 MOROKUMA, K., EU, B. C., and KARPLUS, M., *Journal of Chemical Physics*, **51**, 5193 (1969)

Section 4.4

12 Ref. 8, Chap. 2, 8, 10 and 13
13 SCHATZ, G. C., and KUPPERMANN, A., *Journal of Chemical Physics*, **65**, 4642 and 4668 (1976)
14 TRUHLAR, D. G., and KUPPERMANN, A., *Journal of the American Chemical Society*, **93**, 1840 (1971)
15 MARCUS, R. A., *Journal of Chemical Physics*, **45**, 4493 (1966) and **46**, 959 (1967)
16 JOHNSTON, H. S., and RAPP, D., *Journal of the American Chemical Society*, **83**, 1 (1961)
17 MARCUS, R. A., and COLTRIN, M. E., *Journal of Chemical Physics*, **67**, 2609 (1977)
18 NIKITIN, E. E., *Theory of Thermally Induced Gas Phase Reactions*, Indiana University Press, Bloomington (1966)

Section 4.5

19 ROBINSON, P. J., and HOLBROOK, K. A., *Unimolecular Reactions*, Wiley, London (1972)

20 FORST, W., *Theory of Unimolecular Reactions*, Academic Press, New York (1973)

21 TROE, J., in *International Review of Science. Physical Chemistry*, Series 2, Vol. 9, *Chemical Kinetics*, Ed. Herschbach, D. R., Chap. 1, Butterworths, London (1976)

22 QUACK, M., and TROE, J., in *Specialist Periodical Reports. Gas Kinetics and Energy Transfer*, vol. 2, Eds. Ashmore, P. G., and Donovan, R. J., Chap. 5, Chemical Society, London (1977)

23 MARCUS, R. A., and RICE, O. K., *Journal of Physical and Colloid Chemistry*, **55**, 894 (1951)

24 MARCUS, R. A., *Journal of Chemical Physics*, **20**, 359 (1952); WEIDER, G. M., and MARCUS, R. A., *Journal of Chemical Physics*, **37**, 1835 (1962); MARCUS, R. A., *Journal of Chemical Physics*, **43**, 2658 (1965)

25 KASSEL, L. S., *The Kinetics of Homogeneous Gas Reactions*, Chemical Catalog Co., New York (1932)

26 WAAGE, E. V., and RABINOVITCH, B. S., *Chemical Reviews*, **70**, 377 (1970)

27 QUACK, M., and TROE, J., *Berichte der Bunsengesellschaft für Physikalische Chemie*, **78**, 240 (1974) and **79**, 170 (1975)

28 WHITTEN, G. Z., and RABINOVITCH, B. S., *Journal of Chemical Physics*, **38**, 2466 (1963)

29 TROE, J., *Journal of Chemical Physics*, **66**, 4758 (1977)

30 BUNKER, D. L., and PATTENGILL, M., *Journal of Chemical Physics*, **48**, 772 (1968)

31 TROE, J., *Berichte der Bunsengesellschaft für Physikalische Chemie*, **78**, 478 (1974)

32 TROE, J., *Berichte der Bunsengesellschaft für Physikalische Chemie*, **77**, 665 (1973)

33 QUACK, M., and TROE, J., *Berichte der Bunsengesellschaft für Physikalische Chemie*, **81**, 329 (1977)

34 LIGHT, J. C., *Discussions of the Faraday Society*, **44**, 14 (1967)

35 LEVINE, R. D., and BERNSTEIN, R. B., *Accounts of Chemical Research*, **7**, 393 (1974)

36 LEVINE, R. D., and BERNSTEIN, R. B., in *Dynamics of Molecular Collisions*, Part B, Ed. Miller W. H., Chap. 7, Plenum, New York, (1976)

37 LEVINE, R. D., and BEN-SHAUL, A., in *Chemical and Biochemical Applications of Lasers*, Ed. Moore, C. B., Vol. 2, Chap. 4, Academic Press, New York (1977)

38 SAFRON, S. A., WEINSTEIN, N. D., HERSCHBACH, D. R., and TULLY, J. C., *Chemical Physics Letters*, **12**, 564 (1972)

39 MARCUS, R. A., *Journal of Chemical Physics*, **62**, 1372 (1975)

40 FARRAR, J. M., and LEE, Y. T., *Journal of Chemical Physics*, **65**, 1414 (1976)

41 BREEN, J. E., QUY, R. B., and GLASS, G. P., *Journal of Chemical Physics*, **59**, 556 (1973)

Note added in proof (*see p. 145*)

E. Pollack and P. Pechukas [*Journal of Chemical Physics*, **69**, 1218 (1978)] argue that the best choice of transition state for a particular total energy is a 'classical bound state embodied in the continuum'. These are found by searching for 'trapped trajectories'. As the total energy is raised well beyond the minimum required for reaction, these periodic trajectories no longer traverse the saddlepoint region on the surface.

Part 2:

Comparisons between experiment and theory

Chapter 5

Thermal rate data

5.1 General introduction

In the first four chapters of this book, theoretical aspects of chemical kinetics have been emphasized to the virtual exclusion of all else. Besides describing the simpler rate theories rather fully, an attempt has been made to provide the reader with some insight into refinements of these theories that have recently—or, in many cases, still are—taking place. Now it is time to see how well these theoretical ideas accord with experimental observations. This is the theme in the remainder of this book.

In the present chapter, the empirical data derived from 'conventional' kinetics experiments, i.e., rate constants, activation energies, and pre-exponential factors, are considered. Comparison with theory is carried out at two different levels. In some sections, values of the kinetic parameters for a particular kind of reaction are presented. The 'typical' values of these quantities are related to the general form of the potential in that type of reaction and reasons are suggested to explain any exceptions to the general rules. In addition, the rate data for some individual reactions are examined in greater detail to see whether extensive, but conventional, kinetic results can provide a test of, particularly, transition state theory, and to show how careful comparisons of this kind can furnish limited information about some aspects of the reactive potential.

5.2 Experimental methods[1]

It is certainly *not* the object in this section to review exhaustively all the experimental methods that can be applied to the determination of thermal rate constants for elementary reactions in the gas phase. Rather an attempt is made to describe briefly the main modern methods, emphasizing the strengths, and more particularly the weaknesses, of each technique. The reader will certainly not be able to set up any experiments of his own without a great deal more information. However, by highlighting the possible limits and deficiencies of each technique, it is hoped that the reader will subsequently be able to adopt a

properly critical attitude to reports of rate constant measurements in the literature.

Major errors in rate constant determinations are now very rarely caused by flaws in the observational technique[2]. They arise because nearly always observations are made on only a limited number of the species participating in the chemical changes that are occurring in the reaction mixture. Frequently, the concentration of only a single species is monitored directly. This will not lead to errors as long as the concentration of the observed species is only altered by the single reaction whose kinetics are supposedly being studied, or if a more complicated reaction mechanism correctly identifies all the processes contributing to its removal. Consequently, when evaluating measurements of rate constants, particular attention should be paid to that part of the paper dealing with the assumed mechanism. Preferably, side and secondary, 'interfering', reactions should be eliminated—but not neglected without good cause! Where some allowance has to be made for these reactions, one should examine how dependent the derived rate constant is on those for other reactions in the mechanistic scheme and whether these values are firmly established.

These remarks can perhaps be clarified by considering a specific example. For this we shall take the bimolecular reaction

$$OH + NH_3 \rightarrow H_2O + NH_2 \tag{5.1}$$

which is of that important class in which an atom is transferred from a 'stable' molecule to a free radical. The products are, of course, a new 'stable' molecule and new free radical. When direct methods are used to determine rate constants for such reactions (Sections 5.2.1 to 5.2.3), conditions are chosen so that the initial concentration of the reactant radical, $[OH]_{t=0}$, is very much less than that of the molecule, $[NH_3]_{t=0}$, so that $[OH]_t$, which is what is observed experimentally, decays exponentially. Hence

$$-\mathrm{d}\ln[OH]_t/\mathrm{d}t = k_{1st} \tag{5.2a}$$

and

$$k_1 = k_{1st}/[NH_3] \tag{5.2b}$$

where k_1 is the second-order rate constant for the reaction.

Two important advantages stem from being able to follow the course of the reaction under conditions such that $[OH]_{t=0} \ll [NH_3]$. First, there is no need to know the *absolute* concentrations of the radical. Secondly, one is most likely to eliminate significant interference by other reactions.

It is useful to consider the possible effects of side and secondary reactions quantitatively. For our example, it will be necessary to consider *at least*

$$OH + OH \rightarrow H_2O + O \tag{5.3a}$$

$$OH + OH(+M) \rightarrow H_2O_2(+M) \tag{5.3b}$$

and

$$OH + NH_2 \rightarrow \text{products} \tag{5.4}$$

There may be further possibilities depending on the method used to generate OH radicals and the other products of this process. If the second-order rate constant for (5.3a) and (5.3b) combined is k_3 and that for (5.4) is k_4, the equation allowing for removal of OH by these reactions, as well as the primary reaction, is

$$-\mathrm{d}\ln[\mathrm{OH}]/\mathrm{d}t = k_1[\mathrm{NH_3}] + 2k_3[\mathrm{OH}]_t + k_4[\mathrm{NH_2}]_t \tag{5.5}$$

To a first approximation,

$$[\mathrm{OH}]_t = [\mathrm{OH}]_{t=0}\exp(-k_{1\mathrm{st}}t) \tag{5.6}$$

and

$$\begin{aligned}[\mathrm{NH_2}]_t &= [\mathrm{OH}]_{t=0} - [\mathrm{OH}]_t \\ &= [\mathrm{OH}]_{t=0}\{1 - \exp(-k_{1\mathrm{st}}t)\}\end{aligned} \tag{5.7}$$

so equation (5.5) may be rewritten as

$$\begin{aligned}k_{1\mathrm{st}} &\approx k_1[\mathrm{NH_3}] \\ &\times \left\{1 + \frac{[\mathrm{OH}]_{t=0}}{k_1[\mathrm{NH_3}]}[k_4 + (2k_3 - k_4)\exp(-k_{1\mathrm{st}}t)]\right\}\end{aligned} \tag{5.8}$$

This equation can easily be adapted to other reaction systems. It makes very clear that what is required of the experimental method is that measurements can be made under conditions such that the second factor in the { } braces is $\ll 1$[a]. This can be achieved by a combination of high detection sensitivity, so that the initial concentration of radicals can be made very small, and rapid time resolution. The latter determines how large $k_{1\mathrm{st}} (\approx k_1[\mathrm{NH_3}])$ can be made and still measured. The parameter $k_1[\mathrm{Reac}]/[\mathrm{Rad}]_{t=0}$, where [Reac] and [Rad] replace $[\mathrm{NH_3}]$ and [OH] in our specific example, has the dimensions of a second-order rate constant. It can serve as a very useful indication of the ability of an experimental technique to study a particular reaction in isolation. Ideally this 'resolution-sensitivity index' should exceed Z, the rate constant associated with the occurrence of bimolecular collisions, since some of the interfering reactions are likely to involve two free radicals, and therefore their rate constants may approach this value. It is, of course, true that the right-hand side of equation (5.8) is time-dependent so that a semi-logarithmic plot of radical concentration versus time should curve, if reactions other than the primary one are removing radicals. However, this rarely provides a sensitive test of whether such effects arise.

To conclude, the effects of side and secondary removal of the radical species may arise in any 'direct' method of measuring rate constants, and should

[a] The factor in the { } braces is frequently referred to as the *stoichiometry* of the reaction with respect to removal of the radicals. Its value can sometimes be determined by an extension of the experimental technique, or by knowledge of the rate constants k_3 and k_4, and the appropriate correction applied. It is, however, preferable if the stochiometry is demonstrably unity, so that the primary reaction is observed in isolation.

always be carefully considered. In 'indirect' experiments, the progress of a reaction is not monitored whilst it is actually occurring, but after some measured time the gas mixture is removed from the source of initiation, e.g., a high-temperature vessel or a light source, and analysed. To extract even *relative* rate constants for elementary reactions from such experiments, it is absolutely crucial to know the reaction mechanism.

Each technique has, of course, its own special difficulties or limitations. These are considered under the appropriate headings in the sections that follow. The main two direct techniques that are used to obtain rate constants at relatively low temperatures (up to ~1000 K, but usually much lower) of fast bimolecular reactions, particularly of the atom-transfer type, and of radical + radical association reactions are considered first. In current applications of both of these methods, the radical species are nearly always observed by spectrosocopic techniques of one kind or another.

5.2.1 Discharge–flow systems[3]

The main features of a discharge–flow apparatus suitable for studying elementary gas-phase reactions are shown in *Figure 5.1.* Atoms are created by partial dissociation of a flowing diatomic gas, usually in a microwave discharge. In this way H, N, O and halogen atoms can be produced in yields of ~ 1–10% from the parent molecule, which is usually diluted in He or Ar. These atoms may be converted into different radicals by adding, at an inlet downstream of the discharge, a species which rapidly reacts with the atoms converting them into the radical that is required. Thus OH radicals are produced in the rapid reaction

$$H + NO_2 \rightarrow OH + NO \tag{5.9}$$

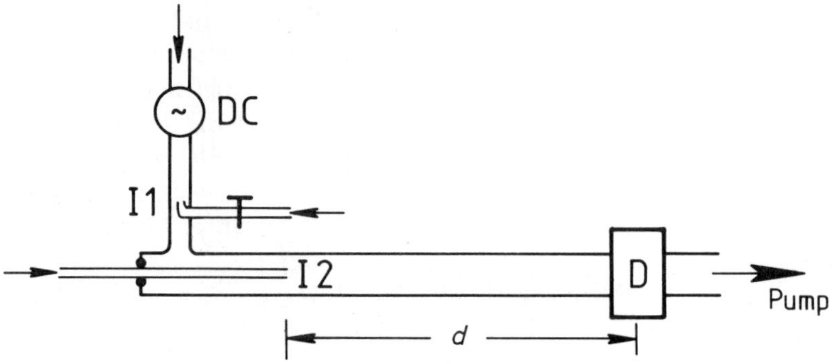

Figure 5.1. Schematic diagram of a discharge–flow apparatus. Atoms are created by dissociating diatomic gas in the discharge cavity (DC), *and may be converted into other free radicals by gas titration at inlet* I1. *The molecular reagent enters through a movable injector* I2 *so that the distance* (*d*) *between its tip and the fixed detector* (D) *can be altered. The arrows indicate the direction of gas flow*

The discharge may also produce metastable molecular species with excess of vibrational or electronic energy. For example, discharged oxygen contains appreciable proportions of $O_2(^1\Delta_g)$, and it is frequently preferable to generate atomic oxygen by discharging N_2 and converting the N atoms into O atoms quantitatively through the fast reaction

$$N + NO \rightarrow N_2 + O \tag{5.10}$$

After their production, the radicals in their diluent enter the main flow tube, which is typically 1 m long and 25 mm internal diameter. Their (relative) concentration is measured at a point near the downstream end of the flow tube. The method depends upon being able to alter the distance between this observation point and the inlet upstream through which the molecular reagent is admitted. This is usually achieved by keeping the detector fixed and using a movable injector.

With the conditions generally employed to study gas-phase reactions, there is a parabolic profile of velocities across the flow tube. As a result, without rapid radial diffusion, species near the walls would have a longer residence time, and therefore experience greater extents of reaction, than those near the axis of the tube. In practice, flow tube experiments can be carried out at pressures and linear flow velocities sufficiently low to ensure that diffusion maintains an essentially flat radial distribution of reacting radicals. On the other hand, they should not be so low that appreciable longitudinal diffusion occurs while the gas passes along the flow tube. With correctly chosen conditions, a 'plug' of gas introduced at the upstream end of the tube should retain its size and form as it is transported down the tube, and the radical concentration which is observed corresponds to that at a time $t = d/\bar{u}$ after reaction has commenced, where d is the distance from inlet I2 to the detector D and $\bar{u}$ is the mean linear velocity of the gas along the tube.

The need to have conditions where the assumption of 'plug-flow' is valid restricts the range of flow rates and total pressures that can be used in flow tube experiments. These limits, and the corrections that become necessary at either end of the allowed range, have been discussed fully elsewhere[3, 4]. Typically, the total pressure in the flow tube is between 0.1 and 1.0 kN m^{-2} (0.75—7.5 torr) and the linear flow velocity is between 1 and 50 m s^{-1}. Values of k_{1st} can be measured in the range ~ 500 to 5 s^{-1}. Since there are limits to the proportion of the total gas flow that should enter through the movable inlet and because very slow linear flow velocities lead to problems with longitudinal diffusion, it is difficult to measure second-order rate constants below about 10^{-16} cm^3 molecule^{-1} s^{-1} without taking special precautions.

Perhaps the major weakness of flow tube experiments is that heterogeneous reactions may occur. In 0.05 s every species present in the reactor makes roughly 10^3 collisions with the walls of the flow tube. In a fixed observation point apparatus of the type illustrated in *Figure 5.1*, the number of collisions that radicals make with the wall, between the point at which they are produced and the detector, is unaffected by the addition of the molecular reactant through the movable injector, unless this forms a major part of the total gas

flow. Consequently, the contribution of the surface-catalysed combination of the radicals to their overall removal should remain the same in all experiments. However, it is not always easy to be certain that the efficiency of this process is unchanged by the addition of the molecular reactant, or that there is no heterogeneous contribution to the primary radical + molecule reaction. To minimize such surface effects, it is customary to coat the wall of the flow tube with materials, such as Teflon or phosphoric acid, which suppress the heterogeneous recombination of most radicals.

A major strength of the discharge–flow method is that it can be applied to reactions of a wide variety of atoms. It has also been used to study the reactions of several diatomic radicals, but only rarely those of larger radicals. A second advantage is that the flow system is in a steady-state so that, however the radicals are observed, the signal can, if necessary, be averaged over several minutes to improve the signal-to-noise ratio. Several different methods have now been used to determine the concentrations of radicals in experiments of this type. A particularly simple method is to view the variation in intensity associated with a chemiluminescent recombination reaction. In the case of atomic oxygen, a small flow of NO is added to the gas stream and the variation in intensity of the 'air afterglow', which is associated with the reaction

$$O + NO(+M) \rightarrow NO_2^*(+M) \rightarrow NO_2 + h\nu \tag{5.11}$$

is observed.

Nowadays, changes in the concentration of the reacting radicals are usually observed spectroscopically. Reactions of the OH radical have recently received a great deal of attention, not least because of their importance in combustion and in atmospheric chemistry. The spectroscopic techniques that have been used to measure concentrations of OH in flow-tube experiments are listed in *Table 5.1*. A value of the resolution-sensitivity index that was referred to in the preceding section has been calculated for each of these methods, assuming $k_{1st} = 200\ s^{-1}$ and a value of $[OH]_{t=0}$ in the middle of the range used in experiments described in the papers that are referenced below the table. Although these figures are only approximate, they do indicate that, for OH, the use of resonance fluorescence or laser magnetic resonance detection should eliminate any need to allow for removal of the radicals by *gas-phase* reactions other than the primary reaction.

When optical techniques are used for observation, usually the concentration of only a single species—in most cases, the radical reactant—is monitored. Mass spectrometric sampling, on the other hand, provides 'universal detection'. In the past, this application of mass spectrometry has been bedevilled with difficulties. The major problem has been to ensure that the sample reaching the ion source of the mass analyser is representative of the reaction mixture and has not been altered either by reactions on the surfaces of the sampling system or by the separation of species of different mass at the sampling orifice. Only quite recently have these problems been overcome and mass spectrometric analyses of discharge-flow systems started to provide reliable rate constants.

Table 5.1 Spectroscopic methods of observing OH radicals in flow tube experiments

Detection method	$(k_{1st}/[OH]_{t=0})$ / cm^3 molecule^{-1} s^{-1}	*Ref.*
Ultraviolet resonance absorption	$\sim 10^{-12}$	*a*
Electron spin resonance	$\sim 10^{-11}$	*b*
Ultraviolet resonance fluorescence	$\sim 5 \times 10^{-9}$	*c*
Laser magnetic resonance	$\sim 5 \times 10^{-8}$	*d*

[a] DEL GRECO, F. P., and KAUFMAN, F., *Discussions of the Faraday Society*, **33**, 128 (1962); PASTRANA V, A., and CARR, Jr., R. W., *International Journal of Chemical Kinetics*, **6**, 587 (1974)
[b] BREEN, J. E., and GLASS, G. P., *Journal of Chemical Physics*, **52**, 1082 (1970); WESTENBERG, A. A., and DE HAAS, N., *Journal of Chemical Physics*, **57**, 5375 (1972)
[c] ANDERSON, J. G., and KAUFMAN, F., *Chemical Physics Letters*, **16**, 375 (1972); HARRIS, G. W., and WAYNE, R. P., *Journal of the Chemical Society, Faraday Transactions*, I, **71**, 610 (1975)
[d] HOWARD, C. J., and EVENSON, K. M., *Journal of Chemical Physics*, **61**, 1943 (1974)

Mass spectrometry can be invaluable in revealing the occurrence of previously unsuspected side reactions or different pathways for the primary reaction. Although it scarcely qualifies as a conventional discharge–flow experiment, an example of this type of observation is provided by a recent study[5] of the reaction between $O(^3P)$ atoms and CS_2. The reagents were in the form of uncollimated beams allowing detection of the primary reaction products under conditions where there was virtually no possibility of secondary reactions. These measurements identified the reaction channels

$$O(^3P) + CS_2 \rightarrow OCS + S \tag{5.12}$$

and

$$O(^3P) + CS_2 \rightarrow CO + S_2 \tag{5.12b}$$

as well as the major atom-transfer route:

$$O(^3P) + CS_2 \rightarrow SO + CS \tag{5.12c}$$

Finally, it should be mentioned that the flow technique has been adapted to the study of ion–molecule reactions, for which these experiments are the main source of thermal rate constants[6]. The ionic reactant is produced upstream of the reactor, either by direct ionization using an electron gun or by a Penning ionization process involving metastable atoms produced in a discharge, e.g.,

$$He(2^3S) + N_2 \rightarrow He + N_2^+ + e \tag{5.13}$$

Inert gas then carries the ions into the main flow tube where the molecular reactant can be injected. The concentrations of the ionic reactant and any ionic products of reaction are observed at the downstream end of the flow tube by mass analysis. Because ions are destroyed very efficiently at surfaces, it is

necessary to use a wider flow tube and a greater flow velocity than in similar experiments involving only neutral species. Consequently, the simple direct relationship between residence time and distance cannot be used to interpret the results, but the laminar flow of gas in the tube must be fully modelled in order to extract rate constants for the ion–molecule reaction.

5.2.2 Flash photolysis[7]

The second major technique for studying the rates of elementary reactions directly is based on the production of the free radical reactant in a short, intense flash of radiation. Although only flash photolysis is considered here, the related techniques of pulse radiolysis[8], where reaction is initiated by a burst of X-rays or high-energy electrons, and modulated photolysis[9] are now also being successfully applied to gas-phase reactions. The application of laser flash photolysis to the study of elementary gas-phase reactions is likely to increase, as powerful lasers for the far and vacuum ultraviolet regions of the spectrum become increasingly available.

In experiments where the radicals are created by flash photolysis, a static gas mixture is prepared which contains the photochemical precursor of the radicals, the molecular reagent, and sufficient chemically inert gas to ensure that the temperature does not rise significantly as heat is released in photochemical and chemical processes. In most early flash photolysis experiments, transient species were observed by recording their absorption spectra photographically using a second flash lamp to provide a background continuum at some preselected delay time after the photolytic flash had been fired. The duration of these light flashes, obtained by discharging energy from a capacitor through an inert gas, depends on the capacitance and voltage used. As a rough guide, using a charging voltage of 20 kV, a flash of 1000 J energy is likely to last about 50 μs, whereas one of 100 J, a typical value for the secondary flash lamp, would last $\leqslant 10\ \mu$s.

By performing several photographic experiments on different samples of the same gas mixture with only the delay time between flashes altered, a record could be built up of the kinetic behaviour of the transient species. The result of an experiment of this kind is shown later in *Figure 6.2*. However, as well as being tedious, this method is also an insensitive way of monitoring species with discrete spectra. In principle, a spectrograph of high resolving power could be used and the extent of absorption at the line centres determined, but in practice a great many individual 'shots' would have to be superimposed to obtain a proper measure of the absorption, because the intensity of the background flash will be dispersed widely across the photographic plate. This has led to the increasing use of *resonance absorption* and *resonance fluorescence* measurements in quantitative kinetic studies.

For measurements by resonance absorption, the secondary flash lamp in 'conventional' flash photolysis experiments is replaced by a lamp that emits the characteristic resonance radiation of the species whose concentration one

wishes to observe. In addition, the combination of spectrograph and photographic plate gives way to a monochromator, which is set at the wavelength of one of the strongest lines in the resonance spectrum, and a photomultiplier. This then provides a continuous record of the variation in concentration of the transient after its formation in the photolytic flash.

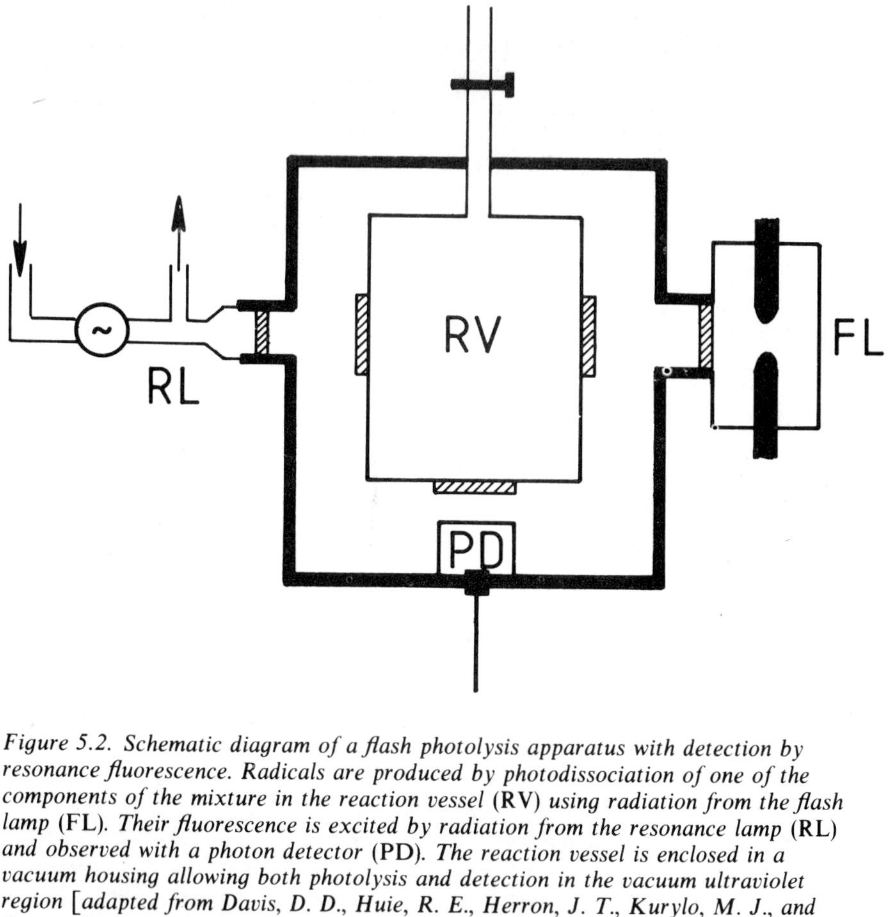

Figure 5.2. Schematic diagram of a flash photolysis apparatus with detection by resonance fluorescence. Radicals are produced by photodissociation of one of the components of the mixture in the reaction vessel (RV) *using radiation from the flash lamp* (FL). *Their fluorescence is excited by radiation from the resonance lamp* (RL) *and observed with a photon detector* (PD). *The reaction vessel is enclosed in a vacuum housing allowing both photolysis and detection in the vacuum ultraviolet region* [*adapted from Davis, D. D., Huie, R. E., Herron, J. T., Kurylo, M. J., and Braun, W.,* Journal of Chemical Physics, **56**, 4868 (1972)]

Even greater sensitivity is obtained by observing not the attenuation of the resonance radiation as it passes through the reaction mixture but the intensity of the flourescence excited by this radiation. The apparatus now changes considerably; *Figure 5.2* provides an illustration of a typical design. Light from the pulsed photolysis lamp and radiation from the resonance lamp illuminate the same portion of the gas mixture and the fluorescence is observed in a

direction perpendicular to the radiation from both the photolysing flash and the resonance lamp. The fluorescence signals are extremely weak, but this may be compensated for by making the decay times (i.e., $1/k_{1st}$) relatively long (≥ 1 ms), and averaging the signals from many rapidly repeated experiments. The high sensitivity of resonance fluorescence detection means that only very small initial concentrations of radicals ($< 10^{12}$ molecule cm^{-3}) need be produced, and the effects of side and secondary reactions are minimized. In addition, the photolysing flash need not be very strong. Consequently its duration is short, so the photochemical production of radicals occurs much more rapidly than their removal by chemical reaction.

Values of the resolution-sensitivity index for the different spectroscopic methods that have been used to observe OH radicals in flash photolysis experiments are compared in *Table 5.2*. With detection by resonance fluorescence, it is clear that OH should not be removed to a significant extent in reactions other than the primary reaction.

Table 5.2 Spectroscopic methods of observing OH in flash photolysis experiments

Detection method	$\dfrac{(k_{1st}/[OH]_{t=0})}{\text{cm}^3\ \text{molecule}^{-1}\ \text{s}^{-1}}$	*Ref.*
Plate photography of ultraviolet absorption	$\sim 5 \times 10^{-11}$	*a*
Ultraviolet resonance absorption	$\sim 2 \times 10^{-10}$	*b*
Ultraviolet resonance fluorescence	$\sim 2 \times 10^{-9}$	*c*

[a] GREINER, N. R., *Journal of Chemical Physics*, **46**, 2795 (1967)
[b] MORLEY, C., and SMITH, I. W. M., *Journal of the Chemical Society, Faraday Transactions* II, **68**, 1016 (1972); SMITH, I. W. M., and ZELLNER, R., *Journal of the Chemical Society, Faraday Transactions* II, **69**, 1617 (1973)
[c] STUHL, F., and NIKI, H., *Journal of Chemical Physics*, **57**, 3671 (1972); DAVIS, D. D., FISCHER, S., and SCHIFF, R., *Journal of Chemical Physics*, **61**, 2213 (1974)

One major advantage of flash photolysis over discharge–flow experiments is that heterogeneous reactions can be eliminated. A second is that measurements can be made over a wide range of total pressures. This is particularly useful in the investigation of association reactions; Troe's group have contrived to study such reactions by flash photolysis at total pressures up to about 10^8 N m^{-2} (1000 atmospheres)[10]. The main problems with quantitative kinetic studies are those associated with finding suitable photochemical sources for the radicals of interest and with minimizing the photolytic production of unwanted reactive species in high concentration. These difficulties become acute for reactions of species that do not have a discrete spectrum; for example, polyatomic free radicals such as HO_2 and CH_3. The sensitivity of detection is low, because the absorption strength is dispersed over a wide range of wavelengths and the unstable upper electronic state will not fluoresce. Now

careful checks are necessary to establish that only the primary reaction is removing the species whose concentration is being observed.

Finally, it is worth noting that the discharge–flow and flash photolysis techniques can be combined to yield rate data on reactions between two different radicals. In this way, Schacke, Schmatjko and Wolfrum[11] studied the reaction

$$O + CN \rightarrow CO + N \tag{5.14}$$

A mixture of oxygen atoms with $(CN)_2$ could be prepared by discharge–flow techniques, since the reaction between atomic oxygen and cyanogen is slow. CN radicals were then generated by flash photolysing part of the $(CN)_2$ and the rate of reaction (5.14) measured by kinetic spectroscopy on these radicals. Further use of this combined method can be expected in the future.

5.2.3. Shock tubes[12, 13]

The discharge–flow and flash photolytic methods have emerged as the most important experimental techniques for kinetic measurements at relatively low temperatures. Shock tube experiments are quite frequently used to measure rate data down to 500 K (and sometimes, at even lower temperatures), thus overlapping the range of temperature covered by the other techniques, but the primary importance of this method lies in its potential for measurements at very high temperatures.

The basic elements of a conventional shock tube apparatus are illustrated in *Figure 5.3*. A thin diaphragm initially separates a short region of the tube containing inert 'driver' gas at high pressure, typically at 10^5—10^6 N m^{-2}, i.e., several atmospheres, from a section several metres long, which contains the potential reactants usually diluted in inert gas and at much lower pressure. When the diaphragm is punctured, pressure waves pass into the low-pressure section of the tube. Because the sound velocity in a gas increases with temperature, the later waves catch up the ones at the front so that they soon unite to form a single sharply defined shock front passing down the shock tube. This shock front compresses the gas containing reactants adiabatically and raises the translational temperature at any point in the shock tube in less than 1 μs.

When observations are made on the *incident* shock wave, that is on the shock wave making its first passage down the tube, temperatures as high as about 5000 K can be obtained. Temperatures as high as 20 000 K can be achieved in the shock wave *reflected* from the end plate of the tube. The wide range of temperature, as well as of pressure, that can be covered in shock tube experiments is one of its major attractions as an experimental technique.

The time resolution of shock tube experiments is set by the time it takes the shocked gas to pass through the region which is observed by each observation station. The shock velocity is greater than the speed of sound and this time is typically about 1 μs. Processes that occur over the range 1 μs to 1 ms are usually observed in shock tube measurements.

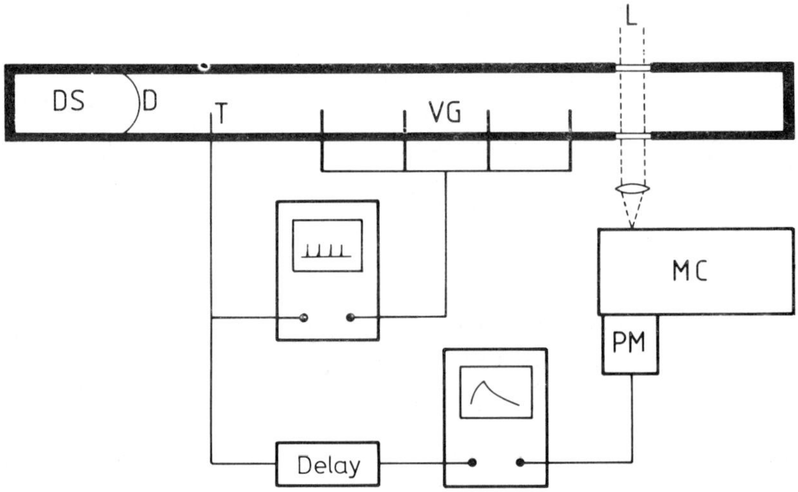

Figure 5.3. Schematic diagram of a shock tube experiment employing optical monitoring. The experimental section of the tube is separated from the driver section (DS) *by a diaphram* (D). *Once the diaphragm is ruptured, the shock wave propagates along the tube passing a trigger* (T) *and velocity gauges* (VG). *Analysing light* (L) *passes through windows flush with the tube walls and is dispersed by a monochromator* (MC), *and the signal from the photomultiplier* (PM) *is displayed* [*adapted from Bradley, J. N.,* Fast Reactions, *Oxford University Press, Oxford* (1975)]

The shock tube method is particularly well adapted to measuring the rates of vibrational energy transfer or relaxation and of thermal decomposition[14]. The study of bimolecular radical + molecule reactions is more difficult, since such a process must be preceded by thermal dissociation of a precursor to produce the appropriate radical. Consequently, the shock-heated mixture must contain at least two 'reactive' constituents and the chemical mechanism can be complicated. A possible way around this difficulty is suggested by some recent experiments[15] in which the low-pressure gas mixture was flowed slowly through the shock tube and contained radicals produced in a discharge.

As with discharge–flow and flash photolysis experiments, a wide range of observational techniques have been used with shock tubes. Both spectroscopic absorption and emission (but *not* resonance fluorescence) have been commonly used. A technique that is particularly well-suited to the study of vibrational relaxation or thermal dissociation is to follow the variation of temperature—or more commonly the related density change—behind the shock front. As was mentioned earlier, the translational heating at the shock front occurs in much less than 1 μs. Subsequently, as energy flows into the vibrational degrees of freedom of molecular species in the shock-heated gas, or is needed to drive endothermic reactions, the translational temperature falls below its initial value. The reason that the reagent is usually diluted is to avoid drastic temperature changes as a result of such processes. However, small

changes can be permitted and measured, and hence the rate of the endothermic processes deduced. Although the processes of energy transfer and thermal dissociation are connected, it is usually possible to find conditions such that the observed change is related to one or other process.

5.2.4 Static thermal and photochemical experiments

Three experimental methods that have been considered so far are all *direct* techniques of studying *fast* elementary reactions. When a reaction occurs over a characteristic time of minutes, as with many dissociation and isomerization reactions at moderate temperatures, less sophisticated techniques can provide accurate and reliable data, as long as the reaction mechanism is known. The interpretation of such experiments is particularly straightforward when the reaction occurs without the intervention of radical intermediates. The reactant, either pure or diluted in inert gas, is put into an accurately thermostatted high-temperature vessel. After a measured time, reaction is 'quenched' by expanding the gas into a vessel at room temperature. Samples are then withdrawn from this container for analysis, usually by gas chromatography or mass spectrometry.

When the reactant molecule decomposes to free radicals, as with ethane and nitric acid [*Table 4.3* (iii)] then secondary reactions will certainly occur. However the dissociation process is likely to be rate-determining, and its rate can be found if the scheme of subsequent reactions is known. Interpretation is greatly facilitated if the overall reaction is allowed to proceed to only a very limited extent; if possible, less than 1 per cent of the original reactant should be allowed to decompose. This is quite possible with modern analytical techniques.

Thermal experiments of this type are restricted to a limited class of rather slow reactions. Furthermore, since these reactions have large activation energies, their rate constants depend strongly on the temperature. As a result the rate of any particular reaction is only of a convenient speed within a fairly small range of temperature, although this can be extended to a limited degree by flowing the reactant fairly slowly through the high-temperature reactor.

Accurate *relative* rate data for radical + molecule reactions can be obtained in 'classical' photolysis experiments with continuously operating, low-intensity, lamps. These are analogous to static thermal experiments; the reaction mixture, which includes a suitable precursor for the radical species, is irradiated for a period of time, samples are then withdrawn and analysed.

One example of experiments of this type is Cvetanovic's study[16] of the reactions of $O(^3P)$ atoms with olefins. The atoms were usually generated by mercury-sensitized decomposition of N_2O:

$$Hg(^1S_0) + h\nu(\lambda = 253.7\ \text{nm}) \rightarrow Hg(^3P_1) \tag{5.15}$$

$$Hg(^3P_1) + N_2O(^1\Sigma) \rightarrow Hg(^1S_0) + N_2(^1\Sigma_g^+) + O(^3P) \qquad (5.16)^a$$

[a] It is interesting to note the different electronic states of the products in (5.16) and

$$O(^3P) + \text{olefin} \rightarrow \text{products} \tag{5.17}$$

A second example is provided by experiments on the reactions of OH radicals using photolysis of H_2O

$$H_2O + h\nu(\lambda \leqslant 185\ \text{nm}) \rightarrow OH + H \tag{5.18}$$

or of N_2O/H_2 mixtures

$$N_2O(^1\Sigma) + h\nu(\lambda \sim 200\ \text{nm}) \rightarrow N_2(^1\Sigma_g^+) + O(^1D) \tag{5.19}^a$$

$$O(^1D) + H_2 \rightarrow OH + H \tag{5.20}$$

to generate OH radicals. The use of N_2O in such experiments is particularly convenient, since the N_2 produced is unreactive and therefore serves as an *internal actinometer*, i.e., a measure of how many radicals have been produced in the initiation process.

The limitations to continuous photolysis experiments are fairly obvious. The chemical mechanism should be simple and must be properly understood, only the selected precursor should undergo photochemical change, and rate constants are only determined relative to that for some reference reaction. If absolute values of the rate constants are sought, one has to rely on the value of the rate constant for the reference reaction determined in a *direct* experiment. However, within these constraints, classical photolysis experiments continue to provide useful kinetic data.

5.3 Bimolecular reactions[17]

Provided with numerical data on reactions of a similar type, the experienced kineticist can frequently make quite good guesses at the values of the kinetic parameters—rate constant, pre-exponential factor, activation energy—for an elementary reaction. Attempting to impart this kind of knowledge or intuition in the space of a few paragraphs is dangerous, since the number of results that can be quoted is necessarily limited. One approach would be to omit exceptions and present only typical data. However, this would give a quite misleading impression about the certainty of this procedure, as well as leaving neglected the reasons why in certain cases quite 'untypical' parameters are observed. In this section, every effort is made to present the range of typical values of the rate constant or pre-exponential factor that is found for reactions of a particular kind, but some exceptions to the general rules are also considered. The size of the typical parameters are discussed in relationship to the type of intermolecular potential that operates between reactants of different kinds.

---

[a] (5.19). If direct photochemical excitation is to an unbound singlet state, then both obey the requirements of *spin conservation* (Section 2.4.3).

5.3.1 Reactions between two ions

The rates at which oppositely charged ions neutralize one another have been determined in a few cases. Whereas association of an atomic ion with an electron, for example

$$H^{+} + e \rightarrow H \tag{5.21}$$

requires the intervention of a third species to remove energy and stabilize the neutral atom, reactions between more complex charged species, for example

$$NO^{+} + SF_6^{-} \rightarrow \text{products} \tag{5.22}$$

are found to have mixed second- and third-order kinetics, i.e.,

$$-d[NO^{+}]/dt = (k_{2nd} + k_{3rd}[M])[NO^{+}][SF_6^{-}] \tag{5.23}$$

The second-order rate constant for this reaction and for some similar reactions are listed in *Table 5.3*. Their values are all very similar and are approximately 10^3 greater than the rate constants associated with hard-sphere collisions between uncharged species at 298 K.

Table 5.3 Second-order rate constants for the mutual neutralization of ions

Reactants	$10^8 k$ / cm^3 molecule^{-1} s^{-1}	*D* / nm	*Ref.*
$NO^{+} + SF_6^{-}$	19	2.1	*a*
$C_6H_6^{+} + SF_6^{-}$	18	2.8	*a*
$Tl_2Br^{+} + TlBr_2^{-}$	5.1	2.15	*b*
$Pb_2I_3^{+} + I^{-}$	3.7	1.2	*b*

[a] CARLTON, T. S., and MAHAN, B. H., *Journal of Chemical Physics*, **40**, 3683 (1964)
[b] FISK, G. A., MAHAN, B. H., and PARKS, E. K., *Journal of Chemical Physics*, **47**, 2649 (1967)

The large rate constants in *Table 5.3* reflect, of course, the strong, long-range, Coulombic potential that acts between ions of opposite charge. If the expression for this potential is inserted into equation (3.20) the following equation is obtained.

$$\varepsilon_T = \tfrac{1}{2}\mu \dot{r}^2 + \varepsilon_T b^2/r^2 + z_1 z_2 e^2/(4\pi\varepsilon_0)r \tag{5.24}$$

Whatever the values of the collision energy, ε_T, and the impact parameter, *b*, there is no centrifugal barrier in the effective potential for this system; at high $\varepsilon_T b^2$, the orbit of the ions is a hyperbolic one.

For mutual neutralization to occur, it is assumed that two ions must approach to at least some critical distance, *D*, at which the system can cross to a potential surface that correlates with neutral products (cf. *Figure 2.6*). This can be looked upon as the distance at which an electron can jump from the negative ion to the positive ion. Using a simple model, *D* can be estimated from

the observed rate coefficient. It is assumed that, for collisions at a given energy, there is a maximum value of the impact parameter at which the ions will just approach to the critical distance. At this point, the radial velocity will be zero, so from equation (5.24) with $\dot{r} = 0$ and $r = D$ one obtains

$$S_r(\varepsilon_T) = \pi b_{max}^2 = \pi D^2(1 - z_1 z_2 e^2/(4\pi\varepsilon_0)D\varepsilon_T) \tag{5.25}$$

The calculated value of the rate constant is found by averaging this expression over the thermal distribution of collision energies. This procedure yields

$$k_{2nd} = \pi D^2(8k_BT/\pi\mu)^{\frac{1}{2}}(1 - z_1 z_2 e^2/(4\pi\varepsilon_0)Dk_BT) \tag{5.26}$$

Substitution of the observed second-order rate coefficients into this equation yields the values of D that are listed in *Table 5.3*.

The neutral products of ion–ion reactions are usually not identified. At the critical distances given in *Table 5.3*, the Coulombic potential energies are still relatively small; thus, at $r = 2$ nm, $\mathscr{V}(r) = -69.5$ kJ mol^{-1} ($\equiv 0.72$ eV) for two singly charged ions of opposite charge. On the other hand, the difference between the ionization potential of NO and the electron affinity of SF_6 is 755 kJ mol^{-1} ($\equiv 7.8$ eV). This difference indicates that the non-ionic surfaces involved in the crossing correlate with neutral species in electronically excited or dissociative states.

Two mechanisms have been suggested for the third-order contribution to the overall rate. The proposal that ion–molecule 'chaperon' species, $(AM)^+$, are involved has been rejected by Mahan[18]. He has obtained satisfactory quantitative agreement with experiment by assuming that the role of the third-body species, M, is to collide with one or other of the ions while they are in a hyperbolic orbit with $b > b_{max}$. If energy is then transferred to M, the ions can fall into a closer orbit, allowing them to approach within the critical distance and neutralize one another. These 'energy-transfer' and 'chaperon' mechanisms are directly analogous to those proposed for the third-order recombination of atomic radicals (Section 5.8.2).

5.3.2 Reactions of ions with neutral species[19]

The reactions of ions with electrically neutral atoms and molecules are important in a wide variety of environments: for example, in electrical discharges, in the upper atmosphere, and in flames. Rate constants have now been measured for many such reactions, particularly using the 'flowing afterglow' technique that was mentioned in Section 5.2.1, and by introducing controlled amounts of neutral gases just downstream from the ion source of a mass-spectrometer. Some representative rate constants for ion–molecule reactions are given in *Table 5.4*. In the first part of the table, rate constants are listed for *charge-transfer* reactions, in which only an electron is transferred between the reactants. In section (ii), rate constants are given for reactions where a neutral or charged atom is transferred.

The ultrasimple Langevin model for ion–molecule reactions was outlined in

Table 5.4 Rate constants for some reactions of positive ions with neutral species at 298 K

Reactions	$10^9 k$ / cm^3 molecule^{-1} s^{-1}
(i) *Charge transfer*	
$H^+ + NO \rightarrow NO^+ + H$	1.9
$N^+ + CO \rightarrow CO^+ + N$	0.5
$O^+ + H_2O \rightarrow H_2O^+ + O$	2.3
$N_2^+ + NO \rightarrow NO^+ + N_2$	0.4
$N_2^+ + O_2 \rightarrow O_2^+ + N_2$	0.06
(ii) *Atom-transfer*	
$H_2^+ + H_2 \rightarrow H_3^+ + H$	1.9
$Ar^+ + H_2 \rightarrow ArH^+ + H$	0.69
$H_2O^+ + H_2O \rightarrow H_3O^+ + OH$	1.7
$O^+ + N_2 \rightarrow NO^+ + N$	1.2×10^{-3}

All data taken from GARVIN, D., *Chemical Kinetics Data Survey IV. Preliminary Tables of Chemical Data for Modelling of the Stratosphere* (N.B.S. report, NBSIR 73–203, 1973)

Section 3.4.3. The principal assumptions that were made in deriving equation (3.53) for the thermal rate constant were:

(1) all 'close' collisions between the reactants lead to reaction;
(2) the rate of close collisions, i.e., those in which the centrifugal barrier is overcome, can be estimated by ignoring specific chemical forces and assuming that only an ion–induced dipole interaction is important; and
(3) the magnitude of the ion–induced dipole potential can be calculated making use of the orientation averaged polarizability (α) of the molecule determined from macroscopic measurements.

Equation (3.74) can be simplified to yield

$$k(T) = (4\pi\alpha e^2/(4\pi\varepsilon_0)^2\mu)^{\frac{1}{2}}\Gamma(\tfrac{1}{2}) \tag{5.27}$$

For typical values of $(\alpha/4\pi\varepsilon_0) = 2.5 \times 10^{-3}$ nm^3 and $\mu = 15$ atomic mass units, this equation yields a rate constant of 0.96×10^{-9} cm^3 molecule^{-1} s^{-1}.

Most of the rate constants in *Table 5.4* and those for many other ion–molecule reactions are *approximately* the size expected from the Langevin relationship. This is true when negative ions instead of positive ions are reactants, and for charge-transfer reactions, as well as atom- and ion-transfer reactions. These rate constants are significantly larger than the values for hard-sphere collisions, but are smaller than those for ion–ion reactions. This indicates that the long-range attractive forces, though weaker than those

between two ions, frequently play an important role. However, one should not conclude that virtually all ion–molecule reactions have rate constants close to the value predicted by the simple treament. The Langevin model provides a useful first basis for comparison, *but nothing more.* To a large extent, the agreement of experimental data with its predictions has been overemphasized, because measuring the rates of rapid ion–molecule reactions has proved to be easier than determining those of slow reactions.

The reaction between O^+ and N_2 provides a rather famous example of an ion–molecule reaction with a rather small rate constant. It is important 'practically' because of the part it plays in atmospheric chemistry above 50 km. With the electronic states of the reactants and products stated, the reaction is

$$O^+(^4S) + N_2(^1\Sigma_g^+) \rightarrow NO^+(^1\Sigma^+) + N(^4S) \qquad (5.28a)$$

Both reactants and products correlate with a linear $^4\Sigma^-$ state of NNO^+, but this is not the ground electronic state of the triatomic species. Large basis set LCAO–MO–CI calculations[20] indicate that the $^4\Sigma_g^-$ state presents a barrier to the atom-transfer reaction. It therefore appears that the reaction may proceed via the $^2\Pi_i$ ground state of NNO^+, the system undergoing two 'spin-forbidden' transitions, one in the reactant region, the other in the product region, where the $^2\Pi_i$ state crosses the $^4\Sigma^-$ potential hypersurface, i.e.,

$$O^+(^4S) + N_2(^1\Sigma_g^+) \rightarrow NNO^+(^2\Pi_i) \rightarrow NO^+(^1\Sigma^+) + N(^4S) \qquad (5.28b)$$

5.3.3 Reactions between two free radicals

The depth of the potential well associated with the dispersion forces between two neutral species is approximately 1 kJ mol^{-1}, i.e., $(\varepsilon_{LJ}/k_B) \sim 120\,K$ (*see Table 2.4*). This energy is smaller than both the average collision energy at normal temperatures and the typical values of activation energies for elementary reactions between uncharged species. Consequently, the long-range forces between reactants are expected to be much less important than in reaction involving charged species.

It was seen earlier that the potential energy curves for the bound states of diatomic molecules approach their dissociation limit asymptotically. The same is usually true for polyatomic molecules; as a particular bond is continuously extended, the potential energy increases monotonically until dissociation occurs. Consequently, when free radicals collide, at least for certain orientations, there is usually no electronic potential energy barrier preventing the formation of an energized collision complex. However, unless this complex breaks down to products that can be distinguished from the reactants, or it is stabilized by a further collision, it will redissociate, and it will not be possible to determine the rate at which collisions between the radical species temporarily form molecules. As a result, it is necessary to look at the rate constants of association reactions in the limit of high [M], if one is to find the appropriate value to compare with collision numbers.

Table 5.5 Rate constants for association reactions at 298 K and in the limit of high [M]

Reactants	*Total degeneracy*	$10^{11}k$ / cm^3 $molecule^{-1}$ s^{-1}	*Ref.*
(i) *Spin-allowed reactions*			
$O+O_2$	27	0.28	*a*
$O+NO$	36	2.8	*a*
$I+NO$	16	0.83	*b*
$OH+NO_2$	8	1.4	*c*
CH_3+NO	8	1.7	*d*
CH_3+CH_3	4	4.0	*e*
$(CH_3)C+(CH_3)_3C$	4	0.40	*e*
(ii) *Spin-disallowed reactions*			
$O+CO$	9	2.0×10^{-5}	*f*
$O+N_2$	9	unobservable ($E_{act} \sim 100$ kJ mol^{-1})	*g*

[a] QUACK, M., and TROE, J., *Berichte der Bunsengesellschaft für Physikalische Chemie*, **81**, 329 (1977)
[b] VAN DEN BERGH H., and TROE, J., *Chemical Physics Letters*, **31**, 351 (1975)
[c] ANASTASI, C., and SMITH, I. W. M., *Journal of the Chemical Society, Faraday Transactions*, II, **72**, 1459 (1976)
[d] VAN DEN BERGH, H., and CALLEAR, A. B., *Transactions of the Faraday Society*, **67**, 2017 (1971)
[e] PARKES, D. A., PAUL, D. M., and QUINN, C. P., *Journal of the Chemical Society, Faraday Transactions*, I, **72**, 1935 (1976); PARKES, D. A., and QUINN, C. P., *Journal of the Chemical Society, Faraday Transactions*, I, **72**, 1952 (1976)
[f] SIMONAITIS, R., and HEICKLEN, J., *Journal of Chemical Physics*, **56**, 2004, (1972)
[g] STUHL, F., and NIKI, H., *Journal of Chemical Physics*, **55**, 3943 (1971)

Table 5.5 lists the limiting high-pressure values of the rate constants for some association reactions involving species[a] of different molecular complexity. For comparison, it is useful to remember (*Tale 3.1*) that the second-order rate constant associated with hard-sphere collisions at 298 K, between species with a collisional reduced mass of 15 atomic mass units and a collision diameter of 0.3 nm, is 1.8×10^{-10} cm^3 $molecule^{-1}$ s^{-1}.

There are two fairly obvious factors which will cause the rates of radical–radical collisions to fall below the values expected on a simple collisional basis. The first arises because the electronic states of radicals, by definition, possess some electronic degeneracy. As a result more than one molecular state will arise when two radicals interact. In the simplest case—as with two $H(^2S_{\frac{1}{2}})$ atoms—there are a bound singlet and non-bound triplet state; only 1 in 4 of the collisions occur on the hypersurface corresponding to the bound state and can lead to the formation of collision complexes. In the second column of *Table 5.5*, the total electronic degeneracy associated with each pair

[a] All the reactants in the first part of *Table 5.5* are free radicals in the sense that they possess unpaired electrons.

of radicals is listed. In some cases, more than one overall electronic state is bound, but in all cases the fraction of bound states will be less than $\frac{1}{2}$. Correction of the observed rate constants by a factor corresponding to the total number of electronic states divided by the number of bound states gives numbers closer to Z_{hs}.

One might also expect that an orientation effect or steric factor would reduce the rate constants below the collision number. In fact, this effect is smaller than might have been expected. For example, if the rate constant for methyl radical recombination is multiplied by four, one obtains a rate constant very close to the value predicted on the basis of reaction between structureless particles, implying that a large fraction of the collisions that take place on the singlet hypersurface lead to formation of an energized C_2H_6 molecule. These comparisons strongly suggest that the transition state in reactions of this type is generally 'loose', so that the recombining radicals are still far apart relative to their final separation and directional forces are rather weak.

Table 5.5 also includes data on the reactions of oxygen atoms with the closed-shell molecules CO and N_2. The rates are much slower than for the other reactions in the *Table 5.5*. The reasons for this were discussed in Section 2.4.2. The electronic states of the separated ground state species do not correlate with the ground state of the triatomic molecule. For reaction to occur, the system must undergo a spin-forbidden transition between triplet and singlet surfaces and these intersect at an energy above that of the separated fragments.

Table 5.6 gives the rate constants for a few atom-transfer reactions between two free radicals. The reactions in the first part of the table can proceed via the

Table 5.6 Rate constants for radical + radical atom-transfer reactions at 298 K

	Reaction	$\frac{10^{11}k}{\text{cm}^3\ \text{molecule}^{-1}\ \text{s}^{-1}}$	*Ref.*
(i)			
	$D + OH \rightarrow OD + H$	13	*a*
	$^{18}O + {}^{16}O^{16}O \rightarrow {}^{18}O^{16}O + {}^{16}O$	0.1	*b*
	$^{18}O + N^{16}O \rightarrow N^{18}O + {}^{16}O$	0.18	*b*
	$O + OH \rightarrow O_2 + H$	4.2	*c*
(ii)			
	$H + OH \rightarrow H_2 + O$	$\sim 3 \times 10^{-6}$	*c*
	$OH + OH \rightarrow H_2O + O$	0.21	*c*
	$C_2H_5 + C_2H_5 \rightarrow C_2H_6 + C_2H_4$	0.18	*d*

[a] MARGITAN, J. J., KAUFMAN, F., and ANDERSON, J. G., *Chemical Physics Letters*, **34**, 485 (1975)
[b] HERRON, J. T., and KLEIN, F. S., *Journal of Chemical Physics*, **40**, 2731 (1964)
[c] HAMPSON, R. F., and GARVIN, D., *Reaction Rate and Photochemical Data for Atmospheric Chemistry–1977* (N.B.S. Special Publication 513, 1978)
[d] *See* ref. *e* under Table 5

bound state of a triatomic species (HDO, O_3, NO_2 or HO_2) and their rate constants are similar to those in *Table 5.5*, as one should expect. The reactions in section (ii) of *Table 5.6* are formally similar to those in the first part of the table, but none of these reactions can proceed via the formation of a collision complex which then dissociates with only one bond being broken. Thus, collisions between $H(^2S_{\frac{1}{2}})$ and $OH(^2\Pi_i)$ can occur on the potential hypersurface corresponding to the $^1A_1$ electronic ground state of H_2O. There can be little doubt that this is the mechanism responsible for the extremely rapid reaction listed first in *Table 5.6*. However, if this route were to lead to $H_2(^1\Sigma_g^+) + O(^3P)$ as products, a spin-forbidden transition to a triplet surface would be necessary as well as the simultaneous breaking of the two OH bonds and the formation of the H—H bond. In fact, the reaction has an activation energy of 41.4 kJ mol^{-1} and presumably proceeds directly on a triplet hypersurface.

The examples that have been quoted should be sufficient to illustrate the main factors that determine the rate of reactions between species with open electronic shells. When interpreting, or attempting to predict, the rates of such reactions, it is of primary importance to establish whether there exists a reaction route via a bound state of the overall system which is allowed by the correlation rules for spin and orbital symmetry. If so, the reaction is likely to proceed at a rate which is no more than two orders-of-magnitude slower than the collision rate, and the rate constant will be relatively insensitive to temperature. If, on the other hand, no such route exists, the kinetic data for the reaction are likely to resemble those for comparable radical + molecule reactions. These are considered next. As there is usually an appreciable activation energy for these reactions, and its value is difficult to estimate, it is much harder to predict the rate constants, even to an order-of-magnitude.

5.3.4 Atom-transfer reactions between a radical and a molecule

For exoergic reactions of this type, the activation energy is usually in the range 0—50 kJ mol^{-1}. With an activation energy of 50 kJ mol^{-1} and $T = 298$ K, the exponential term in the Arrhenius expression for the rate constant is only 1.7×10^{-9}. Clearly the rate constants might vary over many orders-of-magnitude.

In this section, a comparison is only made between the pre-exponential factors for reactions involving species of different molecular complexity, and between these values and Z_{hs}, the collisional rate constant. It is important to appreciate the difficulty of determining pre-exponential factors *accurately*, for reactions where there is an appreciable activation energy. To illustrate this, let us consider a reaction for which rate constants have been measured between 400 and 600 K. The energy of activation would be determined from the slope of the Arrhenius plot, i.e., ln k versus $1/RT$. Let us suppose this yields a value of 40 ± 1 kJ mol^{-1}. If A is then evaluated by substituting this activation energy and the value of ln k at the 'centre-of-gravity' of the Arrhenius plot into the equation

$$\ln A = \ln k + E_{act}/RT \qquad (5.29)$$

the error in A corresponds to approximately 30 per cent. Consequently, the pre-exponential factors determined by different investigators for a particular reaction are usually in much worse agreement than the activation energies or the rate constants themselves. The values of A listed in *Table 5.7* have been arrived at, after some attempt by the author at evaluation.

Table 5.7 Pre-exponential factors (10^{11} A/cm^3 molecule^{-1} s^{-1}) for some bimolecular atom-transfer reactions

Radical	*Molecule*	H_2	HCl	Cl_2	CH_4
H (D)		(5.0)	4	14	10
Cl		8	–	–	5.6
O		2.5	0.25	1.0	3.5
OH		1.8	0.3	–	0.5
CH_3 (CD_3)		0.6	0.16	1.3	(0.1)

A general trend of decreasing pre-exponential factor as the radical becomes progressively more complex is evident in the data in *Table 5.7*. The change when a polyatomic molecular reactant, CH_4, replaces a diatomic reactant is less pronounced even when allowance is made for the fact that there are four equivalent sites for radical attack on CH_4. In any case, the change in the magnitude of the pre-exponential factor with differences in molecular complexity of the reactants and transition state is much less than would be expected on the basis of the simple analysis referred to, but rejected, in Section 4.2.3. Comparison of the numbers in *Table 5.7* with the hard-sphere collision number or with the number obtained including the effect of the Lennard-Jones attraction indicates steric factors ranging from ~ 0.3 to ~ 0.003.

5.3.5 Reactions between two molecules

Finally, reactions are considered briefly that occur between two molecules both of which have closed electronic shells. Reactions of this type are distinguished chiefly by their rarity. The literature contains many examples of proposed reactions that have later been shown to occur by some mechanism other than direct bimolecular reaction. The most celebrated case concerns the reaction between H_2 and I_2, which for two or three decades served as the textbook example of a bimolecular reaction with a pre-exponential factor close to the collision number. In fact, the reaction is now known to proceed via the following mechanism[21]

$$I_2 \rightleftarrows 2I \tag{5.30}$$

$$H_2 + I + I \rightarrow 2HI \tag{5.31}$$

There have been a number of studies in shock tubes which indicate that

bimolecular four-centre reactions involving the simultaneous breaking of two bonds and formation of two new ones can occur at high temperatures. An example is provided by the isotope scrambling reaction[22]

$$^{16}O^{16}O + {}^{18}O^{18}O \rightarrow {}^{16}O^{18}O + {}^{16}O^{18}O \tag{5.32}$$

However, the activation energies of these reactions are high; for reaction (5.32), $E_{act} \approx 170$ kJ mol^{-1}. When unsaturated hydrocarbons are involved and a single molecule is the reaction product four-centre reactions appear to become somewhat easier. Thus the reaction[23]

$$C_2H_4 + HI\,(+M) \rightarrow C_2H_5I\,(+M) \tag{5.33}$$

has an activation energy of 121 kJ mol^{-1}. The pre-exponential factor has been determined as 5.5×10^{-13} cm^3 molecule^{-1} s^{-1}. This relatively low value suggests that, in the transition state, HI is not free to rotate nor is there free rotation around the C—C bond. In other words, the reaction proceeds by way of a 'tight' transition state.

As a general rule, bimolecular, four-centre, reactions have high activation energies and are likely to occur only at high temperatures. As with most types of reaction, exceptions to the general rule have been found. Thus a number of facile four-centre reactions have been studied in molecular beam experiments. These all involve ionic species and proceed by way of collision complexes that owe their stability to 'ionic forces'. Thus the reaction[24]

$$CsCl + KI \rightarrow CsI + KCl \tag{5.34}$$

is presumed to occur via a complex that may be written as $Cs^+ \begin{matrix} Cl^- \\ I^- \end{matrix} K^+$.

Further discussion of such reactions is given in Section 7.5.4.

5.4 The H_3 reactions[25]

The important position that the H_3 system occupies in theoretical chemical kinetics has already been emphasized. All four isotopic variations of the atom-transfer reaction have now had their rate constants determined in direct experiments over a range of temperatures. It is however an instance of Nature's perversity that these reactions, which are electronically so simple, are extremely difficult to study experimentally. Until 1965, the only kinetic data that were available came from indirect measurements, mostly carried out in the 1930s. Since then, however, in an experimental *tour-de-force*, Le Roy and his co-workers have studied all four of these reactions, and those where isotopic exchange takes place have also been studied by Westenberg and de Haas. The scope of these measurements is indicated in *Table 5.8*.

In all of the experiments summarized in *Table* 5.8, the flow method was used. Since, at the low end of the temperature ranges, the reaction rates were rather slow—for example, the rate constant for the $D + H_2$ reaction at 200 K is only

Table 5.8 Experimental kinetic data for the $H + H_2$ reaction and its isotopic variants

Reaction	*Range of* T/K	*Method of analysis**	*Rate expression:*† $\log_{10}(k/\mathrm{cm^3\ molecule^{-1}\ s^{-1}})$	Ref.
$H + \text{para-}H_2 \rightarrow \text{ortho-}H_2 + H$	330—440	GSC	$-8.32_6 - 3.49_0 \times 10^3/T + 3.83_6 \times 10^5/T^2$	*a, b*
$D + \text{ortho-}D_2 \rightarrow \text{para-}D_2 + D$	358—468	TC	$-10.69 - 1.67 \times 10^3/T$	*c*
$D + H_2 \rightarrow HD + H$	274—468	TC	not given	*d*
	167—346	ESR	$-16.56 \times 2\log_{10} T - 1.17 \times 10^3/T$	e
	250—750	ESR	above 450 K: $-10.14 - 1.66 \times 10^3/T$	*f*
$H + D_2 \rightarrow HD + D$	368—468	GSC	$-11.14 - 1.60 \times 10^3/T$	*g, b*
	300—750	ESR	above 450 K: $-10.09 - 2.05 \times 10^3/T$	*f*

* Symbols denote: GSC, low temperature gas–solid chromatography; TC, thermal conductivity analysis; ESR, election spin resonance.

† Rate constants for the $H + \text{para-}H_2$ and $D + \text{ortho-}D_2$ reactions are as given by Le Roy and his co-workers. To obtain the total rate constants, i.e., including $H + \text{para-}H_2 \rightarrow \text{para-}H_2 + H$ and $D + \text{ortho-}D_2 \rightarrow \text{ortho-}D_2 + D$, these should presumably be multiplied by (4/3) and (9/3) respectively.

[a] SCHULZ, W. R., and LE ROY, D. J., *Journal of Chemical Physics*, **42,** 3869 (1985)

[b] QUICKERT, K. A., and LE ROY, D. J., *Journal of Chemical Physics*, **53**, 1325 (1970) give *relative* rate data for the reactions $H + \text{para-}H_2$ and $H + D_2$ between 290 and 690 K.

[c] LE ROY, D. J., RIDLEY, B. A., and QUICKERT, K. A., *Discussions of the Faraday Society*, **44**, 92 (1967)

[d] RIDLEY, B. A., SCHULZ, W. R., and LE ROY, D. J., *Journal of Chemical Physics*, **44,** 3344 (1966),— ○

[e] MITCHELL, D. N., and LE ROY, D. J. *Journal of Chemical Physics*, **58**, 3449 (1973),— □

[f] WESTENBERG, A. A., and DE HAAS, N., *Journal of Chemical Physics*, **47**, 1393 (1967),— ●, ▲

[g] SCHULZ, W. R., and LE ROY, D. J., *Canadian Journal of Chemistry*, **42**, 2480 (1964),— △

Data from the last four references are plotted in *Figure 5.4* using the symbols shown.

1.7×10^{-18} cm^3 molecule^{-1} s^{-1}—great care had to be taken to eliminate effects such as back diffusion of the atoms, which occurs rapidly on account of their small mass. For the isotope exchange reactions, ESR spectroscopy could be used to observe the concentrations of both the H and the D atoms quantitatively. In experiments on the $H + H_2$ and $D + D_2$ reactions, probes which catalytically recombined the atoms, thereby releasing heat, were used to determine the total atom concentrations at different points along the tube. Furthermore, because they effectively removed all atoms, thereby stopping the reaction, the gas downstream could be analysed for ortho-H_2 and para-H_2, or ortho-D_2 and para-D_2, to see how far the atom-transfer reaction had proceeded by the time the catalytic probe was reached.

The range and quality of the experimental data on these reactions and the accuracy of the *ab initio* calculations on the H_3 potential are so good that it is valuable to compare the observed rate constants with those predicted by the conventional TST rate expression. This comparison has been carried out by Koeppl[26] based on the potential of Liu. In calculating the rate constants, motions in the transition state were separated and all internal motions were treated in the rigid rotor–harmonic oscillator approximation. *Figure 5.4* compares, on an Arrhenius plot, the experimental and theoretical rate constants, the latter having been calculated without allowance for quantum mechanical tunnelling. For each reaction lower and upper limits for the rate constants are given, corresponding to the upper and lower bounds on the barrier energy that were estimated by Liu (*Table 2.6*).

At the high end of the temperature range for which a comparison is possible, experimentally determined rate constants and those calculated using a barrier height of 39.7 kJ mol^{-1}, the lower bound estimated by Liu, agree quite well. At lower temperatures, the observed rate constants are greater than the maximum calculated values. This deviation amounts to a factor of about 1.8 for both reactions at 325 K, increasing to about 12 for the $D + H_2$ reaction at 200 K. The most straightforward explanation of these observations is that:

(1) at temperatures $\geqslant$ 500 K, conventional TST theory provides a satisfactory estimate of the rate constants for these reactions;
(2) the energy barrier on the H_3 surface is close to 39.7 kJ mol^{-1}; and
(3) quantum penetration effects cause the deviation between the experimental and calculated rate constants below about 500 K.

However, these conclusions would not appear to be consistent with comparisons between the predictions of quantum scattering calculations and conventional transition state theory. Unfortunately, Schatz and Kuppermann[27] performed calculations only on the $H + H_2$ reaction and, furthermore, they used Porter and Karplus's potential, rather than that of Liu. The deviation between Schatz and Kuppermann's rate constants and those calculated from the standard TST rate expression was 20 at 300 K and 427 at 200 K. Although the effect of tunnelling will be smaller for $D + H_2$ and $H + D_2$, it is unlikely to be very much less, and this suggests that the energy

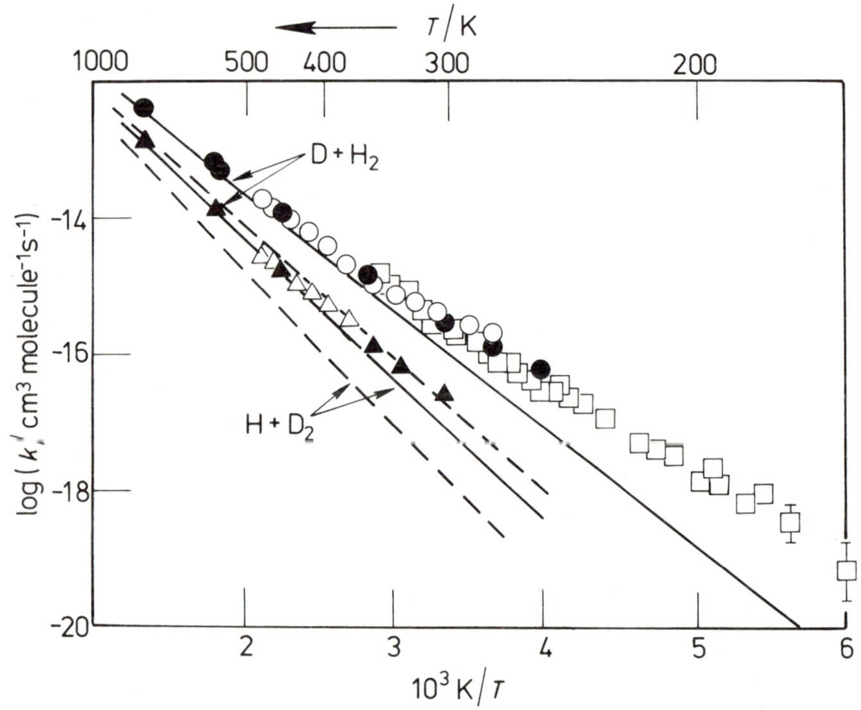

Figure 5.4. Comparison of the experimental rate constants for the reactions $D + H_2 \rightarrow DH + H$ *and* $H + D_2 \rightarrow HD + D$ *with conventional* TST *calculations with no tunnelling corrections. The sources of the points are identified in the list of references under Table 5.8. The continuous and dashed lines represent upper and lower theoretical bounds corresponding to the limits of accuracy for the* H_3 *potential* [*adapted from Koeppl, G. W.*, Journal of Chemical Physics, **59**, 3425 (1973)]

barrier on the H_3 surface may actually be closer to Liu's upper, rather than this lower, limit.

5.5 Non-Arrhenius behaviour

In recent years, the methods for measuring rate constants for elementary reactions have improved considerably in accuracy and precision. Furthermore, it is now possible to determine rate constants for many reactions over a wide range of temperature by reliable techniques. These improvements make it possible to examine just how well the variation of the rate constant with temperature can be expressed in terms of a single (temperature-independent) activation energy.

Comparisons of this kind are usually made within the framework of transition state theory. The case of a bimolecular reaction was considered in Section 4.2.4. There it was shown that, if motion along the reaction coordinate

can be separated and treated classically, then the activation energy is given by the equation:

$$E_{act} = \Delta E_0^{\ddagger} - \tfrac{1}{2}RT - \frac{d \ln Q_{int}^{\ddagger}}{d(1/RT)} + \sum_{reactants} \frac{d \ln Q_{int}}{d(1/RT)} \qquad (5.35)$$

This equation can usually be simplified further by treating the overall rotations of the reactants and of the activated complex classically, so that their partition functions are proportional to T, if they are linear, or to $T^{\frac{3}{2}}$, if they are non-linear. Then

$$E_{act} = \Delta E_0^{\ddagger} + (\Delta n_R^{\ddagger} - 1)\tfrac{1}{2}RT - \frac{d \ln Q_V^{\ddagger}}{d(1/RT)} + \sum_{reactants} \frac{d \ln Q_V}{d(1/RT)} \qquad (5.36)$$

where $\Delta n_R^{\ddagger}$ is the number of overall rotational degrees of freedom in the transition state less those in the separated reactants. As $\Delta n_R^{\ddagger}$ must be less than or equal to unity, E_{act} will be $\leqslant \Delta E_0^{\ddagger}$, unless the third and fourth terms on the right-hand side of equation (5.36) counteract the tendency of the second term to lower E_{act} as the temperature increases.

In estimating the effect of changes in the vibrational partition functions with temperature, it is usually sufficient to consider only the 'new' internal motions in the transition state; these usually have abnormally low frequencies (*see* Section 4.2.3.). Each contributes a term

$$-d \ln Q_s^{\ddagger}/d(1/RT) = RT\{u_s \exp(-u_s)/[1-\exp(-u_s)]\} \qquad (5.37)$$

where $u_s = hc\tilde{\nu}_s/k_BT$, to the right-hand side of equation (5.36). Of course, the size of these terms can only be estimated, unless accurate information is available about the potential in the vicinity of the transition state. It is to be expected, however, that the sum of these factors will usually exceed $(\Delta n_R^{\ddagger} - 1)\frac{1}{2}RT$ so that the energy of activation will increase with temperature. Although this argument has been presented in terms of conventional TST, with motions at the transition state separated and treated according to the rigid rotor–harmonic oscillator approximation, the qualitative conclusions are unlikely to be sensitive to these assumptions.

Curvature in Arrhenius plots can also arise as a result of quantum mechanical tunnelling. Indeed, non-Arrhenius behaviour has been sought chiefly as evidence for tunnelling. Making allowance for this phenomenon within the framework of TST theory introduces an extra term $\{-d \ln \kappa_{qu}(T)/d(1/RT)\}$, on the right-hand side of equation (5.36). The effect of quantum penetration is most marked when the energy barrier and the imaginary frequency associated with motion along the reaction coordinate in the transition state are both large, and when the temperature is low. If tunnelling is appreciable, the energy of activation is lowered, since $\kappa_{qu}(T)$ decreases with increasing temperature, i.e., $d \ln \kappa_{qu}(T)/d(1/RT)$ is positive. This can be clearly seen for the reactions considered in Section 5.3. Thus for the $D + H_2$ reaction, the energy of activation at 167 K, estimated from Mitchell and Le Roy's expression given in *Table 5.8*, is 25.2 kJ mol^{-1}. The

activation energy can then be used to calculate a value of the Arrhenius pre-exponential factor from the equation:

$$\log\{A(T)/\text{cm}^3\ \text{molecule}^{-1}\ \text{s}^{-1}\} = \log\{k(T)/\text{cm}^3\ \text{molecule}^{-1}\ \text{s}^{-1}\} + E_{act}(T)/RT \quad (5.38)$$

This yields $A(T = 167\ \text{K}) = 5.6 \times 10^{-12}\ \text{cm}^3\ \text{molecule}^{-1}\ \text{s}^{-1}$. This value is about two orders-of-magnitude below the rate constant associated with D + H_2 collisions at 167 K and is also more than one order-of-magnitude less than the value of the pre-exponential factor estimated from the TST rate expression without allowance for tunnelling. These figures support the view that, in this instance, the curvature in the Arrhenius plot is caused by quantum penetration effects that are most important at low temperatures, and that this causes the observed energy of activation at these temperatures to be appreciably less than $\Delta E_0^{\ddagger}$.

5.5.1 The reaction: $OH + CH_4 \rightarrow H_2O + CH_3$

The reaction

$$OH + CH_4 \rightarrow H_2O + CH_3 \quad (5.39)$$

provides a further illustration of the point that has just been made. This reaction initiates the oxidation of methane, both in combustion and in the upper atmosphere, and its kinetics have been thoroughly studied over a wide range of temperatures. The rate measurements carried out by Zellner and Steinert[28] are particularly valuable since they covered the range 300—900 K using the single technique of flash photolysis–resonance absorption. The Arrhenius plot, shown in *Figure 5.5*, is clearly curved, and was fitted empirically to the relationship

$$\log(k/\text{cm}^3\ \text{molecule}^{-1}\ \text{s}^{-1}) = -20.24 + 3.08\log(T/\text{K}) - 1.01 \times 10^3\text{K}/2.303T \quad (5.40)$$

Reaction (5.39) clearly has a lower barrier than that in the H_3 system but the origin of the 'non-Arrhenius' behaviour is not immediately apparent. However, if temperature-dependent Arrhenius factors are calculated, then at 300 K one finds

$$E_{act} = 16.1\ \text{kJ mol}^{-1} \quad \text{and} \quad A = 5.3 \times 10^{-12}\ \text{cm}^3\ \text{molecule}^{-1}\ \text{s}^{-1} \quad (5.41a)$$

whilst at 900 K

$$E_{act} = 31.4\ \text{kJ mol}^{-1} \quad \text{and} \quad A = 1.6 \times 10^{-10}\ \text{cm}^3\ \text{molecule}^{-1}\ \text{s}^{-1} \quad (5.41b)$$

For the species in this reaction, through the range of temperature covered by the experiments, the overall rotations may be treated classically and $\Delta n_R^{\ddagger}$ is equal to -2. The most straightforward explanation of the observations is that

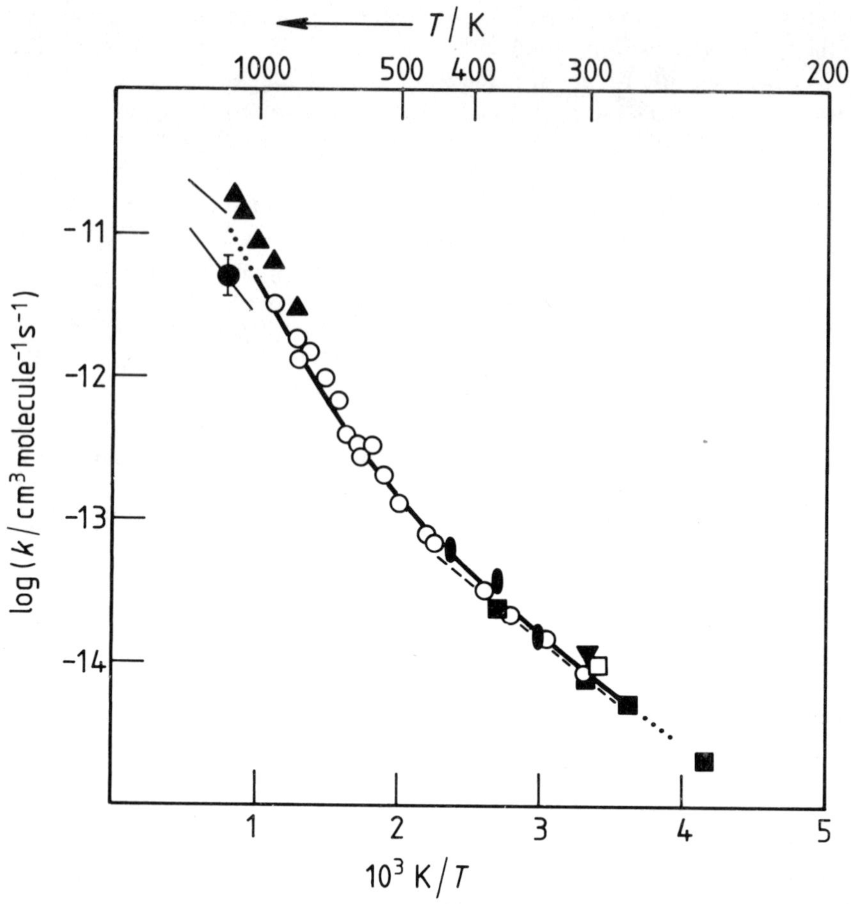

Figure 5.5. Arrhenius plot of the rate constants for the reaction $OH + CH_4 \rightarrow H_2O + CH_3$. *The filled circles joined by the continuous line show the results of Zellner, R., and Steinert W.,* International Journal of Chemical Kinetics, **8**, 397 (1976). *The sources of the other data are given in their paper from which this figure is adapted*

the energy of activation increases with temperature because the third term on the right-hand side of equation (5.36) grows appreciably as the temperature is raised from 300 to 900 K. This growth is associated with an increase in the partition function, $Q_v^{\ddagger}$, owing to the existence of low frequency motions in the transition state. An analysis similar to that carried out in Section 4.2.4 indicates that there are seven such internal modes in the transition state for reaction (5.39).

The alternative explanation for the variation of E_{act} with temperature is that tunnelling is significant at the lower end of the temperature range. This cannot be discounted, but appears not to be the *predominant* effect leading to curvature in the Arrhenius plot. If it were so, then the value of A at 300 K

would be unusually low, whilst the Arrhenius factors at high temperatures would correspond to those predicted by the usual TST rate expression with $\kappa_{qu}(T)$ close to unity. In fact, it is the value of A at 300 K, indicating a 'steric factor' of about 0.01, that is 'normal' and this interpretation suggests that at high temperature E_{act} must be appreciably greater than $\Delta E_0^{\ddagger}$. Finally, the conclusion that tunnelling is *not* the principal origin of the 'non-Arrhenius' behaviour of this reaction is consistent with the relatively minor effect of tunnelling at 300 K in the $D + H_2$ reaction, where the potential barrier to reaction is certainly higher and the imaginary frequency associated with motion along the reaction coordinate is very unlikely to be less.

5.5.2 The reaction: $OH + CO \rightarrow CO_2 + H$

As a final illustration of non-Arrhenius behaviour and of the semi-quantitative application of TST to elementary bimolecular reactions, the reaction:

$$OH + CO \rightarrow CO_2 + H \qquad (5.42)$$

is considered. This is the final step in the production of CO_2 during the combustion of any carbon-containing fuel. Like reaction (5.39), it has been studied recently by a wide range of experimental techniques and over a very wide range of temperature ($220 \leqslant T \leqslant 4000$ K)[29].

For several years the results obtained at temperatures below about 500 K and those determined at temperatures greater than about 1000 K appeared irreconcilable, since the experiments at low temperature indicated very little variation of the rate constant with temperature, whereas those at high temperature suggested an activation energy of about 20 kJ mol^{-1}. It now appears that attempts to encompass all the experimental data by a single Arrhenius expression with temperature-independent parameters, A and E_{act}, were misguided. The rate constants measured recently at temperatures from 300 to 900 K[30] agree well with previous values at either end of the range and show clearly that the energy of activation for this reaction increases continuously with temperature.

The possibility that the Arrhenius plot for this reaction might be strongly curved was first raised by Dryer, Naegeli and Glassman[32]. Since at low temperature the rate constant scarcely depends on temperature, this should cause no surprise, as there can clearly not be a large barrier to reaction, and pre-exponential terms in the TST rate expression should play a major role in determining the variation of $k(T)$. Using a TST model, Dryer *et al.* showed that the Arrhenius plot could curve strongly, but they made no attempt to match absolute calculated values of the rate constant to those observed experimentally.

Smith and Zellner[33] noted that the species HOCO had been observed spectroscopically in a CO matrix at 14 K and that there seemed no reason why this species could not form in bimolecular gas-phase collisions between OH and CO. However, if passage through the transition state leading to formation of this collision complex were rate-determining for reaction (5.42), TST

calculations would suggest that either there would be a stronger dependence of the low-temperature rate constants on temperature, if there were a potential energy barrier to this process, or these rate constants would be larger than the values found experimentally. Good agreement between the calculated and observed values of the rate constant, its dependence on temperature, and the change when OH is replaced by OD could only be obtained by assuming that in the transition state for the reaction all three heavy atoms lie on a straight line. Smith and Zellner therefore suggested that the potential energy along the reaction path of minimum energy varies in the manner shown in *Figure 5.6*. It was then possible to choose a reasonable model for the transition state which led, as *Figure 5.7* shows, to good agreement between the calculated and observed rate constants over the full range of temperature[a]. Striking confirmation of the proposal that an HOCO collision complex may form in collisions between OH and CO, and that at low pressure this complex may

[a] The data of Steinert and Zellner were obtained after Smith and Zellner's TST calculations. Only a small adjustment in the transition state model would be required to improve the agreement with these results, and this would scarcely affect the level of agreement with the other experimental data.

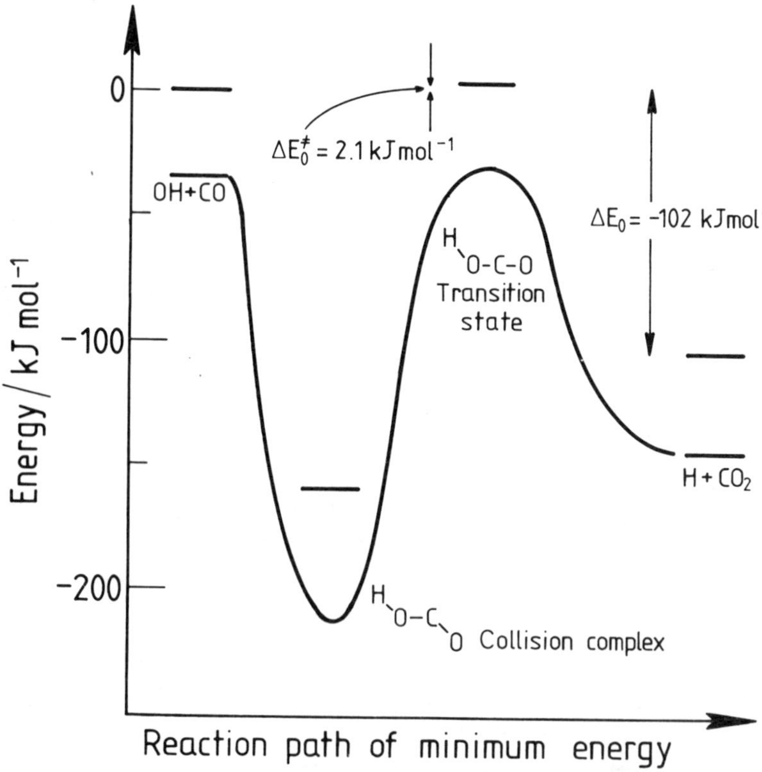

Figure 5.6. Proposed reaction path for the reaction $OH + CO \rightarrow CO_2 + H$

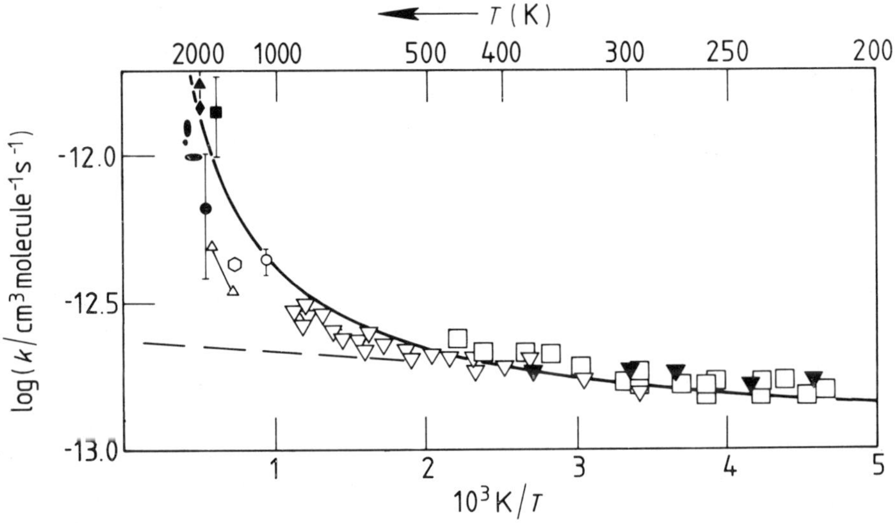

Figure 5.7. Arrhenius plot of the rate constants for the reaction $OH + CO \rightarrow CO_2 + H$. *The experimental data below* 1000 K *are those of Smith and Zellner*[33] (□), *Davis, Fischer and Schiff*[31] (▼), *and Steinert and Zellner*[30] (▽). *The points above* 1000 K *are taken from the review by Wilson*[29]. *The full line shows the result of* TST *calculations based on the model of reaction illustrated in Figure 5.6. The dashed line shows an extrapolation of the Arrhenius expression chosen by Davis* et al. *to represent their low-temperature* ($220 \leqslant T \leqslant 373$ K) *data*

break down at comparable rates to either $OH + CO$ or $CO_2 + H$, has been obtained recently[34, 35]. In experiments at atmospheric pressure, OH and CO react about 2.5 times faster than at the low pressures used in the earlier experiments. It appears that, under these conditions, HOCO can be collisionally stabilized; the subsequent fate of the HOCO radical has not yet been firmly established[35].

5.5.3 Summary and the relationship between 'non-Arrhenius' behaviour and quantum mechanical tunnelling

The examples in this sub-section have been chosen to emphasize a number of points. First, there is no theoretical reason to suppose that rate constants for elementary reactions will be matched by a simple, two-parameter, Arrhenius expression over a wide range of temperature. Secondly, the accuracy and precision of rate constant measurements have now reached a stage where the 'non-Arrhenius' behaviour of elementary bimolecular reactions can be observed, and this behaviour has been established in the case of several reactions. The situation has been reached that whenever an algebraic expression is used to represent the dependence of an observed rate constant with temperature, whether it be of the simple Arrhenius type or something more complicated, it should be accompanied by a clear statement of the

temperature range in which the expressed dependence has been established. If such a formula is then used to calculate the rate constant in a very different temperature regime, the result should be accepted with some caution.

These considerations also have a bearing on how rate data from several different investigations should be evaluated. Until recently, it was customary to assume that experimental measurements would not be sufficiently accurate to uncover non-Arrhenius behaviour, so results that did not correlate with the 'best' Arrhenius expression were rejected. It is now clear that when the measurements cover a wide range of temperature, say from 200 K to several thousand degrees, then such a procedure is unjustified. That this is now realized is illustrated in recent evaluations of the rate constants for reaction (5.42)[29].

The next point to consider in the light of the examples that have been given is the relationship between non-Arrhenius behaviour and quantum penetration effects. Because $\ln \kappa_{qu}(T)$ is not proportional to $1/RT$, tunnelling will lead to curvature in an Arrhenius plot. However, unless such effects are quite large, the experimental measurements very accurate, and the range of temperature over which rate constants have been determined quite extensive, the *apparent* effect of tunnelling may simply be to lower the observed activation energy. Thus, even for the $D + H_2$ reaction, examination of *Figure 5.4* reveals that all the results below about 350 K could be satisfactorily matched by a simple Arrhenius expression with a lower value of E_{act} than that in the high-temperature limit. Consequently, the absence of curvature should not necessarily be construed as evidence for the absence of tunnelling. Conversely, the existence of curvature in the Arrhenius plot for an elementary reaction should not be interpreted in terms of tunnelling before other possibilities have been eliminated. Before seeking *any* theoretical interpretation, one should first examine carefully whether the observation does not result from a systematic effect, such as the removal of the reactant that is being observed in a side or secondary reaction which has a different energy of activation from the primary process.

Finally, the discussion in this sub-section has demonstrated how the application of transition state theory can provide some insights into the details of an elementary reaction, where accurate rate constants are available over a wide temperature range. Thus, the evidence is quite strong that the curvatures in the Arrhenius plots for the reactions of $D + H_2$, $OH + CH_4$, and $OH + CO$ have quite different explanations.

5.6 Kinetic isotope effects[36, 37]

5.6.1 Transition state theory and kinetic isotope effects

Within the limits of the Born–Oppenheimer approximation, the electronic potential energy of a polyatomic system is unaffected by isotopic substitution. Thus the potential energy curves for H_2, HD and D_2 are the same, as are the

potential energy hypersurfaces for H_3, H_2D, HD_2 and D_3. What is altered, however, is the energy of the quantum states which are associated with the relative motion of these different combinations of nuclear particles. This gives rise to isotopic effects of many kinds. A familiar example is the difference in zero-point energies between pairs or groups of isotopically related molecules. For example, the energy of the v = 0 level in H_2, HD and D_2 above the minimum in the potential curve is 2168 cm^{-1} (25.9 kJ mol^{-1}), 1885 cm^{-1} (22.5 kJ mol^{-1}) and 1543 cm^{-1} (18.5 kJ mol^{-1}) respectively. As there are no zero-point effects associated with the relative translational motion of the atoms above the dissociation limit, the dissociation energies measured from the lowest molecular energy state, i.e., D_0, have different values: 432.0 kJ mol^{-1}, 435.4 kJ mol^{-1} and 439.4 kJ mol^{-1}, compared with the single value of 457.9 kJ mol^{-1} for D_e, the difference between the energy at the minimum of the potential curve and the energy when the atoms are completely separated.

Because D_0 corresponds to the minimum internal energy that is required by a non-rotating molecule to dissociate, the difference in zero-point energies between H_2, HD and D_2 is reflected in the different energies of activation for thermal dissociation of these molecules. This provides a particularly simple example of a kinetic isotopic effect. In the case of bimolecular reactions, zero-point effects will arise at all nuclear configurations, but the magnitude of the zero-point energy will differ at different points on the potential hypersurface. Generally, kinetic isotopic effects are treated within the framework of transition state theory, so that attention is particularly focused on the changes in the disposition of the quantum states (or in the partition functions) that take place on passing from reactants to the selected transition state.

It should already be clear that much the largest kinetic isotopic effects occur when H atoms are substituted by D atoms, since the large change in the reduced masses associated with internal motions in the system brings about correspondingly large shifts in the quantum state energies. Only this case will be considered here, and the ratio of the rate constants for a pair of isotopically related reactions will be written as (k_H/k_D).

The value of (k_H/k_D) can be equated to any of several equivalent expressions that can be derived from TST. For example, from the usual rate expressions incorporating partition functions based on normal mode analyses of the reactant and transition state motions, and ignoring, for the moment, the possible effects of tunnelling, one can obtain

$$(k_H/k_D) = \left(\frac{Q^{\ddagger}_{int,H}}{Q^{\ddagger}_{int,D}}\right)\left(\frac{Q_{T,D}\prod_i Q_{int,D}}{Q_{T,H}\prod_i Q_{int,H}}\right)\exp(-\Delta[\Delta\varepsilon_0^{\ddagger}]/k_B T) \qquad (5.43)$$

where the vibrational partition functions are products of terms of the form $(1-\exp[-u_s])^{-1}$ with $u_s = hc\tilde{\nu}_s/k_B T$[*see* equation (4.29)]; $\Delta[\Delta\varepsilon_0^{\ddagger}]$ is equal to $(\Delta\varepsilon^{\ddagger}_{0,H} - \Delta\varepsilon^{\ddagger}_{0,D})$ and is non-zero when the zero-point energies of the reactants and in the transition state are not changed to the same extent on making an

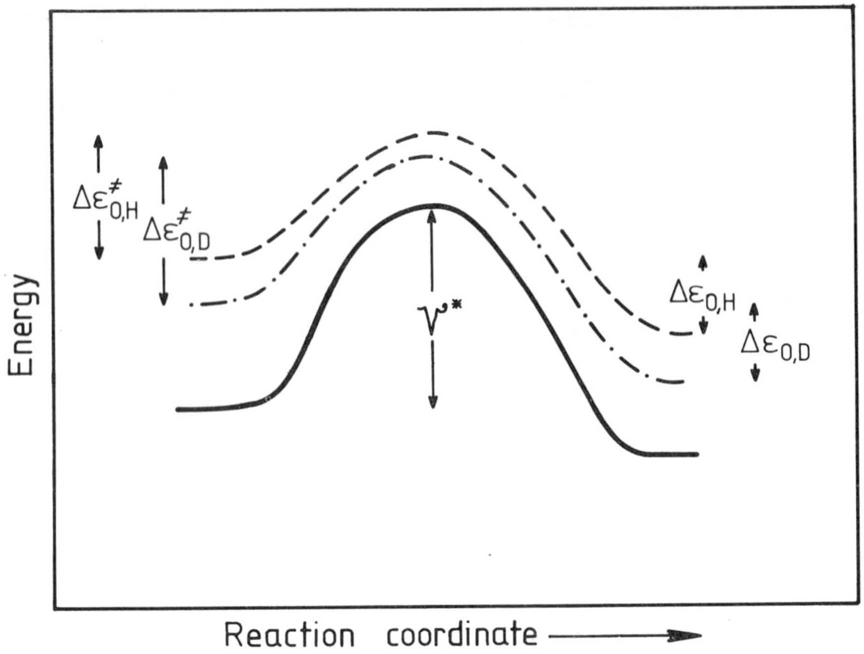

Figure 5.8. Representation of the variation of zero-point energy along the reaction coordinate for a pair of isotopically related reactions

isotopic substitution. Such a case is illustrated in *Figure 5.8*. Alternatively, the vibrational partition functions can be written as $\prod_s \exp[-u_s/2]$, $(1-\exp[-u_s])^{-1}$, so that the exponential term need not be included explicitly [cf. equation (4.32)] and

$$(k_H/k_D) = \left(\frac{Q^{\neq}_{int,H}}{Q^{\neq}_{int,D}}\right)\left(\frac{Q_{T,D}\prod_i Q_{int,D}}{Q_{T,H}\prod_i Q_{int,H}}\right) \tag{5.44}$$

Equations (5.43) and (5.44) show that, according to transition state theory, isotopic substitution will make a large change to the reaction rate when $\Delta(\Delta\varepsilon_0^{\neq})$ is large; that is when the changes in zero-point energy brought about by isotopic substitution in the reactants and in the transition state are quite different. If an isotopic substitution is made at a position that does not directly participate in the reaction, the zero-point energies in the reactants and the transition state will be changed, but to very similar extents, as the frequencies of the motions that are affected do not alter very much in passing from the reactants to the transition state. Only a small change in rate is to be expected for this *secondary* kinetic isotope effect. An example is provided by the pair of reactions

$$CH_3 + HCl \rightarrow CH_4 + Cl \tag{5.45a}$$

and

$$CD_3 + HCl \rightarrow CD_3H + Cl \tag{5.45b}$$

On the other hand, in the reaction

$$CH_3 + DCl \rightarrow CH_3D + Cl \tag{5.45c}$$

the mass of the atom actually being transferred has been changed. Now the changes in zero-point energies in the transition state and reactions will not cancel and an appreciable change in the rate occurs. This constitutes an example of a *primary* kinetic isotope effect.

Another useful equation for the ratio of rate constants for an isotopically related pair of reactions can be derived from the TST expression for the rate constant of a reaction in terms of local bond—rather than normal coordinate—properties. Johnston[36] favours this approach as it affords maximum separation of mass-independent and mass-dependent terms in the rate expression. The ratio of the rate constants is then given by

$$(k_H/k_D) = (\tilde{\nu}^*_{x,H}/\tilde{\nu}^*_{x,D}) \prod^{3N-7} (\Gamma^{\ddagger}_H/\Gamma^{\ddagger}_D) \prod_{\text{reactants}} (\Gamma_D/\Gamma_H) \tag{5.46}$$

where the $\tilde{\nu}^*_x$ are the imaginary frequencies given by a normal model analysis at the transition state, and the Γ are defined by the equation

$$\Gamma_s = (u_s/2)/\sinh(u_s/2) \tag{5.47}$$

where u_s has the same meaning as before. A particularly useful feature of this expression is that it shows that, at the limit of high temperature, since all $\Gamma \rightarrow 1$,

$$(k_H/k_D) \rightarrow (\tilde{\nu}^*_{x,H}/\tilde{\nu}^*_{x,D}) \tag{5.48}$$

This result does not depend on whether quantum tunnelling is allowed for, since the correction factors, κ_{qu}, for tunnelling also go to unity as $T \rightarrow \infty$.

When an attempt is made to determine (k_H/k_D) using any one of equations (5.43), (5.44) and (5.46), the usual obstacle to the fully predictive use of TST presents itself: the data required, in the case of equation (5.46), the real and imaginary frequencies in the transition state, are not available from *a priori* calculations, except for the isotopic variants of the H_3 system. To circumvent this difficulty, a semi-empirical approach is usually adopted. First, rate constants for both or all of the isotopically different reactions are measured over as wide a range of temperature as possible. Then a LEPS, BEBO, or adjusted BEBO potential is found that correctly predicts the temperature dependence of the rate constant for the unsubstituted 'reference' reaction and, hopefully, the absolute magnitude of the rate constants as well. Finally, without further adjustment to the potential, the ratio (k_H/k_D) is calculated and compared with the observed value of this ratio throughout the range of temperature for which experimental results are available.

Unfortunately, this procedure runs up against a major difficulty. As was indicated earlier, kinetic isotope effects are most marked for H or D atom-transfer reactions, particularly when there is an appreciable energy barrier to

the reaction, but these reactions are just the type for which quantum penetration effects are likely to be important. Consequently, each of the expressions for (k_H/k_D) should be multiplied by a factor $(\kappa_{qu,H}/\kappa_{qu,D})$ to allow for the different degree of tunnelling in the isotopically varied reactions. The difficulty of accurately calculating values of κ_{qu} was emphasized in Section 4.4; it is a major source of uncertainty when experimental values of (k_H/k_D) are compared with those calculated using semi-empirical potentials.

In order to evaluate any of the equations for (k_H/k_D) properly, it is necessary to have values for all the bond lengths, angles, and force constants in the transition state. Then a full normal mode analysis—which must usually be carried out by computer—yields the vibrational frequencies, of which one will be imaginary, and they can be inserted into any of equations (5.43), (5.44) or (5.46). When only three atoms are involved in the reaction, this procedure can be carried through easily, all the required data being supplied from the selected semi-empirical potential. When a larger number of atoms are involved, the lengths and force constants associated with the partial bond either side of the atom that is being transferred and the bending force constant about this atom are also supplied from the potential function, but the other transition state parameters are chosen by interpolation between the properties of the reactant and product molecules.

5.6.2 Magnitude of the kinetic isotope effect and the nature of the reactive potential

Before taking an example of a full treatment of kinetic isotope measurements, it is instructive to consider how the magnitude of the effect depends on general features of the potential surface for a reaction. Let us consider first an exoergic reaction, with a low energy barrier located well into the entrance valley of the potential surface, such as

$$\mathrm{Cl + HI \rightarrow HCl + I} \qquad (5.49a)$$

Simple pictorial arguments (cf. Section 3.5.4) suffice to show that the motion along the reaction coordinate at the transition state is essentially that of the relative translation of the approaching reactants. Furthermore, the real stretching frequency in the transition state will be only slightly different from that in HI. To estimate (k_H/k_D), the ratio of rate constants for (5.49a) and for

$$\mathrm{Cl + DI \rightarrow DCl + I} \qquad (5.49b)$$

the partition functions associated with the bending vibrations in the transition state and the rotations in the reagents should be evaluated. The bending frequencies are likely to be very low, so a classical approximation to their partition functions is reasonable, and to depend inversely on the square root of the mass of the light central atom. If this is so, the ratio of partition functions in equation (5.44) is approximately unity. Certainly, because the reduced masses associated with the imaginary frequencies are close to the collisional reduced masses and these are almost identical, the limiting high-temperature value of (k_H/k_D) should, according to equation (5.48), be very close to unity.

It is also quite easy to estimate the magnitude of the isotope effect on the reverse endoergic reaction. Applying the law of detailed balancing (Section 1.2.2), and distinguishing rate constants for reaction in the forward and reverse directions by the same subscripts as previously, the ratio of the equilibrium constants for reactions (5.49a) and (5.49b) is given by

$$(K_H/K_D) = (k_{f,H}/k_{r,H})/(k_{f,D}/k_{r,D}) \tag{5.50}$$

and therefore the required ratio of rate constants is

$$(k_{r,H}/k_{r,D}) = (k_{f,H}/k_{f,D})\,(K_D/K_H) \tag{5.51}$$

The ratio of the equilibrium constants can be calculated using the standard expression from statistical mechanics that was given as equation (4.10). Since $\Delta\mathscr{V}$, the difference in electronic potential energy between products and reactants, is the same for isotopically related reactions, the ratio

$$(K_H/K_D) = \left(\frac{Q_{HCl}Q_I}{Q_{HI}Q_{Cl}}\right)\left(\frac{Q_{DI}Q_{Cl}}{Q_{DCl}Q_I}\right)\exp(-\Delta[\Delta\varepsilon_0]/k_B T) \tag{5.52}$$

where the Q are total partition functions and $\Delta[\Delta\varepsilon_0]$ is equal to $(\Delta\varepsilon_{O,H} - \Delta\varepsilon_{O,D})$, and is non-zero because of differences in the zero-point energies of the products and reagents of the two reactions. The ratio of partition functions in this equation is very close to unity, so that

$$(K_D/K_H) \approx \exp(\Delta[\Delta\varepsilon_0]/k_B T) \tag{5.53}$$

which has a value of 1.6 at 298 K. According to the arguments in the previous paragraph, the ratio $(k_{f,H}/k_{r,D})$ is nearly unity, so that the reverse of reaction (5.49a) should be approximately 1.6 times faster than its isotopic counterpart at 298 K. Once again, the kinetic isotope effect is a relatively minor one.

In practice, the transition state for an atom-transfer reaction usually lies at some intermediate position along the reaction coordinate where the internuclear distances r_{AB} and r_{BC} are both extended relative to the bond lengths in the isolated AB and BC molecules. The kinetic isotopic effect will generally be greater than in either of the limiting cases, but the situation is somewhat more difficult to analyse. One simple example is where A and C are identical, and r_{AB} and r_{BC} are equal in the transition state. Then, motion along the reaction coordinate in the transition state can be identified with the asymmetric displacement of B between A and C, while the motion orthogonal to this in collinear collisions is the symmetric A,C stretch whose frequency is unaffected by isotopic replacement of the B atom. Considering only these frequencies, this type of symmetrical reaction should display a maximum kinetic isotope effect, since changing the B atom from H to D only alters the zero-point energy in the reactants with compensating effect in the transition state. In reality, the doubly degenerate 'bending motions' in the three-centre transition state must also be taken into account. Although the zero-point energies associated with these modes will decrease $\Delta(\Delta\varepsilon_0^{\ddagger})$, the isotope effect is likely to remain quite large.

The pair of reactions,

$$O + HCl \rightarrow OH + Cl, \ \Delta H_0^\circ = +3.8 \text{ kJ mol}^{-1} \qquad (5.54a)$$

and

$$O + DCl \rightarrow OD + Cl, \ \Delta H_0^\circ = +2.6 \text{ kJ mol}^{-1} \qquad (5.54b)$$

provides an example of a system where the transition state is apparently located midway along the reaction coordinate[38]. Thus, if a LEPS potential is adjusted to match the experimentally observed activation energy for reaction (5.55a), which is 25 kJ mol^{-1}, the calculated transition state frequencies are 435.2, 1376.5i and 514.9 cm^{-1} for O–H–Cl and 434.4, 984.3i and 368.5 cm^{-1} for O–D–Cl, where the doubly degenerate bending frequencies are given last. The similarity of the real stretching frequencies, i.e., 435.2 and 434.4 cm^{-1}, and the fact that the ratio of the imaginary frequencies is very close to $2^{\frac{1}{2}}$ are the facts that suggest the 'symmetric' position of the transition state; motion along the reaction coordinate corresponds to motion of the H or D atom between essentially stationary O and Cl atoms, whereas in the real stretching motion the central atom does not move, so the frequency is unaltered by isotopic substitution.

Evaluation of equation (5.46) for this model, with no allowance for tunnelling, leads to a predicted ratio (k_H/k_D) of 4.0 at 400 K (compared with an experimental value of 2.8 ± 1) which increases to 6.9 at 298 K. These values are appreciably larger than those that were estimated for the forward and reverse reactions (5.49). This comparison supports the view that kinetic isotope effects are likely to be most pronounced when the A–B and B–C bond lengths in the transition state are distorted by comparable amounts from their equilibrium values.

5.6.3 The reaction: $CF_3 + CH_4 \rightarrow CF_3H + CH_3$ and its isotopic variations[36]

Some of the most thorough experimental and theoretical studies of kinetic isotope effects have been carried out by Johnston and his co-workers. Johnston gives a detailed description of these in his book and any one considering investigations of this kind is recommended to this source.

A particularly valuable study is that carried out on the reactions between trifluoromethyl radicals and partially deuterated methanes. The reaction

$$CF_3 + CH_4 \rightarrow CF_3H + CH_3 \qquad (5.55a)$$

had been studied earlier in steady-state photolysis experiments and found to have an activation energy of 43 kJ mol^{-1} between 395 and 524 K. Experiments on the kinetic isotope effect were carried out by observing the relative rates of the automatically competitive reactions

$$CF_3 + CHD_3 \rightarrow CF_3H + CD_3 \qquad (5.55b)$$

$$CF_3 + CHD_3 \rightarrow CF_3D + CHD_2 \qquad (5.55c)$$

over an unusually wide range of temperature, 300–1800 K[39, 40].

If the observed ratio of reaction rates for (5.55b) and (5.55c) is multiplied by 3, (k_H/k_D), the ratio of rate constants 'per unit reaction site', is obtained. Because the reactants for both these channels are the same, the TST expression for (k_H/k_D) has a particularly simple form. Based on equation (5.46), but with the possibility of quantum mechanical tunnelling also considered, one obtains

$$(k_H/k_D) = (\tilde{\nu}^*_{x,H}/\tilde{\nu}^*_{x,D})\,(\kappa_{qu,H}/\kappa_{qu,D}) \prod^{3N-7} (\Gamma^{\ddagger}_H/\Gamma^{\ddagger}_D) \qquad (5.56)$$

To evaluate equation (5.56) values are required for the one imaginary and 20 real frequencies in the transition states for the two reactions. Some internuclear distances and force constants were obtained from a three-centre LEPS potential that was chosen to match the experimental activation energy. Any bond not directly involved in the reaction was assumed to have the same length and force constant as in the stable molecule. Bending force constants between a single and a partial bond were given half their value in the stable molecule and rotations about all full and partial bonds were assumed to be free. The calculated values of (k_H/k_D) with $(\kappa_{qu,H}/\kappa_{qu,D}) = 1$ are compared with the experimentally observed ratios in *Figure 5.9*. Johnston also carried out calculations with simplified models of the transition state. The calculated values of (k_H/k_D) changed very little. This confirms that it is changes in the frequencies of the motions which are directly affected by the isotopic replacement that determine the magnitude of kinetic isotope effects.

It is clear from *Figure 5.9* that at low temperatures the values of (k_H/k_D) which are calculated from equation (5.56), with no allowance being made for tunnelling, are much too low. Johnston calculated values of $(\kappa_{qu,H}/\kappa_{qu,D})$ by a method which recognized that one-dimensional tunnelling calculations based on the profile of potential energy along the reaction path of minimum energy would appreciably overestimate the quantum dynamical effects. In his method, a series of cuts are taken through the potential energy surface for collinear collisions parallel to the reaction coordinate at the transition state. An Eckart potential is then fitted to the potential energy profile along each of these cuts and one-dimensional correction factors are evaluated. This method yields tunnelling factors as a function of the value of the symmetric stretching coordinate in the transition state. Finally, these factors are averaged over the classical distribution of phase points with respect to this coordinate in the transition state to yield κ_{qu}. For the $H + H_2$ reaction, this procedure estimates a value of κ_{qu} ($T = 333$ K) that is about one-quarter of the value obtained by the simple means of matching an Eckart potential to the profile of potential energy along the reaction path of minimum electronic energy.

The values of $(\kappa_{qu,H}/\kappa_{qu,D})$ that are obtained using the above procedure bring the calculated and experimental values of (k_H/k_D) for reactions (5.55b) and (5.55c) into excellent agreement. However, a warning-flag to accepting this result too readily was raised by Johnston: although the experimental and calculated values of (k_H/k_D) agree very well, their absolute values, and

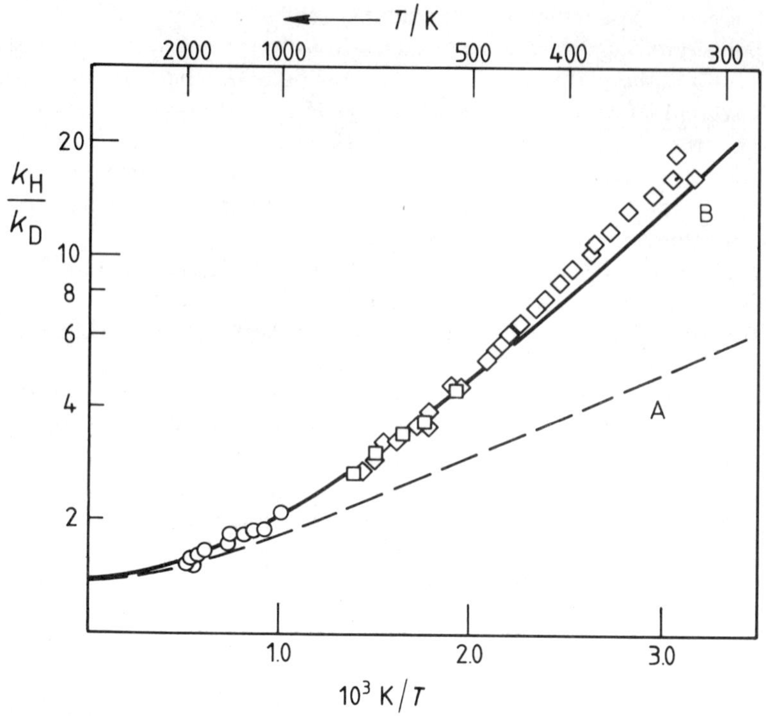

Figure 5.9. Comparison of observed and calculated kinetic isotope effect for reactions (55b) and (55c). The points show experimental data from ref. (39) and ref. (40), and the lines results of TST *calculations* A *without and* B *with allowance for tunnelling* [*adapted from Figures 13–3 and 13–5 in Johnston, H. S.,* Gas Phase Reaction Theory, *Ronald Press, New York* (1966)]

therefore those of A_H and A_D, calculated from TST using the same transition state model, are an order-of-magnitude too low.

The truth of the matter is that there are three stages in calculating the rate constant ratio for isotopically related reactions: the choice of a semi-empirical potential, the use of the standard TST rate expression—derived with the assumption of separable motions in the transition state and evaluated using formulae for rigid rotor–simple harmonic motions—and the estimate of a tunnelling correction factor. Each of these steps is approximate. Consequently, it is difficult to be confident that agreement between the observed and theoretical values of (k_H/k_D) does not result from a lucky cancellation of errors introduced in different parts of the calculation. Conversely, if the experimental and calculated values of (k_H/k_D) do not agree, it can be difficult to identify the cause.

Johnston decided that the pairs of isotopically related reactions which he considered could be divided into three groups. For reactions such as (5.55), where the energy barrier $\mathscr{V}^*$ is appreciably greater than the zero-point energy associated with the vibration along the bond that is broken in the reaction,

agreement between experimental and calculated values of (k_H/k_D) was judged to be excellent—although the failure of calculations based on the same model to predict the absolute values of k_H and k_D should not be forgotten. In Johnston's second category, to which reaction (5.54) belongs, $\mathscr{V}^*$ was not much greater than the zero-point energy of the 'reactive bond' and the agreement between the calculated and observed results was fair. Finally, where $\mathscr{V}^*$ was less than the zero-point energy of the reactive bond, the calculated and experimental values of (k_H/k_D) did not agree.

Whether there is any fundamental connection between (1) the ability of conventional TST calculations, based on semi-empirical potentials and with approximate corrections for tunnelling, to predict the magnitude of kinetic isotope effects successfully, and (2) the relative values of $\mathscr{V}^*$ and the zero-point energy associated with the stretching vibration of the bond that is to be broken in the reaction—as implied by Johnston—seems somewhat dubious. Rather it appears that the lack of agreement between theory and experiment for Johnston's third category of reaction is just one manifestation of a general failure of the standard TST rate expression where the potential barrier to reaction is relatively low. In this situation, the assumption of separability in the transition state and the evaluation of the partition function $Q^{\ddagger}_{int}$ on the basis of rigid rotor–harmonic oscillator motions is certain to lead to some error, since there will be exceptionally small force constants associated with some of the internal motions.

On balance, the Author is inclined to a somewhat gloomier view than Johnston on the success of TST calculations on kinetic isotope effects, since it appears that these are quite generally unable to reproduce both the relative and absolute values of k_H and k_D, using a single model of the transition state. Frequently it is not clear which stage in these calculations is responsible for this failure, although it does not appear that errors in just one step can be responsible for all the discrepancies that are found.

5.7 Unimolecular reactions[41]

5.7.1 Pre-exponential factors

Three different kinds of unimolecular reaction were identified in Section 4.5.1 and activation energies for two examples of each type were given in *Table 4.3*. These reactions are listed again in *Table 5.9*, this time with the Arrhenius pre-exponential factors derived from measurements at high pressure.

In terms of transition state theory, the rate constants for unimolecular reactions in the limit of high [M] are given by

$$k^{\infty}(T) = (k_B T/h)(Q^{\ddagger}_{int}/Q_{int})\exp(-\Delta\varepsilon^{\ddagger}_0/k_B T) \qquad (5.57)$$

If the variation of the partition functions with temperature is ignored, the activation energy corresponds to $(\Delta\varepsilon^{\ddagger}_0 + k_B T)N_A$, and the pre-exponential factor, $A^{\infty}(T)$, is given by

Table 5.9 Pre-exponential factors for some unimolecular reactions

	Reaction[a]	$\Delta S^{\circ b}_{c,298}$ / (J mol^{-1} K^{-1})	$\log\left(\frac{A^{\infty}}{s^{-1}}\right)$	$\log(k_B Te/h)$[c]
(i)	Isomerizations			
	$CH_3NC \rightarrow CH_3CN$	−1.3	13.6	13.45
	cyclo-$C_3H_6 \rightarrow CH_3$—$CH=CH_2$	+29.3	15.45	13.62
(ii)	Dissociation to stable molecules			
	$C_2H_5Cl \rightarrow C_2H_4 + HCl$	+501	14.6	13.6
	cyclo-$C_4H_8 \rightarrow 2C_2H_4$	+544	15.6	13.6
(iii)	Dissociation to free radicals			
	$C_2H_6 \rightarrow 2CH_3$	+527	16.0	13.65
	$HNO_3 \rightarrow OH + NO_2$	+528	15.3	13.75

[a] References for the kinetic data are given in *Table 4.3*.
[b] The standard state is taken as 1 molecule cm^{-3}.
[c] Calculated for T in the middle of the range covered by the experimental measurements.

$$A^{\infty}(T) = (k_B Te/h)(Q^{\ddagger}_{int}/Q_{int}) \tag{5.58}$$

Therefore, major differences between the observed value of $A^{\infty}(T)$ and the calculated value of $(k_B Te/h)$ should reflect differences in the partition functions for the transition state species and the reactant molecule. Because of this connection, the observed magnitude of $A^{\infty}(T)$ provides a clue as to the nature of the transition state in a unimolecular reaction. The usual approach in unimolecular rate theory is to choose a model for the transition state so that the observed high-pressure Arrhenius parameters, E^{∞}_{act} and $A^{\infty}(T)$, match the values that are given by the normal TST rate expressions. The model is then retained in subsequent calculations designed to interpret or predict how the rate of reaction will vary with [M]. Before considering how this is done for two of the reactions listed in *Table 5.9*, it may be helpful to illustrate how semi-quantitative deductions can be made about the nature of the transition state from the observed magnitude of $A^{\infty}(T)$.

The isomerization of CH_3NC has the lowest pre-exponential factor of the reactions that are listed in *Table 5.9* It is, in fact, only slightly larger than $(k_B Te/h)$, which indicates that $Q^{\ddagger}_{int} \approx Q_{int}$, and this in turn suggests that the reaction must proceed via a tightly bound transition state. This reaction is one of those that is discussed more fully later. In contrast to this isomerization reaction, the dissociation of C_2H_6 has a pre-exponential factor more than two orders of magnitude greater than $(k_B Te/h)$. This provides strong evidence that the system passes through a 'loose' transition state, with the two methyl groups already well separated and quite free to rotate relative to one another.

Further insight can be gained by making use of the relationship between the standard entropy change for any system and A_f and A_r, the pre-exponential

factors associated with reaction in the forward and reverse directions. It was shown in Section 1.2.2 that for an isomerization reaction, as there is no difference between the number of molecules on the left and right hand sides of the chemical equation, this relationship has the form:

$$\ln(A_f/A_r) = \Delta S_c^\circ \qquad (5.59)$$

whereas for a dissociation–association reaction,

$$\ln(A_f/A_r) = \Delta S_c^\circ - 1 \qquad (5.60)$$

These equations emphasize that any feature of the transition state which lowers the pre-exponential factor associated with reaction in one direction must have a similar effect for the reverse reaction. For example, the A-factor for $CH_3CN \rightarrow CH_3NC$ is also 'low': $\log(A_r^\infty/s^{-1}) = 13.67$. However, if equation (5.59) is applied to the cyclopropane–propene isomerization, for the reverse reaction one finds $\log(A_r^\infty/s^{-1}) \simeq 14.0$[a], a much lower value than for the forward reaction and one that is close to $\log(k_B Te/h)$. This result strongly suggests that the transition state structure bears a much closer resemblance to that of propene than that of cyclopropane.

Equation (5.60) can be used in a similar fashion to relate the pre-exponential factors for unimolecular dissociation to those for bimolecular association, although this has to be done with some care, since the forward and reverse reactions have often been studied experimentally in different temperature ranges. The high A-factors for C_2H_6 and HNO_3 decomposition are, however, quite consistent with the rate constants that were quoted in *Table 5.5* for the reverse radical recombination reactions. Likewise, the lower value of $A^\infty(T)$ for $C_2H_5Cl \rightarrow C_2H_4 + HCl$ is in agreement with the rather small A-factors that are usually found for bimolecular four-centre reactions (*see* Section 5.3.5).

5.7.2 The reaction: $CH_3NC\ (+M) \rightarrow CH_3CN\ (+M)$[42] (5.61)

Reaction (5.61) and its isotopic analogues were studied very thoroughly some years ago. It is examined here as an example of a unimolecular reaction that proceeds via a 'tight' transition state.

Schneider and Rabinovitch used an extremely straightforward experimental method. Methyl isocyanide, either pure or diluted in inert gas, was admitted at a known pressure into a well-thermostatted vessel held at an accurately controlled temperature. At measured intervals, samples were withdrawn for analysis by gas–liquid chromatography. In this way, first-order rate constants were measured ranging from $10^{-6}\ s^{-1}$ to $2 \times 10^{-3}\ s^{-1}$, the upper limit being determined by the intrinsically poor time resolution of the experimental method, and the lower limit by the sensitivity of the analytical technique and the patience of the experimenter.

[a] For equation (5.59) to be applied *accurately*, values of A_f and ΔS_c° should be available for the same temperature.

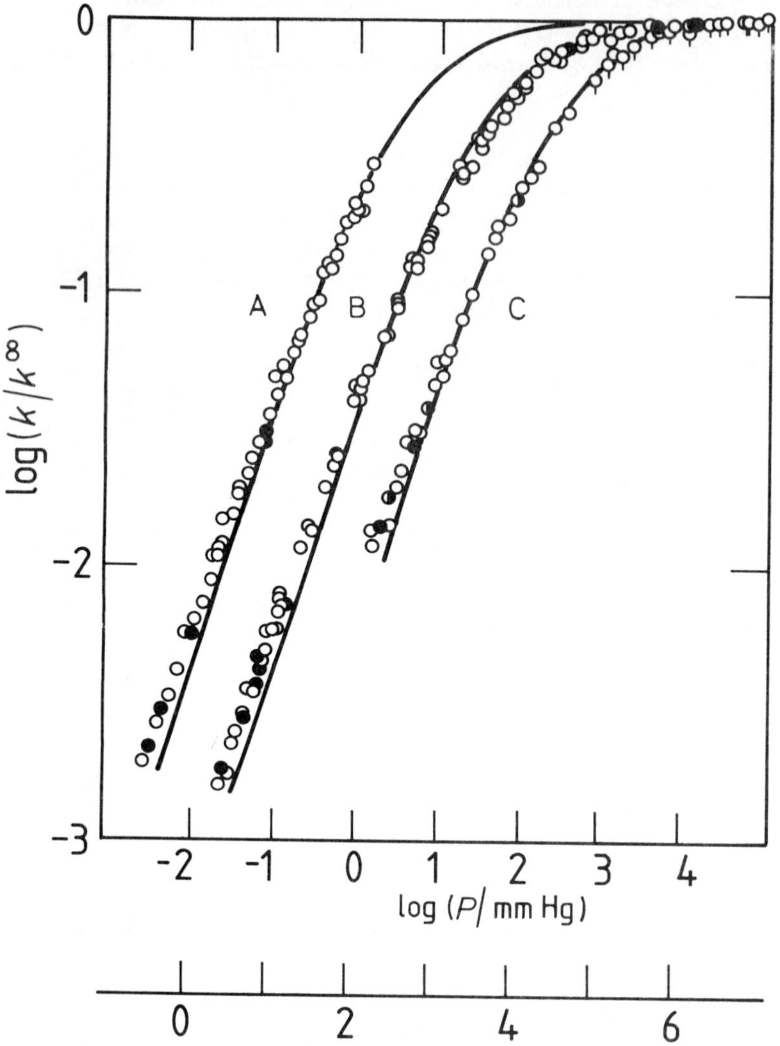

Figure 5.10. Variation of the effective first-order rate constant for the isomerization of CH_3NC *with total pressure at* A *532.9* K, B *503.5* K, *and* C *472.5* K. *For clarity, the data at 532.9* K *have displaced one* log (P/mmHg) *unit to the left and those at 472.5* K *one unit to the right. The vertical marks under some points distinguish results obtained at 472.5* K *from those determined at 503.5* K; *other differences indicate rate constants obtained in different reaction vessels. The curves show the results of* **RRKM** *calculations based on the transition-state model whose properties are summarized in Table 5.10* [*adapted from Schneider, F. W., and Rabinovitch, B. S.*, Journal of the American Chemical Society, **84**, 4215 (1962)]

Experiments were performed at 472.5, 503.5 and 532.9 K, and over a very wide range of total pressures: $5 \times 10^{14} \leqslant [M] \leqslant 2 \times 10^{20}$ molecule cm^{-3}. At 503.5 K, the reaction rate was determined throughout the region in which the reaction order changes from one to two. Data showing the observed variation of k/k^{∞} are given in *Figure 5.10*.

The shape of the observed 'fall-off' curves for this reaction could be reproduced by calculations based on equation (4.105) from RRK theory with s treated as an adjustable parameter. However, the best agreement was obtained with $s = 3.6$, a number much lower than the value of 12 which is obtained by counting all the normal mode vibrations of CH_3NC. This finding, arising from the neglect of zero-point effects, is not peculiar to this reaction. Fall-off curves for unimolecular reactions can usually be matched in this way, indicating that the procedure provides a rather insensitive test of theoretical models.

Schneider and Rabinovitch also carried out detailed RRKM calculations using models of the transition state with the C, C and N atoms in a triangular configuration. Internal quantum states were counted according to the empirical semi-classical procedure of Whitten and Rabinovitch that was described in Section 4.5.5. The theoretical curves in *Figure 5.10.* were determined with the transition state model whose frequencies and moments of inertia are listed in *Table 5.10*. The experimental variation of k/k^{∞} is clearly reproduced rather well and this is also true for data on the $CD_3NC \rightarrow CD_3CN$ isomerization.

Table 5.10 Properties of transition state model for $CH_3NC \rightarrow CH_3CN$

CH_3NC		*Transition state*	
Mode	$\tilde{\nu}$/cm^{-1}	*Mode*	$\tilde{\nu}$/cm^{-1}
CH_3 str.	2966, 2×3014	CH_3 str.	3×2998
N≡C str.	2161	Ring def.	1990
CH_3 def.	1410, 2×1459	CH_3 def.	3×1443
CH_3 rock	2×1041	CH_3 rock	2×1041
C—N str.	945	Ring def.	600
C—N—C bend	2×270	Ring def.	reac. coord.
		Twist	270
Rotational constants	5.22, 2×0.335	Rotational constants	1.45, 0.533, 0.421

However, it is important to note that the level of agreement between experiment and calculation was altered only slightly by quite extensive changes in the model of the transition state. This is encouraging from the point of view of semi-empirical prediction, since it suggests that it might be possible to estimate rate constants for unimolecular reactants at all values of [M], even if only part of the fall-off curve close to the high-pressure limit has been determined. On the other hand, it demonstrates the difficulty of deducing anything more than semi-quantitative information about the transition state in reactions of this kind.

A major factor limiting the interpretation of kinetic data on unimolecular reactions when there is an appreciable activation energy for both the forward and reverse processes is that they can only be studied over a narrow range of temperature, because of the rapid change in reaction rate as the temperature is altered. For example, the rate constant for reaction (5.60), at any value of [M], alters by about two orders-of-magnitude over the 60.4 K range covered in Schneider and Rabinovitch's experiments. This means that it is virtually impossible to obtain any additional insight into the nature of the activated complex—or the validity of conventional transition state theory—by observing the variation of k^{∞} with temperature. This possibility does, however, exist in the case of reactions of type (iii) in *Tables 4.3* and *5.9*, since, as well as measurements on thermal decomposition at high temperatures, rate constants can be determined at low temperatures for the association reactions. The two sets of kinetic data can, of course, be related to one another through the relationships that were established in Section 1.2.2 and referred to in Section 5.7.1.

5.7.3 The reaction: $HNO_3(+M) \rightleftarrows OH + NO_2(+M)$[43,44] (5.62)

This reaction is important in atmospheric chemistry and, partly because of this, its rate has been measured in a number of recent experiments. It also serves as a good example of a reaction where there is apparently little or no maximum in the variation of electronic potential energy along the reaction coordinate. Rate constants are now available over a wide range of temperature and quite a wide range of total gas density, and in this section the observed results are compared with predictions made by the application of unimolecular rate theory.

The experimental data are of two kinds. At low temperatures, it is the radical association reaction (remember that NO_2 has an odd electron) which can be studied directly. In the most extensive measurements[43], OH radicals were produced by flash photolysis of HNO_3 or H_2O and their concentration monitored by resonance absorption. Effective second-order rate constants, defined by

$$k = (-\mathrm{d}\ln[OH]/\mathrm{d}t)/[NO_2] \qquad (5.63)$$

were determined through the ranges: $220 \leqslant T \leqslant 550\,K$ and $3.2 \times 10^{17} \leqslant [M] \leqslant 1.6 \times 10^{19}$ molecule cm^{-3}. Measurements with [M] as low as 3×10^{16} molecule cm^{-3} have also been made using discharge–flow techniques.

At high temperatures, rate constants have been measured for nitric acid dissociation. This reaction has been studied in shock tubes[44] over the ranges: $890 \leqslant T \leqslant 1200$ K and $4.8 \times 10^{18} \leqslant [M] \leqslant 1.9 \times 10^{20}$ molecule cm^{-3}. Some earlier results from static pyrolysis experiments help to fill the gap between the high-temperature shock tube measurements and the low-temperature results on the association reaction.

None of the experiments that have just been mentioned cover as wide a

range of [M] as that achieved by Schneider and Rabinowitch in their measurements on the isomerization of methyl isocyanide. Consequently, in order to discover the rate constants in the limit of high and low [M], i.e., k^0 and k^∞, use was made of the empirical fitting procedure[45] which was described in Section 4.5.6. Fall-off curves were calculated from Kassel integrals of the type to be found on the right-hand side of equation (4.105) using an 'effective' value of s that is derived empirically. For a species M whose third-body collisional efficiency (β) is unity, the appropriate value of s is given by

$$\begin{aligned} s_{\text{eff}} &= \{-\partial \ln Q_V/\partial(1/T)\}/T \\ &= E_V/RT \end{aligned} \tag{5.64}$$

where Q_V and E_V are the vibrational partition function and average vibrational energy of the unimolecular reagent at temperature T. For cases where $\beta < 1$, s_{eff} is further adjusted according to prescriptions given by Troe[45]. Once reduced fall-off curves, i.e., plots of log (k/k^∞) versus log [M], have been calculated, they are matched to the experimental data to yield values for k^0 and k^∞. This procedure allows the limiting rate constants to be determined from experimental rate constants that have been obtained only in the transition region, although obviously the accuracy of this interpolation depends on how closely the observations approach the limiting values.

Figure 5.11 shows values of the second-order rate constant defined by equation (5.63) and measured at 296 K plotted against $[N_2]$. The straight lines show $k^0_{N_2}[N_2]$ and k^∞ as derived from the matching procedure described in the previous paragraph. In an attempt to interpret the observed variation of the rate constant with [M] and T, RRKM calculations have been carried out. First, a model for the transition state was chosen to yield a partition function corresponding to

$$(Q^{\ddagger}/V) = k^\infty(Q_{OH}/V)(Q_{NO_2}/V)/(k_B T/h) \tag{5.65}$$

where k^∞ is the high-pressure limiting rate constant derived for $T = 296$ K. In this equation, $(Q^{\ddagger}/V)$ is the partition function per unit volume for the transition state species with one degree of freedom removed, relative to the zero-point energy level of the separated reactants, OH and NO_2. It was then assumed that $\Delta\varepsilon_0^{\ddagger} = 0$ and the transition state was given the properties summarized in *Table 5.11*. With a collisional efficiency for N_2 of $\beta = 0.3$, the fall-curve shown in *Figure 5.11* was calculated according to the methods of RRKM theory that were outlined in Section 4.5.5.

At first sight, the agreement between the observed rate constants at 296 K and the fall-off curves calculated from RRKM theory is encouraging, but it provides a deceptive picture of the state of our understanding of this reaction—and probably of others of the same kind. *Table 5.12* compares the experimental values of k^∞ at different temperatures with values given by the standard TST expression using the transition state properties summarized in *Table 5.11*. Whereas the measured values of k^∞ fall by a factor of eight in going from 296 to 1100 K, the calculated values increase by more than three times.

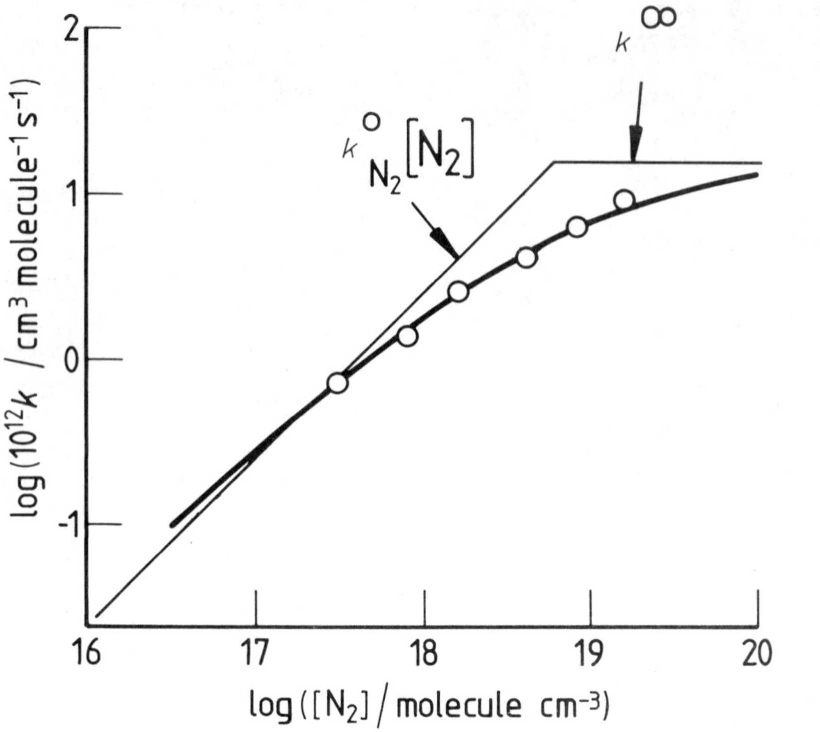

Figure 5.11. Variation of the effective second-order rate constant for $OH + NO_2$ $(+N_2) \rightarrow HNO_3(+N_2)$ *with* $[N_2]$ *at 296* K. *The curve shows the results of* RRKM *calculations based on the transition-state model whose properties are summarized in Table* 5.11

The situation can be improved, by allowing the separation of the two radicals in the transition state to depend on the temperature. Quack and Troe[46] have calculated $k^{\infty}(T)$ for several association reactions involving two radicals, using the method outlined in Section 4.5.6, which is based on finding the configuration of maximum free energy. The variation of both the electronic potential energy and partition function along the reaction path was described in terms of a single parameter which was kept constant for all the reactions studied. There is no adjustment to observed values of $k^{\infty}(T)$ with this method. The third column of *Table 5.12* gives results of calculations on $OH + NO_2$. The absolute value of $k^{\infty}(T)$ at room temperature is in quite good agreement with that from experiment. The calculated rate constants are now virtually independent of temperature. This is an improvement on the TST prediction based on a fixed transition state geometry, but is still some way from matching the observations.

Table 5.11 Properties of transition state model for $HNO_3 \rightleftarrows OH + NO_2$

	$\tilde{\nu}/cm^{-1}$		
Mode	HNO_3	*Transition state*	OH/NO_2
O′H str.	3550	3550	3589
NO antisym. str.	1708	1650	1618
HO′N bend	1331	free rotor	
NO sym. str.	1325	1325	1320
NO_2 def.	879	800	750
NO out-of-plane bend	762	135	
NO′ str.	647	reac. coord.	
ONO′ in-plane bend	579	135	
O′H torsion	(456)	free rotor	
Rotational constants			
internal	18.7	18.7	
overall	0.43, 0.38, 0.206	0.43, 0.065, 0.056	

Table 5.12 Comparison of experimental and calculated values of k^{∞} for $OH + NO_2$ $(+M) \rightarrow HNO_3$ $(+M)$

T/K	10^{11} k^{∞}/cm^3 $molecule^{-1}$ s^{-1}		
	expt.	calc.[c]	calc.[d]
220	2.0[a]	1.45	–
296	1.6[a]	1.6	2.2
550	0.44[a]	2.6	–
1000	0.19[b]	5.2	2.4

[a] ANASTASI, C., and SMITH, I. W. M., *Journal of the Chemical Society, Faraday Transactions*, II, **72**, 1459 (1976)

[b] GLÄNZER, K., AND TROE, J., *Berichte der Bunsengesellschaft für Physikalische Chemie* **78**, 71 (1974)

[c] Using fixed transition state with the properties listed in *Table 5.11*.

[d] QUACK, M., and TROE, J., *Berichte der Bunsengesellschaft für Physikalische Chemie* **81**, 329 (1977)

At present, there are very few reactions involving dissociation to two radicals or the reverse association for which accurate high-pressure rate constants have been measured over a wide range of temperature. Nevertheless, it seems unlikely that the lack of agreement between experiment and theory for reaction (5.62) is a special case. Certainly, transition state theory in its conventional form will fail, when it is applied to loosely bound systems where there is no justification for treating the internal motions as separable small amplitude vibrations. Quack and Troe's method[46] rightly emphasizes two fundamental points. First, the transition state is associated with a free energy

maximum and its location can depend on temperature. Secondly, the standard methods for evaluating internal partition functions are inappropriate for loosely bound transition state species. How well the quantitative predictions of the model will generally fit the experimental facts remains to be seen.

5.8 Atomic recombination

The association reactions that have been considered up to this point produce molecules containing several atoms. In interpreting the kinetics of these reactions, it has been implicitly assumed that, as the two radicals come together, some of the energy which is released can flow into degrees of freedom other than that corresponding to relative motion of the radicals along the direction of the bond which is being formed. Consequently, the newly created molecule does not immediately redissociate, but rather it has a lifetime considerably longer than the period characteristic of a molecular vibration. Furthermore, all 'memory' of the original location of the internal energy is assumed to be lost, so that the distribution of this energy—and hence the distribution of the lifetimes of the energized species with respect to redissociation—is determined simply by statistical factors.

As the number of atoms involved in this kind of reaction becomes less, so does the number of degrees of freedom among which the energy can be distributed. Hence the average lifetime with respect to redissociation becomes shorter, and the changeover between third-order kinetics, when the rate of reaction is proportional to the total gas concentration [M], and second-order kinetics, when the rate is independent of [M], occurs at increasingly high values of [M]. The limit of this trend is obviously reached in the case of recombination of two atoms, denoted in the following paragraphs by A.

When two atoms collide, the details of the trajectory—for example, the distance of their closest approach—depends on the form of the potential energy curve for the electronic state that is involved. However, the nett result does not. As was emphasized in Chapter 3, as long as the dynamics is governed by a single potential curve, i.e., no electronic transitions take place, the two atoms must scatter elastically. If the atoms are to combine, there has to be a mechanism for energy to be 'lost'.

There are three essentially distinct mechanisms by which atomic free radicals can combine. The first possibility is that the atoms approach one another on a potential curve other than that of the electronic ground state of the diatomic molecule and, whilst they are close together, the system emits a photon and falls to a level in a lower electronic state which is stable with respect to dissociation. The second pathway for atomic recombination is generally referred to as the 'energy-transfer' mechanism and it is equivalent to that by which larger radicals combine. For atoms to recombine in this way, the third-body species M must collide with, and remove energy from, a pair of A atoms that are themselves in the act of colliding. The third possible route for atomic recombination is variously referred to as the 'atom-chaperon' or 'bound-

complex' mechanism. In this model, it is proposed that reaction proceeds via the formation of weakly bound AM species and the subsequent bimolecular reaction of these complexes with a 'free' atom A.

The three mechanisms for atomic recombination are discussed, in fairly simple terms, in the next two sections. It is shown that the energy-transfer and bound-complex mechanisms both give rise to third-order kinetic behaviour, i.e., rates of reaction that are proportional to $[A]^2[M]$, and lead to similar order-of-magnitude estimates for the third-order rate constant. Consequently, when some experimental data are presented in Section 5.8.3, emphasis is placed on data that indicate the likely contribution of each of the mechanisms to the total observed reaction rate. Finally, some extensions to the basic models for three-body atomic recombination are reviewed.

5.8.1 Two body radiative recombination[47]

As was seen in Chapter 2, the interaction of atoms which are not in $^1S_0$ states gives rise to more than one electronic state. When two such atoms approach one another the chance of their 'entering' any particular state is determined purely by statistical factors. As a general rule, the dynamics of two-atom collisions is controlled by the form of a single potential energy curve, and the proportion of the total number of collisions that occur on any one of these curves can be calculated by dividing the statistical weight of that state by the sum of the statistical weights associated with all the states that correlate with the separated atoms. Frequently, more than one of the potential curves correlating with the separated atoms have deep minima, i.e., represent bound states. Then one must expect that the energy-transfer and bound-complex mechanisms, which require the participation of a third-body, must operate via these excited electronic states, as well as through the molecular ground state. For the present, however, we are concerned only with the formation of stable molecules in *two-body* collisions during which the system spontaneously radiates and falls into a stable level of a bound state, without the interference of another atom or molecule.

It is easy to demonstrate that the probability of radiative atomic recombination is *always* extremely small. This can be done by comparing the characteristic mean radiative lifetime associated with fully allowed electronic transitions ($\sim 10^{-8}$ s) with the typical duration of a collision between two atoms ($\sim 10^{-13}$ s). This indicates that radiative recombination is only likely to occur about once in every 10^5 collisions. In fact, this estimate is a rough upper bound on the probability as a number of factors can reduce the probability still further.

Detailed mechanisms for two-body radiative recombination can be discussed in terms of the schematic potential energy diagrams in *Figure 5.12*. The first case shown there represents a system where, as well as the bound ground state, a repulsive electronic state correlates with the separated atoms. For the rate of the spontaneous emission process to reach its maximum value, two atoms colliding on the upper state potential must possess a considerable

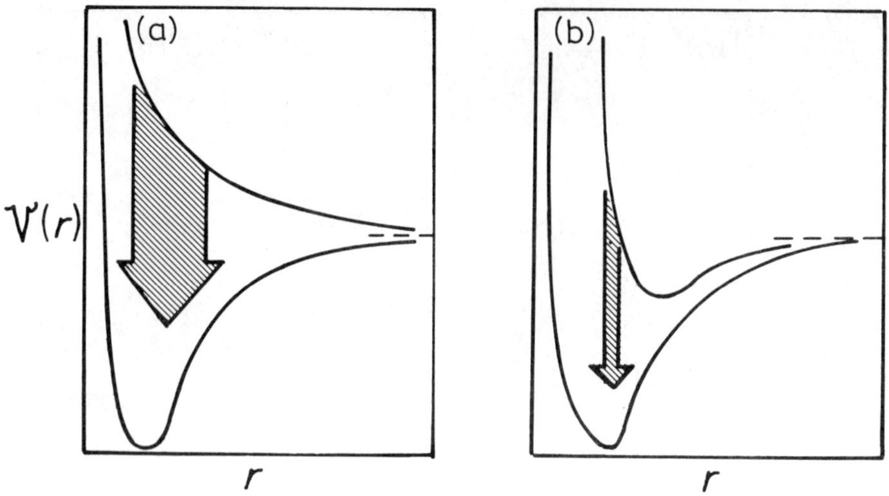

Figure 5.12. Potential-energy diagrams illustrating possible mechanisms for two-body radiative recombination of atoms via the reverse of photodissociation

collision energy to reach a separation approaching the equilibrium internuclear distance of the ground electronic state. Consequently, the thermally averaged probability of recombination via this process will be considerably less than 10^{-5}.

A more favourable situation for radiative recombination is illustrated in *Figure 5.12*(b). Here, an excited bound state exists, with its turning point at the dissociation limit of the state close to the value of r_e for the lower electronic state. Now, in any collisions where the centrifugal barrier is surmounted, the atoms come into close contact, and the thermally averaged probability of radiative recombination will be greater than in the first case—assuming that all other factors are similar in the two cases.

In both the examples that have just been considered, the process of radiative recombination is the reverse of *photodissociation.* A factor which often causes the probability of recombination to be much less than 10^{-5} is that optical transitions between states that correlate with the same pair of atoms are frequently forbidden or, at least, have low electronic transition moments. A good example is provided by O_2 for which potential curves are shown in *Figure 5.13*(a). None of the transitions connecting the $X^3\Sigma_g^-$ ground state of O_2 to the many excited states correlating with $O(^3P)+O(^3P)$ give rise to strong absorptions and radiative recombination plays effectively no part in the recombination of ground state oxygen atoms. On the other hand, the $B^2\Sigma_u^-$—$X^3\Sigma_g^-$ transition is strong. In absorption direct photodissociation is caused by absorption below 175.9 nm producing $O(^1D)+O(^3P)$. Emission arising from radiative recombination via the reverse of this process has been observed in the upper atmosphere.

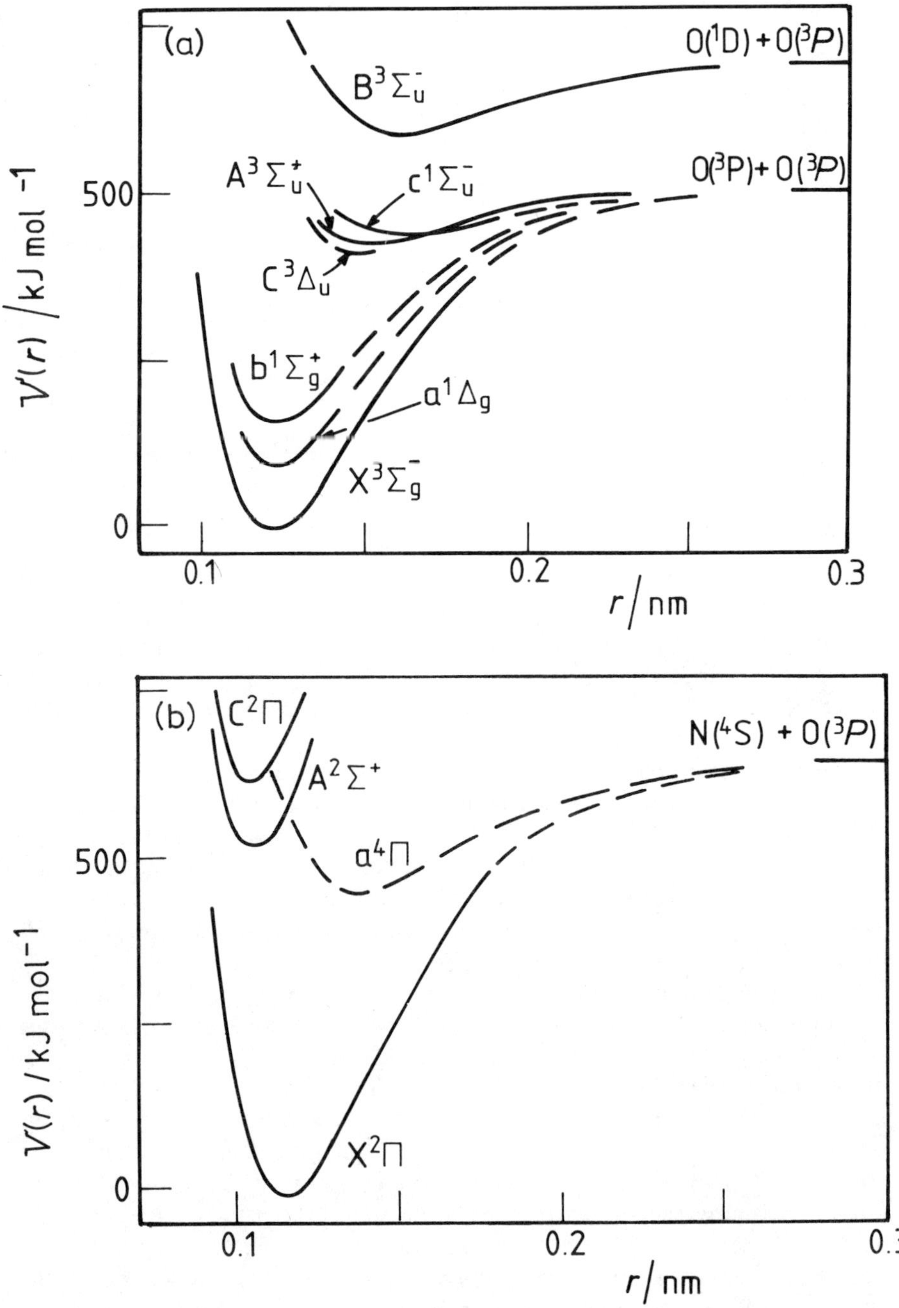

Figure 5.13. Partial potential energy diagram: (a) *for* O_2 *showing curves for all the bound states that correlate with* $O(^3P) + O(^3P)$, *as well as that for* $B^3\Sigma_u^-$ *and* (b) *for* NO, *showing the states that participate in the two-body, preassociative mechanism for radiative recombination*

As well as the reverse of photodissociation, radiative recombination can occur via *inverse predissociation* or *preassociation*. This mechanism requires a juxtaposition of potential energy curves similar to that shown in *Figure 5.13*(b), which is a partial potential energy diagram for NO. Here, the atoms can approach in the $a^4\Pi$ state and then undergo a radiationless transition to $C^2\Pi$ whence they undergo allowed radiative transitions either directly, or via the $A^2\Sigma^+$ state, to the ground electronic state. The overall process can be discussed in terms of the reaction scheme[48, 49]:

$$\begin{array}{l} N(^4S) + O(^3P) \underset{k_r}{\rightleftharpoons} (NO\,a^4\Pi) \overset{k_f}{\rightleftharpoons} NO\,C^2\Pi \\ \qquad\qquad\qquad\qquad\qquad\qquad\quad \downarrow k_{rad} \\ \qquad\qquad\qquad\qquad\qquad\quad NO\,X^2\Pi + h\nu \\ \qquad\qquad\qquad\qquad\quad \text{and } NO\,A^2\Sigma^+ + h\nu \end{array} \tag{5.66}$$

where k_f, k_r and k_{rad} are rate constants for the elementary processes indicated by the arrows. A trivial steady-state treatment is required to derive the total rate of two-body radiative recombination:

$$\text{Rate} = \{k_{rad}k_f/(k_r + k_{rad})\}[N][O]$$

In spite of the fact that the radiationless transition between the $a^4\Pi$ and $C^2\Pi$ states requires the total electron spin to change, $k_r \gg k_{rad}$, so the rate constant for the overall process is very nearly equal to $\{k_{rad}k_f/k_r\}$, which is simply the first-order rate constant for spontaneous emission multiplied by an equilibrium constant. This demonstrates that the rate of recombination via a mechanism of this kind does not depend on the absolute values of k_f or of k_r, as long as $k_r \gg k_{rad}$. Conversely, k_f and k_r cannot be determined by measuring the rate of radiative recombination. In the case of NO, the lowest level in the $C^2\Pi$ state is very close in energy to that of $N(^4S) + O(^3P)$, i.e., to the dissociation limit of the $a^4\Pi$ and $X^2\Pi$ states, and the radiative transitions $C^2\Pi - A^2\Sigma^+$ and $C^2\Pi - X^2\Pi$ are strong. This leads to $\{k_{rad}k_f/k_r\}$ having a value of 1.7 $\times 10^{-17}$ cm^3 molecule^{-1} s^{-1} which is about the maximum value that might be expected on the basis of the ultrasimple arguments that were used earlier to estimate a collisional probability.

To summarize, two-body radiative processes *never* provide an efficient mechanism for recombination. Although observations on the chemiluminescent spectra can provide interesting information concerning the detailed pathways for such processes, this mechanism can only make an appreciable contribution to the overall rate of atomic recombination at low total pressures ($[M] \leqslant 10^{16}$ molecule cm^{-3}).

5.8.2 Three-body recombination of atoms

For a good many years, the non-radiative 'termolecular' recombination of atoms, which can be written as

$$\dot{A} + A + M \rightarrow A_2 + M \tag{5.67}$$

has been considered in terms of the two detailed mechanisms that were referred to earlier: the energy-transfer mechanism and the bound-complex mechanism[50–53]. The difference between them can be discussed in terms of the potential energy surface in *Figure 5.14*. This surface resembles that given in *Figure 2.3*(a) in that its main feature is a single deep valley associated with the chemical bond between the two atoms denoted by A. However, in this case, the surface is drawn in terms of r_{AA} and r_{AM}, rather than r_{AA} and r_{M,A_2} as in *Figure 2.3*(a), and, more importantly, allowance has been made for the existence of weak attractive forces between A and M, giving rise to the shallow valley at the top of the figure, and between A_2 and M, giving rise to the slight hollow close to the bottom right-hand corner. The surface only represents, of course, a cut through the fully dimensional hypersurface, but even if M is polyatomic, since it does not form directional 'chemical bonds' with A, the form of the surface will depend only slightly on the AAM angle or on the internal coordinates, if any, of M.

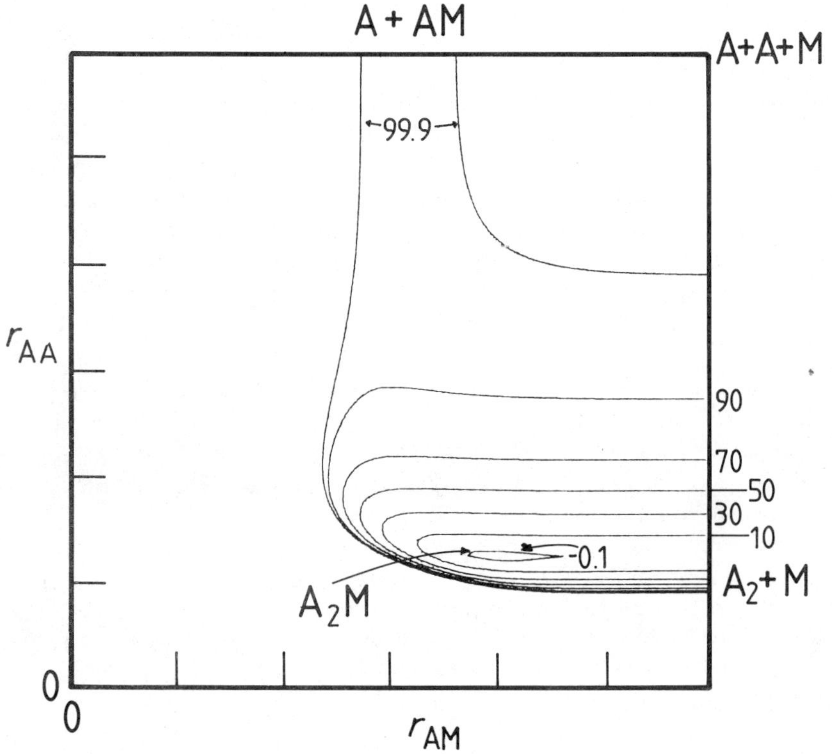

Figure 5.14. Potential surface representing a three-atom system, where A_2 *exists as a chemically bound molecule but the attractive forces between* A *and* M *and between* $A_2 + M$ *are much weaker. The contours are drawn at energies which are expressed as percentages of the dissociation energy of* A_2.

Recombination via the *energy-transfer mechanism* results from trajectories that start in the top right-hand corner of this figure, that is with A, A and M all far apart from one another, and ends with the representative point moving out of the figure along the valley in the lower right-hand corner. The mechanism can be treated in a simplified way in terms of two two-body steps occurring 'simultaneously' or in rapid succession, i.e.,

$$\mathrm{A + A} \underset{k_r}{\overset{k_f}{\rightleftarrows}} \mathrm{A_2^\dagger} \qquad (5.68a)$$

$$\mathrm{A_2^\dagger + M} \xrightarrow{k'} \mathrm{A_2 + M} \qquad (5.68b)$$

Here, $A_2^\dagger$ represents two A atoms in close proximity, which will, in the absence of M, scatter elastically, i.e., they are in continuum states, whereas A_2 represents a molecule that has insufficient energy to dissociate, i.e., it is in a discrete state[a]. This mechanism is clearly analogous to that which brings about the association of larger free radicals, the difference being that k_r is now much greater. Setting $[A_2^\dagger]$ to a steady-state value, an equation is obtained for the overall rate of reaction in terms of the rate constants for the three elementary steps[b]:

$$\begin{aligned} \mathrm{d}[\mathrm{A_2}]/\mathrm{d}t &= \{k'k_f/(k_r + k'[\mathrm{M}])\}[\mathrm{A}]^2[\mathrm{M}] \\ &= k'_{3\mathrm{rd}}[\mathrm{A}]^2[\mathrm{M}] \end{aligned} \qquad (5.69)$$

It is easy to make a rough estimate of k'_{3rd}. The rate constant k_r can be set equal to $\sim 10^{13}\ \mathrm{s^{-1}}$, as it is simply the reciprocal of the characteristic duration of a diatomic collision. This value is much greater than $k'[M]$, at values of $[M]$ that are readily attainable in the gas phase. Consequently,

$$k'_{3\mathrm{rd}} = k'k_f/k_r \qquad (5.70)$$

Now, the second-order rate constants, k_f and k', can both reasonably be set equal to Z, the second-order rate constant for collisions between A and A and between $A_2^\dagger + M$, using either the hard-sphere or hard-sphere plus Lennard-Jones attraction models (*see Table 3.1* for typical values of Z). These substitutions in equation (5.70) yield a value for k'_{3rd} of roughly $10^{-32}\ \mathrm{cm^6\ molecule^{-2}\ s^{-1}}$.

This extremely simple treatment of the energy-transfer mechanism can be improved upon, by recognizing that (k_f/k_r) represents a constant for the equilibrium between pairs of atoms that are (1) close enough for recom-

[a] This ignores the 'in-between' behaviour of the quasi-bound states, shown in *Figure 3.9*, which lie above the dissociation limit but below the centrifugal maximum in $\mathscr{V}_{\mathrm{eff}}(r_{\mathrm{AA}})$.

[b] It can be argued that this treatment includes only those trajectories starting from A + A + M in which r_{AA} becomes 'small' before r_{AM}. One can also include the result of recombination via $\mathrm{A + M \rightleftarrows AM^\dagger}$ and $\mathrm{AM^\dagger + A \rightarrow A_2 + M}$, but, since the treatment that follows is very approximate, this is ignored.

bination to occur, if a collision with M takes place during the transitory lifetime of $A_2^\dagger$, and (2) far apart. This equilibrium constant can be estimated using the equation

$$\frac{k_f}{k_r} = \frac{\int \rho(\varepsilon_T)\exp(-\varepsilon_T/k_B T)\,d\varepsilon_T}{(Q_T/V)} \tag{5.71}$$

where Q_T is the partition function associated with the relative translational motion of a pair of A atoms and $\rho(\varepsilon_T)$ is the density of those states from which A_2 pairs can be transferred to stable states. If these $A_2^\dagger$ states are identified as those orbital states which lie above the centrifugal maximum on the appropriate effective potential curve, $\rho(\varepsilon_T)$ is relatively straightforward to evaluate. Assuming that the potential in the region of the centrifugal maximum is merely the attractive component of the Lennard-Jones expression, then

$$b_{max} = (27\varepsilon_{LJ}/\varepsilon_T)^{1/6}\sigma_{LJ} \tag{5.72}$$

[cf. equation (3.50)], and an expression can be obtained for l_{max}, the maximum value that the orbital quantum number can have, if collisions at a specified energy ε_T are to surmount the centrifugal barrier,

$$l_{max}(l_{max}+1) = \{(2\mu\varepsilon_T)^{\frac{1}{2}} b_{max}/\hbar\}^2 \tag{5.73}$$

[cf. equation (4.51b)]. The total number of orbital states leading to 'close' collisions is given by

$$\sum_0^{l_{max}} (2l+1) = (l_{max}+1)^2, \text{ so}$$

$$\rho(\varepsilon_T) \approx \frac{d}{d\varepsilon_T}\{2\mu\varepsilon_T(27\varepsilon_{LJ}/\varepsilon_T)^{1/3}\sigma_{LJ}^2/\hbar^2\}$$
$$= (4\mu/\hbar^2)(\varepsilon_{LJ}/\varepsilon_T)^{1/3}\sigma_{LJ}^2 \tag{5.74}$$

Substitution of this expression into equation (5.71) yields (with $x = \varepsilon_T/k_B T$):

$$\frac{k_f}{k_r} = \frac{8\pi h\sigma_{LJ}^2}{(2\pi\mu k_B T)^{\frac{1}{2}}}\left(\frac{\varepsilon_{LJ}}{k_B T}\right)^{1/3}\int_0^\infty x^{-1/3}\exp(-x)\,dx$$
$$= \frac{8\pi h\sigma_{LJ}^2}{(2\pi\mu k_B T)^{\frac{1}{2}}}\left(\frac{\varepsilon_{LJ}}{k_B T}\right)^{1/3}\Gamma(2/3) \tag{5.75}$$

With values of $\sigma_{LJ} = 0.3$ nm, $\varepsilon_{LJ}/k_B = 120$ K, and $\mu = 8$ atomic mass units, $(k_f/k_r) = 8.1 \times 10^{-23}$ cm^3 molecule^{-1} and hence $k'_{3rd} \approx 1-2 \times 10^{-32}$ cm^6 molecule^{-2} s^{-1} at 298 K.

Finally, it should be remembered that the separated species usually correlate with more than a single electronic state and so a statistical weight factor should be introduced to allow for this. It is reasonable to assume that each of the bound states of A_2 that correlate with ground state atoms makes a full

contribution to the overall rate of recombination, but they usually amount to less than half of the total number of molecular states. Consequently, the above estimate of k'_{3rd} has generally to be reduced by a factor of between two and four to allow for this effect

The *bound-complex* mechanism is based on the supposition that the second step in atomic recombination involves trajectories on the surface in *Figure 5.14* that originate in the shallow valley representing weakly bound AM species. These bound-complexes are assumed to be formed in three-body ('energy transfer') processes of the type:

$$A + M + M \rightleftarrows AM + M \tag{5.76}$$

The frequency of three-body collisions of this kind is generally much greater than the frequency of collisions between two A atoms and a single M species because $[M] \gg [A]$ under laboratory conditions. Furthermore, since the bound-complexes are very weakly bound and undergo many collisions with M species before encountering an atom A, it is reasonable to assume that they are in thermodynamic equilibrium with the rest of the system, so that their steady-state concentration is given by

$$[AM] = K''[A][M] \tag{5.77}$$

The overall rate of three-body recombination via the bound-complex mechanism is, therefore,

$$\begin{aligned} \text{Rate}'' &= k''[A][AM] \\ &= k''K''[A]^2[M] \\ &= k''_{3rd}[A]^2[M] \end{aligned} \tag{5.78}$$

where k'' is the rate constant for

$$A + AM \rightarrow A_2 + M \tag{5.79}$$

The rate constant k'' is usually assumed to be the collisional rate constant, Z, times a factor that depends on the electronic multiplicities of A, AM and A_2. The equilibrium constant K'' can be found from the usual, statistical thermodynamic equation

$$K'' = \{Q_{AM}/V)/(Q_M/V)(Q_A/V)\} \exp(\varepsilon_{AM}/k_B T) \tag{5.80}$$

where ε_{AM} represents the dissociation energy of the bound-complex AM. If M is polyatomic, the ratio of partition functions can be simplified by recognizing that the attraction between A and M is likely to be much too weak to affect the motions (rotational, as well as vibrational) of M in the AM complex. Consequently, partition functions associated with these degrees of freedom will cancel on the right-hand side of equation (5.79). However, because the attraction between A and M is much weaker than a normal chemical bond, the partition function associated with the remaining internal motions in AM must be evaluated with care. Formulae derived on the assumption of widely spaced vibrational quantum states would certainly be inaccurate.

As a good approximation to K'', where the interaction between A and M can be represented by a Lennard-Jones potential with $(\varepsilon_{LJ}/k_B T) \leqslant 1$, Bunker and Davidson[50] recommended the equation

$$K'' = \pi^{\frac{1}{2}}(\varepsilon_{LJ}/k_B T)^{\frac{3}{2}} \left[\frac{8}{3} + \frac{32\varepsilon_{LJ}}{45 k_B T}\right] \sigma_{LJ}^3 \tag{5.81}$$

If this expression is evaluated for $T = 298$ K with our standard Lennard-Jones 'test' values of $(\varepsilon_{LJ}/k_B) = 120$ K and $\sigma_{LJ} = 0.3$ nm, $K'' = 3.6 \times 10^{-23}$ cm^3 molecule^{-1}. With $k'' = Z$, k''_{3rd} is approximately equal to 10^{-32} cm^6 molecule^{-2} s^{-1}.

What emerges from the simplified discussion in this subsection is that, even if only a Lennard-Jones potential acts between A and M, the contributions of the energy-transfer and bound-complex mechanisms to the rate of atomic recombination at room temperature are similar. Furthermore, a decision cannot be made on which mechanism is predominant on the basis of the observed order of reaction, since both models predict the rate to be proportional to $[A]^2[M]$. However, it should be noted that, whereas k'_{3rd}, the rate constant for recombination via the energy transfer mechanism, is expected to be essentially independent of the nature of M, k''_{3rd}, the rate constant for recombination via the bound-complex, should depend on the strength of the attractive interaction between A and M.

5.8.3 Experimental rate constants

Rate constants for atomic recombination reactions have been determined using both discharge–flow and flash-photolysis methods. Because discharge–flow experiments have to be carried out at fairly low total pressures ($[M] \leqslant 3 \times 10^{17}$ molecule cm^{-3}) and also with quite low concentrations of atoms ($[A] \leqslant 5 \times 10^{15}$ molecule cm^{-3}), the rate of homogeneous recombination is rather slow and the effects of surface-catalysed recombination have to be carefully assessed. Flash photolysis has been used extensively to study the recombination of Br atoms and of I atoms. In these measurements, the atoms are produced by photodissociation of the parent diatomic molecule and the progress of recombination is monitored by following the absorption of light in the Br_2 or I_2 continuum photometrically.

In addition to direct studies of termolecular recombination, there have been investigations of the bimolecular dissociation of diatomic molecules in shock tubes. Rate constants for recombination at high temperatures can be calculated by applying the equation of detailed balancing. For bromine and iodine, flash photolysis experiments on recombination have been extended to high temperatures[54] so that there is an overlap between these results and those obtained from shock tube studies on the reverse reactions. Until recently, there was thought to be a discrepancy of about a factor of three between the rate constants determined directly and those derived using the principle of detailed

balancing[55]. However, the difference has now been resolved, and the two sets of rate constants are consistent with one another[56–59].

Third-order rate constants for a number of atomic recombination reactions, defined by

$$d[A_2]/dt = -\tfrac{1}{2}d[A]/dt = k_{3rd}[A]^2[M] \qquad (5.81)$$

are listed in *Table 5.13*. Although these values have all been quite firmly established, the same cannot be said for the dependence of k_{3rd} on the nature of M—at least for some of the reactions. It is for this reason that rate constants are quoted for different M; there is no single third-body species for which rate constants for all the atomic recombination reactions given in *Table 5.13* have been precisely evaluated. However, where the relative efficiencies of different M have been observed for the same reaction, the variation in k_{3rd} is moderately small as long as M is a noble gas, N_2 or H_2.

Table 5.13 Third-order rate constants and activation energies for some atomic recombination reactions

Reaction	$10^{33}k_{3rd}$ at 298 K / cm^6 molecule^{-2} s^{-2}	E_{act} / kJ mol^{-1}	*Range of* T/K	*Ref.*
$H + H + H_2 \rightarrow H_2 + H_2$	8.3	–	–	*a*
$O + O + N_2 \rightarrow O_2 + N_2$	4.8	−4.0	196, 298	*b*
$N + N + N_2 \rightarrow N_2 + N_2$	10.0	−4.2	100—600	*c*
$N + O + N_2 \rightarrow NO + N_2$	9.2	−1.3	196, 298	*b*
$Cl + Cl + Ar \rightarrow Cl_2 + Ar$	12.3	−7.5	200—500	*d*
$Br + Br + Ar \rightarrow Br_2 + Ar$	6.6	−6.3	298—500	*e*
$Br + Br + N_2 \rightarrow Br_2 + N_2$	9.1	−6.3	298—500	*e*
$I + I + Ar \rightarrow I_2 + Ar$	6.6	−6.7	298—573	*f*

[a] BAULCH, D. L., DRYSDALE, D. D., HORNE, D. G., and LLOYD, A. C., *Evaluated Kinetic Data for High Temperature Reactions*, vol. 1, p. 261, Butterworths, London (1972)
[b] CAMPBELL, I. M., and GRAY, C. N. *Chemical Physics Letters*, **18**, 607 (1973)
[c] BRENNEN, W. and SHANE, E. C., *Journal of Physical Chemistry*, **75**, 1552 (1971)
[d] LLOYD, A. C., *International Journal of Chemical Kinetics*, **3**, 39 (1971)
[e] IP, J. K. K., and BURNS, G., *Journal of Chemical Physics*, **51**, 3414 (1969); DEGRAFF, B. A., and LANG, K. J., *Journal of Physical Chemistry*, **74**, 4181, (1970)
[f] BLAKE, J. A., and BURNS, G., *Journal of Chemical Physics*, **54**, 1480 (1971)

Perhaps the most striking feature of the data in *Table 5.13* is the similarity in all the rate constants. Although the electronic states that are involved and the strengths of the chemical bonds being formed vary considerably, a factor of approximately 2.5 encompasses all the absolute values of k_{3rd} at 298 K. Furthermore, the rate constants are close to the estimate that were made earlier by naïve versions of the energy-transfer and bound-complex mechanisms. For once, the rather unusual situation exists where too many, rather than too few, theories fit the facts.

In order to decide between the two mechanisms for recombination it is

customary to inspect the dependence of k_{3rd} on temperature for a wide variety of third-body gases. Bearing in mind that the rate constant for collisions between hard-spheres with an r^{-6} attraction varies as $T^{1/6}$, versions of the energy-transfer mechanism and bound-complex mechanism based on equations (5.70) and (5.78) predict that $k'_{3rd} \propto T^{-2/3}$ and $k''_{3rd} \propto T^{-4/3}$ (assuming that $32\varepsilon_{LJ}/45k_BT) \ll 8/3$). This means that the third-order rate constant should decrease slightly as the temperature is raised. Although there is clearly no theoretical justification for expecting an Arrhenius expression to match the observed variation of k_{3rd} the rate constants observed over a limited range of temperature can be used to define a *negative* activation energy.

The small negative activation energies in *Table 5.13* reflect a relatively slow variation of k_{3rd} with T. For the recombination reactions of Br atoms and of I atoms, however, it has been possible to obtain rate constants with a much wider variety of third-body species, and then as the data in *Table 5.14* demonstrate, both the absolute values of k_{3rd} and the activation energies show considerably greater variation. These data are taken from a larger compilation published in 1962[51].

Table 5.14 Rate constants and activation energies for the recombination of I atoms

M	$10^{33}k_{3rd}$ at 293 K / cm^6 molecule^{-2} s^{-1}	E_{act}/kJ mol^{-1} $293 \leqslant T \leqslant 523$ K
He	4.1	−1.7
Ar	8.3	−5.4
O_2	19	−6.3
CO_2	37	−7.3
C_6H_6	221	−7.1
$C_6H_5CH_3$	535	−11.3
C_2H_5I	720	−10.0
$C_6H_5(CH_3)_3$	1120	−17.2
I_2	4410	−18.4

All data in this Table are taken from Table 1 in PORTER, G., *Discussions of the Faraday Society*, 33, 198 (1962). The experimental results are from PORTER, G., and SMITH, J. A., *Proceedings of the Royal Society*, ***A*261**, 28 (1961), except for the result for $M = I_2$, which is from BUNKER, D. L., and DAVIDSON, N., *Journal of the American Chemical Society*, **80**, 5085, 5090 (1958).

In an attempt to match the experimental results via a mechanism based on the energy-transfer model, Porter[51] made use of the relationship between k_{3rd} and k_{2nd}, the second-order rate constant for dissociation of I_2, i.e.,

$$k_{3rd} = k_{2nd}K \quad (5.83)$$

where K is the equilibrium constant for $I + I = I_2$. He made the extreme assumption that k_{2nd} could be represented by an expression of the form

$$k_{2nd} = Z(\varepsilon^0/k_B T)^n (1/n!) \exp(-\varepsilon^0/k_B T) \quad (5.84)$$

At first sight, this resembles equation (4.104) for one of the elementary steps in the Lindemann—Hinshelwood mechanism for unimolecular reactions, but now n includes degrees of freedom in M, the collision partner. As Porter recognized, it is physically unreasonable to expect complete *intermolecular* energy transfer during the duration of a single direct collision. In any case, it was impossible to find values of n for most of the third-body species, which simultaneously matched the absolute values of k_{3rd} and their dependence upon temperature.

In contrast to the energy-transfer model, it is clear that the bound-complex mechanism can explain greater differences in the absolute values of k_{3rd} and appreciable negative activation energies, if AM can be reasonably strongly bound, i.e., if ε_{AB}/k_B in equation (5.79) can be larger than values associated with van der Waals forces between noble gases and light diatomic molecules. If it is assumed that the forces holding A and M together in the bound-complex are still non-directional, so that M is free to rotate, and weak, so that the partition function for the A–M stretching vibration can be formulated classically ($Q_V = k_B T/h\nu$), the activation energy is related to ε_{AM} by

$$E_{act} = -N_A(\varepsilon_{AM} - k_B T) \quad (5.85)$$

The values of ε_{AM} for IM complexes that are then needed to explain the dependence of k_{3rd} on T also predict the absolute values of k_{3rd} quite well; they are, however, somewhat greater than the values obtained by applying the combining rules to the values of (ε_{LJ}/k_B) for Xe (next to I in the Periodic Table) and M. It is likely that for certain M, AM is a charge-transfer complex. Further discussion of the relative contributions of the energy-transfer and bound-complex mechanisms, particularly where the AM attraction is not particularly strong, is given in the next section.

5.8.4 Classical trajectories and phase-space theory

The results of experiments on atomic recombination reactions have not only been compared with the relatively simple versions of the energy-transfer and bound-complex models that were considered in the last section. In addition, theoretical calculations have been carried out employing, on the one hand, the method of classical trajectories[56, 57] (Section 3.5.2) and, on the other, the phase space version of transition state theory[58, 59] (Section 4.2.2).

The classical trajectory method has been used both to examine the recombination of atoms via the bound-complex mechanism directly and to study the collisional dissociation of the diatomic molecule. In the latter process, the products may be A + AM or A + A + M, and, since the principle of detailed balancing can be applied to the results, the relative contributions of the energy-transfer and bound-complex mechanisms to the overall rate of the reverse association reaction can be calculated.

A further important general conclusion can be drawn from the results of

trajectory calculations. They show clearly that the diatomic molecules that are formed in either A + AM or A + A + M collisions are produced largely in states quite close to the dissociation limit, whilst the dissociation reaction occurs predominantly from highly excited internal states of A_2. In view of the earlier discussion of reaction dynamics in Section 3.5.4, this finding should cause no surprise. The potential surface shown in *Figure 5.14* represents a limiting case of a highly attractive surface for a highly exoergic atom-transfer reaction. Consequently, the dynamics favour specific energy disposal into the vibration of the newly formed molecule when trajectories run across the hypersurface in the direction of the exoergic reaction, although some broadening of the internal energy distribution doubtless results from secondary encounters. Conversely, the rate of the highly endoergic dissociation of A_2 is promoted selectively by internal excitation, rather than by enhancing the collision energy.

Agreement between the experimental and calculated rate constants, and their dependence on temperature, has been achieved, but only by treating ε_{AM} as an adjustable parameter. The values that were required to get this agreement are significantly larger than estimates based on the application of combining rules to closely related species for which Lennard-Jones potential parameters are available. For example, to obtain agreement with the observed third rate constant for the recombination of Br atoms in Ar, it was necessary to assume[56] that $(\varepsilon_{BrAr}/k_B) = 500$ K which is more than three times the geometric mean of ε_{LJ} for Kr and Ar.

Keck[60] has largely been responsible for pioneering the application of phase-space theories to the problem of atomic recombination. Although the language of his 'variational' approach is somewhat different from that which was employed in Section 4.2.2, the basis of the treatment is essentially the same. Once again a search is made for a critical dividing surface for the AAM system so that the total flux through this surface can be identified as closely as possible with the reactive flux. In the case of atomic recombination, evaluation of the partition functions or phase integrals for the reactants should include those regions of phase space corresponding to AM + A, as well as those for A + A + M. As before, the location of the dividing surface should, in principle, be varied in order to find the position at which the flux in the direction reactants to products is a minimum. This flux should then correspond most closely to the reactive flux but may still overestimate the latter, since a surface may not exist for which all of the flux in one direction coincides with reactive flux, with no trajectories turning back and recrossing the surface. This is particularly a problem when, as in the case of atomic recombination, there is no well-defined potential barrier separating the reactant and product regions of phase space. Consequently—and as was observed earlier (see, for example, *Figure 4.4*)—a theory of this kind provides an upper limit to the true rate of reaction, always assuming that the potential hypersurface is accurately represented, and that the required partition functions or phase integrals are evaluated without recourse to secondary assumptions, such as the separability of internal motions.

In one regard, a theory based on classical phase space is particularly appropriate to atomic recombination reactions. This is because zero-point effects will be virtually absent as the forces in both the AM complexes and the AAM transition state are weak. This fact is made use of in the *modified* version of phase-space theory[58, 59]. In this treatment, one first estimates the total flux through S* by a phase-space calculation of the kind that has just been outlined, then the proportion of this total flux that is reactive is estimated by computing a sample of classical trajectories. These trajectories are not chosen to start from the reactant zone of phase space but, using Monte Carlo sampling procedure, from points on S* itself. For each choice of initial momenta and coordinates on S*, trajectories are run forward and backward in time. When one of this pair of trajectories leads *directly* to separated products, $A_2 + M$, and the other leads to separated reactants, A + A + M or AM + A, then this system makes a contribution to the reactive flux and is therefore 'counted'[61]. In this way, the ratio of reactive to total flux can be determined in a relatively small number of trajectories and the rate constant which is determined from the initial phase-space calculation can be corrected accordingly.

The statistical–dynamical method that has just been described has also been applied, more recently, to bimolecular atom-transfer reactions[62]. In these calculations, the classical phase space at S* was sampled and again trajectories were run forward and back to determine the incidence of multiple crossings of this critical surface. However, in these reactions zero-point effects are much more important than in atomic recombination and the results of the trajectory calculations are equivalent to 'truly' *classical*, rather than *quasiclassical*, trajectory calculations (see Section 3.5.2), since they correspond to a distribution selected, by the potential and the collision dynamics, from the *classical* phase space of the separated reactants. At the present time, it seems that the *quasiclassical* method provides a better estimate of the real behaviour in systems where zero-point effects are significant.

The modified phase-space treatment of atomic recombination makes a second correction to the rate of reaction that is predicted by the unmodified theory. The need for this adjustment arises because, as was pointed out earlier, the trajectory calculations reveal that quite a high proportion of the A_2 molecules formed in A + AM or A + A + M collisions have internal energies close to the dissociation limit. Consequently, there is a significant probability that, rather than being further relaxed in a subsequent collision, these molecules acquire sufficent energy to redissociate. This means that, unavoidably, the observed rate of atomic recombination may be the nett result of reaction in the forward direction proceeding at its equilibrium value, i.e., that with A, AM and M all in thermal equilibrium, but with reaction in the reverse direction occurring at a rate greatly in excess of the equilibrium rate because the internal state distribution of A_2 is heavily weighted towards high levels. Conversely, in high-temperature studies of dissociation, the stationary state populations of levels close to the dissociation limit may be appreciably lower than their equilibrium values on account of depletion by reaction. This effect must again cause the observed rate constant to fall below its value in a fully

equilibrated system. Methods have been devised of estimating these effects and allowances can then be made, either to the phase-space theory results or to those from a classical trajectory study. One interesting feature of these non-equilibrium effects is that they apparently reduce the rate constants for dissociation and association to the same extent at a given temperature; consequently, the ratio remains equal to the equilibrium constant, despite the fact that each rate constant is appreciably less than its value at thermodynamic equilibrium.

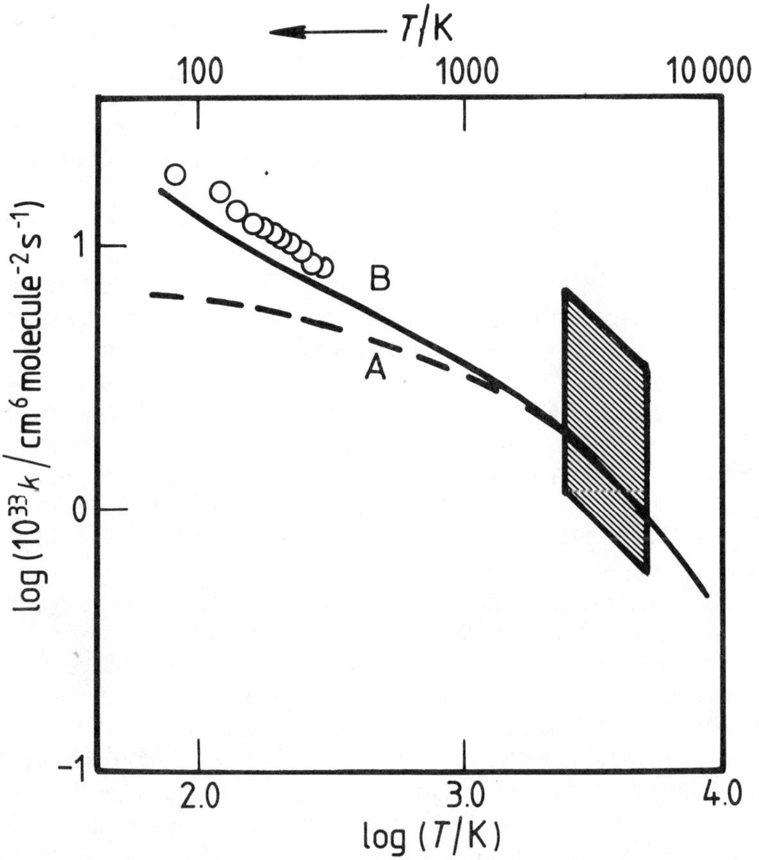

Figure 5.15. Comparison between experimental rate constants for $H + H + H_2 \rightarrow H_2 + H_2$ *and those calculated by modified phase-space theory* A *with no attractive* H, H_2 *interaction, and* B *with* $(\varepsilon_{H,H_2}/k_B) = 38$ K [*adapted from Figure 2 in Shui, V. H., and Appleton, J. P.*, Journal of Chemical Physics, **55**, 3126 (1971); *the experimental points are from Ham, D. O., Trainor, D. W., and Kaufman, F.*, Journal of Chemical Physics, **53**, 495 (1970), *and the shaded area shows the result of an evaluation of high-temperature data by Baulch, D. L., Drysdale, D. D., Horne, D. G.,* and *Lloyd, A. C.*, Evaluated Kinetic Data for High Temperature Reactions, Vol. 1, *Butterworths, London* (1973)]

To illustrate the success of the modified phase-space theory, *Figure 5.15* compares the results of calculations on the reaction

$$H + H + H_2 \rightarrow H_2 + H_2 \qquad (5.86)$$

with the available experimental data. Two sets of calculations were carried out. In those yielding the dashed line, a purely repulsive potential between $H + H_2$ was assumed, whereas in the calculations yielding the continuous line, allowance was made for an attractive interaction between H and H_2 with $\varepsilon_{AM}/k_B = 38$ K. Comparison of the two calculations with the experimental data reveals the importance, even in this system, of the bound-complex mechanism, whose contribution is indicated by the difference between the two calculated curves.

The success of the modified phase-space theory has prompted its advocates to suggest that, by matching the calculated rate constants to those observed experimentally, it is possible to determine the magnitude of the well-depth in the interaction between free-radical atoms and other species. In fact, it should be possible with current molecular beam technology to find accurate intermolecular potentials for many such systems—for example, Br + Ar—by measuring differential cross-sections for elastic scattering and the velocity dependence of the total scattering cross-section. Measurements of this kind have been reported recently[63] for F + rare gas atoms. In fact, molecular beam experiments formed part of the basis for the estimate of $\varepsilon_{AM}/k_B = 38$ K for $H + H_2$ that was used to obtain the theoretical curve shown in *Figure 5.15*. It seems certain that other data of this kind will be forthcoming in the near future and will provide an important cross-check on the agreement between theoretical and experimental results on atomic recombination reactions.

References

Section 5.1

1 BRADLEY, J. N., *Fast Reactions*, Oxford University Press, Oxford (1975)

2 CVETANOVIC, R. J., OVEREND, R. P., and PARASKEVOPOULOS, G., *International Journal of Chemical Kinetics*, symp. 1 on *Chemical Kinetics Data for the Upper and Lower Atmosphere*, 249 (1975); HOWARD, C. J., *Journal of Physical Chemistry*, **83**, 3 (1979)

3 CLYNE, M. A. A., in *Physical Chemistry of Fast Reactions*, Ed. Levitt, B. P., Vol. 1, Chap 4, Plenum, London (1973)

4 POIRIER, R. V., and CARR, R. W., *Journal of Physical Chemistry*, **75**, 1593 (1971)

5 SLAGLE, I. R., GILBERT, J. R., and GUTMAN, D., *Journal of Chemical Physics*, **61**, 704 (1974)

6 FERGUSON, E. E., in *Ion–Molecule Reactions*, Ed. Franklin, J. L., Vol. 2, Plenum, New York (1972)

7 DONOVAN, R. J., and GILLESPIE, H. M., in *Specialist Periodical Reports. Reaction Kinetics*, Ed. Ashmore, P. G., Vol. 1, Chap. 2, Chemical Society, London (1975)

8 GORDON, S., and MULAC, W. A., *International Journal of Chemical Kinetics*, symp. 1 on *Chemical Kinetics Data for the Upper and Lower Atmosphere*, 289 (1975)

9 PAUKERT, T. T., and JOHNSTON, H. S., *Journal of Chemical Physics*, **56**, 2824 (1972)

10 HIPPLER, H., LUTHER, K., and TROE, J., *Berichte der Bunsengesellschaft für Physikalische Chemie*, **77**, 1104 (1973)

11 SCHACKE, H., SCHMATJKO, K. J., and WOLFRUM, J., *Berichte der Bunsengesellschaft für Physikalische Chemie,* **77,** 248 (1973)
12 GREENE, E. F., and TOENNIES, J. P., *Chemical Reactions in Shock Waves,* Arnold, London (1964)
13 BOWMAN, C. T., *Journal of Physical Chemistry,* **83,** 757 (1979)
14 BELFORD, R. L., and STREHLOW, R. A., *Annual Reviews of Physical Chemistry,* **20,** 247 (1969)
15 BREEN, J. E., QUY, R. B., and GLASS, G. P., *Journal of Chemical Physics,* **59,** 556 (1973)
16 CVETANOVIC, R. J., *Advances in Photochemistry,* **1,** 115 (1963)

Section 5.3

17 For recent evaluations of rate data for elementary reactions between neutral species see: (a) HAMPSON, Jr., R. F., and GARVIN, D., *Reaction Rate and Photochemical Data for Atmospheric Chemistry* (N.B.S. Technical note 513, 1978), and (b) *Evaluated Kinetic Data for High Temperature Reactions,* Vol. 1, Baulch, D. L., Drysdale, D. D., Horne, D. G., and Lloyd, A. C., Butterworths, London (1972); Vol. 2, Baulch, D. L., Drysdale, D. D., and Horne, D. G., Butterworths, London (1973); Vol. 3, Baulch, D. L., Drysdale, D. D., Duxbury, J., and Grant, S., Butterworths, London (1976)
18 MAHAN, B. H., *Journal of Chemical Physics,* **48,** 2629 (1968)
19 *Ion-Molecule Reactions,* Ed. Franklin, J. L., Vol. 1 and 2, Plenum, New York (1972), especially Chap. 5 in Vol. 1 by M. Henchmann
20 KAUFMAN, J. J., in *The Excited State in Chemical Physics,* Ed. McGowan, J. W., John Wiley, New York (1975), Vol. XXVIII in the series *Advances in Chemical Physics,* 113
21 SULLIVAN, J. H., *Journal of Chemical Physics,* **46,** 73 (1967)
22 CARROLL, H. F., and BAUER, S. H., *Journal of the American Chemical Society,* **91,** 7727 (1969)
23 BENSON, S. W., and BOSE, A. N., *Journal of Chemical Physics,* **37,** 2935 (1962)
24 KING, D. L., and HERSCHBACH, D. R., *Faraday Discussions of the Chemical Society,* **55,** 331 (1973)

Section 5.4

25 TRUHLAR, D. G., and WYATT, R. E., *Annual Reviews of Physical Chemistry,* **27,** 1 (1976)
26 KOEPPL, G. W., *Journal of Chemical Physics,* **59,** 3425 (1973)
27 SCHATZ, G. C., and KUPPERMANN, A., *Journal of Chemical Physics,* **65,** 4642 and 4668 (1976)

Section 5.5

28 ZELLNER, R., and STEINERT, W., *International Journal of Chemical Kinetics,* **8,** 397 (1976)
29 For reviews see: BAULCH, D. L., and DRYSDALE, D. D., *Combustion and Flame,* **23,** 215 (1974); WILSON, Jr., W. E., *Journal of Physical Chemistry Reference Data,* **1,** 535 (1972)
30 STEINERT, W., and ZELLNER, R., *Second European Symposium on Combustion,* (1975)
31 DAVIS, D. D., FISCHER, S., and SCHIFF, R., *Journal of Chemical Physics,* **61,** 2213 (1974)
32 DRYER, F., NAEGELI, D., and GLASSMAN, I., *Combustion and Flame,* **17,** 270 (1971)
33 SMITH, I. W. M., and ZELLNER, R., *Journal of the Chemical Society, Faraday Transactions, II,* **69,** 1617 (1973)
34 SMITH, I. W. M., *Chemical Physics Letters,* **49,** 112 (1977), and references therein
35 BIERMANN, H. W., ZETZSCH, C., and STUHL, F., *Berichte der Bunsengesellschaft für Physikalische Chemie,* **82,** 633 (1978)

Section 5.6

36 JOHNSTON, H. S., *Gas Phase Reaction Rate Theory,* Chap. 13, Arnold, New York (1966)
37 KLEIN, F. S., *Annual Reviews of Physical Chemistry,* **26,** 191 (1976)
38 BROWN, R. D. H., and SMITH, I. W. M., *International Journal of Chemical Kinetics,* **10,** 1 (1978)

39 SHARP, T. E., and JOHNSTON, H. S., *Journal of Chemical Physics*, **37**, 1541 (1962)
40 JOHNSTON, H. S., and TSCHUIKOW-ROUX, *Journal of Chemical Physics*, **36**, 463 (1962)

Section 5.7

41 ROBINSON, P. J., in *Specialist Periodical Reports. Reaction Kinetics*, Ed. Ashmore, P. G., Vol. 1, Chap. 3, Chemical Society, London (1975); see also the references at the start of Section 4.5.
42 SCHNEIDER, F. W., and RABINOWITCH, B. S., *Journal of the American Chemical Society* **84**, 4215 (1962) and **85**, 2365 (1963)
43 ANASTASI, C., and SMITH, I. W. M., *Journal of the Chemical Society, Faraday Transactions, II*, **72**, 1459 (1976)
44 GLÄNZER, K., and TROE, J., *Berichte der Bunsengesellschaft für Physikalische Chemie*, **78**, 71 (1974)
45 TROE, J., *Berichte der Bunsengesellschaft für Physikalische Chemie*, **78**, 478 (1974)
46 QUACK, M., and TROE, J., *Berichte der Bunsengesellschaft für Physikalische Chemie*, **81**, 329 (1977)

Section 5.8

47 GOLDE, M. F., and THRUSH, B. A., *Reports on Progress in Physics*, **36**, 1285 (1973)
48 CALLEAR, A. B., and PILLING, M. J., *Transactions of the Faraday Society*, **66**, 1618 (1970)
49 MANDELMAN, M., CARRINGTON, T., and YOUNG, R. A., *Journal of Chemical Physics*, **58**, 84 (1973)
50 BUNKER, D. L., and DAVIDSON, N., *Journal of American Chemical Society*, **80**, 5090 (1958)
51 PORTER, G., *Discussions of the Faraday Society*, **33**, 198 (1962)
52 BUNKER, D. L., *Theories of Elementary Gas Reaction Rates*, Chap. 4, Pergamon, Oxford (1966)
53 Chap. 14 in ref. 36.
54 IP, J. K. K., and BURNS, G., *Journal of Chemical Physics*, **51**, 3414 (1969) and **56**, 3155 (1972)
55 IP, J. K. K., and BURNS, G., *Journal of Chemical Physics*, **51**, 3425 (1969)
56 WONG, W. H., and BURNS, G., *Journal of Chemical Physics*, **58**, 4459 (1973)
57 WONG, W. H., and BURNS, G., *Journal of Chemical Physics*, **59**, 2974 (1973)
58 SHUI, V. H., and APPLETON, J. P., *Journal of Chemical Physics*, **55**, 3126 (1971)
59 SHUI, V. H., *Journal of Chemical Physics*, **57**, 1704 (1972)
60 KECK, J. C., *Advances in Chemical Physics*, XIII, 85 (1967)
61 ANDERSON, J. B., *Journal of Chemical Physics*, **58**, 4684 (1973)
62 JAFFE, R. L., HENRY, J. M., and ANDERSON, J. B., *Journal of Chemical Physics*, **59**, 1128 (1973)
63 BECKER, C. H., CASAVECCHIA, P., and LEE, Y. T., *Journal of Chemical Physics*, **69**, 2377 and **70**, 2986 (1978)

Chapter 6

Detailed rate constants

6.1 General introduction

The rate constant for a bimolecular reaction describes the result of collisions between thermally equilibrated reagents without any specification of the states or relative velocities of the reagents or products. In terms of the nomenclature that was summarized in *Figure 1.4*, it is related to the fully detailed rate constants for processes connecting defined reactant and product states (denoted by n and n′) by:

$$k(T) = \sum_{\mathrm{n}} f_{\mathrm{n}} \sum_{\mathrm{n}'} k(\mathrm{n}' \mid \mathrm{n}; T) \qquad (6.1)$$

In this chapter, we shall chiefly be concerned with studies that provide, at least partly, detailed rate constants.

At the present time, these experiments can be separated into two categories which are almost mutually exclusive. First, there are investigations where the aim is to discover the distribution of product species over the internal states which are accessible energetically. These experiments are usually carried out with thermal reagents, in which case they yield absolute, or relative, values of $k(\mathrm{n}' \mid ; T)$. The distribution of product translational energies can only be measured in molecular beam experiments. Discussion of the techniques and results of reactive scattering measurements is held over to Chapter 7. In the second type of experiment, the Boltzmann distribution of reactant energies is perturbed and the reaction rate is measured for different states of the reagents in order to find values of $k(\mid \mathrm{n}; T)$. As well as disturbing the distribution over internal quantum states, it is possible, even in 'bulb' experiments, to produce reactants with a non-thermal velocity distribution and hence to discover something about the excitation function for a reaction.

The methods that are used to determine detailed kinetic data in 'bulb' experiments are based on the techniques that were described in Chapter 5. A general difficulty, which does not arise when only the thermal rate is being measured, is that it is necessary to account for the effect of relaxation processes, which can destroy the non-Boltzmann distribution that one is trying to measure or create. Although they are related, the problems in the two categories of experiment are somewhat different. When a product state

distribution is being studied, relaxation can occur only in collisions *after* the excited species has been formed. By working at a low total pressure, or with sufficient time resolution, relaxation can be eliminated or allowed for. On the other hand, when a reagent is selectively excited, it may be removed in both inelastic and reactive collisions with the same species at comparable rates. The relative importance of these two competing processes can only be discovered by observing the formation of a chemical product and relating its concentration to that of the energized species.

6.2 Experimental methods: excitation of products[1—4]

6.2.1 Steady-state chemiluminescence[5,6]

The excited products of a chemical reaction can frequently lose all or part of their excess energy by radiating spontaneously. If this occurs, it may be possible to deduce how this product is distributed over its excited states from the emitted spectrum. Experiments of this kind are usually carried out in a steady-state mode, the reagents being mixed in a discharge–flow system. The distribution of reaction products over excited states is determined in two stages. For the first, one needs to know the transition probabilities for individual lines in the emission, as well as how the sensitivity of detection varies with wavelength. Then the relative excited state populations can be calculated. Having done this, it may be necessary to allow for relaxation processes, and it is in this second stage that most error or uncertainty is likely to be found.

Electronically excited species are formed in numerous association reactions, such as

$$NO + O(+M) \rightarrow NO_2(\tilde{A}^2B_1) \rightarrow NO_2(\tilde{X}^2A_1) + h\nu \qquad (6.2)$$

as well as in some atom-transfer reactions, like

$$NO + O_3 \rightarrow NO_2(\tilde{A}^2B_1) \rightarrow NO_2(\tilde{X}^2A_1) + h\nu \qquad (6.3)$$

Usually, electronically excited molecules emit visible or ultraviolet chemiluminescence. Some of these systems have been studied for many years. For example, there are now over a thousand papers on 'active' nitrogen, most of them being concerned with the electronic transitions which are emitted when nitrogen is passed through an electrical discharge. Impetus to research on electronic chemiluminescence has been provided recently by the search for lasers that produce visible or ultraviolet radiation and which derive their energy from exothermic chemical reactions.

Of course, observations of chemiluminescence cannot themselves provide any information about the production of unexcited species, and frequently it is difficult to estimate accurately the relative rates of reaction into excited and unexcited electronic states. On the other hand, because the radiative lifetimes of electronically excited species are short—for a fully allowed transition, 10 ns

is typical—it is frequently possible to carry out experiments at sufficiently low total pressures for most excited molecules to radiate before they undergo any collisions. Then the spectrum reveals a distribution over rovibrational levels in the emitting state that is completely unrelaxed, and therefore characteristic of the mode of formation. Observations of this kind have done much to unravel the complex network of non-adiabatic radiationless transitions that connect the excited states of many simple molecules[6].

The study of chemiluminescence from *vibrationally* excited molecules is dealt with here in greater detail, since the dynamics of electronically *adiabatic* reactions are emphasized in this book. Observations of infrared chemiluminescence (IRC) have provided much of the information that is available concerning the release of energy in exoergic reactions. The experiments are similar to those in which electronic emission is observed. The reagents are mixed, preferably at low total pressures(~ 0.01 N m^{-2} or 0.1 mTorr), in a vessel which is equipped to gather as much as possible of the emission. In the White cell illustrated in *Figure 6.1*, the light is gathered using adjustable mirrors at each end of the cell. The collected radiation exits through a gap between the two mirrors at one end and passes into a spectrometer. Interferometers are increasingly being used to record the emission spectra, since they are much more sensitive than grating spectrometers providing the same resolution.

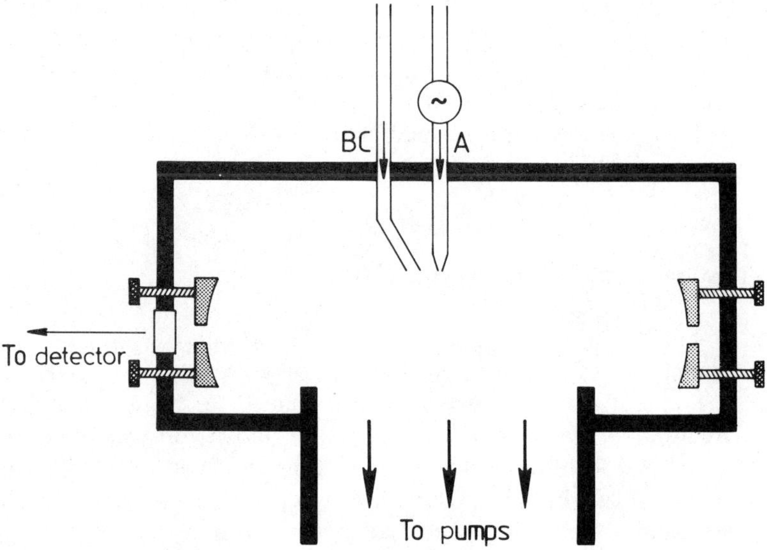

Figure 6.1. Schematic diagram of an apparatus for chemiluminescence experiments. The atomic reagent (A) *is produced in a microwave discharge and mixes with the molecular reagent* (BC) *below the inlet nozzles. The reaction zone may be surrounded with liquid* N_2 *cooled walls (not shown) and the vessel is rapidly pumped. The radiation is collected by the mirrors and passes into the spectrometer through the window at one end of the cell [adapted from Polanyi, J. C., and Woodall, K. B.,* Journal of Chemical Physics, **57**, 1574 (1972)]

Obviously, IRC experiments yield no information about the excitation of either homonuclear diatomic molecules or those modes in polyatomic molecules which are not infrared-active. The great majority of studies have been on reactions producing diatomic hydrides, particularly HF, HCl and OH, because their infrared spectra are easy to resolve and their fundamental bands lie at wavelengths less than 3.5 μm, for which the most sensitive infrared detectors are available. However, recent developments have heralded the much wider application of the IRC method. For example, emission has been observed out to 16 μm from molecules as large as cyclooctanone[7].

A general steady-state equation can be formulated to relate the populations in individual levels ($N_{n'}$), which are derived from the emission spectrum, to the relative rates of chemical reaction producing molecules in these levels ($R_{n'}$). For concentrations in specified *vibrational* levels, this equation is

$$R_{v'} = N_{v'}\tau^{-1} + N_{v'} \sum_{m=1}^{m=v'} A_{v', v'-m} - \sum_{m=1}^{m_{max}} A_{v'+m, v'} N_{v'+m}$$

$$+ \text{ gas-phase collisional deactivation processes}$$
$$+ \text{ wall deactivation} \tag{6.4}$$

where τ is the mean time each molecule spends in the field of view of the spectrometer, and the second and third terms on the right-hand side represent the rates of spontaneous emission *from* the level v to all lower levels and *to* the level v from all higher levels, the A's being Einstein coefficients. In principle, values of $R_{v'}$ can be found by solving the equation for each vibrational level. In general, however, $R_{v'}$ will be the small difference between two large terms, for example, $N_{v'} \sum_{m=1}^{m=v'} A_{v', v'-m}$ and $\sum_{m=1}^{m_{max}} A_{v'+m, v'} N_{v'+m}$, so this procedure is hazardous. It is better to conduct experiments at a low total pressure so that excited molecules reach the walls in a time (τ_w) much shorter than their radiative lifetime, and before they have suffered more than a few collisions. This is usually referred to as the 'arrested relaxation' method. If complete deactivation occurs at the walls, the $R_{v'}$ are directly proportional to the $N_{v'}$, since equation (6.4) reduces to

$$R_{v'} = N_{v'}\tau_w^{-1} \tag{6.5}$$

For equation (6.5) to be valid, the reaction products must undergo some collisions in the region where the 'sprays' of reactants meet. Otherwise, τ_w^{-1} depends on the initial translational energy of the excited molecule and hence, because of energy balance, on the internal states of the products. Evidence that vibrational relaxation is insignificant comes from observing the line intensities within each vibrational band. If these indicate a non-Boltzmann rotational distribution, it should be safe to assume that the observed vibrational distribution is the nascent one, since rotational relaxation is usually much more rapid than vibrational relaxation in collisions between the same pair of molecules.

To summarize, IRC experiments have already provided extremely detailed

information about the release of energy in fast exoergic reactions producing diatomic hydrides. The relative rate of reaction into the lowest vibrational level is not measured directly, but it can generally be inferred from the distribution of relative rates into higher levels. Although the method has been applied much less to product molecules other than the diatomic hydrides, this deficiency is certain to be remedied in the near future.

6.2.2 Kinetic spectroscopy

An inherent weakness of the experiments that were described in the previous section is that they lack resolution in 'real time'. In one variation of the IRC technique, this is introduced by observing emission at different points along a flow tube. In these 'measured relaxation' experiments, $R_{v'}$ are determined by extrapolating the observed vibrational distributions—which become successively more relaxed—back to the point where the reagents mix. However, the time resolution of this method is only ~ 1—10 ms compared with a value of ~ 10 μs which can be obtained by conventional flash photolysis techniques.

There have been only a few attempts to observe the product states of flash-initiated reactions by time-resolved infrared chemiluminescence. Some measurements have been made using kinetic absorption spectroscopy, the vibrational distribution of a product being deduced from its banded electronic spectrum observed shortly after the chemical reaction has been initiated. The reaction

$$O(^3P) + CS_2 \rightarrow SO + CS \qquad (6.6)$$

has been investigated in this way, and *Figure 6.2* shows part of the photographic record that was obtained[8]. The $O(^3P)$ atoms were created by photolysis of NO_2 at wavelengths above 300 nm; absorption bands from vibrationally excited levels of both CS and SO are clearly visible at short delays, although SO quickly disappears through a rapid reaction with undissociated NO_2.

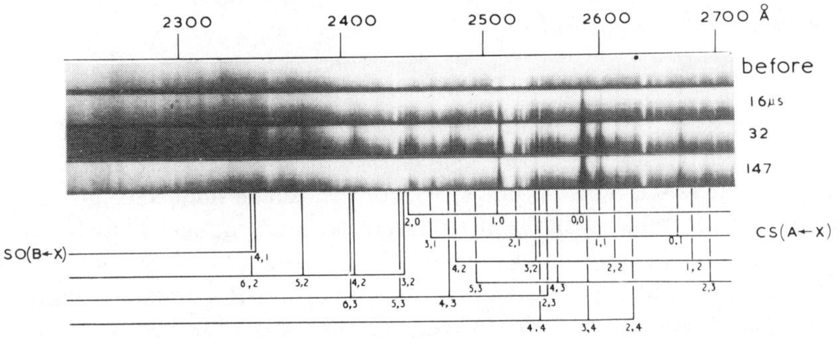

Figure 6.2. Formation of vibrationally excited SO *and* CS *in* $O + CS \rightarrow SO + CS$. *A mixture containing* 8.1×10^{15} *molecule* cm^{-3} NO_2, 4.1×10^{16} *molecule* cm^{-3} CS_2 *and* 2.6×10^{18} *molecule* cm^{-3} Ar *was irradiated with the output from a 2500* J *flash* [*adapted from Smith, I. W. M.*, Discussions of the Faraday Society, **44,** 194 (1976)]

Although this technique was used in pioneering work around 1960, its application has been limited, because several conditions must be satisfied simultaneously if the experiments are to provide anything other than semi-quantitative information. The main requirements are:

(1) a convenient photochemical precursor, say ABC, for the reactive atoms (A), with BC preferably unreactive;
(2) a reagent (XYZ) which, if it is not ABC, should be unaffected by the flash;
(3) products of $A + XYZ \rightarrow AX + YZ$ that possess banded spectra that are unobscured by other absorptions—for example, that of ABC;
(4) if ABC and XYZ are different, they must not react.

The difficulties are compounded by the insensitivity of photographically recording discrete spectra at moderate resolution. Excited molecules have not been observed via resonance fluorescence or resonance absorption in experiments of this kind, but this development will surely take place as tunable laser sources for ultraviolet wavelengths become increasingly available.

6.2.3 Chemical lasers[9,10]

Although there have been rather few studies of *spontaneous* emission from the vibrationally excited products of flash-initiated reactions, observations of *stimulated* emission have recently provided data about the product state distributions from some elementary chemical and photochemical processes. To explain the principle of these experiments, it is easiest to begin by considering a model system of just two levels of energy ε_n, ε_m and degeneracy g_n, g_m, which are resonant at the frequency $\nu_{nm} = (\varepsilon_n - \varepsilon_m)/h$. If radiation of this frequency is passed through a thermal sample, since $(N_n/g_n) < (N_m/g_m)$, it will be attenuated, because the rate of (stimulated) absorption, which is proportional to (N_m/g_m), will exceed that of stimulated emission, which is related via the same proportionality constant (the Einstein coefficient B) to (N_n/g_n). If, however, the medium is perturbed so that $(N_n/g_n) > (N_m/g_m)$, the radiation will be amplified. This is the basis of *laser action.*

Because the products of exoergic chemical reactions are not initially in equilibrium with their surroundings, the potential for converting chemical energy into laser photons clearly exists. However, even with the rovibrational states of a diatomic molecule, we are not dealing with a simple two-level system. Each fundamental vibrational transition $v \rightarrow v-1$, is accompanied by a change in j, either $j' (= j'' + 1) \rightarrow j''$ in an *R*-branch transition or $j' (= j'' - 1) \rightarrow j''$ in a *P*-branch transition. For small molecules, any rotational (and translational) excitation is degraded much more rapidly than vibrational excitation. Consequently, with a chemically inert diluent present, rotational equilibration can usually be assumed, and the inversion condition:

$$\frac{N_{v,j'}}{(2j'+1)} > \frac{N_{v-1,j''}}{(2j''+1)} \tag{6.7}$$

becomes

$$\frac{N_v(2j'+1)\exp(-\varepsilon_{j'}/k_B T)}{(2j'+1)Q_{R,v}} > \frac{N_{v-1}(2j''+1)\exp(-\varepsilon_{j''}/k_B T)}{(2j''+1)Q_{R,v-1}}$$

i.e.,

$$\frac{N_v \exp(-\varepsilon_{j'}/k_B T)}{Q_{R,v}} > \frac{N_{v-1}\exp(-\varepsilon_{j''}/k_B T)}{Q_{R,v-1}} \tag{6.8}$$

where $\varepsilon_{j'}$ and $\varepsilon_{j''}$ are the rotational energies of the upper and lower levels, and $Q_{R,v}$ and $Q_{R,v-1}$ are rotational partition functions for the two vibrational levels.

The amplification brought about by an inversion of population is usually expressed as the *gain coefficient* (α_v), which is defined in terms of the increase in intensity of the radiation at a particular frequency, per unit length of the medium,

$$dI_v = I_v \alpha_v \, dl \tag{6.9}$$

For a spectral line whose profile is determined by Doppler broadening, the gain coefficient at the centre of the line is given by

$$\alpha^0_{v,j'\to v-1,j''} = (1/8\pi k_B T)(m/2\pi k_B T)^{\frac{1}{2}}\lambda^{0^3} F_{j',j''} A_{v,v-1} \times \{N_v B_v \exp(-\varepsilon_{j'}/k_B T) - N_{v-1} B_{v-1} \exp(-\varepsilon_{j''}/k_B T)\} \tag{6.10}$$

where $\lambda^0 = c/\nu^0$ is the wavelength at the line centre, $F_{j',j''}$ is a line strength factor, which is, to a good approximation, equal to j″ for a *P*-branch line and j″ + 1 for an *R*-branch transition, $A_{v,v-1}$ is the spontaneous emission coefficient for the band, and B_v and B_{v-1} are rotational constants in J molecule^{-1}. The concentrations N_v and N_{v-1} can be evaluated if α^0 can be measured accurately for different lines within the same band.

With a tunable, narrow linewidth, continuous wave (cw), infrared laser, absolute values of α^0 could be determined for many lines in the vibrational spectrum of a molecule, just after it had been formed in a reaction initiated by flash photolysis. Unfortunately, suitable 'universal' lasers are not available at the present time, so 'small signal gain' measurements have been confined to a few special cases. The basis of one such experiment is illustrated in *Figure 6.3*(a). A cw laser that emits many lines in the fundamental bands of CO can be constructed fairly easily. Using a reflectance diffraction grating as one component of the optical cavity, individual lines can be selected. This radiation is passed through the reaction mixture and the amplification (or attenuation) is observed, as excited CO is produced and subsequently relaxes. The experiment

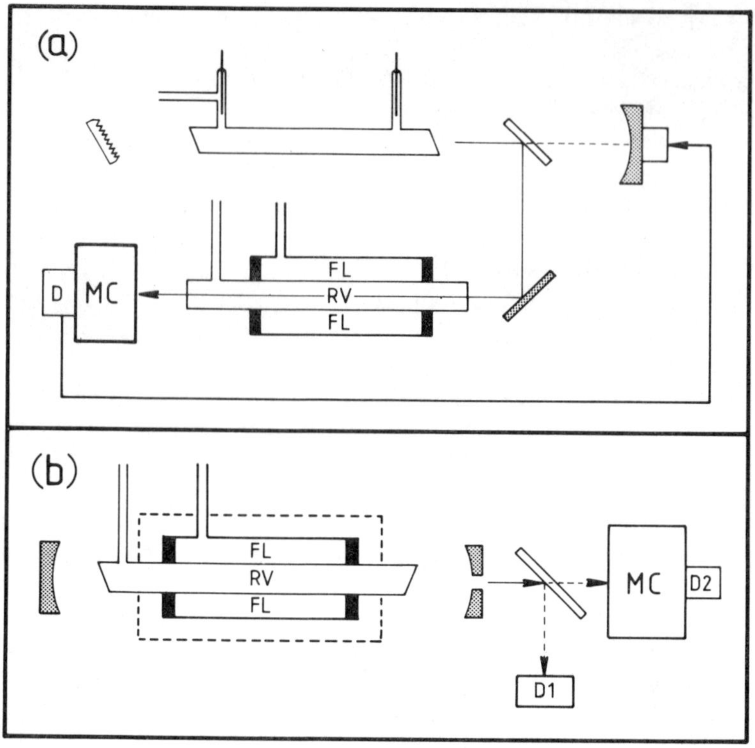

Figure 6.3. Laser methods for studying product distributions. Reaction is initiated using a flash lamp (FL) *concentric with the reaction vessel* (RV). *In* (a) *a selected, stabilized line from the cw laser is sent through* RV *and the attenuation or amplification is measured as a function of time. In* (b), *the reaction system is the laser medium; the total output and that from individual transitions, selected with the monochromator* (MC), *are observed with fast detectors* (D1 *and* D2). *The reaction vessel assembly can be enclosed in an oven* (*represented by the dashed line*)

is repeated using different laser lines and equation (6.10) is used to calculate the CO vibrational populations at various times. This method has, for example, been used to find values of $R_{v'}$ for the reaction[11]

$$O(^3P) + CS \rightarrow CO(v') + S \qquad (6.11)$$

In the experiment that has just been described, the reaction mixture serves as a laser *amplifier*, since an external laser provides the primary radiation. Information about product state distributions can also be obtained by using the reaction mixture as the medium for a laser *oscillator*. The reaction vessel is now placed within an optical cavity, which, in the simplest case, is formed by two parallel mirrors as shown in *Figure 6.3*(b). If oscillation is to occur, besides an inversion of population, the optical gain created by the inversion must exceed the losses suffered as the radiation travels back and forth along the axis of the optical cavity. These losses can arise from imperfections in the optical

components, as well as from the need to couple radiation out of the cavity; for example, by making one of the mirrors partly transmitting.

It might seem as if laser oscillation can only occur when the reaction creates a *total* inversion between neighbouring vibrational levels, i.e., $N_v > N_{v-1}$. In fact this is not so. For *P*-branch lines $\varepsilon_{j''} > \varepsilon_{j'}$, so that α^0 can be positive even when N_v/N_{v-1} is appreciably less than unity. Oscillation can frequently be achieved on such a *partial* inversion, if care is taken to minimize the cavity losses; for example, by mounting the laser mirrors within the reaction vessel.

Several techniques have been used to extract product state distributions from the output of chemical lasers. Not all of these are described here. The most straightforward but most limited method employs the apparatus shown in *Figure 6.3*(b). Reaction is initiated by flash photolysis of a preprepared gas mixture, and one determines which line is the first to emit laser radiation. Since the cavity is comprised of two mirrors, the cavity losses are the same for all potential laser transitions, so it is reasonable to assume that the one which oscillates first has the highest gain. The relative magnitudes of α^0 for lines within a given band depend on the ratio N_v/N_{v-1}. This single observation allows one to calculate upper and lower limits for N_v/N_{v-1}, for the vibrational levels connected by the observed first laser transition. In the 'equal-gain temperature technique', this method is extended by making measurements at different temperatures. At a certain temperature, two lines with adjacent values of j′ within the same band will reach threshold simultaneously and before all other transitions. The expressions for the gain on these two transitions can then be equated, and the value of N_v/N_{v-1} calculated for the pair of vibrational levels connected by these transitions.

In using equations (6.8) and (6.10) to interpret the results of chemical laser experiments, it is assumed that rotational relaxation is rapid, but that vibrational relaxation does not occur in the few microseconds before laser radiation is first emitted, so that N_v/N_{v-1} corresponds to R_v/R_{v-1}. Because the stimulated emission process itself disturbs the vibrational distribution, no information can be obtained by observing later laser transitions. One way around this limitation is to use the 'grating selection method'. One of the laser mirrors is replaced by a diffraction grating, so that the cavity losses are small only for a selected narrow band of wavelengths. The time that it takes a wide range of individual transitions to reach threshold can then be observed, in the absence of other stimulated emission processes. These delay times are related to the gain coefficients and precise values of N_v/N_{v-1} can be obtained for all the vibrational levels that are significantly populated by the reaction.

Because stimulated emission is generally more sensitive in detecting excited molecules than low-resolution spectrography, flash-initiated chemical laser studies have been rather more successful than kinetic spectroscopy in determining accurate vibrational distributions. However, they have, so far, only been applied to hydrogen halides, although it is additionally possible to investigate photoelimination processes, such as

$$CH_2{=}CHCl + h\nu \rightarrow CH{\equiv}CH + HCl(v') \qquad (6.12)$$

One advantage which laser methods possess over IRC experiments is that $R_{v=0}$ is found directly, if $N_{v=1}/N_{v=0}$ is observed experimentally. On the other hand, no information is obtained about the distribution of product species over rotational levels.

6.3 Energy disposal in chemical reactions[2–4]

6.3.1 The reaction: $F + H_2 \rightarrow HF + H$[12–18]

The $H + H_2$ reaction and its isotopic variants were accorded special treatment in Chapter 5. It was emphasized that their importance lies in the possibility—now essentially realized—of obtaining rate constants from *a priori* calculations. However, because $(-\Delta H_0^\circ + E_{act})$ is appreciably less than the excitation energy of $H_2(v = 1)$, the overwhelming majority of the molecules formed in the $H + H_2$ reaction, under thermal conditions, must be in $v' = 0$. The energy released after the potential barrier to reaction has been crossed can only be shared between product rotation and relative translation, and the rotational distribution has not yet been observed. Because of these factors, these reactions do not provide a fertile testing ground for comparisons between experimental and theoretical estimates of detailed rate constants. Consequently, in the area of reaction dynamics, the leading position has been usurped by the reactions shown in *Table 6.1*.

Table 6.1

	ΔH_0°/kJ mol^{-1}	
$F + H_2 \rightarrow HF + H$	−133.5	(6.13)
$F + D_2 \rightarrow DF + D$	−133.0	(6.14)
$F + HD \nearrow HF + D$	−130.1	(6.15a)
$F + HD \searrow DF + H$	−137.2	(6.15b)

From a *theoretical* standpoint, this system remains sufficiently simple electronically for *ab initio* calculations of the lowest potential energy hypersurface to be performed with respectable accuracy; the surface shown in *Figure 2.4* is the result of one such calculation[19]. Unfortunately, the very feature which makes studies of energy disposal informative, namely the existence of a fairly large number of energetically accessible product states, also complicates any quantum scattering calculations, since all of these states, and more, must be included in the basis set. On the other hand, there is good reason to suppose that quasiclassical trajectories provide an adequate description of the collision dynamics and the results of several such studies will be mentioned later in this section.

From an *experimental* standpoint, the reactions (6.13) to (6.15) in Table 6.1 are especially suitable for spectroscopic investigation, because HF and DF

Table 6.2 Energy disposal in the $F + H_2$, D_2, HD reactions [a,b]

Reaction	Method	$k(v' \mid ;T)/k(T)$					$\langle f_{V'} \rangle$	$\langle f_{R'} \rangle$	$\langle f_{T'} \rangle$	$-\lambda_{V'}$	$\frac{\Delta S_{V'}}{\text{J mol}^{-1}\,\text{K}^{-1}}$
		$v' = 0$[c]	1	2	3	4					
$F + H_2 \rightarrow$	IRC	(0.05)	0.17	0.53	0.25	–	0.63	0.08	0.29	6.0	10.9
$HF + H$	IRC	(0.03)	0.15	0.53	0.29	–	0.65	0.08	0.27	6.5	12.1
	CL	0.03	0.15	0.50	0.32	–	0.66	–	–	6.6	12.6
$F + D_2 \rightarrow$	IRC	(0.03)	0.10	0.24	0.37	0.26	0.62	0.08	0.30	6.0	10.0
$DF + D$	IRC	(0.02)	0.07	0.23	0.43	0.25	0.64	0.08	0.28	6.6	11.3
	CL	0.04	0.11	0.24	0.44	0.17	0.59	–	–	5.4	8.8
$F + HD \rightarrow$	IRC	(0.08)	0.19	0.64	0.09	–	0.56	0.13	0.31	4.8	8.4
$HF + D$	CL	0.04	0.21	0.65	0.10	–	0.58	–	–	4.8	9.2
$F + HD \rightarrow$	IRC	(0.03)	0.08	0.22	0.42	0.25	0.62	0.07	0.31	6.1	10.0
$DF + H$	CL	0.06	0.12	0.27	0.43	0.12	0.53	–	–	4.4	6.3

[a] All the data are taken from HOLMES, B. E., and SETSER, D. W., in *Physical Chemistry of Fast Reactions*: vol 2: *Reaction Dynamics*, Ed. Smith, I. W. M., Plenum Press, New York and London (1979)

[b] The average energy available for disposal is $\langle E \rangle = -\Delta H_0^\circ + E_{act} + 5/2RT$.

[c] The values in parentheses are obtained using a surprisal analysis on the values of $k(v' \mid ;T)$ with, $v' > 0$.

have strong vibrational transitions giving rise to easily resolved bands at short infrared wavelengths. Consequently, these reactions, particularly (6.13), have been thoroughly studied by both infrared chemiluminescence (IRC) and chemical laser (CL) techniques. Whereas IRC experiments[12, 13], conducted under conditions of arrested relaxation, have yielded product distributions over j′, as well as v′, for $v' > 0$, measurements using the grating selection version[15] of the CL method have established the complete distribution over v′, including $v' = 0$. Reaction (6.14) has also been studied in molecular beam experiments, but discussion of these results is deferred until Section 7.5.1.

Information about product energy distributions can be presented in several ways. The first part of Table 6.2 lists the fractions of HF or DF produced in specified vibrational levels, when reaction occurs between thermally equilibrated reagents at $T = 300$ K. The second part gives the fraction of energy which, on average, is released as vibration ($\langle f_{V'} \rangle$), as rotation ($\langle f_{R'} \rangle$), and as translation ($\langle f_{T'} \rangle$). The agreement between the results from IRC and CL experiments is excellent.

No indication is given in Table 6.2 of how molecules are distributed over the rotational states within each vibrational level. *Figure 6.4* illustrates one way in which these results can be displayed. Relative values of $k(v', j' |; T)$ are plotted as vertical bars as a function of $f_{V'}$ and $f_{R'}$. Frequently, this diagram is simplified by assuming that the detailed rate constants are points lying on a continuous surface, whose topography can then be represented by contour lines. The result is shown in *Figure 6.5*. Comparison of the two triangular plots indicates that differences in the distributions from the two isotopically related reactions are essentially predictable. They may be thought of as arising from the product quantum states sampling different parts of a (hypothetical) classical energy distribution represented by the contours. This increases one's

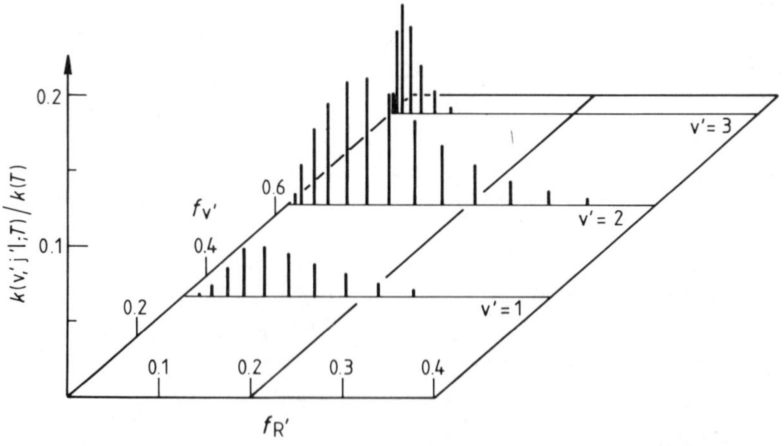

Figure 6.4. 'Stick' diagram showing the distribution over v′, j′ states of the HF *produced in* $F + H_2 \rightarrow HF + H$. *These results were obtained from* IRC *experiments, so no information was obtained about the populations in* $v' = 0$

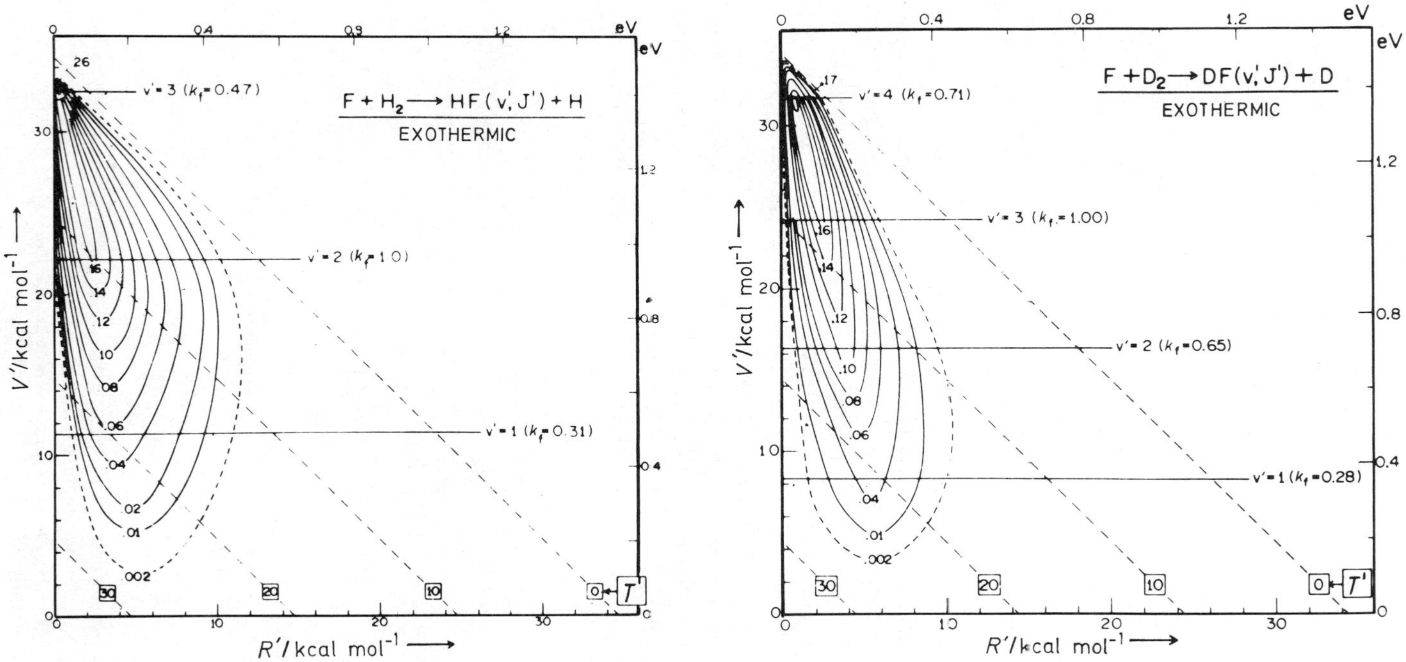

Figure 6.5. Triangular plots showing product energy distributions. The contour lines join equal values of the detailed rate constants $k_f(v', j'|; T)$, *ignoring quantization and with normalization to* $k_f(v'|; T) = 1.0$. *Vibrational and rotational energies* (V' *and* R') *are plotted along rectlinear axes* (1 eV ≡ 23.05 *kcal/mole* ≡ 96.5 kJ mol^{-1}). *Because the energy released is approximately constant, values of the final relative translational energy can be represented by the dashed lines [reproduced, with permission, from Polanyi, J. C., and Woodall, K. B.,* Journal of Chemical Physics, **57**, 1574 (1972)]

confidence in the use of quasiclassical trajectories to investigate reaction dynamics, since this is exactly the procedure which is then applied to find a quantum state distribution from the calculated classical one.

A diagrammatic representation of detailed rate constants is appealing, because a wealth of information is reduced to a single picture. Alternatively, the data can be compressed by comparing detailed rate constants with a prior distribution predicted solely on the basis of phase-space arguments. Figure 6.6 shows that the surprisal associated with the vibrational distributions from the $F + H_2$, D_2, HD reactions varies nearly linearly with $f_{v'}$[13, 15] and that $\lambda_{v'}$, given by the slopes of these lines (equation 4.129), is almost the same for each reaction. These plots not only serve to compact the observed results, but they also provide a basis for obtaining information that cannot be measured directly. For example, surprisal plots are the source of the bracketed values of $k(v' = 0|; T)/k(T)$ that are given in Table 6.2.

When a surprisal plot is linear, the deviation of the distribution from prior expectation can be characterized in terms of the values of $\lambda^0_{v'}$ and $\lambda_{v'}$ in equation (4.129). The average value of the deviation can also be calculated and is known as the entropy of the distribution. The *entropy deficiency* of a vibrational distribution is defined by[20]

$$\Delta S_{v'} = S^0_{v'} - S_{v'}$$
$$= R\sum P(f_{v'})\ln[P(f_{v'})/P^0(f_{v'})] \qquad (6.16)$$

The greater the specificity of energy disposal into a particular degree of freedom, then the larger is the associated entropy deficiency. For the $F + H_2$ reaction, the vibrational entropy deficiency is 12.7 J mol^{-1} K^{-1}.

For reactions (6.13) and (6.15), $\langle f_{R'} \rangle$ is only 0.08, corresponding to $\langle E_{R'} \rangle \approx 12$ kJ mol^{-1}. It was pointed out in Section 3.54 that the need to conserve total angular momentum throughout a molecular collision generally restricts the yield of product rotational energy. In these reactions, as the potential hypersurface is repulsive, rotational excitation must result from the release of energy as the products separate. The extent to which this can channel energy into rotation is limited by two special features. The first is that systems passing through the transition state have a nearly linear configuration so the repulsion produces little torque on the molecular product. Secondly, the initial and final orbital momenta, i.e.,

$$1 = \mu v b = (2\varepsilon_T \mu)^{\frac{1}{2}} b$$

and

$$1' = \mu' v' b' = (2\varepsilon_{T'} \mu')^{\frac{1}{2}} b' \qquad (6.17)$$

are relatively small, because both μ and μ' are small. To some extent, these factors are opposed by the ability of HF and DF, with their low moments of inertia, to possess large amounts of rotational *energy* for relatively modest values of j'.

Figures 6.4 and *6.5* show clearly that the distributions over j' depend

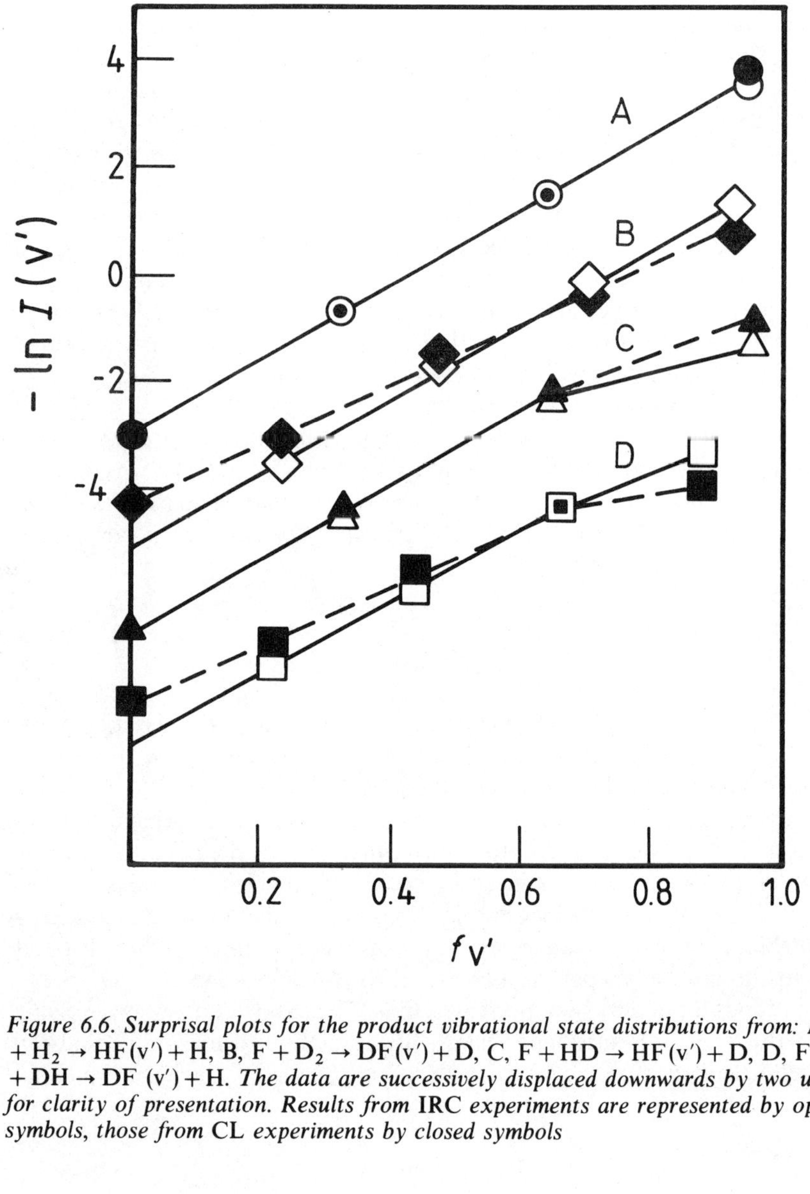

Figure 6.6. Surprisal plots for the product vibrational state distributions from: A, $F + H_2 \rightarrow HF(v') + H$, B, $F + D_2 \rightarrow DF(v') + D$, C, $F + HD \rightarrow HF(v') + D$, D, $F + DH \rightarrow DF(v') + H$. *The data are successively displaced downwards by two units for clarity of presentation. Results from* IRC *experiments are represented by open symbols, those from* CL *experiments by closed symbols*

strongly on v′. In view of the connection between the value of j′ and that of the relative translational momentum of the products

$$j' = J - \mu' v' b' \tag{6.18}$$

this result is scarcely surprising. It suggests the concept of a *conditional* distribution[20], which is used in the information–theoretic analysis. With knowledge of the product vibrational distribution, it is natural to express the

variation of the surprisal associated with rotational distributions in a specified v′ level in terms of the reduced variable $f_{R'}/(1-f_{v'})$.

If the reduced energy of the most populated rotational state in each product vibrational level is denoted by $f_{R'}^{mp}$, it is found that[21]

$$f_{R'}^{mp}/(1-f_{v'}) \approx 0.21 \tag{6.19}$$

for all v′ levels of the HF and DF produced in reactions (6.13) and (6.14). In addition, the conditional surprisal of each rotational distribution can be inspected. The details of this examination will not be considered here. Suffice it to say that the actual and prior distributions are similar, as long as in calculating the prior distribution proper allowance is made for the need to conserve angular momentum as well as total energy.

The dependence of the product energy distributions on reagent collision energy and on reagent rotational state has also been studied in IRC[13, 14] and CL[16,17] experiments. There is not, at the present time, complete agreement between the results obtained by the two techniques. On balance, it would seem that the data from the chemiluminescence studies are to be preferred, since they are less liable to the complications resulting from collisions in which energy is transferred. In these experiments[13] the temperature of the H_2 inlet was varied between 77 and 1315 K. The H_2 rotational distributions corresponded to these temperatures, whilst collision energies were obtained varying from about 100 to 1200 K[a]. The rotational distribution of the reagent molecules has also been altered by carrying out experiments using para-H_2, as well as normal-H_2[14]. Enhancing the reagent translational temperature led to enhanced excitation of the translational and rotational degrees of freedom of the products. The effect of changing the rotational states of H_2 was rather more complicated. It appears that the fraction of energy entering product vibration, $\langle f_{V'} \rangle$, is less from $F + H_2$ (v = 0, j = 1) than from $F + H_2$ (v = 0, j = 0) but then increases for $F + H_2$ (v = 0, j = 2).

The very detailed experimental data which have been obtained for the family of reactions represented by equations (6.13) to (6.15) have inspired a large number of quasiclassical trajectory studies[22]. Some of these were carried out before the accurate, *ab initio*, potential energy calculations. The fact that the main characteristics of the potential were identified by matching the calculated and experimental energy distributions strengthens one's faith in this procedure for obtaining information about the potential for reactive systems. There is now general agreement over several features of the FH_2 hypersurface. First, the barrier to reaction is lowest for collinear configurations of the three atoms. Secondly, this barrier is located at the head of the entrance valley on the potential surface (*see Figure 2.4*), before the reaction path of minimum energy

[a] The effective temperature associated with the relative translation motion is given by $(m_1 T_2 + m_2 T_1)/(m_1 + m_2)$, where T_1 and T_2 are the (different) temperatures of the two reagents and m_1 and m_2 their masses.

'turns the corner'. Thirdly, the surface is highly repulsive: at the point on the *a priori* reaction path of minimum energy where $(r_{HH}/r_{FH}) = (r_{e,HH}/r_{e,FH})$, the potential energy is actually higher than for separated $F + H_2$.

Although the *surface* is highly repulsive, predominantly 'mixed' *energy release* takes place. This is because, for $F + H_2$, D_2, HD, the F atom is heavy compared with the other two. Consequently, once a trajectory has crossed the barrier, the release of energy as repulsion rapidly forces the two light atoms apart, so that much of the total energy is released before the trajectory first reaches a point where $r_{FH} = r_{e,FH}$. This channels a high proportion of the energy released as repulsion between the products into the vibration of the HF molecule.

The product's vibrational energy distribution has served as the main reference point for comparisons between theory and experiment, the potential function being adjusted until the calculated value of $\langle f_{V'} \rangle$ corresponds to the observed value. The calculated vibrational *distribution* has usually been somewhat too narrow. Calculations on a surface designed to match the *ab initio* energy points gave too much vibrational excitation and this result led to the suggestion that this surface may fall too steeply in the important region between the barrier and the exit valley[22]. This example nicely illustrates how dynamical calculations can provide a constructive link between the detailed experimental results for a particular system and features of the potential hypersurface. More use will be made of this kind of comparison, as the quality and range of both the calculated and experimental data improve.

The rotational energy distributions from quasiclassical trajectories are also in fair agreement with those obtained experimentally, although the former do extend to higher j′, particularly for low vibrational excitation. On a triangular plot representing the calculated values of $k(v', j'|; T)$, the 'ridge' of high probability is, therefore, directed more nearly parallel to the hypotenuse than that shown in *Figure 6.5*. In addition, the value of $\langle f_{R'} \rangle$, although still quite small, is about 20–50 per cent above the experimental value.

As mentioned earlier, the $F + D_2$ reaction has been studied in molecular beam experiments[23]. The reaction products are not scattered symmetrically about the angle $\theta = 90$ degrees. This observation, demonstrating that reaction occurs in *direct* molecular collisions, is entirely consistent with the absence of any significantly deep minima on the calculated potential energy hypersurface and with the specificity of energy disposal, as measured, for example, by the large vibrational entropy deficiency. In the sections which follow, some other examples will be considered of energy disposal, first in reactions which occur directly, and then in systems where the collision dynamics are dominated by complex formation. Even if molecular beam experiments have not been carried out, it is usually possible to classify a reaction on the basis of either the energy disposal studies themselves or general chemical knowledge. There are difficulties over this classification in only a few cases. They can arise either if the lifetime of a complex is of the order of a rotational period or when, interestingly, reaction can occur via more than one potential hypersurface, yielding a complex as well as a direct pathway for reaction.

6.3.2 Direct reactions involving three atoms

Although the number of individual reactions which have been studied by IRC and CL techniques is now quite large, they can be divided into a rather small number of categories. Much the largest of these groups can be represented by the equation:

$$X + HY \rightarrow HX + Y \tag{6.20}$$

where X is a halogen atom—usually F—and Y can be either another halogen atom (subsequently denoted by X′) or a larger group. Reactions (6.13) to (6.15) might be considered to belong to this group, but this would ignore the important dynamical differences resulting from Y and HY being much heavier than H (or D) and H_2 (or D_2, HD). The $X + HX'$ reactions are especially important; not only because, with just three atoms, they are the simplest reactions of this type, but also because it is possible to study the influence of reagent vibration on both the total rate and the rates to specific product states. These experiments are considered in Sections 6.6.2 and 6.7.

Several of the

$$X + HX' \rightarrow HX + X' \tag{6.21}$$

reactions are sufficiently energetic to produce the atomic halogen in the excited $^2P_{\frac{1}{2}}$ component of the electronic ground state[4]. Uncertainty about the importance of this process could confuse the interpretation of IRC experiments. However, it seems that $I(^2P_{\frac{1}{2}})$ atoms are not formed in significant amounts when F, Cl or Br atoms react with HI. Weak $Br(^2P_{\frac{1}{2}} - {}^2P_{\frac{3}{2}})$ emission has been detected from the F + HBr reaction and a branching ratio for production of $Br^*(^2P_{\frac{1}{2}})$: $Br(^2P_{\frac{3}{2}})$ of 0.11 has been deduced[24]. Certainly, it seems that $\langle f_{El'} \rangle$ is small for all these reactions.

As the data in Table 6.3 show, $\langle f_{V'} \rangle$, $\langle f_{R'} \rangle$ and $\langle f_{T'} \rangle$ are virtually identical for the Cl + HI and Cl + DI reactions[25]. Furthermore, the relative values of $k(v'|; T)$ when plotted against $f_{v'}$ fall close to a common curve. This agreement supplied the first indication that classical mechanics can provide a satisfactory description of the dynamics of reactive collisions.

As with the reactions in *Table 6.1*, the specific disposal of energy into product vibration in the $X + HX'$ reactions is apparently associated with mixed energy release on a predominantly repulsive surface[26]. Energy is again liberated rapidly once the system has passed through the transition state region, forcing the H atom towards X′ as the H—X′ bond is still extended. However, in contrast to the situation in the $F + H_2$ reaction, X′, being relatively massive, fails to move away rapidly as HX′ is formed. Consequently, as HX′ starts to vibrate, secondary encounters may occur between H and X. Some energy may be transferred, broadening the final vibrational distribution.

The values of $\langle f_{R'} \rangle$ in *Table 6.3* indicate that the extent of rotational excitation in the products of the $X + HX'$ reactions is rather large. *Figure 6.7* shows that the population of rovibrational levels can reach nearly to the energetically accessible limit. Analysis of the distributions from F + HCl and

Table 6.3 Energy disposal in X + HX′ reactions[a]

Reaction	$\langle f_{V'} \rangle$	$\langle f_{R'} \rangle$	$\langle f_{T'} \rangle$	$-\lambda_{v'}$	$\Delta S_{V'}$/J mol^{-1} K^{-1}
$Cl + HI \rightarrow HCl + I$	0.68	0.13	0.19	6.9	13.4
$Cl + DI \rightarrow DCl + I$	0.68	0.13	0.19	6.9	13.4
$Cl + HBr \rightarrow HCl + Br$	0.43	–	–	~3.5	4.6
$Br + HI \rightarrow HBr + I$	0.52	–	–	~4.3	6.7
$F + HCl \rightarrow HF + Cl$	0.53	0.21	0.26	~5	6.3
$F + HBr \rightarrow HF + Br$	0.58	–	–	5.1	7.5
$F + HI \rightarrow HF + I$	0.60	–	–	5.1	7.5

[a] All the data are taken from Holmes and Setser's review. The results are all derived from IRC measurements, with $k(v' = 0|;T)$ estimated from a surprisal analysis.

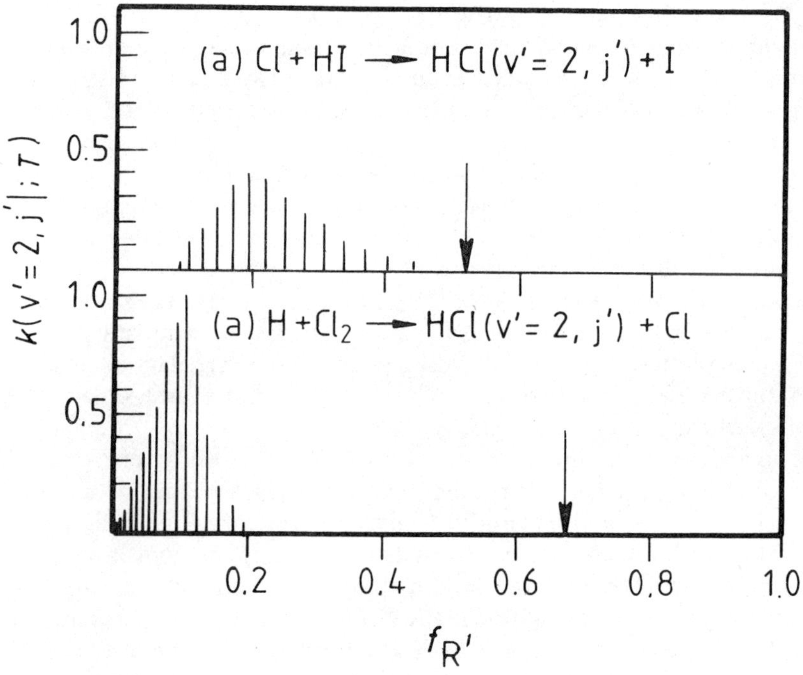

Figure 6.7. Comparison of rotational state distributions. The arrows indicate the positions of the highest energetically accessible levels. The detailed rate constants are normalized to their maximum values: i.e., $k(v' = 4, j' = 5|; T)$ *for* $Cl + HI$ *and* $k(v' = 2, j' = 13|; T)$ *for* $H + Cl_2$

Cl + HI show that these reactions partition the 'available' energy $(1 - f_{v'})\varepsilon_{total}$ almost statistically between the rotational and relative translational degrees of freedom. This behaviour is in marked contrast to that found for F + H_2, D_2, HD and the reasons for this difference are not difficult to identify. One factor is that, on average, the initial orbital momentum is much larger in X + HX′ reactive collisions. As they have large cross sections, these reactions can occur in collisions with rather high impact parameters. Moreover, μ is much greater than the reduced mass associated with F + H_2 collisions. A second important factor is that the force acting between H and X as the products separate will almost entirely increase the momentum of H, rather than X, because of the large mass difference. The component of this force perpendicular to the line joining H to the centre-of-mass of HX (which is, of course, located close to X and some distance from H) will produce a torque exciting HX's rotation. The nett result of these factors is to minimize the dynamical constraints which usually limit the extent of rotational excitation.

Quite different dynamical behaviour from that just described is exhibited by reactions of the type

$$H + XX' \rightarrow HX + X' \qquad (6.22)$$

(where X and X′ may be the same or different halogen atoms). One finding is that the vibrational surprisal plots are markedly non-linear. The vibrational energy yields are also quite low: thus, $\langle f_{V'} \rangle = 0.58$ for XX′ = F_2 and 0.37 for XX′ = Cl_2[27, 28]. These appear to be manifestations of the 'light atom anomaly', that is, of unusual collision dynamics when the attacking atom is much lighter than the other two. One effect is that in a reactive encounter the light H atom is able to get in close to X, before the repulsive forces have had time to separate the relatively massive X and X′ atoms. This results in rather little mixed energy release, i.e., the trajectories tend to follow a rather rectilinear path on the potential surface. Energy which is liberated after the separation of a product molecule has reached its final equilibrium value is *always* channelled very inefficiently into that molecule's vibration. This is especially so with the combination of masses in these reactions, since X′ is pressing on the heavy end of the HX molecule as the products separate. Thus, even if the energy were released impulsively between X and X′ with all three atoms in a line, energy and momentum conservation would limit the fraction of energy deposited in the vibration of HX to $(m_H m_{X'})/(m_H + m_X)(m_X + m_{X'})$. With both X and X′ = ^{35}Cl, this would amount to only (1/72).

Even in non-linear geometries, X–X′ repulsion will produce very little torque on the HX molecule. Moreover, the initial orbital momentum is much less than in X + HX′ collisions. These factors explain the low degree of rotational excitation found in these reactions. Now, as *Figure 6.7* shows, the highest rotational level populated lies some way below the energetic limit.

There have not been many detailed studies of energy disposal in reactions where the 'new' bond formed is that of a non-hydride diatomic molecule. However, the vibrational distributions of CO formed in the reactions of O(^3P) atoms with CS and with CN have been characterized. The results provide

fascinating information about the potential hypersurfaces involved and the role played by spin conservation in reactive collisions.

For the $O(^3P) + CS$ reaction there is only one electronically adiabatic exoergic pathway:

$$O(^3P) + CS(^1\Sigma^+) \rightarrow CO(^1\Sigma^+) + S(^3P);\ \Delta H_0^\circ = -356\ \text{kJ mol}^{-1} \quad (6.23)$$

The reaction is very rapid with $k(T = 300\ \text{K}) = 2.2 \times 10^{-11}\ \text{cm}^3\ \text{molecule}^{-1}\ \text{s}^{-1}$, and the energy released is channelled preferentially into the CO vibration: $\langle f_{V'} \rangle \approx 0.81$ and $\Delta S_{V'} \approx 18.5\ \text{J mol}^{-1}\ \text{K}^{-1}$ [11]. This suggests that (6.23) proceeds directly across an attractive surface with little or no barrier to the reaction. It is *not* consistent with reaction via the $OCS(^1\Sigma^+)$ ground state, which is 'spin-forbidden', since formation of such a collision complex would cause the energy to be shared much less specifically than is observed. Reaction presumably takes place across a triplet hypersurface, apparently without any significant potential hollows in it.

The results for the $O(^3P) + CN$ reaction reveal an even more interesting situation[29]. Once again, the overall reaction is fast—$k(T = 300\ \text{K}) = 2.1 \times 10^{-11}\ \text{cm}^3\ \text{molecule}^{-1}\ \text{s}^{-1}$—but, in contrast to CS, CN is itself a free radical and there are now two exoergic electronically adiabatic pathways:

$$O(^3P) + CN(^2\Pi) \rightarrow CO(^1\Sigma^+) + N(^4S);\ \Delta H_0^\circ = -322\ \text{kJ mol}^{-1} \quad (6.24a)$$

and

$$O(^3P) + CN(^2\Pi) \rightarrow CO(^1\Sigma^+) + N(^2D);\ \Delta E_0 = -91\ \text{kJ mol}^{-1} \quad 6.24b)$$

Channel (a), yielding products both in their electronic ground states, cannot proceed via the $^2\Pi$ lowest state of the NCO radical, whereas (b) can. The observed CO vibrational distribution, which is shown in *Figure 6.8*, is bimodal. It appears that the relatively high populations in the lowest v′ levels (which decrease rapidly) come from (6.24b), the energy being released almost statistically in collisions involving the $NCO(^2\Pi)$ complex, whilst the peaked distribution in higher levels arises from direct reaction according to equation (6.24a). If the contributions from each pathway are unscrambled, two linear surprisal plots are obtained; that for (a) having a much steeper slope (i.e., larger value of $-\lambda_{v'}$) than that for (b). This picture has been beautifully corroborated by measuring the relative yields of $N(^2D)$ and $N(^4S)$ directly by resonance fluorescence. The observations confirm that reaction (6.24a) proceeds five times faster than (6.24b).

6.3.3 Direct reactions involving more atoms[4]

We now consider briefly the reactions represented earlier by equation (6.20) in the case where R is a molecular radical. Infrared chemiluminescence has now been observed from many such reactions. Whenever HR is a saturated molecule, energy is fed into the HF vibration preferentially, indicating that

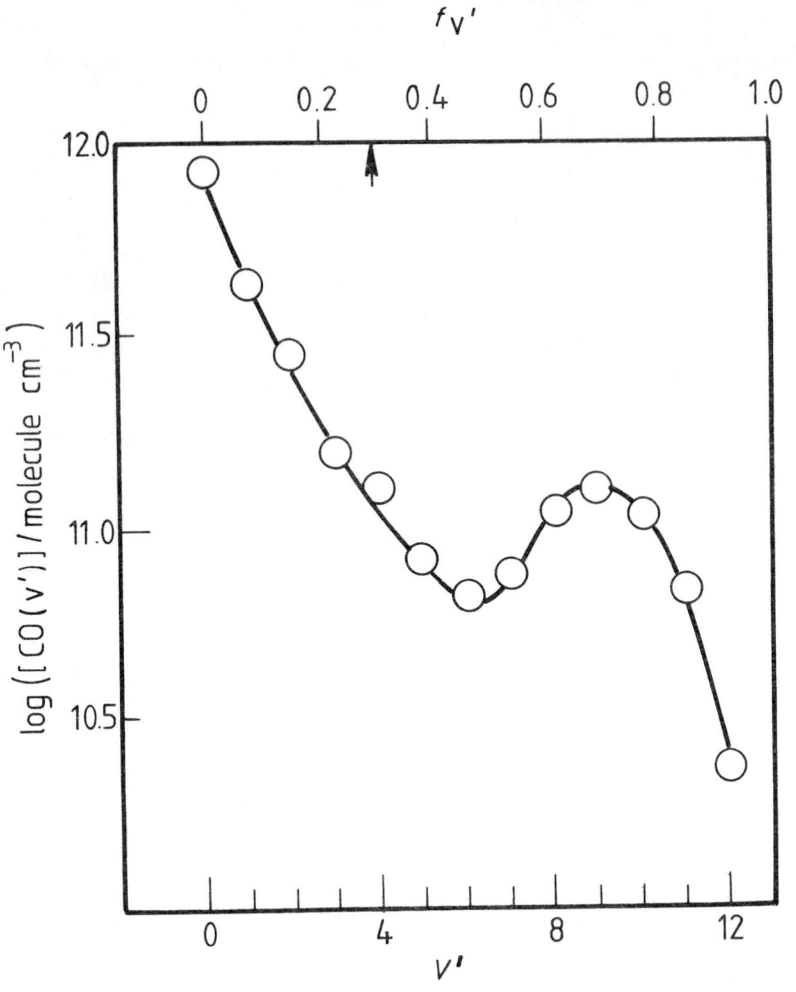

Figure 6.8. Distribution of CO(v′) *from the reaction of* O (^3P) *with* CN. $f_{v'}$ *is calculated assuming* CO + N *are produced in their electronic ground states, whilst the arrow marks the approximate energetic limit for formation of* CO (v′) + N(^2D). *The activation energies for both pathways are assumed to be zero* [*adapted from Schmatjko, K. J., and Wolfrum, J.*, 16th International Symposium on Combustion (1977)]

reaction occurs in direct collisions. As long as the liberated radical R cannot stabilize itself by strengthening its π-bonding, the extent of HF vibrational excitation remains remarkably constant, with $\langle f_{V'} \rangle \approx 0.57 \pm 0.05$, close to the value for the F + HCl reaction. The similarity of $\langle f_{V'} \rangle$ for the triatomic and polyatomic reaction systems is clear evidence that R can absorb little energy in its internal degrees of freedom.

The picture is somewhat altered when the product R is a 'stabilized radical' such as $C_6H_5CH_2$ or HCO. Now the fractional yield of HF vibrational energy is reduced, the highest populated v′ level is below the thermochemical limit by an amount roughly equal to the radical's stabilization energy, and the surprisal plots based on three-atom models are non-linear, unless, in calculating the prior detailed rates, it is assumed that the stabilization energy is not available for distribution. Chemiluminescence from the polyatomic radicals has not been observed. Nevertheless, indirect evidence suggests that the stabilization energy probably goes into the radical's internal modes. Some excitation would be expected as the radical relaxes to a (slightly) more tightly bound structure; this would correspond to the attractive release of energy, not into the newly formed bond, but into the altered bonds of the other product.

The reactions of H atoms with non-hydride triatomic molecules form an interesting sequence. The limited information which is available on energy disposal in these reactions is gathered together in Table 6.4. Although the reactions do not all appear to occur directly, it is sensible to consider their collision dynamics together, in view of their related electronic structure.

Table 6.4 Data on H + XYZ reactions

	ΔH°_0/kJ mol^{-1}			
Reaction	*complex*	*products*	$\langle f_{v'} \rangle$	*Ref.*
$H + CO_2 \rightarrow OH + CO$	~ −50	+102.0	probably low	*a*
$H + NO_2 \rightarrow OH + NO$	−326.0	−123.5	0.23, 0.40	*b, c*
$H + O_3 \rightarrow OH + O_2$	not bound	−322.7	0.90	*c*
$H + NOCl \rightarrow HCl + NO$	not bound	−272.0	0.50	*d*

[a] TRAINOR, D. W., and VON ROSENBERG, Jr., C. W., *Chemical Physics Letters*, **29**, 35 (1974); these authors have failed to detect infrared chemiluminescence from CO_2 formed in the reverse, exothermic reaction and conclude that very little of the exothermicity is released into product vibration

[b] SPENCER, J. E., and GLASS, G. P., *Chemical Physics*, **15**, 35 (1976); SILVER, J. A., DIMPFL, W. L., BROPHY, J. H., and KINSEY, J. L., *Journal of Chemical Physics*, **65**, 1811 (1976)

[c] POLANYI, J. C., and SLOAN, J. J., *International Journal of Chemical Kinetics, Supplement*, **1**, 51 (1975)

[d] NAZAR, M. A., POLANYI, J. C., and SKRLAC, W. J., *Chemical Physics Letters*, **29**, 473 (1974)

The first member of the series

$$H + CO_2 \rightarrow OH + CO \tag{6.25}$$

is endothermic. The proposed profile of potential energy along the reaction path of minimum energy is discussed in Section 5.5.2, where it is suggested that the transition state for the reverse reaction is located after a potential minimum corresponding to the HOCO radical. The unusual form for the reaction path arises because, whereas CO_2 is linear, combination of the molecule with an H atom releases an additional electron into the π-electron

system, with the result that the most stable structure now has the three heavy atoms in a bent configuration, in accordance with Walsh's rules[30]. Attempts to observe infrared chemiluminescence from CO_2, produced in the reverse reaction to (6.25), have been unsuccessful, and it was concluded that the energy yield into the asymmetric ν_3 vibration of CO_2 must be very low. This result appears to be consistent with a 'late' barrier, since then the reaction energy would be released predominantly as repulsion between CO_2 and H as they separate.

In contrast to reaction (6.25), that between H and NO_2

$$H + NO_2 \rightarrow OH + NO \tag{6.26}$$

is exothermic and very fast. The extra electron, compared with CO_2, means that NO_2 is bent. Furthermore, in this system, the reagents and products are all free radicals, in the sense of having unpaired electrons. Interaction of $H(^2S)$ with $NO_2(^2A_1)$ gives rise to two electronic states, $^1A'$ and $^3A'$, whereas $OH(^2\Pi) + NO(^2\Pi)$ correlate with four states, $^1A'$, $^3A'$, $^1A''$ and $^3A''$. Clearly, there are two adiabatic pathways for reaction (6.26) and, of these, the $^1A'$ hypersurface corresponds to the electronic ground state of the nitrous acid molecule. Furthermore, the association reaction

$$OH + NO(+M) \rightarrow HONO(+M);\ \Delta H_0^\circ = -202\ \text{kJ mol}^{-1} \tag{6.27}$$

shows only a slight dependence on temperature, so there is little or no potential maximum separating HONO from OH + NO. Consequently, if reaction (6.26) proceeded exclusively across the $^1A'$ potential via an HONO collision complex, it should result in energy being shared statistically between its products, OH + NO. In practice, it appears that roughly 40 per cent of the energy released enters the OH vibration and that there is little or no internal excitation of the NO. It therefore seems likely that reaction (6.26) proceeds mainly in *direct* collisions over the $^3A'$ hypersurface that connects $H(^2S) + NO_2(^2A_1)$ adiabatically to $OH(^2\Pi) + NO(^2\Pi)$.

The reaction

$$H + O_3 \rightarrow OH + O_2 \tag{6.28}$$

is also rapid, but it gives a remarkably high yield of vibrational energy: OH is produced almost exclusively in levels $v' = 7$, 8 and 9, and $\langle f_{v'} \rangle \sim 0.90$. It seems that, in making the step from $H + NO_2$ to $H + O_3$, the intermediate configurations of the four-atom system become unstable with respect to the pair of diatomic products—but only just[a]. Consequently, the $H + O_3$ reaction proceeds across a highly attractive hypersurface leading to high excitation in the newly formed bond. On the other hand, the reaction of H with NOCl, which has, superficially, a similar electronic structure to O_3, the HCl product is

[a] In his book *Thermochemical Kinetics* (Wiley, New York, 1968), S. W. Benson lists ΔH_f° for HOOO of 0 ± 8 kJ mol^{-1}. This yields a bond strength for HO—OO of ~ 40 kJ mol^{-1}. This value is consistent with a Cl—OO bond strength of (25 ± 8) kJ mol^{-1} and with similarities between the kinetics and thermodynamics of OH and Cl radicals.

less highly excited than the OH from reaction (6.28). The fractional yield, $\langle f_{V'}(HCl) \rangle$, is comparable with those in the reactions of H atoms with diatomic halogens and indicates that the hypersurface for this reaction is not highly attractive.

The reactions of H atoms with NOF, FO_2, ClO_2, F_2O or Cl_2O have not yet been investigated under conditions of 'arrested relaxation'. On the basis of an extrapolation from the systems that have been discussed in the last few paragraphs, it would seem that the hypersurfaces are unlikely to be highly attractive. Bearing in mind the special effects of the light-atom anomaly, it seems unlikely that these reactions will channel a very high percentage of their energy into the vibration of the hydride product. The $H + SCl_2$ reaction has been studied; $\langle f_{V'} \rangle = 0.45$ [31].

The atom-exchange reactions which have been discussed in this section involve more than three atoms and the product not containing the newly formed bond is a molecule, rather than an atom. It appears that in direct reactions, unless the equilibrium geometry of this fragment is appreciably altered, it will remain almost unexcited internally. Repulsive forces are inefficient at exciting molecular vibrations, unless they are more impulsive than usually appears to be the case. This picture can change if there is a hollow of significant depth on the potential hypersurface. Collision complexes can then form providing the opportunity for intramolecular energy transfer during the lifetime of the complex.

6.3.4 Reactions proceeding *via* collision complexes

Some examples are now considered of energy disposal where the reaction dynamics are clearly dominated by the formation of long-lived collision complexes, which decompose to products when sufficient energy accumulates in the internal mode or combination of modes which carries the system through the transition state. For experiments on energy disposal, the complexes are formed with energy above the minimum which they require to dissociate, either in a prior bimolecular association reaction or by photochemical excitation. Although the internal energy which is acquired in these ways is initially localized in the excited molecule, it does, according to theories of unimolecular processes, quickly become randomized throughout all internal modes.

The methods which are used to produce the energized complexes are essentially those of *chemical* and *photochemical activation*. It will be shown in Section 6.5 that the rate at which collision complexes with defined internal energies decompose can be estimated by measuring the yield of product as a function of total gas pressure. However, in the experiments that are presently under review, the aim is to discover the distribution over internal states of the reaction products; the total pressure is too low for stabilizing collisions to occur, so that effectively all the complexes decompose.

In discussing the dynamics of reactions proceeding via long-lived complexes, it is useful to discriminate between the molecular motions and energy

partitioning, before and after passage through the transition state. According to RRKM theory, the distribution over states of the activated complexes is determined solely by statistical factors and by the need to conserve total energy and angular momentum. Any energy which is released after the transition state region and as the final products separate will not be shared statistically if the dynamics in this region are direct. Because of the possibility of this late, non-statistical, release of energy, the observance of non-statistical state distributions among the products of a reaction proceeding via a collision complex is *not*, in itself, sufficient evidence to state that RRKM theory is breaking down.

The points made in the previous paragraph can be illustrated by referring to the schematic potential energy profiles for unimolecular reactions which are given in *Figure 4.7*. Where there is essentially no potential barrier to the reverse reaction, as in *Figure 4.7*(c), dissociation proceeds via a rather 'loose' transition state, and a negligible amount of energy is released after the system passes through the transition state. This type of behaviour is characteristic of unimolecular decompositions where a single bond breaks and two free radicals are produced. It is expected that energy will be partitioned statistically, as long as due account is taken of the conservation laws.

At the other extreme, dissociation to two molecules with closed electronic shells occurs via a 'tight' transition state, where the profile of potential energy along the reaction coordinate, as in *Figure 4.7*(b), passes through a well-defined maximum. Such reactions are exemplified by the elimination of hydrogen halides from substituted alkanes or alkenes via four-centre transition states. The energy disposal in several reactions of this type has been investigated; the results for two of them are summarized in Table 6.5.

The vibrational distribution in HF produced in the reaction

$$CH_3 + CF_3 \rightarrow (CH_3CF_3)^{\dagger} \rightarrow CH_2{=}CF_2 + HF(v'); \quad \Delta H_0^{\circ} = -301 \text{ kJ mol}^{-1} \tag{6.29}$$

has been studied by both IRC and CL methods. Although in both cases the

Table 6.5 Energy disposal in unimolecular processes

Reaction	*Method*	$k(v' \mid; T)/k(T)$ $v' = 0$	1	2	3	4	$\langle f_{V'}(HX) \rangle$	*Ref.*
$CH_3 + CF_3 \rightarrow (CH_3{-}CF_3)^{\dagger}$	IRC	(0.49)	0.32	0.14	0.04	0.01	0.12	*a*
$\rightarrow CH_2{=}CF_2 + HF$	CL	0.45	0.36	0.19			0.11	*b*
$CH_2{=}CHCl + h\nu \rightarrow$ $CH{\equiv}CH + HCl$	CL	0.26	0.22	0.20	0.18	0.14	0.10	*c*

[a] CLOUGH, P. N., POLANYI, J. C., and TAUGUCHI, R. T., *Canadian Journal of Chemistry*, **48**, 2919 (1970)
[b] BERRY, M. J., and PIMENTEL, G. C., *Journal of Chemical Physics*, **49**, 5190 (1968); SIRKIN, E. R., and BERRY, M. J., *Journal of Quantum Electronics*, **QE-10**, 701 (1974)
[c] BERRY, M. J., *Journal of Chemical Physics*, **61**, 3114 (1974)

radicals are produced photochemically, the $(CH_3CF_3)^{\dagger}$ complex is formed in the bimolecular reaction shown, so this constitutes an example of *chemical* activation. The IRC experiments have provided information about the relative values of $k(v'|; T)$ for $v' = 1$—4, whereas the CL measurements have done the same for $v' = 0$—2. Both sets of data yield vibrational surprisal plots which are linear and they merge together quite well. The values of $k(v'|; T)$ fall monotonically with increasing v′ and $\langle f_{v'} \rangle$ is only about 0.12. Although this degree of vibrational excitation is much less than that found for direct abstraction reactions forming HF, it is greater than would be expected on purely statistical grounds. This is confirmed, for example, by the finding that the surprisal plots are characterized by a negative value of $\lambda_{v'}$.

It is not difficult to see why the HF produced in reaction (6.29) should acquire more energy than is predicted on a prior statistical basis. Thermochemical data and the experimentally observed activation energy for elimination of HF from CH_3CF_3 at high pressure indicate that the energetics of the reaction are as shown in *Figure 6.9*. Only about 125 kJ mol^{-1} of energy is available for (statistical) distribution among the internal modes in the transition state. The remaining 176 kJ mol^{-1} is released, as the products separate, in a way which depends on the nature of the potential energy

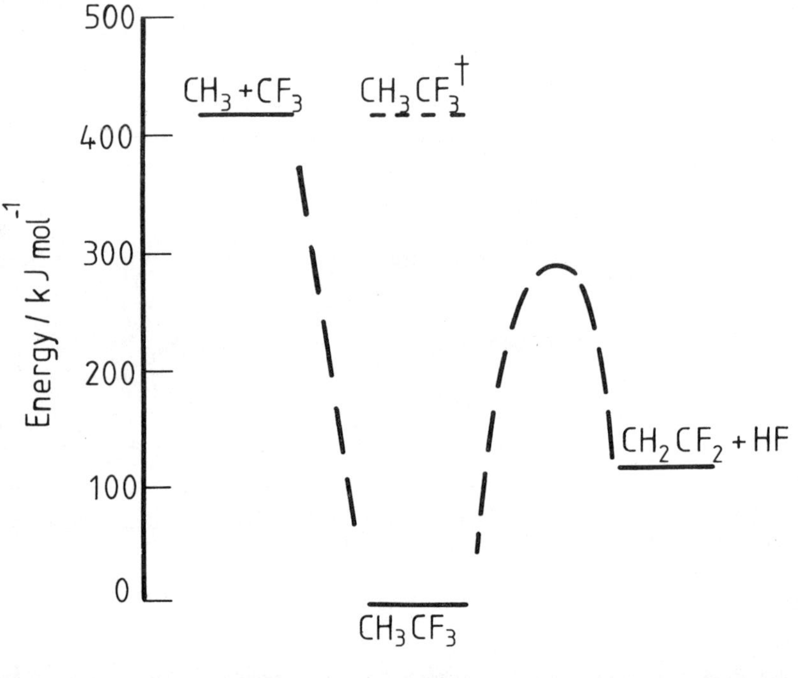

Figure 6.9. Energetics of the reaction $CH_3 + CF_3 \rightarrow (CH_3.CF_3)^{\dagger} \rightarrow CH_2 = CF_2 + HF$

hypersurface in the post-transition state region. The dynamics in this last stage, the reactive event, will usually be direct.

Although the amount of HF excitation exceeds statistical expectation, it is, nevertheless, quite slight. There seem to be two possible reasons why the yield is so small. The first is that the product HF is virtually formed by the time the system passes through the transition state; that is, the distance between these two atoms is only slightly greater than $r_{e,HF}$ and so there is little attractive energy release as the products separate. If the separation in the transition state was much greater than $r_{e,HF}$, and *if the products separated instantaneously*, a higher vibrational energy yield would be expected. The second possible explanation of the observed results is that, although r_{HF} is large in the transition state, the various modes of motion are still strongly coupled as the products separate, so that excitation of the HF product is reduced.

Energy disposal in the photodissociation of chlorinated ethylenes has been studied in very detailed chemical laser experiments using the grating selection technique[32]. The results for

$$CH_2{=}CHCl + h\nu \rightarrow CH{\equiv}CH + HCl \qquad (6.30)$$

are summarized in Table 6.5. $CH_2{=}CHCl$ absorbs in a broad band centred at about 182 nm, and the average energy absorbed is 557 kJ mol^{-1}. The thermal dissociation, for which $\Delta H_0^\circ = 120$ kJ mol^{-1}, has an activation energy in the limit of high pressure of 326 kJ mol^{-1}. These data indicate that the energetics of the reaction are not unlike those for dissociation of CH_3CF_3, *if* photochemical decomposition proceeds via the potential energy hypersurface of the ground electronic state. For this condition to be met, it is necessary for the electronic excitation energy acquired by photon absorption to be transformed rapidly to internal kinetic energy via a process of *internal conversion.*

Unfortunately, the interpretation of photochemical activation experiments is frequently obscured by doubts about the nature of the primary photochemical process. Photodissociation of molecules may occur via one or more potential hypersurfaces. Furthermore, decomposition may be dynamically direct or occur only once sufficient internal energy, after initially being randomized, finds its way into the appropriate mode or combination of modes. For polyatomic molecules, direct photodissociation is fairly unlikely, since sufficient energy would have to be deposited into the correct internal modes during the absorption process.

Although the primary photochemical act is not definitely established, the results for chlorinated ethylenes suggest strongly that *direct* photodissociation does not occur. They can be explained satisfactorily assuming internal conversion followed by unimolecular decomposition on the ground state hypersurface. For reaction (6.30), $\langle f_{V'} \rangle$ is 0.11. Once again, the degree of excitation in the eliminated hydrogen halide is somewhat greater than that expected on the basis of statistical sharing of the energy among all the product degrees of freedom. It has been suggested that slight differences between the

vibrational distribution of HCl produced in (6.30) and those from photolysis of other chlorinated ethylenes, may reflect some contribution from α–α elimination, i.e.,

$$CH_2{=}CHCl + h\nu \rightarrow CH_2{=}C\colon + HCl \tag{6.31}$$

which is not possible for $CH_2{=}CCl_2$. Berry[32] considers that the small extent of HCl excitation is an indication that there is still considerable interaction in the region after the transition state, so that energy leaks out of the HCl vibration as the products separate.

There are large barriers in the exit channels on the potential hypersurfaces for both reactions (6.29) and (6.30). It has already been mentioned that such barriers are usually absent when dissociation leads to two radicals. However, there have been no studies of energy disposal in these reactions up to the present time. A series of very interesting experiments have been carried out on the addition reactions of fluorine atoms to alkenes and substituted alkenes[33]. The free radicals which are formed initially have usually sufficient energy to decompose to several different pairs of products but, in accord with theories of unimolecular reactions, the most exoergic channel is overwhelmingly predominant; that is, the weakest bond in the complex breaks. Reactions yielding Cl or Br atoms, H atoms, and CH_3 radicals have all been studied. The disposal of energy in these reactions was first studied by molecular beam techniques and detailed discussion of their collision dynamics is deferred until Section 7.5.3. All that will be said at this stage is that the results do show clearly how, in reactions proceeding via long-lived collision complexes, the product energy distributions become increasingly non-statistical as the height of any exit channel barrier, relative to the energy of the separated products, increases. If proper account is taken of angular momentum effects, it appears that the product distributions usually correspond to statistical expectation in the limit of zero barrier.

6.3.5 Implications of microscopic reversibility[34,35]

The second half of this chapter is primarily concerned with the dependence of reaction rate on selective excitation of the reagents; that is on the *selectivity* of the energy requirements for reaction, rather than the *specificity* of energy disposal. It was shown in Section 1.3.2 that the principle of microscopic reversibility leads to relationships between the parameters that define the probability or rate of reaction proceeding in opposite directions at various levels of detail. When partially or fully detailed rate constants are available for reaction in one direction, these equations can be used to extract similarly detailed rate constants for the reverse reaction. Our discussion can start with equation (1.31), i.e.,

$$p^2 g_{\mathrm{n}} \sigma_{\mathrm{r}}(\mathrm{n}'|\mathrm{n}; v, \theta) = p'^2 g_{\mathrm{n}'} \sigma_{\mathrm{r}}(\mathrm{n}|\mathrm{n}'; v', \theta) \tag{6.32}$$

This relates the differential reaction cross-sections connecting fully defined

states n and n′ of the 'reagents' and 'products', in collisions at relative velocity $v(\mathrm{n}' \leftarrow \mathrm{n})$ or $v'(\mathrm{n} \leftarrow \mathrm{n}')$, in which species are scattered at the angle θ. The symbols g_{n} and $g_{\mathrm{n}'}$ denote the degeneracies of the n and n′ levels. In the case of an atom-transfer reaction involving only three atoms, e.g.,

$$\mathrm{A} + \mathrm{BC}(\mathrm{v}, \mathrm{j}) \rightleftarrows \mathrm{AB}(\mathrm{v}', \mathrm{j}') + \mathrm{C} \tag{6.33}$$

and, if the only degeneracies are those of BC's and AB's rotational levels,

$$(g_{\mathrm{n}'}/g_{\mathrm{n}}) = (2\mathrm{j}' + 1)/(2\mathrm{j} + 1) \tag{6.34}$$

Then, since

$$\varepsilon_{\mathrm{T}} = p^2/2\mu = \tfrac{1}{2}\mu v^2 \quad \text{and} \quad \varepsilon'_{\mathrm{T}} = p'^2/2\mu' = \tfrac{1}{2}\mu' v'^2 \tag{6.35}$$

the ratio of total reaction cross-sections is given by

$$\frac{S_{\mathrm{r}}(\mathrm{v}', \mathrm{j}' | \mathrm{v}, \mathrm{j}; \varepsilon_{\mathrm{T}})}{S_{\mathrm{r}}(\mathrm{v}, \mathrm{j} | \mathrm{v}', \mathrm{j}'; \varepsilon'_{\mathrm{T}})} = \left(\frac{2\mathrm{j}' + 1}{2\mathrm{j} + 1}\right)\left(\frac{\mu'}{\mu}\right)\left(\frac{\varepsilon'_{\mathrm{T}}}{\varepsilon_{\mathrm{T}}}\right) \tag{6.36}$$

The corresponding equation for the detailed rate coefficients is

$$\frac{k(\mathrm{v}', \mathrm{j}' | \mathrm{v}, \mathrm{j}; \varepsilon_{\mathrm{T}})}{k'(\mathrm{v}, \mathrm{j} | \mathrm{v}', \mathrm{j}'; \varepsilon'_{\mathrm{T}})} = \frac{v S_{\mathrm{r}}(\mathrm{v}', \mathrm{j}' | \mathrm{v}, \mathrm{j}; \varepsilon_{\mathrm{T}})}{v' S_{\mathrm{r}}(\mathrm{v}, \mathrm{j} | \mathrm{v}', \mathrm{j}'; \varepsilon'_{\mathrm{T}})} = \frac{(2\mathrm{j}' + 1)}{(2\mathrm{j} + 1)}\left(\frac{\mu'}{\mu}\right)^{\frac{3}{2}}\left(\frac{\varepsilon'_{\mathrm{T}}}{\varepsilon_{\mathrm{T}}}\right)^{\frac{1}{2}} \tag{6.37}$$

Detailed rate constants which describe the results of collisions between species in defined internal states occurring with a full thermal spread of velocities or collision energies can be obtained by integrating the right-hand side of equation (6.37) according to equation (1.36). Bearing in mind that for a fully specified pair of states, v′, j′ and v, j, $\Delta\varepsilon_{\mathrm{int}} = \varepsilon_{\mathrm{T}} - \varepsilon'_{\mathrm{T}}$, this yields

$$\frac{k(\mathrm{v}', \mathrm{j}' | \hat{\mathrm{v}}, \hat{\mathrm{j}}; \bar{\varepsilon}_{\mathrm{T}})}{k(\hat{\mathrm{v}}, \hat{\mathrm{j}} | \mathrm{v}', \mathrm{j}'; \bar{\varepsilon}'_{\mathrm{T}})} = \left(\frac{2\mathrm{j}' + 1}{2\mathrm{j} + 1}\right)\left(\frac{\mu'}{\mu}\right)^{\frac{3}{2}} \exp(-\Delta\varepsilon_{\mathrm{int}}/k_{\mathrm{B}}T) \tag{6.38}$$

In most experiments, only partially detailed information is obtained. For example, IRC experiments under conditions of 'arrested relaxation' can provide distributions of products over rovibrational states, but usually the reagents are not specially selected. In this situation the rate constants which are measured are defined by

$$k(\mathrm{v}', \mathrm{j}' |; T) = \sum_{\mathrm{v}, \mathrm{j}} f(\mathrm{v}, \mathrm{j}; T)\, k(\mathrm{v}', \mathrm{j}' | \mathrm{v}, \mathrm{j}; T) \tag{6.39}$$

where $f(\mathrm{v}, \mathrm{j}; T)$ is the Boltzmann distribution function, giving the fraction of BC molecules in a particular v, j level. In practice, it has been an approximate form of (6.37) that has been used most frequently to extract detailed rate constants for endoergic reactions from observations on the reverse exoergic process. Anlauf *et al.*[36] argue that, provided the distribution of $S_{\mathrm{r}}(\mathrm{v}', \mathrm{j}' | \mathrm{v}, \mathrm{j}; \varepsilon_{\mathrm{T}})$

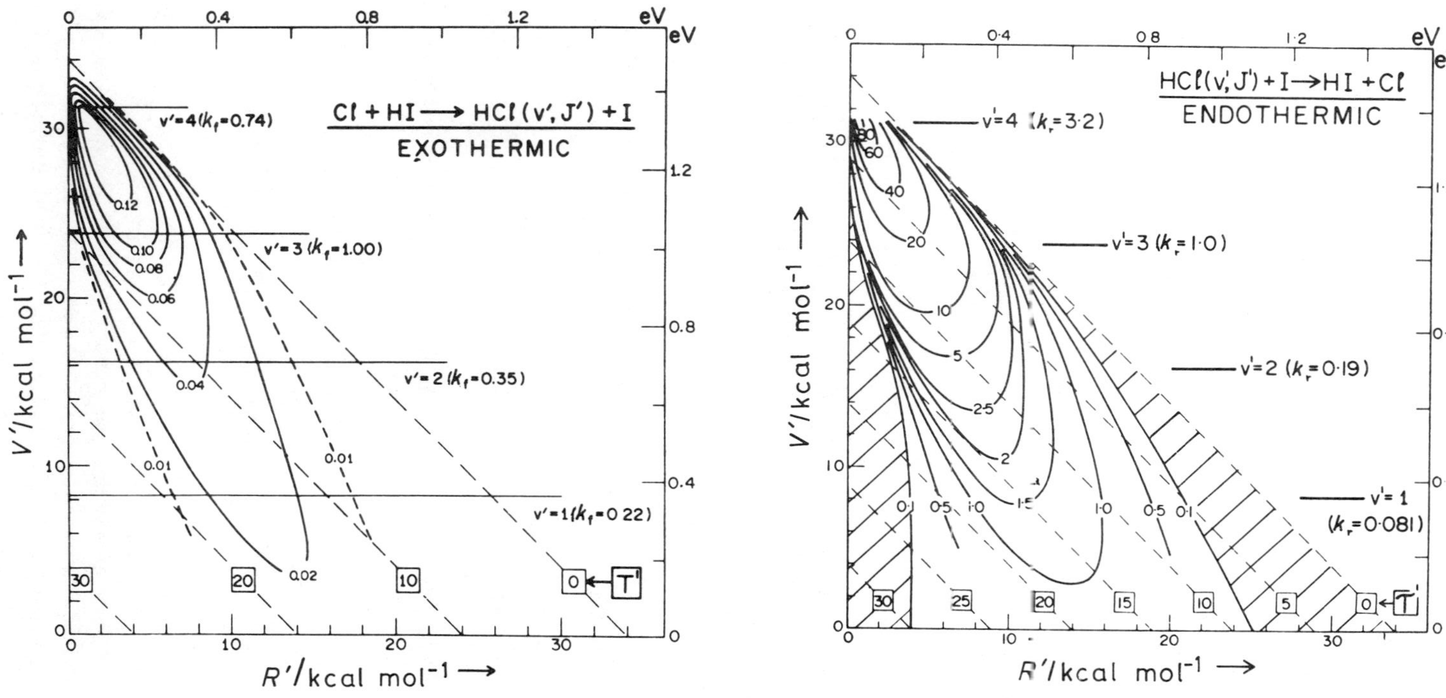
Cl + HI ⟶ HCl (v′, J′) + I
EXOTHERMIC
v′=4 (k_f=0.74)
v′=3 (k_f=1.00)
v′=2 (k_f=0.35)
v′=1 (k_f=0 22)
0.12
0.10
0.08
0.06
0.04
0.02
0.01
0.01
30
20
10
0
T′
V′/kcal mol⁻¹ ⟶
R′/kcal mol⁻¹ ⟶
eV
0
0.4
0.8
1.2
10
20
30
HCl(v′, J′) + I ⟶ HI + Cl
ENDOTHERMIC
v′=4 (k_r=3·2)
v′=3 (k_r=1·0)
v′=2 (k_r=0·19)
v′=1
(k_r=0·081)
60
40
20
10
5
2·5
2
1·5
1·0
0·5
0·1
30
25
20
15
10
5
0
T′
V′/kcal mol⁻¹ ⟶
R′/kcal mol⁻¹ ⟶
eV
0·4
0·8
1·2

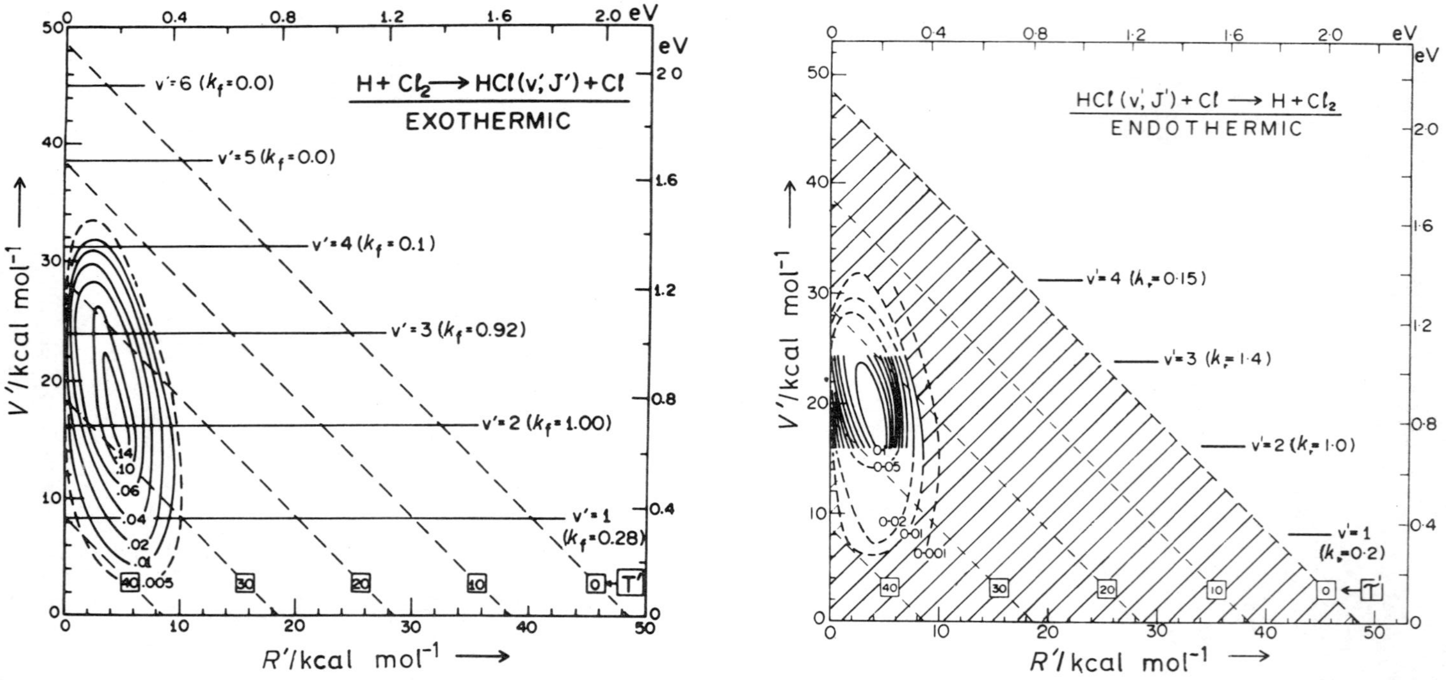

Figure 6.10. Comparison of triangular plots for exoergic and endoergic reactions. The notation is the same as that in Figure 6.5 [*reproduced, with permission, from Anlauf, K. G.*, et al., Journal of Chemical Physics, **51,** 5716 (1969); *Maylotte, D. H.*, et al., Journal of Chemical Physics, **57,** 1547 (1972); *Anlauf, K. G.*, et al., Journal of Chemical Physics, **57,** 1561 (1972)]

is only moderately dependent on ε_T, then equation (6.37) can be rewritten as

$$\frac{k(v', j'|v, j; \varepsilon_T)}{k(v, j|v', j' \varepsilon'_T)} = \frac{k(v', j'|\hat{v}, \hat{j}; \bar{\varepsilon}'_T)}{k(\hat{v}, \hat{j}|v', j'; \varepsilon'_T)}$$

$$= \frac{k(v', j'|; T)}{k(\hat{v}, \hat{j}|v', j'; \bar{\varepsilon}'_T)} \qquad (6.40)$$

where $\hat{v}(=0)$ and $\hat{j}$ are the most populated vibrational and rotational levels in BC and the average translational energy of the collisions leading to exothermic reaction is

$$\bar{\varepsilon}_T = (E_{act} + \tfrac{3}{2}RT)/N_A \qquad (6.41)$$

The total energy available to the products of this reaction is, on average,

$$\bar{\varepsilon}_{total} = (-\Delta H_0^\circ + E_{act} + 5/2RT)/N_A \qquad (6.42)$$

the last two terms in the numerator corresponding to the mean translational plus rotational energy of A + BC collision pairs that react ($\hat{v} = 0$). Finally, $\bar{\varepsilon}'_T$ is given by energy balance as

$$\bar{\varepsilon}'_T = \bar{\varepsilon}_T - (\varepsilon_{v',j'} - \varepsilon_{\hat{v},\hat{j}}) \qquad (6.43)$$

Combining equation (6.37) with (6.40) to (6.43) yields

$$k(\hat{v}, \hat{j}|v', j'; \bar{\varepsilon}'_T) = \left(\frac{2\hat{j}+1}{2j'+1}\right)\left(\frac{\mu}{\mu'}\right)^{\frac{3}{2}}\left(\frac{\bar{\varepsilon}_T}{\bar{\varepsilon}'_T}\right)^{\frac{1}{2}} k(v', j'|; T) \qquad (6.44)$$

allowing detailed rate constants for reaction out of specified states to be calculated, if the detailed rate constants $k(v', J'|; T)$ have been observed.

The above, approximate, analysis makes the assumption that the total energy in all reactive collisions is the same. It yields a set of detailed rate constants which can then be displayed on the kind of triangular plot that was used in Section 6.2.1 to show the product energy distribution from the $F + H_2$ reaction. *Figure 6.10* shows plots for two reactions. As we have already seen, the Cl + HI reaction leads to high vibrational, and moderate rotational, excitation of HCl. In *Figure 6.10*, this is indicated by the position of the peak in the contour line plot. The diagram for the reverse reaction has a similarly placed maximum illustrating that, for total HCl + I energies defined by equation (6.42), the endoergic reaction will be most effectively promoted by placing most of the energy in HCl vibration. The $H + Cl_2$ reaction, because of the light-atom anomaly, leads to less internal excitation of the HCl product. For the reverse reaction, the optimum distribution of energy is ~ 50 per cent in the HCl vibration, with most of the rest in relative translation. Further excitation of HCl at the expense of relative translational energy reduces the likelihood of reaction. This appears to be a further manifestation of a light-atom anomaly.

Although the approximate treatment is useful from the point of view of diagrammatic display on a triangular plot, partially detailed rate constants for

opposing reactions can be *accurately* related to one another. Rearrangement of equation (6.38) and division of both sides by Q_{int}, the partition function over internal states (v, j) of BC, yields:

$$\left\{\frac{(2\text{j}+1)\exp(-\varepsilon_{\text{v,j}}/kT)}{Q_{\text{int}}}\right\} k(\text{v}',\text{j}'|\text{v},\text{j};T)$$

$$= \frac{(2\text{j}'+1)}{Q_{\text{int}}}\left(\frac{\mu'}{\mu}\right)^{\frac{3}{2}} \exp(-\Delta\varepsilon_{\text{v}'\text{j}',0}/kT)\,k(\text{v},\text{j}|\text{v}',\text{j}';T) \tag{6.45}$$

where $\Delta\varepsilon_{\text{v}'\text{j}',0}$ is the difference in internal energy between AB(v′, j′) and BC(v = 0, j = 0). Summing both sides of this equation over v and j, one obtains

$$\frac{k(\text{v}',\text{j}'|;T)}{k(|\text{v}',\text{j}';T)} = \frac{(2\text{j}'+1)}{Q_{\text{int}}}\left(\frac{\mu'}{\mu}\right)^{\frac{3}{2}} \exp(-\Delta\varepsilon_{\text{v}'\text{j}',0}/k_{\text{B}}T) \tag{6.46}$$

Repeating this procedure for j′ leads to

$$\frac{k(\text{v}'|;T)}{k(|\text{v}';T)} = \left(\frac{Q'_{\text{R,v}'}}{Q_{\text{int}}}\right)\left(\frac{\mu'}{\mu}\right)^{\frac{3}{2}} \exp(-\Delta\varepsilon_{\text{v}',0}/k_{\text{B}}T) \tag{6.47}$$

where $Q'_{\text{R,v}'}$ is the rotational partition function for AB(v′) and $\Delta\varepsilon_{\text{v}',0}$ is the energy difference between AB(v′, j′ = 0) and BC(v = 0, j = 0).

Equations (6.46) and (6.47) are extremely valuable since they permit one to calculate the partially detailed rate constants for reaction *from* selected reactant levels to all product states, i.e., $k(|\text{v}',\text{j}';T)$ or $k(|\text{v}';T)$, when the rate constants associated with formation of these species in the reverse, thermal reaction, i.e., $k(\text{v}',\text{j}'|;T)$ or $k(\text{v}'|;T)$, are known. In practice, it is nearly always easier to obtain information about the specificity of energy disposal in the exothermic reaction; these equations can then be used to discover something about the selective energy requirements for the endothermic reaction. This powerful method has one important limitation: it can only provide data for those levels with energies close to, or below, the minimum required for reaction.

Based on a knowledge of energy disposal in different types of exoergic reaction, it is possible to make some generalizations about the energy requirements for endoergic reactions. First, excitation of the reactant vibration will be probably most effective in promoting highly endoergic atom-transfer reactions which involve only three atoms and proceed directly. Secondly, if the reagents are more complex, vibrational excitation of the bond which is broken in the reaction is likely to be most effective. Finally, if reaction proceeds via long-lived collision complexes, the selective nature of the energy requirements are likely to be lowered, especially if there is no potential barrier preventing the formation of the complexes.

6.4 Experimental methods: excitation of reagents

It has already been pointed out that, when a reactive system is not at chemical equilibrium, the distribution of reagents over internal states and collision energies may differ from that predicted by the Boltzmann laws. Non-equilibrium effects of this kind are frequently present *in unimolecular reactions*. Only in the limit of high gas density, when intermolecular energy transfer becomes very rapid, do such effects disappear. Nevertheless, at all pressures, the thermal kinetics of unimolecular reactions are dominated by the behaviour of species with internal energies close to ε^0, the critical value for reaction to occur. Consequently, conventional experiments, even if these elucidate the fall-off characteristics of the reaction, do not reveal much about how $k(\varepsilon)$ varies with values of $\varepsilon > \varepsilon^0$. In order to test theories of unimolecular reactions thoroughly, it is desirable to determine the specific rate constant at internal energies well in excess of the critical energy. These measurements require techniques for selectively exciting the reagent.

Under thermal conditions, the great majority of *bimolecular collisions* involve species in their ground electronic state and in $v = 0$, and the temperature-dependence of the rate constant largely reflects the result of increasing the average collision energy. The ease of translational energy transfer ensures that non-equilibrium effects are negligible. However, the values of the thermal rate constants largely reflect those of the collision cross-section close to the threshold energy. Once again, there is considerable interest in discovering how the reaction depends on initial energy, both of relative translation and in internal modes of the reagents.

Photochemical methods provide the most general means of disturbing the Boltzmann distribution over energy states, and hence of studying the kinetics of processes involving species in defined excited states. The application of photochemical excitation is not restricted to the production of atoms or molecules in selected internal states, since photodissociation, especially of diatomic molecules with monochromatic radiation, can be used to produce species with large and narrowly defined translational energies.

6.4.1 Continuous photochemical activation

The fundamental limitation in continuous photolysis experiments, as was pointed out in Section 5.2.4, is that they are, at best, only capable of determining *relative* rate constants. A classic example is provided by steady-state studies of electronic fluorescence[37, 38]. In the simplest situation, the species which are excited by photon absorption

$$A + h\nu \rightarrow A^* \tag{6.48}$$

decay only by spontaneous emission

$$A^* \rightarrow A + h\nu \tag{6.49}$$

or in collisions with a single component of the gas mixture

$$A^* + Q \rightarrow A + Q \qquad (6.50)$$

The ratio I^0/I, that is the intensity of fluorescence without Q present to that with Q present, is equal to the ratio of the steady-state concentrations of A* under these conditions. Then, if the rate of excitation is unaffected by including Q in the gas mixture,

$$I^0/I = 1 + (k_Q[Q]/k_{rad}) \qquad (6.51)$$

where k_Q and k_{rad} are the second- and first-order rate constants for (6.49) and (6.50). This is the celebrated Stern–Volmer relationship. There are two fundamental conditions which have to be satisfied if the equation is to be used in its simplest form. First, it is essential that the rate of excitation does not change when [Q] is varied, as a result, for example, of the absorption line or lines being pressure-broadened. Secondly, effects due to imprisonment or trapping of the emission should be avoided. Otherwise, the effective value of k_{rad} will not correspond to the true spontaneous lifetime of A*. If these difficulties can be overcome, and if the value of k_{rad} can be determined in some separate experiment, then steady-state quenching measurements can supply accurate rate constants. However, they provide no information about the mechanism of the process removing the excited species; for this, it is necessary to observe the products of the quenching process.

The analyses of most continuous photolysis experiments are similar to that which has just been described. When the quenching of fluorescence is observed, the radiative lifetime of the excited species provides a convenient 'internal clock'. In photochemical studies of unimolecular reactions[39], and in experiments where translationally 'hot' atoms are generated photochemically[40], reaction and collisional deactivation of the energized species compete and it may be possible to deduce the specific reaction rate by studying the results of this competition. Unfortunately, the interpretation of the observations is less certain than one might hope. The reason is the same in both cases. If a species is excited to an energy well past threshold, the 'strong collision assumption' may not be valid; that is, the first non-reactive collision may not remove sufficient energy to reduce the probability of reaction at some later time to zero. As a result a rather complicated analysis of the competition between reaction and energy transfer may be called for.

An additional difficulty in photochemical activation studies of unimolecular reactions, already referred to in Section 5.2.4, is that details of the primary photochemistry may not be fully understood[41]. When *diatomic* molecules are excited to dissociative states, the atoms separate in about 10^{-13} s and the quantum yield is unity at all pressures that can be conveniently reached. Energy over and above that required to dissociate the molecule must appear as energy of relative translational motion. With *polyatomic* molecules, there is likely to be a time lapse between excitation and any unimolecular chemical change, and this may allow stabilizing collisions to occur. Consequently, the quantum yield (ϕ) of the reactive channel may depend on the total pressure.

To interpret the results fully one needs to know whether radiationless transitions occur between different electronic states, and particularly whether internal conversion to vibrationally excited levels of the electronic ground state occurs. The competition between reaction and collisional stabilization can be represented by the equations[39]

$$A + h\nu \rightarrow A_i^* \qquad (6.52)$$

$$A_i^* \xrightarrow{k(\varepsilon_i)} \text{products} \qquad (6.53)$$

$$A_i^* + M \rightarrow A_j^* + M \qquad (6.54)$$

$$A_j^* \xrightarrow{k(\varepsilon_j)} \text{products} \qquad (6.55)$$

$$A_j^* + M \rightarrow A + M \qquad (6.56)$$

The subscripts i and j distinguish the states that are excited initially from those, still above threshold, that are populated by partial (stepwise) deactivation. The strength of the photochemical activation method resides in the possibility that photolysis can be carried out with monochromatic radiation at any wavelength absorbed by the molecule under investigation. Thus, ε_i can usually be varied over quite a wide range, but, at the same time, it can be well-defined in any individual experiment. (The spread in ε_i mainly comes from the distribution of internal energies in the absorbing species.)

In the simplest model of photochemical activation, it is assumed that reaction from states other than those which are excited initially is unimportant. That is either A_i^* molecules are deactivated directly to states below threshold, or at least they are transferred to states from which the reaction probability is negligibly small. In this case, a simple steady-state treatment yields an equation analogous to the Stern–Volmer relationship, i.e.,

$$\phi^{-1} = 1 + k_M[M]/k(\varepsilon_i) \qquad (6.57)$$

where k_M is the second-order rate constant associated with 'quenching' collisions. Clearly, k_M or $k(\varepsilon_i)$ can be evaluated, if the value of the other is assumed, and if the quantum yield is shown to vary in the way predicted by this equation. For example, if k_M is taken to be Z_M, i.e., the rate constant associated with collisions between A and M, then $k(\varepsilon_i)$ can be determined. In practice, this strong collision model is an oversimplification. More complex treatments are considered briefly in Section 6.5.

Photochemical studies of bimolecular reactions of translationally 'hot' atoms are similar—both in method and difficulties—to photochemical activation experiments on unimolecular reactions. The technique has mainly been used to study reactions of H (or D) atoms generated by photolysis of hydrides. The relative translational energy of the separating fragments is given by energy balance, i.e.,

$$\varepsilon'_T = h\nu - D_0 - \varepsilon'_{int} \qquad (6.58)$$

where D_0 is the dissociation energy of the parent molecule, ε'_{int} is any internal energy carried by the fragments, and any initial internal energy of the parent is

ignored. The situation is simplest when a diatomic is photolysed, since ε'_{int} can only be electronic energy. This has led to the hydrogen halides (HX), which absorb continuously over wide ranges in the far ultraviolet region, being used extensively as sources of hot H atoms[40]. Unfortunately, more than one dissociative state is involved, and they do not all correlate with ground state atoms. This means that photolysis with monochromatic radiation can produce atoms with two separate, though well-defined, initial energies, one corresponding to production of $H(^2S) + X(^2P_{\frac{3}{2}})$ and the other to production of $H(^2S) + X(^2P_{\frac{1}{2}})$.

Because momentum, as well as energy, must be conserved in the photodissociation of HX, the two equations

$$\varepsilon_T = \tfrac{1}{2}m_H v'_H + \tfrac{1}{2}m_X v'^2_X \qquad (6.59)$$

define the translational energy of the H atoms:

$$\tfrac{1}{2}m_H v_H^2 = \varepsilon'_T/(1 + m_H/m_X) \qquad (6.61)$$

The H atom carries away the lion's share of the energy, because the halogen atom (X) is much heavier. Consequently, photolysis of HBr at 184.9 nm, which is believed to produce $Br(^2P_{\frac{3}{2}})$ atoms almost exclusively, yields H atoms with the equivalent of 281 kJ mol^{-1} of energy.

A second advantage of using HX (X = Br or I) as the source of hot H atoms is that the parent molecule *normally* reacts rapidly with H atoms:

$$H + HX \rightarrow H_2 + X \qquad (6.62)$$

As a result, atoms that lose translational energy in non-reactive collisions, and eventually become thermalized, are 'scavenged'. Small concentrations of HX are photolysed in a large excess of a second species that will react with H atoms—but only in a process with a threshold energy much greater than $k_B T$. This has to yield a distinguishable product from (6.62), say

$$H + DR \rightarrow HD + R \qquad (6.63)$$

Then the ratio of product yields, $[HD]/[H_2]$, is measured for different wavelengths of photolysing light and as a function of added inert gas, usually referred to as the 'moderator'.

When H atoms make their very first collisions with DR there will clearly be a cross-section for reaction which is much larger than the average value in thermal collisions. The energy in these first collisions will be $\frac{1}{2}\mu v^2$ where $\mu = (m_H m_{DR})/(m_H + m_{DR})$ and $\boldsymbol{v} = \boldsymbol{v}_H - \boldsymbol{v}_{DR}$. Because the values of $\boldsymbol{v}_{DR}$ will be thermally distributed, some definition of the collision energy is lost. However, this blurring is small if $\mu_H \ll \mu_{DR}$ so that $\boldsymbol{v}_H \gg \bar{\boldsymbol{v}}_{DR}$.

The simplest analytical approach parallels that used for unimolecular reactions: one assumes that hot atoms either react in their first collision or sufficient energy is removed to render them unreactive in subsequent collisions. In practice, this oversimplification is invalid. More realistic treatments, which take account of the fact that hot atoms generally make

several collisions before they react or fall below threshold, are referred to in Section 6.6.1.

6.4.2 Pulsed photochemical activation

Pulsed photochemical experiments on excited species in the gas phase are almost entirely limited to studies of the kinetics of atoms and *small* molecules in selected internal states. In principle, a large molecule can be excited with a short pulse of radiation and the formation of its dissociation products monitored as a function of time. In practice, it is impossible to ensure that the excited molecules survive for times that can be resolved, without their making collisions either with other gas molecules or with the walls of the container. The only exception has been in experiments where a beam of molecules is crossed by a powerful, pulsed, laser beam[41]. The velocities of the dissociative fragments are measured by timing their arrival at a detector some distance from the interaction zone, and their internal energies can then be evaluated using equation (6.58). An indication of the lifetime of the photoexcited molecules is given by the angular distribution of the fragments. If it is symmetrical (see Section 7.3.2), it shows that the excited species survive for at least a rotational period before breaking up, thereby 'losing their memory' of any specificity in the mode of formation, resulting from the laser beam's being polarized.

In non-beam experiments, variations in the concentration of an excited species can be followed spectroscopically. In order to obtain kinetic data, it is essential that the exciting light pulse is at least as short as the lifetime of the excited species. In the limit where excitation is achieved in a time much shorter than the decay time, and if the kinetics can be described by the simple scheme represented by equations (6.48) to (6.50), then

$$[A^*]_t = [A^*]_{t=0}\exp\{-(k_{rad}+k_Q[Q])\} \tag{6.64}$$

Time-resolved measurements have been made on the fluorescence from a number of short-lived atomic states[42]. As with steady-state quenching experiments, effects can arise owing to the imprisonment of the emitted radiation. This is illustrated in *Figure 6.11*, which shows how the effective value of k_{rad} decreases dramatically as [K] rises above 5×10^{10} molecules cm^{-3}. At first sight, this increase in the effective radiative lifetime might not appear to affect direct measurements. However, at concentrations of A which cause appreciable trapping, the effective value of k_{rad} depends on the magnitude of $k_Q[Q]$. As $k_Q[Q]$ increases, a smaller fraction of A* species decay by radiative transitions, the effect of radiation trapping diminishes, and k_{rad} eventually corresponds to the reciprocal of the mean radiative lifetime.

Metastable electronically excited species and vibrationally excited molecules have longer radiative lifetimes. Frequently, in studies of their kinetics, radiative processes can be neglected altogether, although it may be necessary to consider whether heterogeneous process contribute to the overall relaxation.

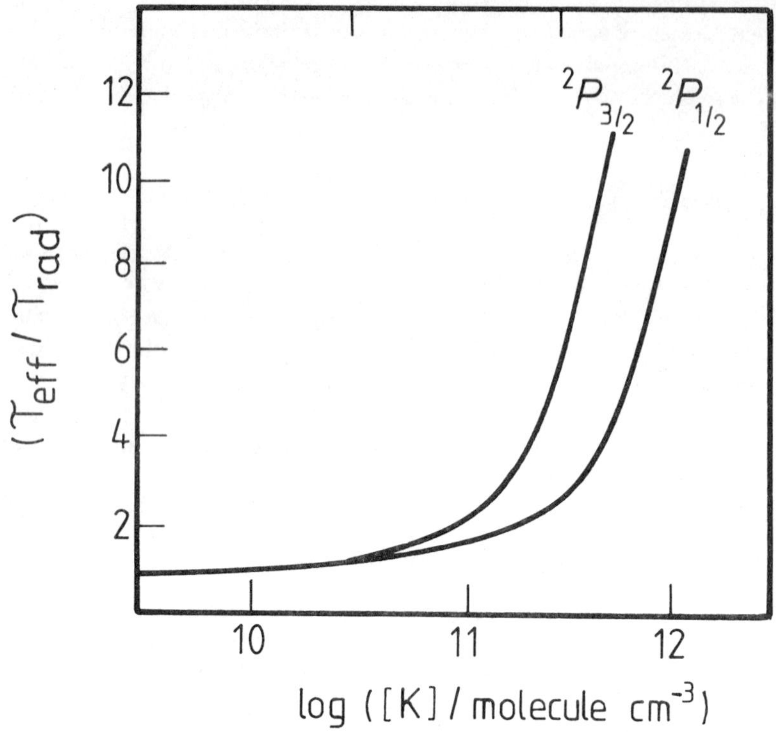

Figure 6.11. Variation of effective radiative lifetimes (τ_{eff}) *for the* $^2P_{\frac{3}{2},\frac{1}{2}}$–$^2S_{\frac{1}{2}}$ *resonance transitions in potassium; the true radiative lifetime,* $\tau_{rad} = 2.8 \times 10^{-8}$ s [*adapted from Copley, G., and Krause, L.*, Canadian Journal of Physics, **47**, 533 (1969)]

Quenching of metastable atoms has been studied extensively using flash photolysis of a suitable precursor to create the atoms and resonance absorption or fluorescence to follow their relaxation[43].

In contrast to *electronic* photochemistry, *vibrational* photochemistry has had to await the development of powerful lasers capable of exciting molecules in their comparatively weak infrared transitions. In the past decade, however, collisional energy transfer from vibrationally excited molecules has been investigated widely. In these experiments, molecules are excited by a short laser pulse and the consequent infrared fluorescence is observed as a function of time in gas mixtures of differing composition. The technique has two principal limitations. The first arises because the laser must emit frequencies absorbed by the gas. The lack of powerful, tunable, infrared lasers has meant that these studies have largely been confined to molecules, such as HF and HCl, for which the corresponding molecular laser can be constructed[35, 44]. Because these lasers cannot be used to excite molecules in overtone transitions, the method

has been almost totally restricted to molecules in their first level of excitation.

Finally, it is re-emphasized that experiments in which only the concentration of the excited state species is observed fail to provide any direct evidence regarding the mechanism of the collisional processes responsible for their removal.

6.4.3 Chemical activation

The term chemical activation is usually applied to studies of reactions like (6.29), in which the yield of a unimolecular reaction product is measured relative to that of the stabilized adduct, at different total pressures. The treatment of the results follows the same lines as the analysis of photochemical activation studies. However, there is an additional limitation: rather rarely is it possible to find more than one bimolecular association reaction to form the energized complex. Consequently, the initial energy of the collision complex cannot be varied as in photochemical experiments.

Some other experiments can be classified under the heading chemical activation, if it is interpreted very broadly. Stretching the term to its limits, one could include studies of the reactions of atoms produced in nuclear recoil reactions[45], such He^3(n, p)T. The energy of the tritium ions is initially about 200 keV ($\sim$ 19 GJ mol^{-1}). By the time they have been moderated into the chemical range, the ions have been neutralized. The hot atoms continue to undergo energy-degrading collisions until they react, either with a species whose reaction with tritium has an appreciable threshold energy or with a scavenger. Such experiments played an important role in the development of hot-atom chemistry. However, they cannot yield accurate excitation functions and are now performed less frequently.

Exoergic bimolecular reactions yield excited products, but it is not easy to use these molecules as reagents for subsequent reactions and thereby study how the rate of the second reaction depends on the degree of excitation of the intermediate. One difficulty is that only the product vibrational excitation survives many collisions. A second is that the products of the first reaction are not created in a single state. The most successful studies of successive bimolecular reactions have involved measurements of depletion of the infrared chemiluminescence from products of a primary reaction, when these are crossed with a stream of secondary reagents. As with so many of the techniques discussed in this section, it is the excited reagent (for the secondary reaction) which is observed directly, so once again it is difficult to distinguish whether it is reaction or non-reactive energy transfer, which is primarily responsible for any reduction in chemiluminescence.

6.5 Selective excitation: unimolecular processes[39,46,47]

6.5.1 Tests of energy randomization

A central assumption in the usual, statistical based, theories of unimolecular reactions is that internal energy, whether it be acquired by thermal, chemical, or photochemical action, is rapidly randomized, so that any 'mode specificity' is lost 'immediately'. Several experiments have been devised to test this postulate (*see* p. 308).

One check relies on the production of a common collision complex by different primary processes. For example, 'energized' methylcyclopropane is formed both by the addition of triplet methylene radicals to propene

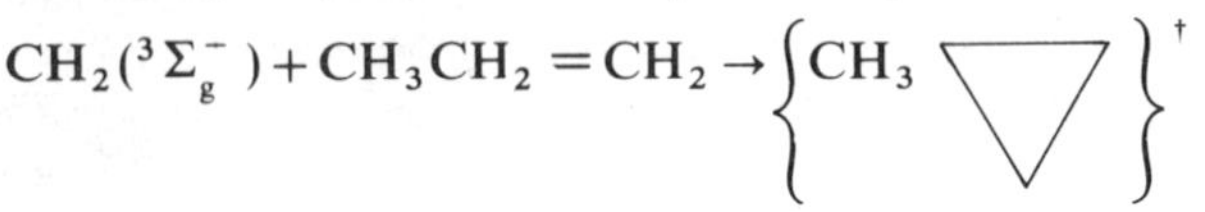

(6.65)

and by the insertion reaction of singlet methylene with cyclopropane

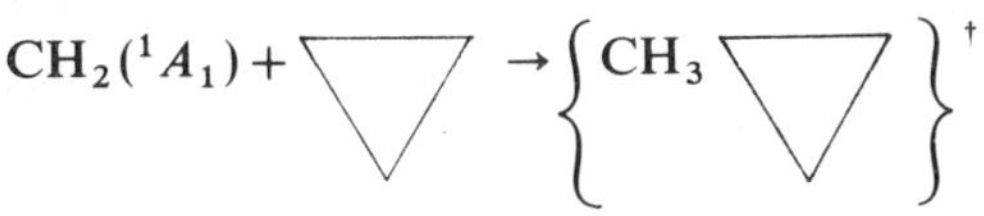

(6.66)

Unless the adduct is stabilized in collisions, it isomerizes, yielding one of several butene isomers. The relative importance of the various pathways for loss of the energized methylcyclopropane molecules formed in reactions (6.65) and (6.66) have been compared[48]. Only minor differences were found, and these could be attributed to the slight difference in internal energy between the adducts produced in the two primary bimolecular steps. These observations were consistent with rapid randomization of the internal excitation resulting from the exothermicity of reactions (6.65) and (6.66), and therefore with the behaviour of the energized molecules depending only on their total energy and angular momentum.

Some more recent experiments[49] provide a more sensitive test for the completeness of energy randomization. Partially deuterated hexafluorobicyclopropyl (HBC) molecules were created by the reactions:

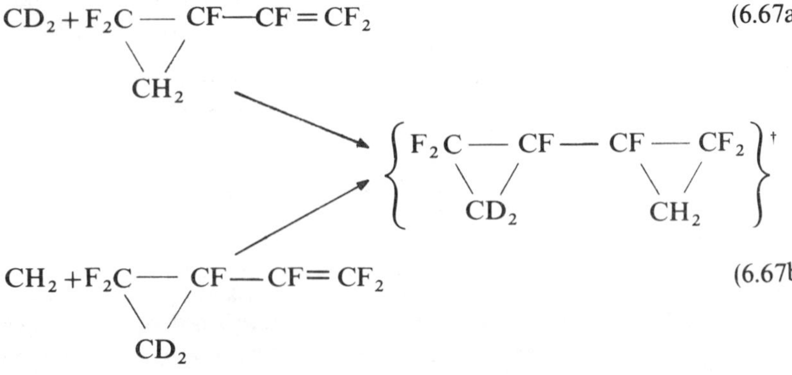

(6.67a)

(6.67b)

CF_2 can be eliminated from either side of the energized adduct and the relative yields of $CD_2{=}CF{-}CF{-}CH_2{-}CF_2$ and $CH_2{=}CF{-}CF{-}CD_2{-}CF_2$ determined accurately from their fragment mass spectra. At low *total* pressures, *all* the $HBC^{\dagger}$ species decompose, irrespective of whether the internal energy becomes randomized. Under these conditions, the yields indicated a slight preference for elimination of CF_2 from the side of the adduct directly involved in the initial addition. This preference, which is of a similar magnitude for both reactions (6.67a) and (6.67b), was interpreted as being the result of decomposition of the $HBC^{\dagger}$ molecules prior to randomization of their internal energy. As the data in *Figure 6.12* show, approximately 3 per cent of the energized adducts appeared to dissociate in this way. As the pressure is raised, an increasing proportion of those $HBC^{\dagger}$ species in which the internal energy has been randomized can be stabilized by collision and do not dissociate. However, to deactivate the adducts which decompose 'immediately' would require very high pressures indeed. Consequently, the proportion of $HBC^{\dagger}$ molecules which decompose before energy randomization increases with pressure, and the relative yields of the two products of decomposition reflect this.

A rather different test of energy randomization is provided by experiments

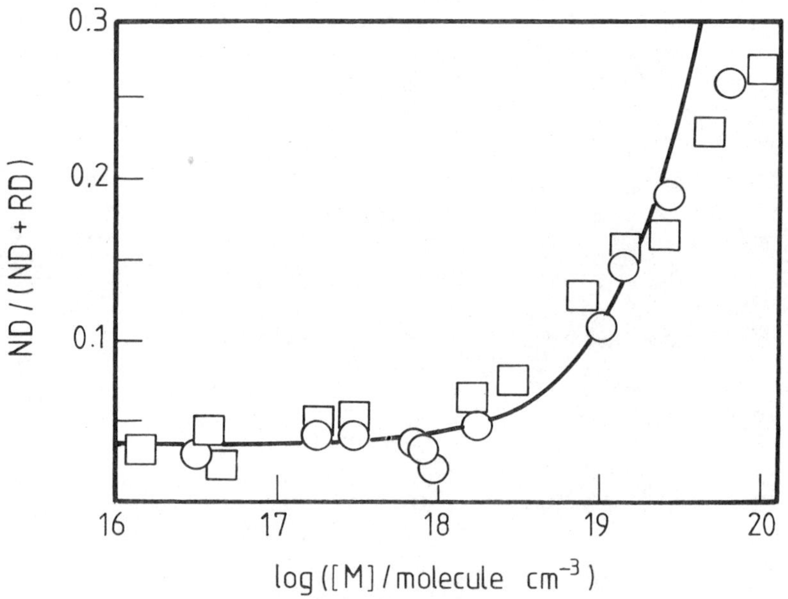

Figure 6.12. Ratio of non-random decomposition (ND) *to total decomposition* (ND + RD) *of hexafluorobicyclopropyl* † *radical formed in reactions* (6.67a) ○ *and* (6.67b) □ [*adapted from Rynbrandt, J. D., and Rabinovitch, B. S.,* Journal of Physical Chemistry, **75**, 2164 (1971)]. 2.49×10^{-19} molecules cm^{-3} = 1 atm.

on bimolecular vibrational energy transfer in systems where the dynamics can be dominated by the formation of collision complexes[35, 50, 51]. Recent experiments have demonstrated that the relaxations of O_2 by O, of NO by O and Cl, and of OH by H, NO and NO_2 are all unusually rapid. In all these cases, a collision complex can form, and breakdown of a randomized complex should lead predominantly to molecules in their lowest vibrational levels, since these products are associated with the greatest volume of phase space. According to such a model, if energy randomization is complete, the rate constant for vibrational deactivation should be close to the second-order rate constant for the combination reaction in the limit of high pressure.

6.5.2 Photodissociation of NO_2[47, 52]

Nitrogen dioxide absorbs, albeit rather weakly, over a wide range of wavelengths in the visible and ultraviolet parts of the spectrum. The photochemical threshold for dissociation of NO_2

$$NO_2 \rightarrow NO + O; \qquad \Delta H_0^\circ = 300.5 \text{ kJ mol}^{-1} \tag{6.68}$$

is at 397.9 nm. At low pressures, excitation at shorter wavelengths causes dissociation, whereas irradiation with light of longer wavelengths gives rise to fluorescence[53]. A predissociative mechanism apparently allows dissociation to occur at, or very close to, the energetically allowed threshold, although the behaviour close to this threshold is somewhat obscured because of the distribution of internal energies in the absorbing molecules.

There is interest in the NO_2 fluorescence for a number of reasons: not least, the recognition that this emission is closely related to the chemiluminescence which is observed when NO reacts with O atoms or with O_3. The interpretation of the results on all these systems is complicated by the profusion of excited states that exist in the pertinent range of energy. Thus, it has been calculated that there are five electronic states of NO_2 between about 200 and 400 kJ mol^{-1} above the ground state, although only two of these ($^2B_1$ and $^2B_2$) can be excited from the ground state ($^2A_1$) in fully allowed radiative transitions. Lifetime measurements under collision-free conditions[54] suggest that it is the $^2B_1$ state which is primarily excited, but that these molecules decay by radiationless transitions into a dense manifold of levels which have mixed $^2B_2$ and $^2A_1$ character and a rather long radiative lifetime. Although the coupling between different electronic states does complicate some issues, it also bestows advantages, deriving from the fact that the kinetics of photodissociation and thermal decomposition are both determined by the behaviour of molecules in the mixed levels close to and above the dissociation limit of the ground state. As a result, this system serves as a valuable prototype of unimolecular reactions in which a small molecule dissociates to two free radical fragments. A further advantage is that information about collisional energy transfer from the optically excited states can be obtained from steady-state measurements on the fluorescence excited with wavelengths > 397.9 nm[53].

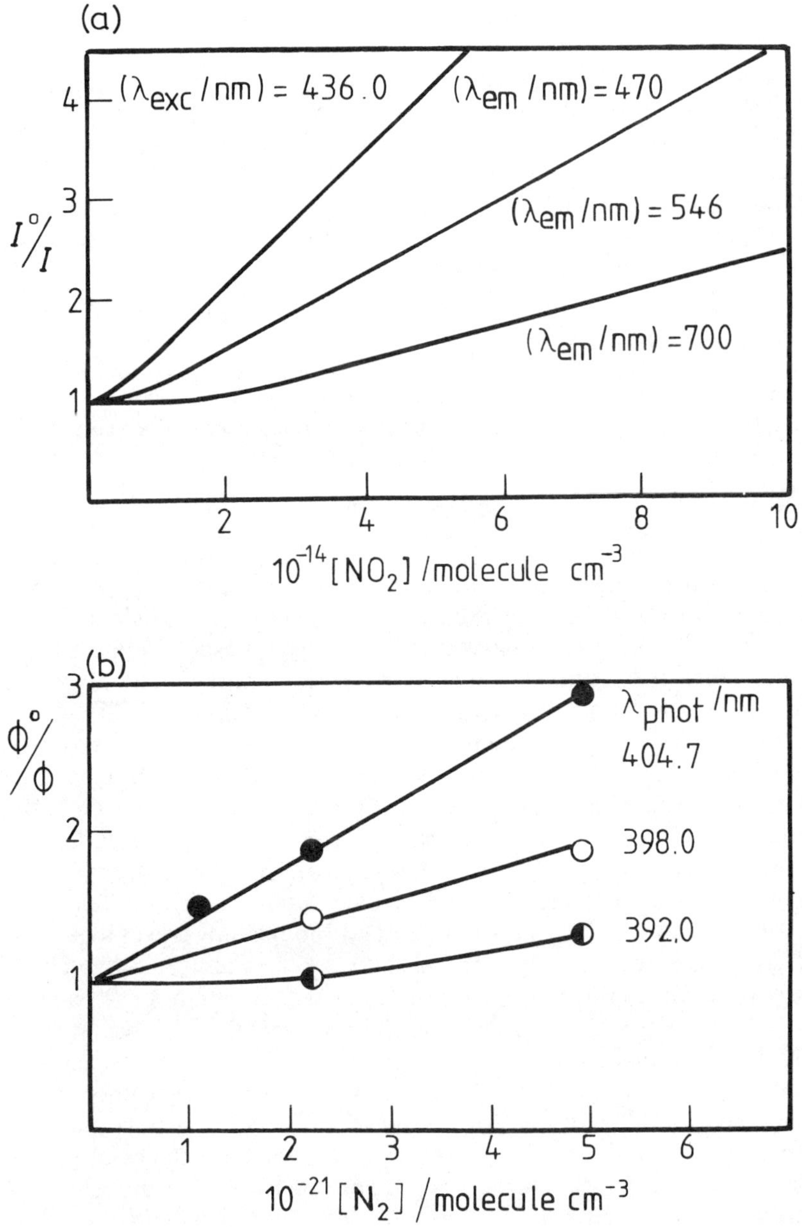

Figure 6.13. Stern–Volmer plots for (a) NO_2 *fluorescence, excited with* λ_{exc} = *436.0*/nm, *and* (b) NO_2 *photolysis, for various wavelengths* $\leqslant$ 404.7 nm [*adopted from Keyser, L. F.*, et al., Journal of Chemical Physics, **54,** 355 (1971), *and Gaedtke, H., and Troe, J.*, Berichte der Bunsengesellschaft für Physikalische Chemie, **79,** 184 (1975)]

Some results of a Stern–Volmer analysis of quenching measurements are shown in *Figure 6.13*(a). The plots are non-linear and the variation of intensity with $[NO_2]$ is steepest for emission at wavelengths closest to the excitation wavelength. The longer the wavelength of fluorescence the higher is the proportion of emission emanating from levels populated by collisional relaxation of higher levels. The observations show that vibrational relaxation predominates over total electronic quenching and occurs by stepwise de-excitation down the manifold of levels: in simple terms, the 'strong collision assumption' is invalid.

The lifetime of NO_2 states giving rise to fluorescence are long enough ($\tau_{rad} \sim 50\ \mu s$) to be measured directly. Analysis of the plots in *Figure 6.13*(a) indicate that collisions with unexcited NO_2 molecules remove roughly 12 kJ mol^{-1} of energy on each collision. The second panel in *Figure 6.13* compares what happens when the source wavelength is short enough to dissociate NO_2. Now, pressures in the range of 100 atmospheres are required if collisional deactivation is to compete with the predominant low pressure

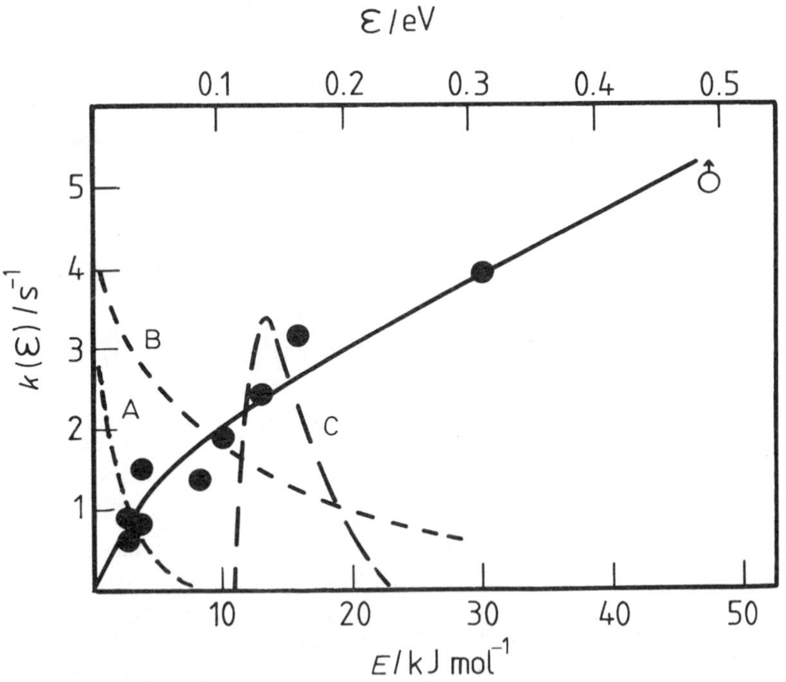

Figure 6.14. Specific rate constants for dissociation of NO_2 *from photolysis experiments (●) and calculated (–). The dashed curves indicate f(ε), the distribution of internal energies with* A *thermal activation at 300* K, **B** *thermal activation at 1500* K, *and* C *photochemical activation at 300* K [*adapted from Gaedtke H.*, et al., Chemical Physics Letters, **16,** 177 (1972)]

process, which is photodissociation. From these results, using the information on collisional deactivation gleaned from fluorescence quenching measurements, $k(\varepsilon)$ has been determined for a variety of photolysis wavelengths.

Figure 6.14 shows that the values of $k(\varepsilon)$ which are found experimentally agree rather well with the results of a statistical calculation based on equation (4.111). However, two factors were allowed for over and above those usually included in an RRKM calculation. First, the transition state was not assumed to have a fixed nuclear geometry. It was found that, as $(\varepsilon/\varepsilon^0)$ was increased, the minimum density of states appeared at a decreasingly small ON---O distance. Consequently, the transition state was redefined for each value of (ε, J), although its structure depended only very slightly on J.·In addition, the anharmonicity of vibrations in the reactant and transition state was recognized. As was mentioned in Section 4.5.5, this is important for processes involving *small* molecules, when the critical internal energy corresponds to an excitation of several quanta per vibrational mode.

Figure 6.14 also illustrates the advantage of photochemical studies of dissociation over thermal rate measurements. In thermal measurements at low temperature, the kinetics are dominated by the form of $k(\varepsilon)$ just above threshold. Raising the temperature probes $k(\varepsilon)$ over a wider range of values, but only at the expense of a considerable loss of definition. Photolysis experiments do not provide data for exactly defined internal energies owing to the different internal energies possessed by absorbing molecules. Nevertheless, the spread can be kept quite narrow, and the position of curve C in *Figure 6.14* can be altered by changing the wavelength of the photolysis source.

6.5.3 Photoisomerization of cycloheptatriene[55—58]

Thermal studies of the isomerization of cycloheptatriene (CHT)

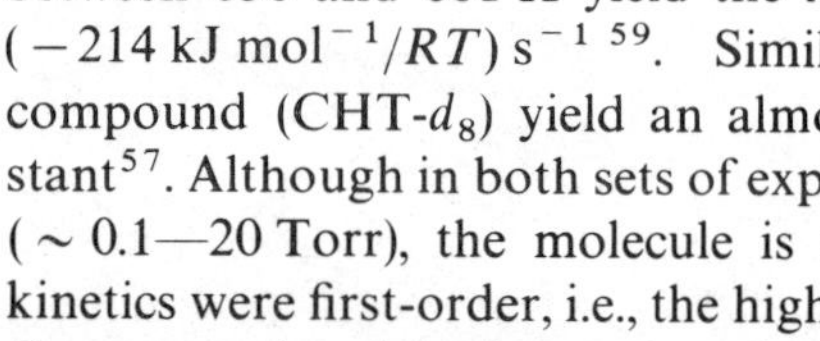

$$; \Delta H^\circ_{298} = -136\ \mathrm{kJ\ mol^{-1}} \qquad (6.69)$$

between 636 and 681 K yield the Arrhenius expression $k = 3.5 \times 10^{13} \exp(-214\ \mathrm{kJ\ mol^{-1}}/RT)\ \mathrm{s^{-1}}$ [59]. Similar experiments of the perdeutero-compound (CHT-d_8) yield an almost identical equation for the rate constant[57]. Although in both sets of experiments the total pressure was quite low ($\sim$ 0.1—20 Torr), the molecule is sufficiently complex to ensure that the kinetics were first-order, i.e., the high-pressure limiting rate had been reached. Consequently, the observed activation energy corresponds to E^0 for the reaction; the relatively low value of the pre-exponential factor (cf. data in *Table 5.9*) indicates, as one would expect, that the reaction proceeds via a rather tightly bound transition state.

Cycloheptatriene absorbs strongly below about 320 nm(≡ 375 kJ mol^{-1}). The excited molecules apparently undergo rapid internal conversion, thereby producing molecules in the electronic ground state with internal energy well in excess of the minimum required to isomerize. Thorough studies of this photoisomerization have been carried out, using a variety of sources to provide monochromatic radiation at wavelengths from 313.0 nm(≡ 383 kJ mol^{-1}) to 228.8 nm(≡ 524 kJ mol^{-1}). Although any molecule excited by one of these sources acquires internal energy greatly in excess of ε^0, the energized species, because of its large number of vibrational modes, can survive for a relatively long time. This means that collisional quenching of photochemically activated CHT molecules can be studied using pressures in the convenient range between 1 Torr and 1 atmosphere (in contrast to the experiments on NO_2 referred to earlier).

Plots of ϕ^{-1} against [M], the concentration of the gas responsible for collisional deactivation, are linear to a good degree of approximation. However, the slopes of these lines depend quite strongly on the identity of M.

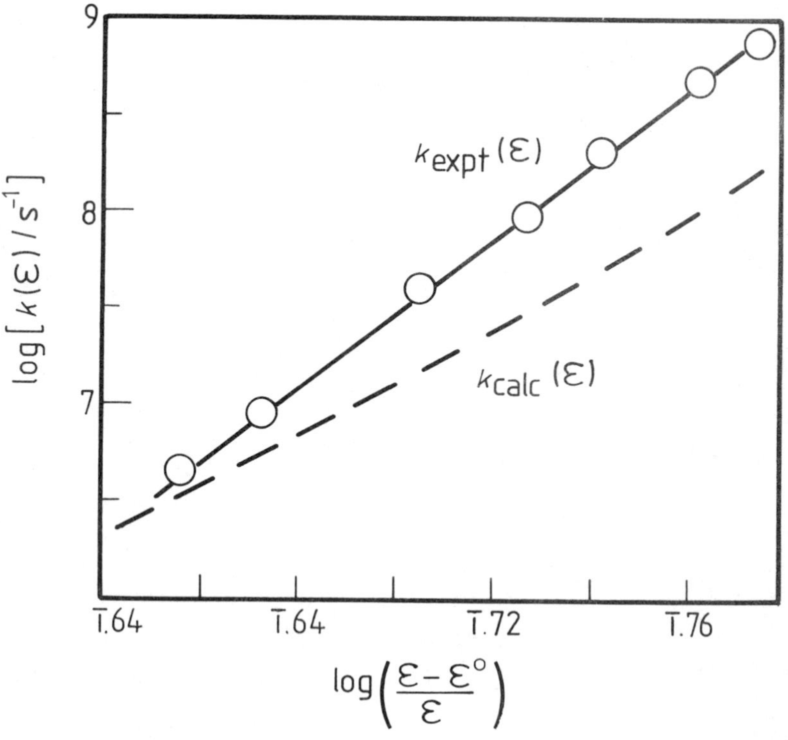

Figure 6.15. Comparison of $k_{expt}(\varepsilon)$ and $k_{calc}(\varepsilon)$ for photoisomerization of cycloheptatriene [*adapted from Atkinson, R., and Thrush, B. A.*, Proceedings of the Royal Society, **A316**, 131 (1970)]

CHT itself is one of the most effective quenching gases and, if it is assumed that excited molecules are completely deactivated at every collision, values of $k(\varepsilon)$ can be calculated for different wavelengths of excitation. As *Figure 6.15* shows, the resulting values of $k(\varepsilon)$ are appreciably larger than those calculated from equation (4.111), using the Whitten–Rabinovitch method to estimate the density of vibrational levels. A similar result is found for CHT-d_8. In these *large* molecules, the neglect of anharmonicity is unimportant, since ε corresponds, on average, to less than one quantum of excitation per vibration. The disagreement illustrated in *Figure 6.15* could reflect inadequacies in the theoretical treatment. On the other hand, it could result from a breakdown of the strong collision assumption. This would scarcely be surprising, since the internal energy in the photoexcited molecules is much greater than ε^0. In fact, more effective deactivating species than CHT, such as n-C_8H_{18}, have been discovered.

The results of experiments on the photoisomerization of CHT can provide information about the efficiency with which various M species remove energy from highly excited CHT molecules. A variety of analytical models can be used to extract this information. In these treatments, it is first necessary to make some assumption about the form of $k(\varepsilon)$. Then an adjustable function $P(\varepsilon'|\varepsilon)$ must be chosen to describe the distribution of final internal energies (ε') in CHT following collisions with M of CHT molecules with an initial internal energy (ε). The data in *Table 6.6* were derived using a form for $P(\varepsilon'|\varepsilon)$ which is based on a statistical description of the product energy distribution from CHT(ε) + M collisions. The values of (ΔE) that were deduced are not

Table 6.6 Average energies removed from CHT with ε = 461 kJ mol^{-1} in collisions with M at 293 K[a]

	$-\langle\Delta\varepsilon\rangle$/kJ mol^{-1}	
M	*with* $k_{expt}(\varepsilon)^b$	*with* $k_{calc}(\varepsilon)^c$
CHT	30	9
He	1.2	0.4
Xe	5.2	1.8
H_2	2.3	0.6
N_2	2.6	0.7
CO_2	8.4	3.2
CH_4	7.4	2.8
SF_6	8.1	3.1
n-C_4F_{10}	20.	6.7
n-C_8H_{18}	–	17.

[a] Data from LUU, S. H., and TROE, J., *Berichte der Bunsengesellschaft für Physikalische Chemie*, **78**, 766 (1974)
[b] Calculated on the basis that the strong collision assumption is valid for $CHT^{\dagger}$ + n-C_8H_{18} collisions
[c] Calculated on the basis that $k(\varepsilon)$ has the value given by an RRKM treatment using the Whitten–Rabinovitch modification

apparently very sensitive to the detailed form of this function. Two values of $\langle \Delta E \rangle$ are given for each M. These were calculated using the same $P(\varepsilon'|\varepsilon)$ but different assumptions about $k(\varepsilon)$, as explained in the footnotes to the table. They probably represent upper and lower limits to the average amount of energy actually transferred in these collisions.

6.6 Selective excitation: bimolecular reactions[46]

6.6.1 Translational excitation of reagents[40, 60]

Before considering the form of some experimentally determined excitation functions, it may be as well to recapitulate briefly on the predictions of simple theoretical models. According to these, a bimolecular reaction in which an atom is transferred from a saturated molecule to a free radical is usually characterized by a potential with a significant energy barrier. Consequently, the excitation function should exhibit a more or less well-defined threshold, ε_T^0, and rise monotonically at higher collision energies, until it eventually levels out at some asymptotic value. Averaged over a thermal distribution of collision energies, this leads to rate constants which increase with temperature in a manner which can usually be described by the Arrhenius equation. More sophisticated arguments suggest that the potential barrier usually occupies an 'early' position on the surface for an exoergic reaction and that translational energy is especially effective in promoting reaction, whereas endoergic reactions have late barriers and are selectively enhanced by vibrational excitation of the bond which is broken in the reaction.

The form of $S_r(\varepsilon_T)$ may be quite different for ion–molecule reactions. Now the relatively strong long-range attractive forces, which depend little on orientation, may play an important role in the dynamics, allowing close contact in any collisions which surmount the centrifugal barrier on the effective potential. If this is so, there should be no threshold, and $S_r(\varepsilon_T)$ should decrease as the collision energy is raised—at least until it reaches the 'hard-sphere' collision value. This variation of $S_r(\varepsilon_T)$ is very different from that described in the first paragraph. Between these two extremes, one might expect to discover other forms of $S_r(\varepsilon_T)$; for example, reactions between two radicals, where there is no potential energy barrier for favourable orientations, may constitute an intermediate class.

Although there have been numerous experiments on reactions of photochemically generated hot H (or D) atoms, a reasonably accurate excitation function has been determined in only one case[61]. Several attempts have been made to study isotopic variations of the $H + H_2$ reaction. One study of the photolysis of DI in the presence of a large excess of H_2 led to the first determination of a phenomenological threshold energy for the $D + H_2$ reaction[62]. Unfortunately, hot-atom studies of the H_3 reactions, which are theoretically so interesting, are bedevilled by even greater difficulties than is usual in experiments of this kind. In particular, the spread of collision energies

for a specified deuterium atom laboratory velocity is unusually wide, because of the broad thermal distribution of velocities in the light target molecules of H_2. This is especially frustrating in the important range of collision energies close to threshold. A second problem is that the relative rates of the reactions

$$H + DX \rightarrow HD + X \qquad (6.70a)$$

and

$$H + DX \rightarrow HX + D \qquad (6.70b)$$

are not well known. Nevertheless, it was possible to derive a threshold energy for the $D + H_2$ reaction of approximately 24 kJ mol^{-1} and show that $S_r(\varepsilon_T)$ increases monotonically to reach a value of approximately 0.03 nm^2 at $\varepsilon_T \sim 140$ kJ mol^{-1}, which was the highest collision energy that was studied.

The problems associated with interpreting the results of photochemical hot atom experiments have largely been overcome in measurements on the reaction[61]

$$H + n\text{-}C_4D_{10} \rightarrow HD + C_4D_9 \qquad (6.71)$$

Abstraction of the secondary D-atoms is approximately 40 kJ mol^{-1} exoergic and the reaction has an activation energy of about 30 kJ mol^{-1}. Hot H atoms were produced with initial energies between 34 and 198 kJ mol^{-1}, by photolysis of HBr using several monochromatic sources. As C_4D_{10} is much heavier than H, the relative energies of the 'first collisions' are close to the laboratory energies of the hot atoms and almost monoenergetic for a given photolysis wavelength, since the spread of C_4D_{10} thermal velocities is quite narrow. Subsequent reactions of the C_4D_9 radicals did not complicate the analysis and reaction cross-sections were obtained by measuring relative yields of HD and H_2.

The excitation function for reaction (6.71), which is shown in *Figure 6.16*, has two specially interesting features. The first is that $S_r(\varepsilon_T)$ passes through a maximum at a collision energy only about 3 times greater than the threshold at 34 kJ mol^{-1}. Trajectory calculations on systems of three atoms indicate that this can happen at energies where new reactive pathways, such as collisional dissociation, open up, but the collision energy in this case is far too small for that. It appears that the decrease in $S_r(\varepsilon_T)$ may be the result of collisions in which the trajectories traverse the transition state region twice. As the collision energy becomes large, an increasing proportion of collisions which cross the energy barrier may fail to turn the corner leading to the exit valley and separated products. These trajectories are reflected from the repulsive wall at the far side of the exit valley and recross the barrier.

A second interesting aspect of the results shown in *Figure 6.16* is the steep rise in $S_r(\varepsilon_T)$ immediately after ε_T^0. The increase predicted by a normalized line-of-centre model is appreciably slower. It may be that relative translational energy is especially effective at promoting reaction because the potential barrier is located in the approach portion of the energy surface. This position would be consistent with the exothermicity of reaction (6.71).

Excitation functions for a few neutral–neutral reactions and numerous

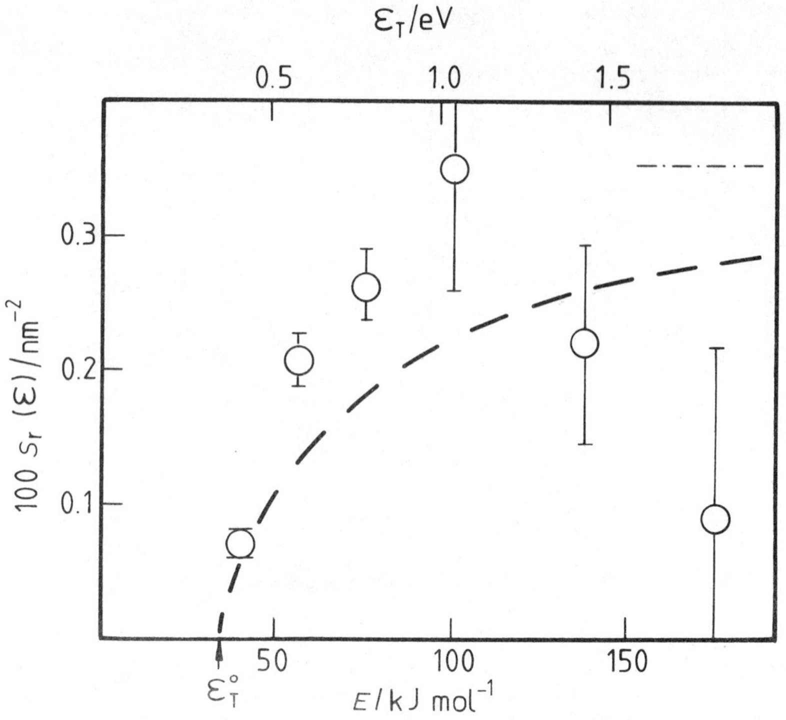

Figure 6.16. Reaction cross-sections for $H + n\text{-}C_4D_{10} \rightarrow HD + C_4D_9$-(sec). *The dashed line shows the excitation function for a simple collision theory model using the observed threshold and an asymptotic limit equal to the maximum observed value of* $S_r(\varepsilon)$ [*adapted from Gann, R. G.*, et al., Journal of Chemical Physics, **54**, 2304 (1971)]

ion–molecule reactions have been measured in crossed molecular beam experiments. It is appropriate to consider some examples of these measurements here, even though the application of molecular beam experiments to reaction dynamics is really the subject of Chapter 7. Particularly detailed results have been obtained for the reactions:

$$K + CH_3I \rightarrow KI + CH_3; \quad \Delta H_0^\circ = -92 \text{ kJ mol}^{-1} \qquad (6.72)$$

and

$$K + HCl \rightarrow KCl + H; \quad \Delta H_0^\circ = +6 \text{ kJ mol}^{-1} \qquad (6.73)$$

The excitation functions for these reactions are shown in *Figure 6.17*. They make an interesting comparison.

Reaction (6.72) is believed to occur via an 'electron-jump' mechanism[63]. If the reagents approach to a distance of about 0.32 nm, the difference between the ionization potential of K and the electron affinity of CH_3I is equal to the

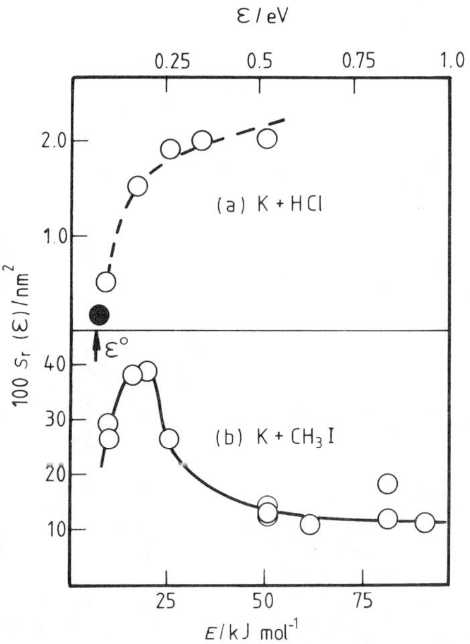

Figure 6.17. Comparison of the excitation function for (a) $K + HCl \rightarrow KCl + H$ *and* (b) $K + CH_3I \rightarrow KI + CH_3$ [*adapted from Pruett J. G.*, et al., Journal of Chemical Physics, **63**, 1173 (1975) *and Gersh, M. E., and Bernstein, R. B.*, Journal of Chemical Physics, **56**, 6131 (1972)]

Coulombic attraction between two ions of opposite charge. Therefore, an electron can be transferred 'instantaneously' to CH_3I from the K atoms, or, put another way, the system has a finite chance of switching from a surface which is essentially covalent to one dominated by the Coulombic attraction.

The form of the excitation function for reaction (6.73) has been explained in two ways. Trajectory calculations on three particles moving on an adiabatic surface, which has covalent character at large reactant separations and ionic character when the three particles are close together, have reproduced the behaviour of the observed excitation function remarkably well[64]. The decrease in $S_r(\varepsilon_T)$ at higher energies is associated with the increasing number of trajectories which recross the surface in phase space connecting the points where the ionic and covalent potential energies are equal. The possibility that the system does not remain on the lowest potential but continues through the avoided crossing region on the non-ionic potential cannot be treated properly in trajectory calculations. Qualitatively, the probability of non-adiabatic transitions should increase to higher ε_T and further improve the agreement between experiment and calculation.

A simpler heuristic explanation for the observed variation of $S_r(\varepsilon_T)$ has also been proposed[65]. This is based on a recognition that the excitation function at low ε_T is similar to that predicted by 'simple collision theory', whereas at higher ε_T it resembles that found when the potential is dominated by long-range

attraction. A model potential combining a low orientation-independent barrier of 4.2 kJ mol^{-1} at $r_{K–CH_3I}$ of 0.45 nm, with an attractive Coulombic attraction at smaller separations, reproduces the observations remarkably well. At higher collision energies, most of the collisions with impact parameters $b < 0.45$ nm surmount the 4.2 kJ mol^{-1} barrier but an increasing fraction scatter off the centrifugal barrier without reacting. The simplicity of this model is appealing, and it emphasizes the position of this reaction intermediate between atom-transfer reactions with large barriers and ion–molecule reactions with long-range attractive forces. In contrast to the conclusions from the trajectory calculations, however, this model predicts that it is large impact parameter collisions which at high energy fail to react in spite of crossing the potential energy barrier.

In contrast to that with CH_3I, the reaction of K atom with HCl is slightly endothermic. The variation of reaction cross-sections is matched quite well by an excitation function of the form derived from simple collision theory; $S_r(\varepsilon_T) = \pi D^2(1 - \varepsilon_T^0/\varepsilon_T)$ with $\pi D^2 = 0.0226$ nm^2 and $\varepsilon_T^0 \approx 6.4$ kJ mol^{-1}[166]. However,

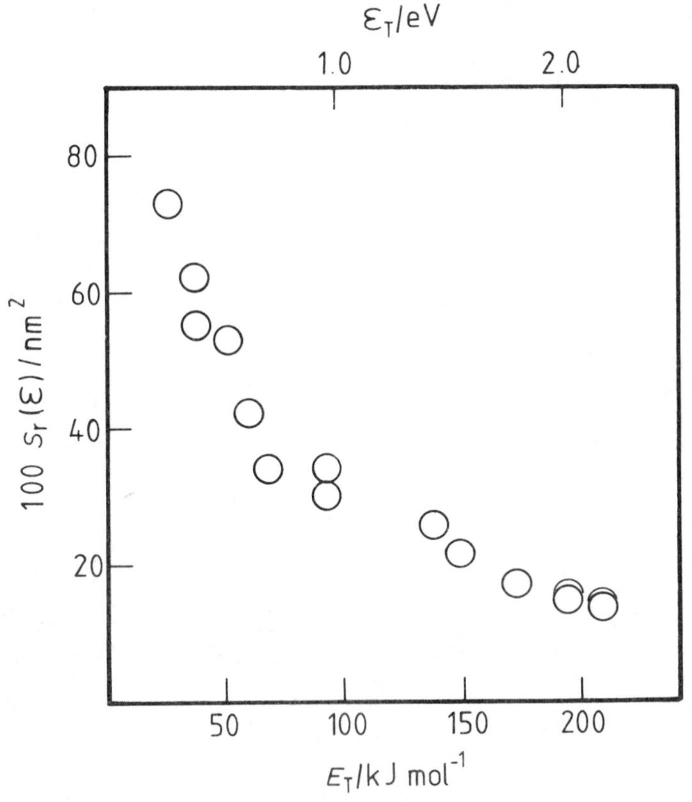

Figure 6.18. Excitation function for the reaction $Ar^+ + D_2 \rightarrow ArD^+ + D$ [*data from Giese, C. F., and Maier, W. B.,* Journal of Chemical Physics, **39**, 739 (1963)]

the agreement must be regarded as fortuitous, since the value of D is unreasonably small for an idealized hard-sphere model. Furthermore, in a parallel series of beam experiments, the effect was observed of exciting a proportion of the HCl to v = 1, using a pulsed HCl chemical laser[67]. In these experiments, using beams with a thermal spread of translational energies, the velocity-averaged reaction cross-section increased from $\langle S_r(\text{v} = 0) \rangle \sim 1.5 \times 10^{-3}$ nm^2 to $\langle S_r(\text{v} = 1) \rangle \sim 0.2$ nm^2. The latter value is about an order-of-magnitude greater than that for K + HCl(v = 0) collisions with the excess of energy present totally as relative translation of the reagents.

The preference for reactant vibrational excitation which these experiments reveal is consistent with reaction across a potential surface with a late energy barrier. On the other hand, it might seem that the lack of *very* selective energy requirements for reaction and the fairly steep rise of $S_r(\varepsilon_T)$ for K + HCl(v = 0) is difficult to reconcile with a late barrier position. However, it should be remembered that reaction (6.73) is only slightly endothermic and that the mass combination is unusual, the ejected H atom being much lighter than the other two. The scaled and skewed surface for collinear collisions has axes that are nearly rectlinear but the entrance valley is considerably elongated relative to the exit valley. This may reduce the difference between the effects of vibrational and translational reactant excitation.

Finally, *Figure 6.18* shows a characteristic excitation function for an ion–molecule reaction at moderate collision energies[68]. The form of $S_r(\varepsilon)$ contrasts sharply with those for reactions between neutral species, and the monotonic fall to higher collision energies is consistent with the Langevin analysis based on long-range attraction. It is re-emphasized, however, that this model is oversimplified once $\varepsilon_T \geqslant 50$ kJ mol^{-1}. An additional factor leading to the decrease in $S_r(\varepsilon_T)$ at high collision energies will be the increasing fraction of encounters in which the reactants, having come into hard close contact, bounce back along the entrance valley on the potential hypersurface, with or without energy being transferred[69].

6.6.2 Internal excitation of reagents

Relatively little of what is known about the influence of vibrational or rotational excitation on the rates of bimolecular reactions has been deduced directly from experiment. The difficulties associated with such experiments were referred to in Section 6.4.2. In Section 6.3.5, it was pointed out that values of $k(|\text{v}', \text{j}'; T)$ for an endoergic reaction can be accurately calculated, if values of $k(\text{v}', \text{j}'|; T)$, i.e., the detailed rate constants for the reverse exoergic reaction *into* specified product levels, are known. Before illustrating this, it is useful to show how the information–theoretic analysis of the detailed rate constants which are obtained in this and other ways can provide a basis for saying how *selectively* a reaction is promoted by excitation of a particular degree of freedom in the reagents[70]. However, it is first necessary to modify slightly the definition of surprisal that is given in Section 4.6.2.

A case will be considered where both the translational and rotational

degrees of freedom have thermal distributions, and the vibrational surprisal associated with the partially detailed rate constants is given by

$$I(\mathrm{v}'; T) = -\ln[k(|\mathrm{v}'; T)/k^0(|\mathrm{v}'; T)] \tag{6.74}$$

It can be shown that this equation corresponds to (4.126) at a given temperature, and $P^0(\mathrm{v}')$ is the limit of $k^0(\mathrm{v}'|; T)/k^0(T)$ when $(-\Delta\varepsilon_0/k_BT) \gg 0$. Because the prior rates, as well as the actual rates, must conform to equation (6.47), $I(\mathrm{v}'; T)$ is independent of whether the forward or reverse reaction is being considered, so

$$\begin{aligned} I(\mathrm{v}'; T) &= -\ln[k(|\mathrm{v}'; T)/k^0(|\mathrm{v}'; T) \\ &= -\ln(k(\mathrm{v}'|; T)/k^0(\mathrm{v}'|; T)] \\ &= -\ln[P(\mathrm{v}')/P^0(\mathrm{v}')] \end{aligned} \tag{6.75}$$

When the surprisal plot is linear, one can write

$$I(\mathrm{v}'; T) = I_0(T) + \Lambda_{\mathrm{v}'}\varepsilon_{\mathrm{v}'}/k_BT \tag{6.76}$$

so that

$$k(|\mathrm{v}'; T) = k^0(|\mathrm{v}'; T)\exp[-I_0 - \Lambda_{\mathrm{v}'}\varepsilon_{\mathrm{v}'}/k_BT] \tag{6.77}$$

where $\Lambda_{\mathrm{v}'}$ is related to $\lambda_{\mathrm{v}'}$, the parameter introduced in equation (4.128), by

$$\Lambda_{\mathrm{v}'} = \lambda_{\mathrm{v}'}(k_BT/\bar{\varepsilon}_{\mathrm{total}}) \tag{6.78}$$

A reaction is enhanced preferentially or selectively by reactant vibrational excitation if $k(|\mathrm{v}'; T)$ increases more rapidly with v′ than $k^0(|\mathrm{v}'; T)$: that is, if the acceleration is not simply associated with the increased volume of phase space accessible to the reaction products. For the reverse of the direct exothermic reactions which were considered in Section 6.3, $\Lambda_{\mathrm{v}'}$ is negative. Vibrational excitation enhances these endothermic reactions *selectively*.

Figure 6.19 illustrates the use of the equations that were derived in the previous paragraph. It compares a surprisal plot of the CO(v′) distribution from the exothermic reaction (6.23) with the partially detailed rate constants for the reverse reaction. The values of $\log[k(|\mathrm{v}'; T)]$ fall on a steeply rising line. This is neither very surprising nor very revealing. The prior rate constants for the forward and reverse reactions are related through equation (6.47), i.e.,

$$k^0(|\mathrm{v}'; T) = k^0(\mathrm{v}'|; T)\left(\frac{Q_{\mathrm{int}}}{Q'_{\mathrm{R,v}'}}\right)\left(\frac{\mu}{\mu'}\right)^{\frac{3}{2}}\exp(\Delta\varepsilon_{\mathrm{v}',0}/k_BT) \tag{6.79}$$

Substitution into (6.77) shows clearly that a large part of the increase in $k(|\mathrm{v}'; T)$ is merely related to the steep increase in the prior rate constants as $\Delta\varepsilon_{\mathrm{v}',0}$ is reduced. A much smaller fraction of the increase is related to specially selective energy requirements. This is one reason why detailed information about state-selected kinetics is always easiest to obtain by studying the product state distribution of the exoergic process and by using the equations derived from the principle of microscopic reversibility. There is, as shown earlier, a limitation to this procedure: the highest values of v and v′ for which detailed rate constants can be found are those which are energetically accessible in

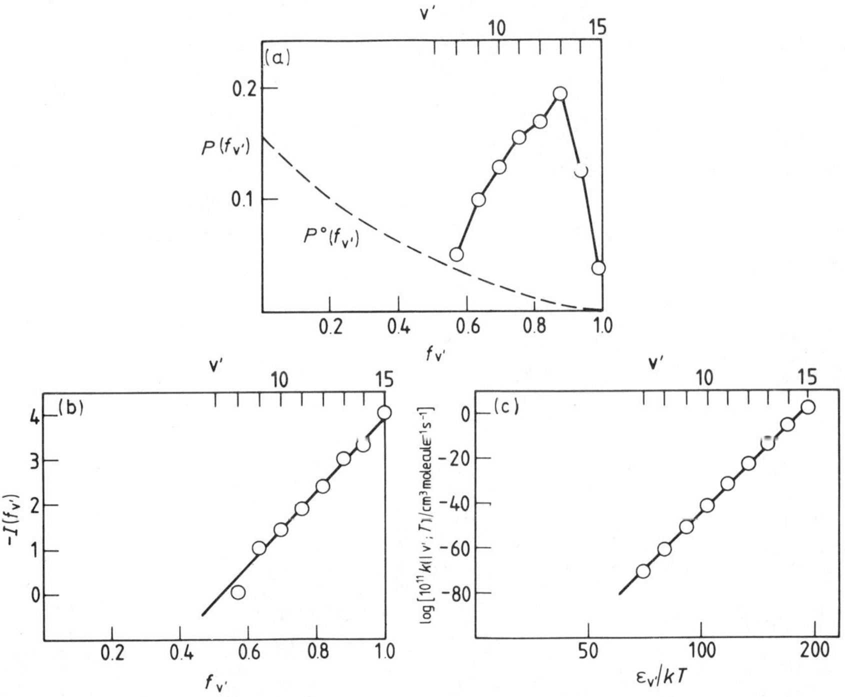

Figure 6.19. Information-theoretic analysis of detailed rate constants for the reaction $O + CS \rightleftarrows CO(v') + S$. *Panel* (a) *compares* $P(f_{v'})$, *the observed distribution over* v′ *from the exothermic reaction, to* $P^\circ(f_{v'})$, *the distribution expected on prior grounds. Panel* (b) *shows the surprisal associated with this distribution. In panel* (c), *the values of* $k(|v';T)$, *calculated from detailed balancing, are plotted against* $\varepsilon_{v'}/k_B T$ *for* $T = 300$ K [*reproduced , with permission, from Smith, I. W. M.*, Specialist Periodical Reports, Gas Kinetics and Energy Transfer, vol. 2, *Eds. Ashmore, P. G., and Donovan, R. J., Chemical Society, London* (1977)]

measurements on the exoergic reaction. If these experiments are carried out at room temperature, $v_{max} = 0$ and v'_{max} is set by $(-\Delta H_0^\circ + E_{act} + 5/2RT)$ for the reaction. To go further, the reagents must be disturbed from a thermal distribution. This is easier to do in a trajectory calculation than in an experiment.

Some of the results of trajectory studies which were specifically designed to explore the energy requirements for reaction were discussed in Section 3.5.4. The relationship between selective energy consumption and barrier location was described. Broadly speaking, as the potential barrier is moved to a progressively later position along the reaction path of minimum energy, then the more 'useful' vibrational, rather than translational, energy is in promoting reaction. The information–theoretic approach supplies a method for reporting the selectivity of the energy requirements quantitatively. For direct endoergic reactions with late barriers, the calculated values of $k(|v';T)$ are found to rise more steeply than the prior rate constants as long as $\Delta\varepsilon_{v',0}$ in equation (6.79) is

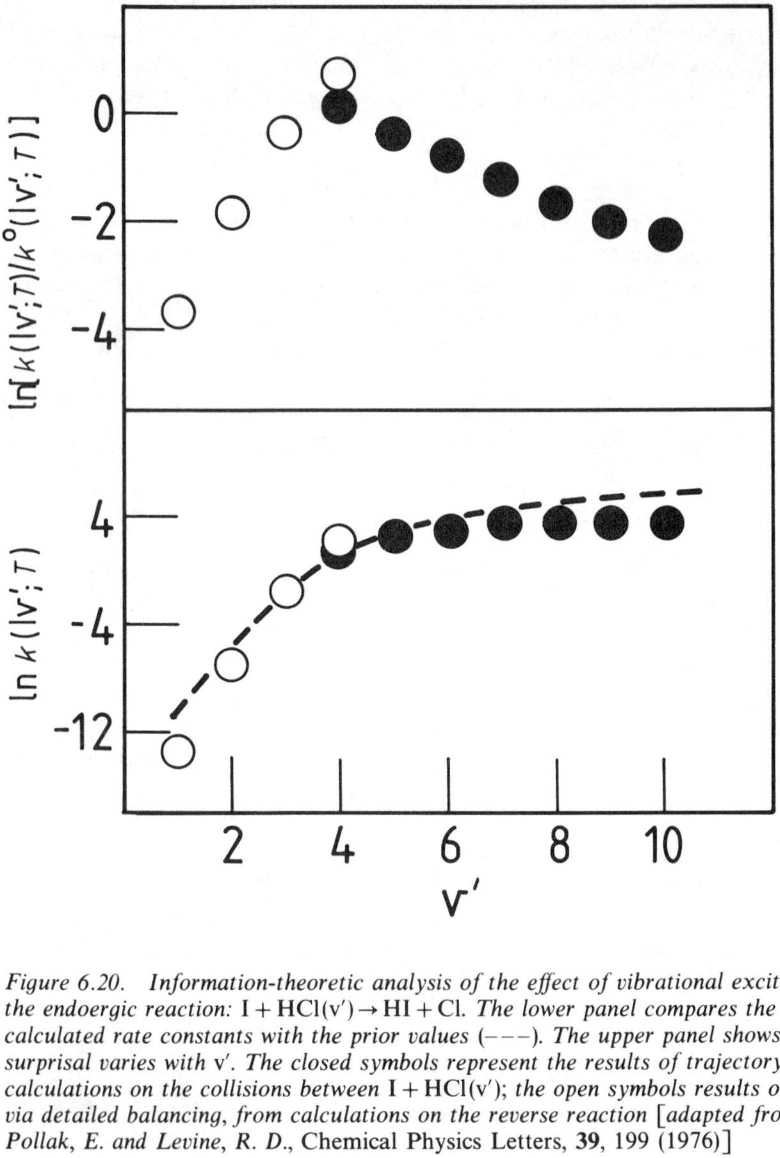

Figure 6.20. Information-theoretic analysis of the effect of vibrational excitation on the endoergic reaction: $I + HCl(v') \rightarrow HI + Cl$. *The lower panel compares the calculated rate constants with the prior values (---). The upper panel shows how the surprisal varies with v'. The closed symbols represent the results of trajectory calculations on the collisions between* $I + HCl(v')$; *the open symbols results obtained, via detailed balancing, from calculations on the reverse reaction [adapted from Pollak, E. and Levine, R. D.,* Chemical Physics Letters, **39**, 199 (1976)]

negative. This behaviour is exemplified, in *Figure 6.20*, by some results for the reaction[70]

$$I + HCl(v') \rightarrow HI + Cl; \Delta H^\circ_0 = +133.5 \text{ kJ mol}^{-1} \qquad (6.80)$$

However, once $\Delta\varepsilon_{v',0}$ becomes positive—which, for reaction (6.80), occurs when $v' \geqslant 4$—the prior rate constants increase more rapidly than the actual calculated values. Vibrational excitation is still promoting the reaction

selectively, but the extent of this selectivity decreases with further increases in v′. For ergoneutral reactions, such as $H + H_2(v')$, $\Delta\varepsilon_{v',0} \geqslant 0$ for all values of v′. These reactions seem to be selectively enhanced by reactant vibrational energy but the magnitude of the selectivity falls monotonically, a result which is consistent with those for endoergic reactions[71]. The general conclusion is that differences between the effects of vibrational and translational excitation are most marked at total energies close to threshold.

Quite recently, direct studies have been begun of the chemical kinetics of molecules excited vibrationally by infrared lasers. Here, consideration is given only to systems where molecules are prepared in a well-defined state as a result of absorbing single photons from relatively low power lasers. To exemplify investigations of this kind, we consider experiments on the reactions

$$O + HCl(v') \rightarrow OH + Cl;\ \Delta H_0^\circ = +3.8\ \text{kJ mol}^{-1} \qquad (6.81a)$$

$$Br + HCl(v') \rightarrow HBr + Cl;\ \Delta H_0^\circ = +65.6\ \text{kJ mol}^{-1} \qquad (6.82a)$$

whose energetics are shown in *Figure 6.21.*

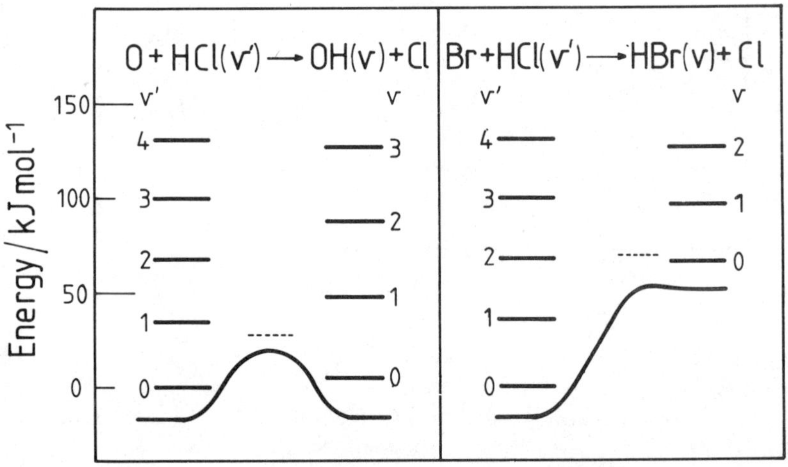

Figure 6.21. Energetics of the reaction of oxygen and bromine atoms with HCl *in different vibrational levels. The dotted lines indicate approximately the zero-point energies in the transition state*

The thermal reaction between O atoms and HCl is quite slow. The activation energy is approximately 25 kJ mol^{-1} and $k(T = 298\ \text{K}) = 1.2 \times 10^{-16}$ cm^3 molecule^{-1} s^{-1}. By exciting HCl to v′ = 1, using a pulsed chemical laser, and making measurements on the decay of the subsequent infrared fluorescence, it has been established that the *total* rate constant for removal of HCl(v′ = 1), by (6.81b)

$$O + HCl(v') \rightarrow O + HCl(v'' < v') \qquad (6.81b)$$

as well as by (6.81a), is 9.3×10^{-13} cm^3 molecule^{-1} s^{-1} [72, 73]. This is almost four orders-of-magnitude larger than the thermal rate constant, $k_a(v' = 0)$, at the same temperature. Clearly, experiments to determine the relative contributions to $k_{a+b}(v' = 1)$ of the reactive (6.81a) and non-reactive (6.81b) processes are crucial. They are also difficult to perform. Present indications are that $k_a(v' = 1)$ is only between 250 and 750 times $k_a(v' = 0)$[73, 74], which means that chemical reaction only accounts for a small fraction of the total loss of HCl($v' = 1$). Interestingly, the experimental value of $k_a(v' = 1)$ is quite consistent with that calculated from quasiclassical trajectories using a LEPS potential[75]. The barrier on this potential was positioned roughly halfway along the reaction path of minimum energy and its height was chosen to yield an activation energy in agreement with the experimental value. The trajectory study also indicated that $k_a(v')$ should increase a further 13 times as v' is increased from 1 to 2. Experiments on HCl($v' = 2$) yield $k_{a+b}(v' = 2) = 5.3 \times 10^{-12}$ cm^3 molecule^{-1} s^{-1} [73]. It appears that chemical reaction plays a larger, though not necessarily major, role in the removal of HCl excited to $v' = 2$.

The observations show that non-reactive relaxation of HCl(v') by O atoms is several orders-of-magnitude faster than by atoms which are not 'potentially reactive'. This may be the result of the different shapes of the adiabatic potentials for these systems. However, the major mechanism appears to be one involving electronically non-adiabatic transitions between the (three) electronic states which arise when O(^3P) and HCl interact[73].

A similar non-adiabatic mechanism appears to be responsible for the relaxation of HCl($v' = 1$) by Br atoms[76]:

$$\mathrm{Br + HCl(v') \rightarrow Br + HCl(v'' < v')} \qquad (6.82b)$$

The reactive channel (6.82a) remains strongly disfavoured energetically until $v' = 2$. Results of experiments on Br + HCl($v' = 2$) reveal an interesting discrepancy. The thermal rate constant and the product vibrational distribution for reaction (6.82a) in the reverse, exoergic, direction have both been measured[77, 78]. Since the Boltzmann laws ensure that the reacting HCl molecules in these experiments are overwhelmingly in $v = 0$, these measurements together yield: $k_a(v' = 2|v = 0; T) = 1.5 \times 10^{-12}$ cm^3 molecule^{-1} s^{-1}. Straightforward application of the principle of detailed balancing, which requires the use of equation (6.47), allows one to derive: $k_a(v = 0|v' = 2; T) = 4.0 \times 10^{-12}$ cm^3 molecule^{-1} s^{-1} for the reaction as written in equation (6.82a). However, direct measurements on the *total* rate of loss of HCl($v' = 2$) in collisions with Br atoms indicate $k_{a+b}(|v' = 2; T) = 1.8 \times 10^{-12}$ cm^3 molecule^{-1} s^{-1} [76].

This example has been quoted to emphasize how the equations based on the principle of microscopic reversibility can be used to examine critically experimental results in a few favoured cases. A second general point which emerges from our brief inspection of the O + HCl(v') and Br + HCl(v') systems is that reaction and non-reactive relaxation may frequently occur in parallel processes at competing rates. This is especially likely if either reactant has

orbital angular momentum, since then multiple potential hypersurfaces exist, and relaxation may occur by an electronically non-adiabatic transition between different electronic states.

For a final example of the influence of internal energy on chemical reactivity, we turn to the field of ion–molecule reactions. Systems involving ions offer special opportunities for the control of reactant energy, since by applying electric fields the translational velocity of the ions can be varied and well-defined. Less obviously, selection of the vibrational distribution is possible if ions are produced by photoionization. By controlling the monochromatic wavelength producing the ions and knowing the Franck–Condon factors for the transition from the neutral to the ionized molecule, it is possible to create diatomic ions with known, and variable, vibrational distributions for studies of their subsequent reactions with neutral species[79].

For those exoergic ion–molecule reactions where the simple Langevin model is reasonably satisfactory, one might expect that the rate of collisions surmounting the centrifugal barrier—and therefore the overall reaction rate—would be altered very little by internal excitation of the reagents. If, however, more than one distinct set of products are possible, extra reactant energy might alter the relative importance of these channels. If no product channel is strongly exoergic and the reaction proceeds via a collision complex, the break-up of the complex to the original reagents may become important. This process would become more favoured at high total energies, so the rate might decrease slightly as the reactant excitation is increased.

The qualitative behaviour which has just been described has been found for several exoergic ion–molecule reactions. For example, for the reaction

$$H_2 + H_2^+(v) \rightarrow H_3^+ + H;\ \Delta H_0^\circ = -37\ \text{kJ mol}^{-1} \tag{6.83}$$

the reaction cross-section falls gradually as v increases from 0 to 4, the value for H_2^+(v = 4) being 0.79 times that for H_2^+(v = 0).

Extensive data exist for the endoergic reaction

$$He + H_2^+(v') \rightarrow HHe^+ + H;\ \Delta H_0^\circ = +74.5\ \text{kJ mol}^{-1} \tag{6.84}$$

The experimental results, which are summarized in Table 6.7, show that vibrational excitation of H_2^+ is much more effective than relative translational energy in promoting reaction. *Ab initio* calculations[80] indicate a well depth of only about 10–20 kJ mol^{-1} for H_2He^+ relative to the asymptotic limit for H_2^+ + He. This is much less than the experimental collision energies, so the influence of the potential well is small, reaction proceeds directly, and any explanation based on the Langevin model is inappropriate. The *ab initio* surface shows the late barrier characteristic of endoergic reactions, and the selective enhancement by reactant vibrational energy is consistent with this. In addition, it has been suggested[81] that the acceleration of reaction as v′ is increased is also influenced by an increase in attraction between the reactants when the H_2^+ bond is extended.

Table 6.7 Relative probability for the reaction $He + H_2^+ (v') \rightarrow HHe^+ + H$ as a function of 'total' reagent energy and of v'

	Relative probability			
v'	*Total energy* = 1.0 eV 96.5 kJ mol^{-1}	2.0 eV 193. kJ mol^{-1}	3.0 eV 289.5 kJ mol^{-1}	4.0 eV 386. kJ mol^{-1}
0	0.06	0.10	0.13	0.17
1	0.49	0.35	0.31	0.25
2	1.95	0.93	0.55	0.34
3	–	1.70	0.99	0.56
4	–	2.35	1.22	0.68
5	–	2.49	1.70	0.89

6.7 State-to-state rate constants

The ultimate goal of the experiments which have been described in this chapter is the provision of information at a level of detail and accuracy similar to that enjoyed by the spectroscopist. Thus, for many *radiative* transitions between states of diatomic molecules, the frequencies and relative intensities are known accurately for spectral lines connecting many pairs of v, j levels. These results can be used to construct accurate potential curves. Similarly, the reaction dynamicist seeks to measure a large matrix of detailed rate constants which connect fully defined reactant and product states by *chemical reaction*, in the belief that it should then be possible to map the potential energy hypersurface for the reaction with reasonable accuracy.

The reaction systems which have been considered in this chapter include examples of those for which the most detailed kinetic data are available at the present time. Nevertheless, the experiments fall some way short of supplying the fine detail which the experimentalist is striving for. Indeed, this failure is implied by the division of the present chapter into separate sections on energy disposal and on selective excitation. Some few experiments have investigated how the product distribution changes when a 'non-Boltzmann' distribution of reagents are mixed. Two examples of investigations of this kind are considered next to indicate the current 'state-of-the-art' in this area of endeavour.

For the first example, we return to reaction (6.13) between F atoms and H_2. This system provides one with an unusual opportunity to investigate the role of reactant rotation, about which very little has so far been said. This opportunity arises because H_2 has widely spaced rotational levels and exists in ortho and para forms[82]. Careful measurements have been made of the relative rates of reaction of F atoms with para-H_2 and with normal H_2 (consisting of three parts ortho to one part para). At 175 K, para-H_2 was found to react 1.025 times faster than normal H_2, whilst no difference could be detected in the rates at 273 or 298 K. Infrared chemiluminescence and chemical laser studies have also been carried out to investigate how the distribution over product HF states is modified when normal H_2 is replaced by para-H_2. The CL experiments[18] indicated that the ratios $k(v' = 3|; T)$ and $k(v' = 2|; T)/k(v' = 1|; T)$ were larger from F + para-H_2. The IRC chemiluminescence measurements[83] provided somewhat more refined data. The fractional yield of vibrational energy, $\langle f_{v'} \rangle$, was determined as 0.70 from $F + H_2(j = 0)$, 0.67 from $F + H_2(j = 1)$, and 0.69 from $F + H_2(j = 2)$. The extent of product rotational excitation remains essentially unchanged. These relatively minor effects of varying the rotational state of a molecular reagent are consistent with expectations based on QCL trajectory calculations.

A more general technique has been developed in Polanyi's laboratory to investigate how 'excess' reactant energy is disposed of in chemical reactions. These experiments, in which chemical activation provides a non-thermal distribution of reactant, can best be described by referring to a specific example. A very recent study of the reverse reaction to (6.81a)[84], i.e.,

$$OH(v) + Cl \rightarrow HCl(v') + O; \ \Delta H_0^\circ = -3.8 \text{ kJ mol}^{-1} \qquad (6.85)$$

is ideal for this purpose. In the experiments, the OH reagent for (6.85) was formed vibrationally excited in one or other of two 'pre-reactions': either the reaction between H and NO_2 or that between H and O_3. The product distributions from these reactions are discussed in Section 6.3.3.

Infrared chemiluminescence from OH was observed a short way downstream from where the primary reaction occurred and in a region where the gas flow containing excited OH could be crossed by a gas stream containing Cl atoms. The infrared OH spectra were observed with and without the Cl atoms present and, from the 'chemiluminescence depletion', relative rates of reaction from the observed OH levels could be determined. Furthermore, the distribution of excited HCl formed in reaction(6.85) could be found from its chemiluminescent spectrum.

These sophisticated experiments indicate that the total rate of reaction (6.85) increased by roughly an order-of-magnitude as v rises from 1 to 10. Furthermore, the results indicate a very high degree of vibrational adiabaticity; that is, the OH vibrational energy is converted into HCl vibrational energy with close to 100 per cent efficiency. This conclusion is consistent with the results of quasiclassical trajectory studies, which reveal that *high* vibrational excitation tends to lead to trajectories which 'cut the corner' on the potential surface causing high mixed-energy release and high vibrational excitation in the product. This feature is accentuated in a system where a light atom is transferred between two much heavier atoms. As is pointed out in Section 3.5.4, the skewed and scaled surface for collinear collisions has reactant and product valleys at an acute angle to one another. As a result the representative mass point is very likely to enter the product valley from the side and in a direction almost perpendicular to its floor. This will clearly result in high vibrational excitation as the products separate.

Up to the present time, the observation of infrared chemiluminescence under conditions of 'arrested relaxation' has proved to be the most successful technique for measuring detailed rate constants. In these experiments, two 'sprays', or uncollimated beams, each containing one of the reagents, are mixed. The total pressure in the mixing zone is typically 1—2 N m^{-2} or 10 mTorr, which is sufficient to ensure that any excited chemical products make several collisions before they leave the volume observed by the spectrometer. The occurrence of collisions is important because, if the translational energy of the products were not moderated, the time molecules spent in the observation region would depend on their translational energy and therefore, in turn, on their internal state. Consequently, the observed steady-state distribution would be distorted to higher internal states from that actually formed in the reaction, and equation (6.5) could not be used to calculate the detailed rates. In practice, infrared chemiluminescence has not been observed in genuine crossed beam experiments. However, the ability to work under conditions where only single collisions occur in the region where the reagents mix bestows a number of immense advantages. In particular, the

distributions of reaction products over both space and relative translational velocity can, at least in principle, be observed. The results which can be obtained in crossed molecular beam studies of chemical reactions are considered in the next, final, chapter of this book.

References

Section 6.2

1 BRADLEY, J. N., *Fast Reactions*, Oxford University Press, Oxford (1975)
2 SMITH, I. W. M., in *The Excited State in Chemical Physics*, Ed. McGowan, J. W., John Wiley, New York (1975), vol XXVIII in the series *Advances in Chemical Physics*, 1
3 FARRAR, J. M., and LEE, Y. T., *Annual Review of Physical Chemistry*, **25**, 357 (1974)
4 HOLMES, B. E., and SETSER, D. W., in *Physical Chemistry of Fast Reactions*: Vol. 2, *Reaction dynamics*, Ed. Smith, I. W. M., Plenum, New York (1979)
5 CARRINGTON, T., and POLANYI, J. C., in *International Review of Science. Physical Chemistry*, series 1, vol. 9, *Chemical Kinetics*, Ed. Polanyi, J. C., Chap. 5, Butterworths, London (1972)
6 GOLDE, M. F., and THRUSH, B. A., *Reports on Progress in Physics*, **36**, 1285 (1973)
7 MOEHLMANN, J. G., and McDONALD, J. D., *Journal of Chemical Physics*, **59**, 6683 (1973)
8 SMITH, I. W. M., *Discussions of the Faraday Society*, **44**, 194 (1967)
9 BERRY, M. J., in *Molecular Energy Transfer*, Eds. Jortner, J., and Levine, R. D., Wiley, New York (1975)
10 BERRY, M. J., *Annual Review of Physical Chemistry*, **26**, 259 (1975)
11 HANCOCK, G., MORLEY, C., and SMITH, I. W. M., *Chemical Physics Letters*, **12**, 193 (1971); DJEU, N., *Journal of Chemical Physcis*, **60**, 4109 (1974)

Section 6.3

12 POLANYI, J. C., and WOODALL, K. B., *Journal of Chemical Physics*, **57**, 1574 (1972)
13 PERRY, D. S., and POLANYI, J. C., *Chemical Physics*, **12**, 419 (1976)
14 DOUGLAS, D. J., and POLANYI, J. C., *Chemical Physics*, **16**, 1 (1976)
15 BERRY, M. J., *Journal of Chemical Physics*, **59**, 6229 (1973)
16 COOMBE, R. D., and PIMENTEL, G. C., *Journal of Chemical Physics*, **59**, 251 (1973)
17 COOMBE, R. D., and PIMENTEL, G. C., *Journal of Chemical Physics*, **59**, 1535 (1973)
18 KROGH, O. D., STONE, D. K., and PIMENTEL, G. C., *Journal of Chemical Physics*, **66**, 368 (1977)
19 BENDER, C. F., PEARSON, P. K., O'NEILL, S. V., and SCHAEFER III, H. F., *Journal of Chemical Physics*, **56**, 4626 (1972)
20 LEVINE, R. D., and BERNSTEIN, R. B., *Accounts of Chemical Research*, **7**, 393 (1974)
21 LEVINE, R. D., JOHNSON, B. R., and BERNSTEIN, R. B., *Chemical Physics Letters*, **19**, 1 (1973)
22 POLANYI, J. C., and SCHREIBER, J. L., *Faraday Discussions of the Chemical Society*, **62**, 267 (1977)
23 SCHAFER, T. D., SISKA, P. E., PARSON, J. M., TULLY, F. P., WONG, Y. C., and LEE, Y. T., *Journal of Chemical Physics*, **53**, 3385 (1970)
24 SUNG, J. P., and SETSER, D. W., *Chemical Physics Letters*, **48**, 413 (1977)
25 ANLAUF, K. G., POLANYI, J. C., WONG, W. H., and WOODALL, K. B., *Journal of Chemical Physics*, **49**, 5189 (1968); MAYLOTTE, D. H., POLANYI, J. C., and WOODALL, K. B., *Journal of Chemical Physics*, **57**, 1547 (1972)
26 PARR, C. A., POLANYI, J. C., and WONG, W. H., *Journal of Chemical Physics*, **58**, 5 (1973)
27 POLANYI, J. C., SLOAN, J. J., and WANNER, J., *Chemical Physics*, **13**, 1 (1976)
28 ANLAUF, K. G., HORNE, D. S., MACDONALD, R. G., POLANYI, J. C., and WOODALL, K. B., *Journal of Chemical Physics*, **57**, 1561 (1972)
29 SCHMATJKO, K. J., and WOLFRUM, J., *Sixteenth International Symposium on Combustion*, The Combustion Institute (1977), and WOLFUM, J., *Berichte der Bunsengesellschaft für Physikalische Chemie*, **81**, 114 (1977)

30 WALSH, A. D., *Journal of the Chemical Society*, 2260—2332 (1953)
31 HEYDTMANN, H., and POLANYI, J. C., *Applied Optics*, **10,** 1738 (1971)
32 BERRY, M. J., *Journal of Chemical Physics*, **61,** 3114 (1974)
33 MOEHLMANN, J. G., and McDONALD, J. D., *Journal of Chemical Physics*, **62,** 3052 (1975)
34 KAPLAN, H., LEVINE, R. D., and MANZ, J., *Chemical Physics*, **12,** 447 (1976)
35 SMITH, I. W. M., in *Specialist Periodical Reports. Gas Kinetics and Energy Transfer*, vol. 2, Eds. ASHMORE, P. G. and DONOVAN, R. J. Vol. 2, Chap. 1, Chemical Society, London (1977)
36 ANLAUF, K. G., MAYLOTTE, D. H., POLANYI, J. C., and BERNSTEIN, R. B., *Journal of Chemical Physics*, **51,** 5716 (1969)

Section 6.4

37 STEINFELD, J. I., *Accounts of Chemical Research*, **3,** 313 (1970)
38 MITCHELL, A. C. G., and ZEMANSKY, M. W., *Resonance Radiation and Excited Atoms*, Cambridge University Press, Cambridge (1961)
39 TROE, J., in *Physical Chemistry. An Advanced Treatise*, Vol. VIB, Eds. Eyring, H., Henderson, D., and Jost, W., Chap. 11, Academic Press, New York (1975)
40 OLDERSHAW, G., in ref. 35, Chap. 3
41 SIMONS, J. P., in ref. 35, Chap. 2
42 KRAUSE, L., in ref. 2, 267
43 HUSAIN, D., *Berichte der Bunsengesellschaft für Physikalische Chemie*, **81,** 168 (1977)
44 WEITZ, E., and FLYNN, G., *Annual Review of Physical Chemistry*, **25,** 275 (1974)
45 WOLFGANG, R., *Progress in Reaction Kinetics*, **3,** 97 (1965)

Section 6.5

46 SMITH, I. W. M., in *Physical Chemistry of Fast Reactions*, vol. 2, *Reaction dynamics*, Ed. Smith, I. W. M., Plenum, New York (1979); MOORE, C. B., and SMITH, I. W. M., *Faraday Discussions of the Chemical Society*, **67,** (1979)
47 TROE, J., in ref. 35, Chap. 5.
48 BUTLER, J. N., and KISTIAKOWSKY, G. B., *Journal of the American Chemical Society*, **82,** 759 (1960)
49 RYNBRANDT, J. D., and RABINOVITCH, B. S., *Journal of Physical Chemistry*, **75,** 2164 (1971)
50 SMITH, I. W. M., *Accounts of Chemical Research*, **9,** 161 (1976)
51 QUACK, M., and TROE, J., *Berichte der Bunsengesellschaft für Physikalische Chemie*, **81,** 160 (1977)
52 GAEDTKE, H., and TROE, J., *Berichte der Bunsengesellschaft für Physikalische Chemie*, **77,** 24 (1973) and **79,** 184 (1975)
53 KEYSER, L. F., LEVINE, S. Z., and KAUFMAN, F., *Journal of Chemical Physics*, **54,** 355 (1971)
54 PAECH, F., SCHMIEDL, R., and DEMTRÖDER, W., *Journal of Chemical Physics*, **63,** 4369 (1975)
55 ATKINSON, R., and THRUSH, B. A., *Proceedings of the Royal Society*, *A*, **316,** 123 (1970)
56 ATKINSON, R., and THRUSH, B. A., *Proceedings of the Royal Society*, *A*, **316,** 131 (1970)
57 ORCHARD, S. W., and THRUSH, B. A., *Proceedings of the Royal Society*, *A*, **329,** 233 (1972)
58 LUU, S. H., and TROE, J., *Berichte der Bunsengesellschaft für Physikalische Chemie*, **77,** 325 (1973) and **78,** 766 (1974)
59 KLUMP, K. N., and CHESICK, J. P., *Journal of the American Chemical Society*, **85,** 130 (1963)

Section 6.6

60 DUBRIN, J., *Annual Review of Physical Chemistry*, **24,** 97 (1973)
61 GANN, R. G., OLLISON, W. M., and DUBRIN, J., *Journal of Chemical Physics*, **54,** 2304 (1971)

62 KUPPERMANN, A., *Nobel Symposium,* no. 5, Ed. Claesson, S., Interscience, New York (1967) p. 131

63 GERSH, M. E., and BERNSTEIN, R. B., *Journal of Chemical Physics,* **56,** 6131 (1972); BERNSTEIN, R. B., and RULIS, A. M., *Faraday Discussions of the Chemical Society,* **55,** 293 (1973)

64 LaBUDDE, R. A., KUNTZ, P. J., BERNSTEIN, R. B., and LEVINE, R. D., *Journal of Chemical Physics,* **59,** 6286 (1973)

65 HARRIS, R, M., and HERSCHBACH, D. R., *Faraday Discussions of the Chemical Society,* **55,** 121 (1973)

66 PRUETT, J. G., GRABINER, F. R., and BROOKS, P. R., *Journal of Chemical Physics,* **63,** 1173 (1975)

67 ODIORNE, T. J., BROOKS, P. R., and KASPER, J. V. V., *Journal of Chemical Physics,* **55,** 1980 (1971)

68 GERSH, C. F., and MAIER, W. B., *Journal of Chemical Physics,* **39,** 739 (1963)

69 MAHAN, B. H., *Journal of Chemical Education,* **51,** 308 and 377 (1974)

70 POLLAK, E., and LEVINE, R. D., *Chemical Physics Letters,* **39,** 199 (1976)

71 SMITH, I. W. M., *Chemical Physics Letters,* **47,** 219 (1977)

72 BROWN, R. D. H., GLASS, G. P., and SMITH, I. W. M., *Chemical Physics Letters,* **32,** 517 (1975)

73 MACDONALD, R. G., and MOORE, C. B., *Journal of Chemical Physics,* **68,** 513 (1978)

74 KARNY, Z., KATZ, B., and SZOKE, A., *Chemical Physics Letters,* **35,** 100 (1975)

75 BROWN, R. D. H., and SMITH, I. W. M., *International Journal of Chemical Kinetics,* **10,** 1 (1978)

76 LEONE, S. R., MACDONALD, R. G., and MOORE, C. B., *Journal of Chemical Physics,* **63,** 4735 (1975)

77 BERGMANN, K., and MOORE, C. B., *Journal of Chemical Physics,* **63,** 643 (1975)

78 DOUGLAS, D. J., POLANYI, J. C., and SLOAN, J. J., *Chemical Physics,* **13,** 15 (1976)

79 CHUPKA, W. A., in *Ion–Molecule Reactions,* Vol. 1, Ed. Franklin, J. L., Chap. 3, Plenum Press, New York (1972)

80 BROWN, P. J., and HAYES, E. F., *Journal of Chemical Physics,* **55,** 922 (1971)

81 KUNTZ, P. J., and WHITTON, W. N., *Chemical Physics Letters,* **34,** 340 (1975)

Section 6.7

82 KLEIN, F. S., and PERSKY, A., *Journal of Chemical Physics,* **61,** 2472 (1974)

83 DOUGLAS, D. J., and POLANYI, J. C., *Chemical Physics,* **16,** 1 (1976)

84 BLACKWELL, B. A., POLANYI, J. C., and SLOAN, J. J., *Chemical Physics,* **24,** 25 (1977)

Note added in proof

The question of at what energy vibrations in a molecule are strongly coupled so that the dynamics depend only on internal energy and angular momentum is of current interest, largely in response to new experiments where molecules are excited vibrationally by absorption of multiple quanta from powerful infrared lasers or by single photon absorptions in high overtone bands. For some experiments and discussion, *see Faraday Discussions of the Chemical Society,* **67** (1979)

Chapter 7

Reactive scattering

7.1 Introduction[1–6]

In 'bulb' experiments, no attempt is made to define the directions at which the potential reagents for a collisional reaction approach one another. Consequently, there is no possibility of discovering whether the products of the collisions are scattered isotropically or preferentially at some angles relative to the original direction of approach. However 'detailed' the observation from a 'bulb' experiment, it is necessarily one integrated over all scattering angles, and the parameter used to describe the result must lie in the lower right-hand quarter of *Figure 1.4* (*see* p. 365).

In scattering experiments, one forms two spatially well-defined beams which intersect. Within this crossing zone, and only there, molecular collisions take place. They can lead to three kinds of result:

(1) *elastic scattering*, in which the total translational energy and the magnitude of the linear momentum are conserved;
(2) *inelastic scattering*, in which energy is transferred between different degrees of freedom;
(3) *reactive scattering*, in which molecular rearrangements occur.

In measurements on reactive scattering, which are our chief concern, the object is to detect the amounts of reaction product scattered on to a detector as its position is varied in relation to the direction of the incident beams. The intensity of the product flux reaching the detector is measured over a sufficiently wide range of positions to allow one to build up a complete picture of the angular scattering pattern.

If a molecular beam is to remain spatially well-defined over distances of several centimetres, it is necessary to minimize the frequency of collisions between species in the beam. This can be done either by using beams of very low particle density, or by ensuring that all the particles have nearly the same velocity so that they keep their positions relative to one another. In addition, of course, it is necessary to keep the background pressure in the vacuum chamber through which the beam is travelling as low as possible.

Crossed molecular beam experiments are technologically demanding.

Ideally, sources are required which can produce intense beams of a wide range of species, radicals as well as molecules, with well-defined velocities which can be varied at will. Then a sensitive detector, which can discriminate between different species, is needed to detect the extremely low fluxes of product molecules which are likely to reach it in many situations. Furthermore, improving the two 'ends' of the experiment cannot be considered entirely separately. If one is successful in producing intense well-collimated beams of the potential reagents, then the intensity of products scattered in any particular direction will rise accordingly. However, the use of intense beam sources requires the utilization of larger pumps. These are not only needed to keep down the background pressure in the main scattering chamber. Mass spectrometry is now the main technique for observing scattered products. It has the immense advantage of being able to detect all species and discriminate between those of different masses, it, therefore, provides 'universal detection'. However, to use this detector to its full, it is necessary to keep the background pressure in the ionization source extremely low. In spite of the considerable experimental difficulties, there has recently been tremendous progress in the application of molecular beam techniques to chemical reactions. Section 7.2 provides a guide to these improvements. As in Chapters 5 and 6, this experimental section is not intended to provide a full technical description of the experimental method.

Before describing crossed molecular beam experiments in any greater detail, it is as well to draw attention to a possible source of uncertainty in their interpretation. In the crudest crossed-beam experiment, the velocities in both beams have thermal distributions and scattered reaction products are observed without discriminating between species with different velocities or internal energies. A differential reaction cross-section, averaged over initial velocities and internal states can be derived from this experiment. However, the results do *not* provide a direct measure of the quantity denoted by $\bar{\sigma}_r(\theta)$ in *Figure 1.4*, because the scattering angle θ referred to there is one measured relative to the centre-of-mass (CM) of the two-particle system as origin. The observations measure the corresponding quantity in a system of coordinates which has the position of the scattering zone in the apparatus as its origin. This quantity in the LAB frame-of-reference will be referred to as $\bar{\sigma}_r(\Theta)$, to distinguish it from $\bar{\sigma}_r(\theta)$ in CM coordinates. The relationship between pairs of differential scattering cross-sections in these two systems of coordinates and a description of the procedures which are used to transform the LAB quantities to the corresponding CM quantities are given in Section 7.3. For the moment, it is sufficient to point out that there is no certain procedure for transforming $\bar{\sigma}_r(\Theta)$ to $\bar{\sigma}_r(\theta)$. This is because a detector placed at any angle Θ can observe scattered product with a range of u', θ values[a]. Consequently, not only does the simplest beam experiment supply fairly limited information about the

[a] The symbols v'_X and u'_X distinguish the final velocity of a scattered species in the LAB frame-of-reference (v'_X) and that measured relative to the centre-of-mass of the colliding pair (u'_X). The relationship between these quantities is explained in Section 7.3.

collision dynamics, but it is also subject to a good deal of analytical uncertainty.

These two factors, coupled with technical advances, have led to much more sophisticated experiments. In these, the velocity in one or both of the well-collimated incident beams is selected, and can be varied. In addition, and just as important, the distribution of LAB velocities $f(v'_X)$ of the observed product (X) is measured at a wide range of LAB scattering angles. This means that the distribution $f(v'_x, \Theta)$ is determined directly, and it is then relatively straightforward to find the corresponding quantities in the CM frame-of-reference, i.e., $f(u'_X, \theta)$. If these distribution functions are measured for a variety of initial relative velocities, then the differential reaction cross-sections $\sigma_r(v'_x \mid v_x; \theta)$ can be calculated.

It is more difficult to observe reactive scattering from species in defined internal states—or to monitor the internal states of products at particular scattering angles. Of course, if v is defined and the distribution of v' is measured, the equation of energy conservation provides one, via a trivial calculation, with a function describing the distribution of products over *total* internal energies. Therefore, a crossed-beam experiment with both *velocity selection* of the reagents and *velocity analysis* of the products goes close to providing values of $\sigma_r(n' \mid n; v, \theta)$ which, it was suggested in Chapter 1, describe the ultimate level of detail in experiments on the dynamics of collisional reactions.

7.2 Experimental methods[1—7]

7.2.1 'Primitive' experiments

The essential elements of a crossed-beam apparatus for investigating reactive scattering are:

(1) sources for the two beams;
(2) a detector for one or both of the products of reactive collisions; and
(3) some means of varying the relative positions of the detector and the beams.

Finally, it is necessary that the main chamber of the apparatus can be evacuated—with the beams in operation—to a very low background pressure; certainly, less than about 10^{-5} N m^{-2}($\leqslant 10^{-7}$ Torr).

The pioneering experiments on reactive scattering were conducted with apparatus which was remarkably unsophisticated. However, this early work was confined almost entirely to reactions of alkali metals (M) producing alkali-metal halides (MX), such as

$$K + HBr \rightarrow KBr + H; \quad \Delta H_0^\circ = -17\ \text{kJ mol}^{-1} \tag{7.1}$$

There were a number of factors which led to this state of affairs. The most important was the discovery of a remarkably efficient, and simple, detector for

alkali-metal atoms and their salts. It relies on the fact that these atoms and molecules can be ionized with almost 100 per cent efficiency when they come into contact with a heated wire of appropriate work function. Crucially, the relative ionization efficiencies for M and MX depend critically on the material of the wire or its pretreatment. This allowed the development of *differential* surface ionization detectors[1]. One such detector consists of two filaments: one of tungsten, or rhenium, when heated to about 1900 K, produces M^+ from M and MX with almost equal efficiency, whilst the second of platinum–8 per cent tungsten alloy, after suitable pretreatment, is only sensitive to M. The flux of reaction product can be determined by observing the difference between the signals supplied by these two filaments.

The beams in the early experiments were produced by collimating the effusive flow from a small opening in the side of a vessel containing the reagent. The orifice frequently had a rectangular slit cross-section (e.g., 0.02×10 mm). A second similar slit, placed a short distance in front of the first, allowed through only those molecules travelling in the required direction. In the earliest experiments, it was presumed that the pressure in the source should be low enough to guarantee effusive flow (requiring the mean free path to be more than the dimensions of the orifice) and to avoid many collisions in the regions between the slits. Consequently, the beams used were of very low intensity. Typically, only about 10^{14} molecules per second crossed a plane perpendicular to the beam, maintaining a steady-state concentration of M atoms of roughly 10^{10} molecule cm^{-3} in the region where the beams intersected. It was only discovered rather later that these sources could be operated at much higher pressure, yielding more intense beams.

The second major disadvantage of effusive sources is that there is a thermal spread of velocities in the resultant beams. In some of the early studies, the collision energy was more closely defined by selecting the velocity in one or, later, both beams. This was done with a mechanical velocity selector consisting of a series of slotted discs mounted on a common axis. When it was rotated at a constant velocity, only molecules with velocities within a narrow range could successfully thread their way through the slots and reach the crossing zone. The selected velocity could easily be changed by altering the speed of rotation. However, there was a price to pay for this type of velocity selection: only a fraction of the original rather weak beam reached the scattering region, so the incident flux and eventual signal were appreciably reduced. In addition, of course, the choice of incident velocity was restricted to the range which was thermally accessible.

The limitation to thermal collision energies was a second major reason for the early emphasis on reactions of alkali-metal atoms; many of these have cross-sections $> 0.1\ nm^2$ (i.e., $> 10\,Å^2$), with the result that as much as 0.1—1 per cent of the M atoms entering the crossing zone react to produce MX. The necessity for a relatively large reaction cross-section at accessible collision energies remains a major requirement in studies of reactive scattering. The final factor which determined that reactions producing alkali-metal salts were the first to be studied using crossed-beam techniques was simply that the

reactants and products have low vapour pressures at 77 K. Consequently, they could be cryogenically pumped by immersing the scattering chamber in liquid nitrogen. The other pumping requirements were then really quite modest. As we shall see, despite the limitation of these initial studies to an apparently exotic set of reactions, they were able to identify the major categories of reactive scattering and to lay a highly successful foundation for developments in the 'post-alkali-metal atom age'.

7.2.2 The 'supermachines'

The term 'supermachine' has been coined for any apparatus for scattering experiments which incorporates two major improvements over the relatively primitive systems used by the pioneers and described above. These are mass spectrometric, 'universal', detection and supersonic beam sources. Changes in the equipment are, however, considerably more extensive than this simple statement might suggest. This is because a vastly improved vacuum technology is required, incorporating an array of large diffusion pumps which differentiately evacuate separate chambers of what is now a much larger apparatus. The need for these pumps follows inevitably from the new features of the apparatus. Supersonic nozzles provide reagent flows 2—3 orders-of-magnitude greater than effusive sources, but the increased flow of frequently non-condensable gas must be removed. Secondly, the pressure in the electron bombardment region of the mass spectrometer must be kept extremely low to reduce interfering background signals.

Figure 7.1 shows a simplified picture of a nozzle beam source for scattering studies. The pressure of the gas in the nozzle chamber is several orders-of-magnitude greater than that for effusive flow. The gas behaves as a continuous fluid expanding adiabatically into a vacuum[7]. At the 'throat' of the gas stream, the velocity of the fluid reaches the local sound velocity; thereafter, the flow diverges and becomes supersonic. As long as the gas behaves as a fluid, the random relative motions of the molecules, which are responsible for the temperature, are converted into ordered motion along the required direction. Eventually the density falls so far that collisions cannot maintain this exchange, and the temperature is then frozen at the value it has reached. The random transverse temperature may be only about 5 per cent of the source temperature and the spread of velocities along the beam direction can be $\leqslant 10$ per cent of the mean value.

Nozzle beams pass through a skimmer, designed to meet hydrodynamic criteria, and through one or more differentially pumped vacuum sections before reaching the main scattering chamber. As well as providing beams of greatly enhanced intensity and narrow velocity spreads, supersonic nozzle sources allow one to vary the translational energy up to quite high values by altering the source temperature. For heavy molecules, the range of energies can be further increased by 'seeding' the source with a much lighter species. The heavy species, which originally constitutes a small fraction of the gas, is swept along in the expansion by the lighter constituent and reaches kinetic energies

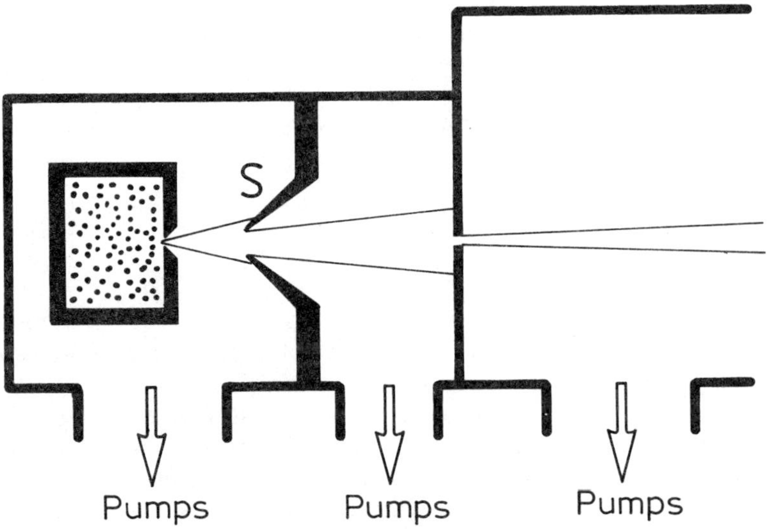

Figure 7.1. Schematic diagram of a nozzle beam source. Molecules emerge from the high-pressure oven, pass through the skimmer (S), *and are collimated, before entering the main chamber. Each chamber is independently pumped*

much greater than those in the oven. Furthermore, the heavy species is concentrated along the axis of the beam, whilst the light gas tends to scatter out of it and is then pumped away. In this manner, neutral beams with energies of several electron-volts (i.e., ~ 500—1000 kJ mol^{-1}) have been obtained, although it should be re-emphasized that such energies can only be achieved for heavy molecules, like HI and CH_3I.

With the advent of universal detection, a good deal of attention has been focused on the provision of beam sources for different free radical species. Successful experiments have been carried out on halogen atoms, H and D atoms, O atoms and methyl radicals, as well as on a wide range of metal atoms. Of the radicals extensively studied by 'bulb' kineticists, only N atoms and OH appear to have eluded the beamists so far.

Radicals for crossed-beam studies, like those for 'bulb' flow studies, are created by either thermal or electrical dissociation of a parent molecule. In either case, it is usually desirable to keep the source pressure quite low and, of course, the radical may constitute only a fairly small fraction of the flow leaving the source chamber. However, it should be appreciated that it is generally unnecessary to use nozzle sources to produce *both* beams. If the molecular reagent (as distinct from the radical reagent) is created in an intense, high energy, beam, it will probably be sufficient to take radicals from an effusive source. Even if the latter beam is not velocity-selected, the *relative* velocity of collisions between the radicals and molecules is likely to be well-defined, because the thermal spread in the radical beam will be quite small

compared with the high, and narrowly specified, velocity in the nozzle beam.

The detection of reactively scattered species by mass spectrometry was somewhat delayed by an overpessimistic view of its usefulness. This arose because it was recognized that electron bombardment was unlikely to ionize more than 1 in every 1000 molecules entering the ion source. Although this efficiency is much lower than that of surface ionization of alkali metals and their salts, the comparison ignores other factors which essentially close the apparent gap between the two detection methods. Thus, it neglects the fact that to measure the extent of reactive scattering by differential surface ionization the signal due to non-reactively scattered M must be subtracted from that due to (M + MX). Unless the reaction cross-section is very large, a small difference has to be calculated between two similar signals. Furthermore, filament noise means that ion-counting techniques cannot be used with surface ionization detectors.[4]

As mentioned earlier, a crucial element in the successful application of a mass spectrometer is the quality of the vacuum in the ion source. A partial pressure of less than 10^{-11} N m^2 ($< 10^{-13}$ Torr) is required. This is not difficult to achieve if the pumping is designed properly, and as long as un-ionized product molecules enter another differentially pumped chamber before they encounter a surface. The ions which are created enter a quadrupole mass filter in which they pass through an oscillating quadrupole field. Only ions of the selected mass/charge ratio can traverse this system; the others execute unstable trajectories and do not reach the ion counter.

In current studies using universal detection, product fluxes are typically 10^4—10^6 molecule s^{-1}. However, respectable data can be obtained with fluxes as low as about 10^3 molecule s^{-1}, which corresponds to counts of only a few ions per second[4].

The marriage between nozzle beam sources and sensitive mass spectrometric detection has done more than allow a wider range of reactions to be investigated by molecular beam techniques. It has also permitted differential and total cross-sections for some of these reactions to be measured as a function of collision energy. Finally, the distribution of product translational velocities has been observed for a number of systems. As was mentioned earlier, these measurements are necessary, if the distribution of product molecules over relative translational velocity and scattering angle in the CM system, $f(v', \theta)$, is to be determined unequivocally.

Most early experiments incorporating velocity analysis employed the mechanical device described in Section 7.2.1, but it was now inserted between the scattering centre and the detector. However, as was pointed out earlier, these velocity selectors suffer from the disadvantage of rejecting a large proportion of the molecules whenever they are operating, without any recompense in terms of being able to use lock-in detection. Most velocity analyses are now carried out using a time-of-flight technique. A single chopper is located only about 1 cm in front of the entrance aperture to the mass spectrometer and is rotated at high speed. When 'open', it passes all the product molecules, but the times which they take to reach the detector depend

on their translational velocities. An optical reference is supplied and one accumulates from the observed signals the continuous spectrum of velocities present in scattered product flux.

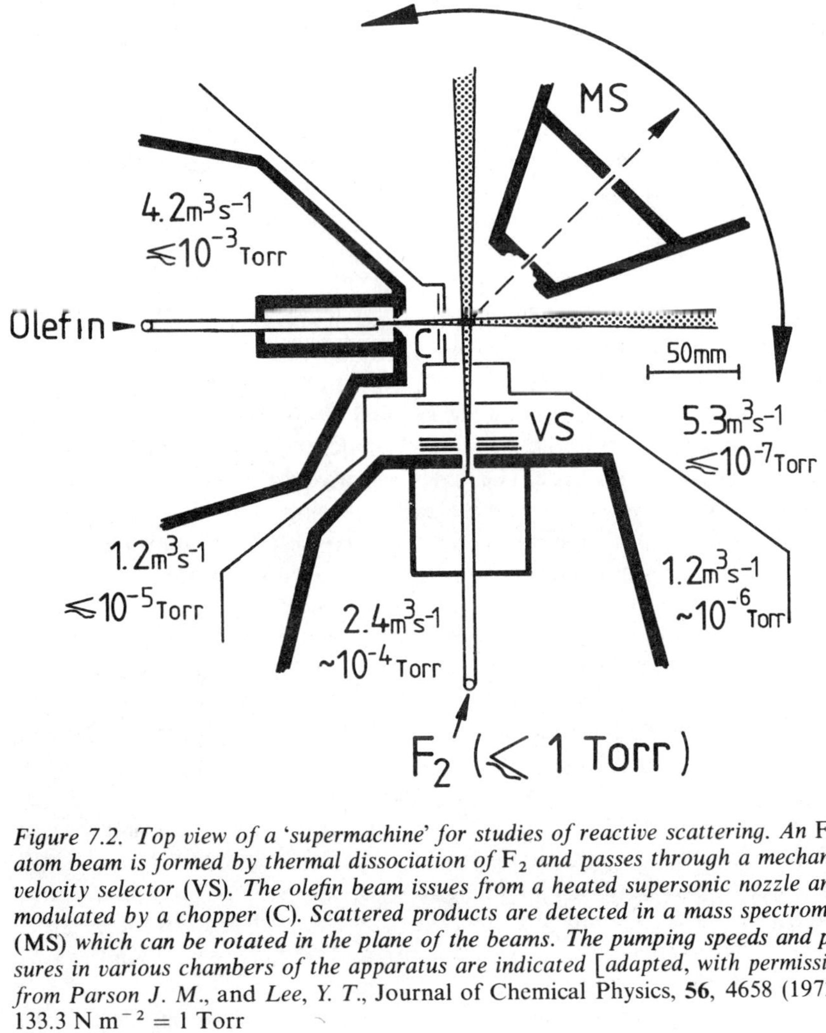

Figure 7.2. Top view of a 'supermachine' for studies of reactive scattering. An F *atom beam is formed by thermal dissociation of* F_2 *and passes through a mechanical velocity selector* (VS). *The olefin beam issues from a heated supersonic nozzle and is modulated by a chopper* (C). *Scattered products are detected in a mass spectrometer* (MS) *which can be rotated in the plane of the beams. The pumping speeds and pressures in various chambers of the apparatus are indicated* [*adapted, with permission, from Parson J. M.*, and *Lee, Y. T.*, Journal of Chemical Physics, **56**, 4658 (1972)]. 133.3 N m^{-2} = 1 Torr

Figure 7.2 provides a schematic drawing of a molecular beam 'supermachine', indicating the major features (apart from pumping) and some dimensions. At the end of this section, *Table 7.1* summarizes the wide range of reactions which have now been investigated by crossed-beam techniques. Individual categories of reaction have been studied to different depths of detail. The symbols in each box provide some guide to this.

Table 7.1. Main categories of reaction studied in reactive scattering experiments

Reagents →	M	M_2	M′	X	$(X_2)_2$	MX			
	Li, Na, K, Rb, Cs	K_2, Cs_2, Rb_2	Mg, Ca, Sr, Ba	F, Cl, Br	$(Cl_2)_2$, $(Br_2)_2$	*e.g.* KCl	H, D, T	CH_3	O
M_2	VS	–	–	P	–	–	VA	–	–
MX	VS; V_r	–	–	–	–	VS	–	–	–
H_2, D_2	–	–	–	(VS/VA)	–	–	VA	–	–
X_2, XX′	(VS/VA)	VS	P	VA	(VS/VA)	(VS/VA)	VA	VA	(VS/VA)
HX	(VS/VA); v, j; V_r	–	P; (v′, j′ | v, j)	–	(VS/VA)	–	P	–	–
O_2	–	–	P	–	–	–	–	–	–
N_2O, NO_2, SO_2	P	–	P; (v′, j′ | v, j)	–	–	–	VA; v′, j′	–	–
CS_2	–	–	–	–	–	–	–	–	(VS/VA)
RX[a]	(VS/VA); V_r; Ω	(VS/VA)	P	VS	–	–	–	–	(VS/VA)
olefins[b]	–	–	–	(VS/VA)	–	–	–	–	–

[a] RX = alkyl or substituted alkyl halide
[b] also halogenated olefins and aromatic molecules

The symbols indicate what is the most detailed beam experiment carried out on each category of reaction. P indicates that only 'primitive' beam experiments have been performed: that is, measurements of reactive scattering from crossed beams containing a thermal distribution of molecules. VS = velocity selection; VA = velocity analysis; (VS/VA) both velocity selection and analysis in the same experiment; V_r indicates that an excitation function has been determined. v, j = selection of initial molecular state by laser excitation; v′, j′ = analysis of final state distribution by laser induced fluorescence (these experiments frequently employ a single beam + low pressure scattering gas). Ω indicates that measurements have been made with partial orientation of the molecular reagent.

7.2.3 Hyperthermal sources, state selection, state analysis

What has been described so far may now be regarded as the standard experiment on reactive scattering: that is the measurement of differential reaction cross-sections and product velocity distributions as a function of collision energy. For some purposes, it is desirable to have even higher beam energies than those provided by seeded nozzle sources. Very high translational energies can be reached with charged species, by simply accelerating the ions produced in an electrical discharge, or from the surface of a hot wire, in an electric field. If required, the charge can then be neutralized by passing the ions through a volume containing gas at low pressure. If this gas is suitably chosen, charge exchange can occur over large impact parameters, so that little momentum is transferred, and the species which were initially charged retain their high velocity. Any ions left in the emerging beam can be deflected away by a suitable field. Hyperthermal beams of this kind have been principally used to probe the repulsive portion of the intermolecular potential between non-reacting species[8], such as pairs of noble-gas atoms, rather than in studies of chemical reactions.

Ion–molecule reactions can be investigated in crossed-beam experiments[9]. The intensity of the ion beam is much lower than that used in studies of neutral–neutral reactions, but reaction cross-sections are usually larger and 100 per cent of the charged product can be detected by mass spectrometry. The major problem in these experiments is to extend the range of collision energies *downwards*, because, as the LAB velocity in a beam of ions is reduced, it becomes progressively more difficult to avoid the beam diverging as a result of Coulombic repulsions. Experiments using 'merging beams' have gone a long way to overcoming this problem. Both beams have high, but similar, LAB velocities, so that the *relative* velocity which determines the collision energy is quite small. In this way, the result of collisions with energies as low as ~ 0.1 eV (10 kJ mol^{-1}) have been investigated.

There have been relatively few measurements on reactive scattering with internal state selection of one of the reagents. During the expansion from a nozzle source, the translational cooling is accompanied by a parallel cooling in the internal degrees of freedom. As rotational relaxation is rapid, the cooling of reactant rotation is likely to be complete and molecules in the beam will have a very low rotational temperature. The vibrational degrees of freedom in small molecules relax much more slowly than rotation, so the vibrational distribution may be anywhere between the distribution present initially in the high-temperature oven and that corresponding to the low translational temperature in the beam. One thing which does occur, and has been exploited, is the formation of 'clusters'. For example, in a nozzle beam expansion of alkali metals (M), there is a high yield of M_2 dimers, and the reactions of these rather weakly bound molecules have been quite extensively studied.

In the previous chapter, it was pointed out that internally excited species can be produced by thermal, photochemical, or chemical activation. All of these methods have been employed in crossed-beam experiments, but only to a very

limited extent. Only photochemical excitation provides a means of populating a single selected state. However, if the absorption cross-section for the transition connecting the unexcited and excited states is large, the chance of the excited species making a collision before it decays spontaneously is very slight. Consequently, the exciting source must be very intense and the excited species must be fairly long-lived. The elegant experiment of Brooks and his colleagues[10] using an HCl chemical laser in a study of the reaction

$$K + HCl(v') \rightarrow KCl + H; \quad \Delta H_0^\circ = +6\ kJ\ mol^{-1} \qquad (7.2)$$

was mentioned in Section 6.6.1. It is likely that as lasers increase in range and power they will be increasingly used for selective excitation in scattering experiments (*see* p. 365).

The use of thermal and chemical methods of producing an excited vibrational distribution in a potential reagent was mentioned in Section 6.6.2, in connection with infrared chemiluminescence experiments under 'near-beam' conditions. A few examples of similar beam experiments have been reported, but obviously the problems of detection can become acute. Both thermal sources and electrical discharges can be used to supply metastable species: for example, vibrationally excited H_2 or N_2, or electronically excited metastables such as $Hg(6^3P_2)$[11]. The exploitation of chemical activation requires a double-scattering experiment[12]. In a first scattering region an exothermic 'prereaction' occurs forming a highly excited distribution of reaction product. A slit is placed where this species scatters most strongly, thereby producing a collimated beam which then serves as one of the two crossed beams in a second scattering process.

Just as it is difficult to use photochemical techniques to excite sufficient *reagents* for scattering studies, so it is difficult to use spectroscopic techniques to observe the distribution of *products* over internal states. In the case of highly exoergic reactions which produce *electronically* excited species, for example

$$Cl + Na_2 \rightarrow NaCl + Na^*(3^2P_{\frac{1}{2},\frac{3}{2}}); \quad \Delta H_0^\circ = -135\ kJ\ mol^{-1} \qquad (7.3)$$

it is usually possible to observe chemiluminescence from the crossing zone. However, one learns nothing about the angular distribution of the electronically excited product, and it is not easy to estimate the ratio of the cross-sections for production of excited and ground state product. The lower emission coefficients for transitions between vibrational levels and the lower sensitivity of infrared detectors prevents observation of infrared chemiluminescence from vibrationally excited molecules produced under 'genuine' crossed-beam conditions. However, two spectroscopic techniques have had some success in observing product state distributions and these will be described briefly to bring this section to a close.

The first spectroscopic technique utilizes a molecular beam resonance spectrometer[13, 14]. One of the few reactions studied in this way is

$$Cs + SF_6 \rightarrow CsF + SF_5; \quad \Delta H_0^\circ \sim -170\ kJ\ mol^{-1} \qquad (7.4)$$

The CsF scattered at a particular angle is collimated and sent into the

spectrometer. This instrument consists basically of a surface ionization detector preceded by an arrangement of three electrical field regions. The quadrupole fields *A* and *B* ensure that only molecules in a particular (j, m_j) state undergo the correct curved trajectory and fall on the detector. Between regions *A* and *B* the molecules pass through a d.c. field where they are irradiated with radiofrequency radiation. If this radiation strikes a resonance and induces a transition, the molecules whose state is changed will not be focused on to the detector by field *B* and the signal falls. Because the Stark splitting is not the same for the same j level in different vibrational states, relative vibrational populations can be determined. However, the method suffers from the disadvantage that it can be applied only to strongly polar molecules and, even then, not all j states can be observed. Consequently, only a partial distribution is measured.

The second spectroscopic technique of laser-induced fluorescence[15–17] promises to have wider applications. The internal state distribution of reaction products is determined through their *excitation spectra.* The excitation spectrum is the total fluorescence which is observed as a tunable dye laser is swept through the frequencies at which the product molecule undergoes an electronic absorption to a bound—and therefore fluorescing—upper state. The method is relatively simple, has a high sensitivity, is capable of providing the entire (v′, j′) distribution, and is likely to find increasing use as the availability and range of tunable lasers increases. A number of reactions of Ba atoms, such as

$$Ba + O_2 \rightarrow BaO(v', j') + O; \quad \Delta H_0^\circ = -63 \text{ kJ mol}^{-1} \tag{7.5}$$

have been studied. The determination of accurate distributions requires a good knowledge of the spectroscopic constants and Franck–Condon factors for the transitions involved. In the earliest experiments of this kind[15, 16] the exciting laser was directed into the zone where the reagent beams intersected but already primitive angular distributions for products in specified states are being provided.[17].

7.3 Interpretation of scattering experiments[1, 3]

The object in any crossed-beam experiment, whether it be on elastic, inelastic, or reactive scattering, is to discover the spectrum of recoil velocity vectors which carry the scattered species away from the centre-of-mass. The centre-of-mass in an individual collision proceeds at a constant velocity through LAB space, undisturbed by the molecular encounter. The relationship between any LAB velocity ($\boldsymbol{v}_X$) and the equivalent CM velocity ($\boldsymbol{u}_X$) is

$$\boldsymbol{v}_X = \boldsymbol{u}_X + \boldsymbol{C} \tag{7.6}$$

where $\boldsymbol{C}$ is the constant vector describing the motion of the centre-of-mass.

The kinematic analysis, which is required to reconstruct the CM scattering pattern (i.e., CM differential cross-sections) from the observed LAB distribution, is similar for elastic, inelastic, and reactive scattering. Elastic scattering is simplest and is considered first. The treatment is an extension of that given for two-particle classical collisions in Chapter 3.

7.3.1 Elastic scattering

Several features of the LAB–CM transformation can best be illustrated by means of a Newton vector diagram. The one in *Figure 7.3* represents a collision between particles A and B with LAB velocities, $\mathbf{v}_A$ and $\mathbf{v}_B$, at right-angles to each other; this is the customary arrangement in crossed-beam experiments. The relative translational velocity and the vector describing the motion of the centre-of-mass of the two particle system are given by equations (7.7) and (7.8):

$$\mathbf{v} = \mathbf{v}_A - \mathbf{v}_B \qquad (7.7)$$

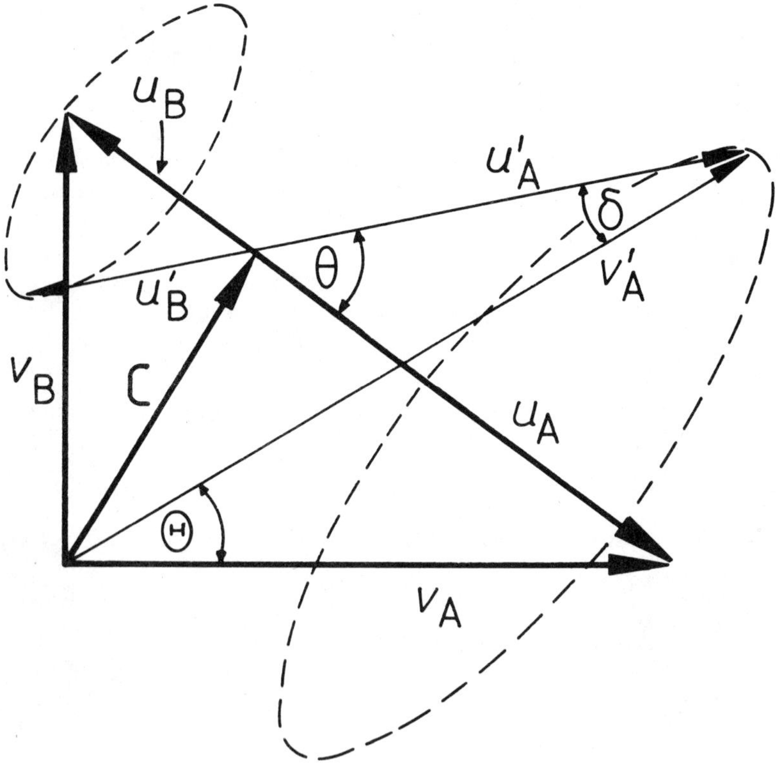

Figure 7.3. Newton diagram for elastic *scattering with* $m_B = 2m_A$ *and* $v_A/v_B = (m_B/m_A)^{1/2}$. A *and* B *scattered at* CM *angle* θ *will be found at* LAB *positions on the perimeters of the bases of two cones; these are shown as dotted lines*

$$C = \frac{m_A \boldsymbol{v}_A + m_B \boldsymbol{v}_B}{(m_A + m_B)} \tag{7.8}$$

Furthermore, because linear momentum must be conserved, $\boldsymbol{C}$ divides $\boldsymbol{v}$ into individual components $\boldsymbol{u}_A$ and $\boldsymbol{u}_B$, i.e., CM velocities of A and B, according to

$$\boldsymbol{u}_A = \boldsymbol{v}_A - \boldsymbol{C} = \frac{m_B \boldsymbol{v}}{(m_A + m_B)} \tag{7.9a}$$

and

$$\boldsymbol{u}_B = \boldsymbol{v}_B - \boldsymbol{C} = \frac{-m_A \boldsymbol{v}}{m_A + m_B} \tag{7.9b}$$

In an elastic collision, the direction of $\boldsymbol{v}$ but not its magnitude (or u_A and u_B) is altered. Since the scattering in collisions that have the same impact parameter must be cylindrically symmetric about $\boldsymbol{v}$ (*see* Figure 3.3), the vectors like $\boldsymbol{v}'$, representing the relative velocity of A and B after collisions must terminate on the perimeters of the bases of the two cones, drawn with $u_A = u'_A$ and $u_B = u'_B$, shown in *Figure 7.3*. Because of this symmetry, imposed by the twin requirements of energy and momentum conservation, it is not necessary to observe scattering out of the plane containing the two beams, or even over a complete range of 'in-plane' angles.

However, although the differential cross-sections in the CM system must be identical at θ and $-\theta$, the same is not true of the corresponding LAB quantities. The LAB differential cross-section, which is the quantity that is determined in an experiment, is defined by an equation similar to (3.29) for $\sigma_{CM}(\theta)$, i.e.,

$$\sigma_{LAB}(\Theta) = dS/d\Omega \tag{7.10}$$

where $d\Omega$ is an element of LAB solid angle. Consequently,

$$\sigma_{CM}(\theta) = \sigma_{LAB}(\Theta)\{d\Omega/d\omega\} \tag{7.11}$$

For elastic collisions $(d\Omega/d\omega)$ can be shown, by straightforward geometrical arguments, to be given by

$$(d\Omega/d\omega) = (u'_A/v'_A)^2 \cos\delta \tag{7.12}$$

where u'_A and v'_A are the CM and LAB velocities of the observed species. Furthermore, if $\boldsymbol{v}$ is sharply defined, i.e., if both beams have a narrow spread of velocities and are well-collimated, u'_A is known and $\cos\delta$ and v'_A both have single, easily calculated values for any particular angle Θ. Then the conversion of $\sigma(\Theta)$ to $\sigma(\theta)$ is trivial. In practice, as long as the velocity of *one* beam is selected, the loss of resolution resulting from the spread in the $\boldsymbol{v}$ and $\boldsymbol{C}$ vector does not cause any great difficulties.

7.3.2 Reactive scattering

Reactive collisions can also be represented by a Newton diagram. Now, however, $\boldsymbol{v}'$, the relative velocity associated with scattering of the reaction

products, may have any value, limited only—in principle, at least—by energy conservation. The vectors $\boldsymbol{v}'$ are again pivoted at the tip of $\boldsymbol{C}$, but the lengths of the segments are determined by momentum balance between two new species, the chemical reaction products. If the reaction is represented by

$$\mathrm{A + BC \rightarrow AB + C} \qquad (7.13)$$

the following equations describe the relationship in each reactive collision between the final relative velocity of AB + C and the individual CM and LAB velocities of the two products:

$$\boldsymbol{v}' = \boldsymbol{u}'_{AB} - \boldsymbol{u}'_{C} \qquad (7.14)$$

$$\boldsymbol{u}'_{AB} = \boldsymbol{v}'_{AB} - \boldsymbol{C} = \frac{m_C \boldsymbol{v}'}{m_{AB} + m_C} \qquad (7.15a)$$

$$\boldsymbol{u}'_{C} = \boldsymbol{v}'_{C} - \boldsymbol{C} = \frac{-m_{AB} \boldsymbol{v}'}{m_{AB} + m_C} \qquad (7.15b)$$

Since $v' = (2\varepsilon'_T/\mu')^{\frac{1}{2}}$, the in-plane scattering of AB (or C) at various values of ε'_T can be represented on the Newton diagram by concentric circles of radius u'_{AB} (or u'_C), as shown in *Figure 7.4*.

With an exoergic reaction, v' (and therefore u'_{AB} and u'_C) can have a wide range of values. A detector positioned at a particular LAB angle Θ will usually observe products, say AB, with a distribution of LAB velocities, each of which corresponds to different values of u'_{AB} and θ. Consequently, the LAB–CM transformation of the scattering observations cannot be carried out accurately, unless the distribution of LAB velocities is measured for each position of the detector. If this is done, the transformation can be made using the equation:

$$\sigma_r(v'_{AB}, \Theta) = \sigma_r(u'_{AB}, \theta)(v'_{AB}/u'_{AB})^2 \qquad (7.16)$$

The results are normally represented as contours on a polar diagram, the contours representing the relative magnitude of the differential reaction cross-section, or more often the product flux, at the position on the plot defining the values of u'_{AB} and θ. Several diagrams of this kind are presented later in this chapter.

As will be seen in Section 7.4.1, qualitative features in the LAB scattering pattern can provide important clues about the reaction dynamics. Contributions to $\sigma_r(v'_{AB}, \Theta)$ with low values of u'_{AB} (and therefore small ε'_T) are strongly weighted by the Jacobian factor, $(v'_{AB}/u'_{AB})^2$, in equation (7.16). This can distort the LAB distribution and simplified arguments, if pushed too far, may lead to errors.

Figure 7.4 shows that those AB molecules which are produced with sufficient translational energy for u'_{AB} to exceed C are scattered over all 4π steradians in the LAB. On the other hand, molecules with $u'_{AB} < C$ are confined to a much smaller range of LAB angles. Such constraints can arise through mass factors. Reaction (7.1) between K atoms and HBr provides an interesting example.

The K + HBr reaction is only moderately exoergic and the maximum energy

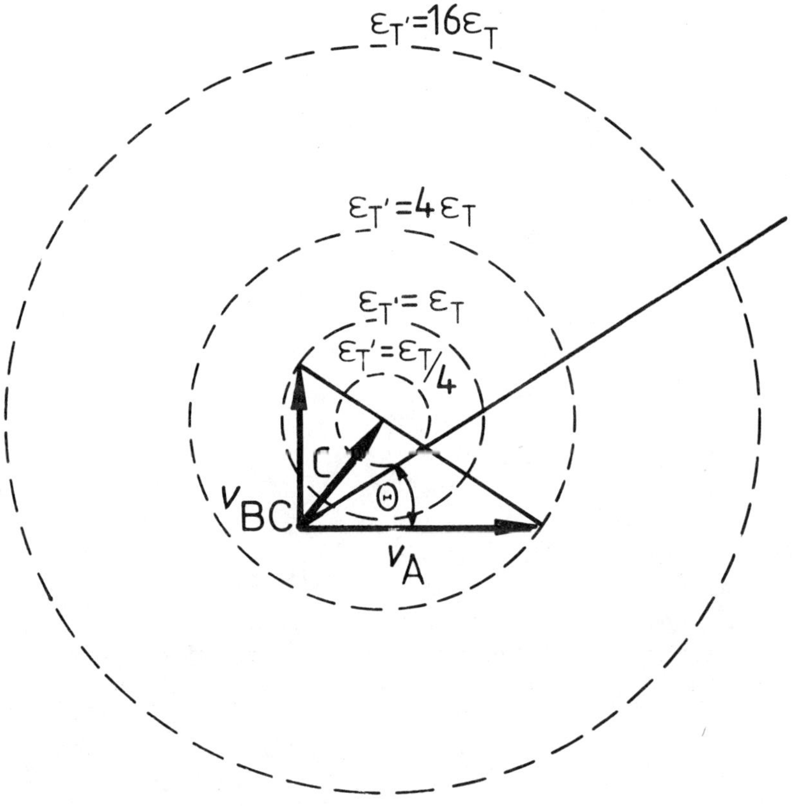

Figure 7.4. Newton diagram for reactive scattering from A + BC → AB + C *with* $m_A = m_B = m_C$ *and* $v_A/v_{BC} = (m_{BC}/m_A)^{1/2}$. *The in-plane* u'_{AB} *vectors terminate on circles of radius* u'_{AB}. *A detector placed at* LAB *angle* Θ *detects* AB *molecules with all energies* $\varepsilon'_T > (\varepsilon'_T/4)$; *each value of* ε'_T *corresponds to a different* θ

which can be released as relative translational motion of KBr + H is only about 0.23 eV or 22 kJ mol^{-1}. Furthermore, however much energy is released as relative translation, it must be shared very unevenly between the two products on account of their very different masses. *Figure 7.5* shows Newton diagrams for this reaction with circles drawn for the values of u'_{KBr} and u'_H corresponding to ε'_T taking its maximum value. These diagrams illustrate several interesting points. First, the KBr must be scattered within a small range of LAB angles ($\Theta_{min} \leqslant \Theta \leqslant \Theta_{max}$) close to the direction of $\boldsymbol{C}$. This makes it relatively easy to detect—recall that this reaction was the first to be studied by crossed-beam methods—but difficult to obtain a well-resolved picture of the KBr scattering. On the other hand, the H atoms are scattered through all LAB angles. Furthermore, for nearly all ε'_T, $\boldsymbol{v}'_H$ must nearly equal $\boldsymbol{u}'_H$ and the LAB scattering

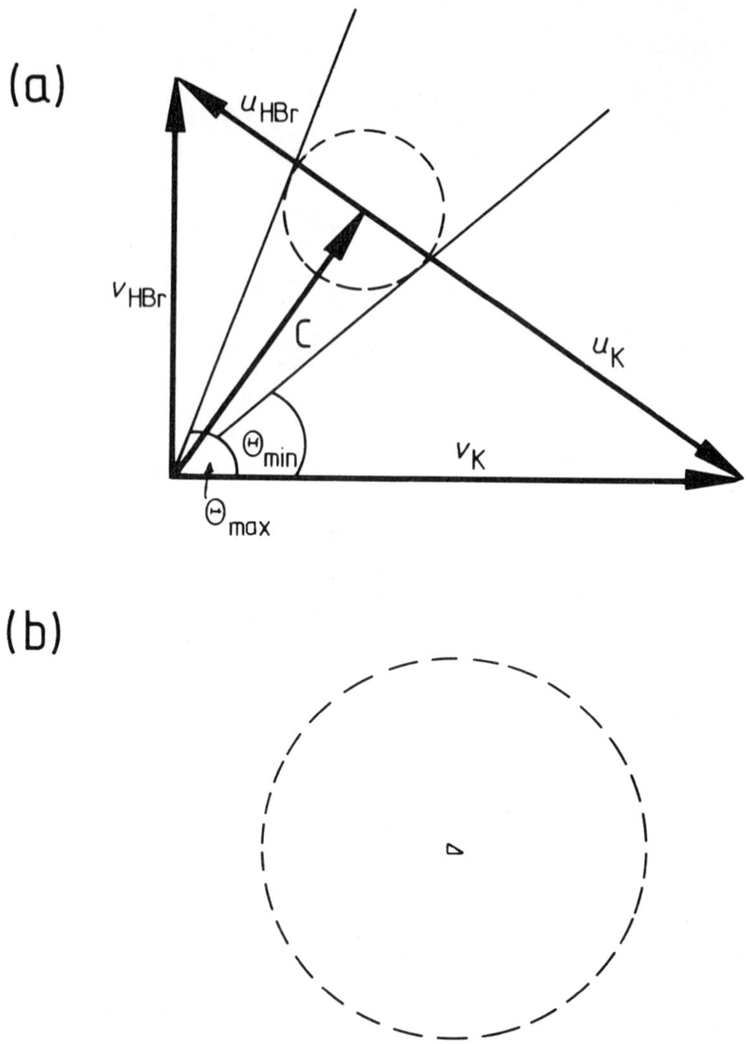

Figure 7.5. Newton diagrams for K + HBr → KBr + H. *The radius of the circle in* (a) *corresponds to* u'_{KBr} *if* all *the energy is released as translational energy. Scattering of* KBr *is restricted to a narrow range of* LAB *angles. The diagram in* (b) *is drawn at (1/50)th of the scale in* (a)*; the circle now represents the maximum possible value of* u'_H

distribution, which is in principle easy to resolve, corresponds closely to the CM distribution.

7.3.3 Angular momentum and reactive scattering

In order to understand reactive scattering fully, it is necessary to consider the

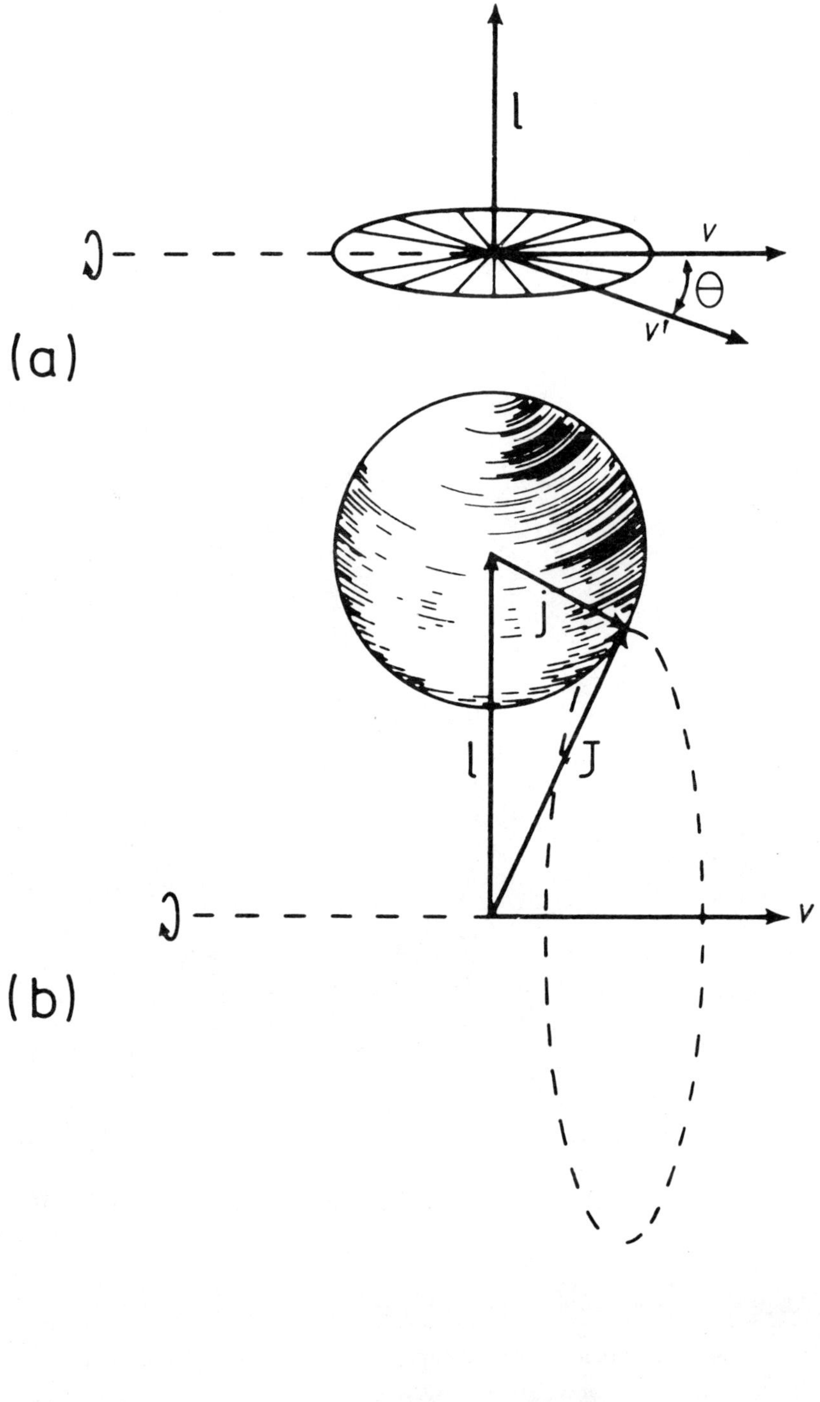
l
v
Θ
v'
(a)
j
l
J
v
(b)

features which arise because angular momentum must be conserved in molecular collisions. Once again, it is useful to begin by briefly considering two-particle systems when there is only *orbital* angular momentum present. Then

$$\mu v b = \mathrm{l} = \mathrm{l}' = \mu' v' b' \tag{7.17}$$

The vectors $\boldsymbol{l}$ and $\boldsymbol{l}'$, representing the orbital angular momenta, are perpendicular to $\boldsymbol{v}$ and $\boldsymbol{v}'$. Consequently, in any single collision, $\boldsymbol{v}$ and $\boldsymbol{v}'$ must lie in the same plane, as illustrated in *Figure 7.6*(a). It is useful to consider a hypothetical case in which, for specified values of $\boldsymbol{l}$ and $\boldsymbol{v}$, the $\boldsymbol{v}'$ vectors are evenly distributed in the $\boldsymbol{v}$, $\boldsymbol{v}'$ plane. This distribution is represented by the evenly spaced spokes drawn in the circle in *Figure 7.6*(a). Of course, in any real experiment, all azimuthal orientations of $\boldsymbol{l}$ and $\boldsymbol{v}$ are equally probable. Then one can imagine constructing the three-dimensional distribution of $\boldsymbol{v}'$ vectors by superimposing many circles like that shown all with the common chord $\boldsymbol{v}$. This clearly leads to a distribution per unit solid angle which is symmetric about the plane defined by $\theta = 90$ degrees, but strongly peaked at $\theta = 0$ and 180 degrees. The same conclusion is reached by considering equation (3.29), i.e.,

$$\sigma(\theta) = (2\pi \sin\theta)^{-1} (\mathrm{d}S/\mathrm{d}\theta) \tag{7.18}$$

and supposing $(\mathrm{d}S/\mathrm{d}\theta)$ is constant.

The presumption that $(\mathrm{d}S/\mathrm{d}\theta)$ is the same for all values of θ is characteristic of the scattering from a collision complex which can survive for at least several rotational periods. When this long-lived complex dissociates, the products are sprayed out randomly in the plane perpendicular to $\boldsymbol{l}$. Two atoms cannot collide in this way, since they must separate far too rapidly. However, equation (7.18) indicates that strong peaks will occur in the CM scattering at the two poles defined by $\theta = 0$ and 180 degrees, unless, as $\theta \to 0$ or $\to 180$ degrees $(\mathrm{d}S/\mathrm{d}\theta) \to 0$ more rapidly than $\sin\theta \to 0$.

For any system comprising three or more atoms, the rotational momenta of the molecules must be included in any statement of angular momentum conservation. For a reaction like (7.13),

$$\begin{aligned}\mathbf{J} &= \mathbf{l} + \mathbf{j} \\ &= \mathbf{l}' + \mathbf{j}' = \mathbf{J}'\end{aligned} \tag{7.19}$$

Figure 7.6. (a) ***Representation of isotropic, in-plane, scattering in two-particle*** *collisions. The complete distribution of* $\boldsymbol{v}'$ *vectors is obtained by rotation around* $\boldsymbol{v}$. (b) *Collisions between* A *and* BC, *where* BC *can possess rotational momentum. Since* $\mathbf{j}$ *can be randomly oriented relative to* $\mathbf{l}$, *the total angular momentum vectors* (**J**) *for a particular* $\mathbf{l}$ *terminate on the surface of a sphere. However, all azimuthal orientations of* $\mathbf{l}$ *are equally probable, so the* **J** *vectors are distributed with cylindrical symmetry around* $\boldsymbol{v}$

where $\mathbf{j}$ and $\mathbf{j}'$ are the rotational momenta of BC and AB and $\mathbf{J}$ and $\mathbf{J}'$ are now the *total* angular momenta of A + BC and AB + C, respectively. $\mathbf{l}$ and $\mathbf{l}'$ must again be aligned perpendicular to $\mathbf{v}$ and $\mathbf{v}'$ but the $\mathbf{j}$ (and $\mathbf{j}'$) vectors can be oriented randomly as indicated in *Figure 7.6*(b). The $\mathbf{l}$ vectors must again be distributed evenly around $\mathbf{v}$. Clearly, for any particular value of $\mathbf{j}$, the $\mathbf{J}$ vectors must also be distributed evenly over azimuthal angles; so summed over all $\mathbf{j}$, the $\mathbf{J}$ vectors have cylindrical symmetry around $\mathbf{v}$. However, because it is now $\mathbf{J}$ and $\mathbf{J}'$, rather than $\mathbf{l}$ and $\mathbf{l}'$, that must be equal, the $\mathbf{v}'$ vectors are not limited, as in elastic scattering, to the plane perpendicular to $\mathbf{l}$. Nor are they constrained to the plane perpendicular to $\mathbf{J}$, since the $\mathbf{j}'$ vectors contribute to $\mathbf{J}'$ and these can be randomly oriented. Nevertheless, because all azimuthal orientations of $\mathbf{J}$ to $\mathbf{v}$ are equally probable, any reactive scattering defined by particular values of $\mathbf{v}'$, $\mathbf{l}'$ and $\mathbf{j}'$ is equally likely at any azimuthal angle. Consequently, reactive scattering is cylindrically symmetric about $\mathbf{v}$.

Average values of j and l can be estimated using the equations:

$$\bar{j} = (2I_{BC}\bar{\varepsilon}_R)^{\frac{1}{2}} = (2\mu_{BC}\bar{\varepsilon}_R)^{\frac{1}{2}}\, r_{BC} \qquad (7.20)$$

$$\bar{l} = \mu \bar{V} \bar{b} = (2\mu\bar{\varepsilon}_T)^{\frac{1}{2}}\, \bar{b}$$
$$\approx 2/3\,(2\mu\varepsilon_T S_r/\pi)^{\frac{1}{2}} \qquad (7.21)$$

For reactions studied in molecular beam experiments, S_r, the total reaction cross-section, is usually large. Hence $\bar{b}$ is appreciably greater than r_{BC}, with the result $\bar{l} \gg \bar{j}$ and $\bar{J} \approx \bar{l}$. Average values of l′ and j′ cannot be estimated in this direct and simple fashion, and their values are usually constrained only weakly by the conservation laws: (1) since $|\mathbf{l}' + \mathbf{j}'|$ may be much less than either l′ or j individually, and (2) because any restriction due to energy conservation is slight if the reaction is appreciably exothermic.

Arguments based on consideration of angular momentum conservation were evoked in Section 6.3.2 to explain differences in product rotational excitation that have been observed in infrared chemiluminescence experiments. A further example of the special effects which can arise when H atoms are involved in a reaction is provided by reaction (7.1). Because S_r is quite large ($\sim 0.3\ \text{nm}^2$) and $\mu \gg \mu_{HBr}$, $\bar{l} \gg \bar{j}$ and so $\bar{J} \approx \bar{l}$. On the other hand, since $\mu' \ll \mu_{KBr}$, $\bar{j}'$ is likely to be much larger than $\bar{l}'$. Consequently,

$$\bar{l} \approx \bar{j}' \qquad (7.22)$$

This leads to two important results. First, the rotational motion of AB molecules scattered in-plane will be strongly polarized. Secondly, if the j′ distribution of KBr were observed, the distribution of l vectors and hence $P_r(b)$ could be determined.

The results of Monte Carlo trajectory calculations suggest that equation (7.22) may approximately hold, even when it does not follow from the conservation requirements. It does not appear that l′ and j′ can both have values much greater than l but opposite signs in any significant number of encounters.

7.4 Reactions of metal atoms and dimers

7.4.1 Direct reactions and complex formation

Although early molecular beam experiments were limited to reactions of alkali-metal atoms (M) and were, by today's standards, primitive, they were able to demonstrate clearly the occurrence of atom-transfer reactions by both direct dynamics and via the formation of collision complexes. Furthermore, it was observed that, in direct reactions, the molecular product (MX) containing the newly formed bond could be scattered into either the *forward* hemisphere ($\theta < 90$ degrees) or the *backward* hemisphere ($\theta > 90$ degrees) relative to the direction of the incident atom beam defining $\theta = 0$ degrees. The following reactions, all studied in the alkali-metal-only era of reactive scattering studies, exemplify the three distinct kinds of dynamical behaviour:

$$K + Br_2 \rightarrow KBr + Br; \quad \Delta H_0^\circ = -190\ \text{kJ mol}^{-1} \tag{7.23}$$

$$K + CH_3I \rightarrow CH_3; \quad \Delta H_0^\circ = -100\ \text{kJ mol}^{-1} \tag{7.24}$$

$$Cs + RbCl \rightarrow CsCl + Rb; \quad \Delta H_0^\circ \sim +20\ \text{kJ mol}^{-1} \tag{7.25}$$

It was soon clear that reactions of alkali-metal atoms with halogens are characterized by very large reaction cross-sections, i.e., $S_r \sim 2$—$4\ \text{nm}^2$, even at thermal collision energies. In practice, it is difficult to measure a *total* reaction cross-section. The differential cross-section, $\sigma_r(\theta)$, must be determined accurately over a wide range of scattering angles, so that the expression $S_r = \int \sigma_r(\theta)\, d\theta$ can be integrated. A second technique has occasionally been applied to the determination of S_r. Rather than measure the intensity of the scattered reaction product, differential cross-sections were found for the *non-reactively* scattered M[18]. These were compared with the scattering cross-sections for a roughly similar, but totally non-reactive, system. For example, the scattering of K from HBr was compared with that of K from Kr. For K + Kr, the differential cross-sections varied with scattering angle in the manner shown in *Figure 3.8*. For K + HBr, the scattering was similar at small θ, but the non-reactive differential cross-sections were smaller at large θ, indicating that chemical reaction was removing K atoms in collisions with small impact parameters. A quantitative comparison yielded a form for the opacity function, $P_r(b)$, and the total reaction- cross-section could be evaluated using the equation:

$$S_r = \int P_r(b)\, 2\pi b\, db \tag{7.26}$$

Experiments of this kind showed that $\overline{S}_r$ for K + HBr was, at approximately $0.3\ \text{nm}^2$, about an order-of-magnitude smaller than the thermally averaged reaction cross-section for reaction (7.23).

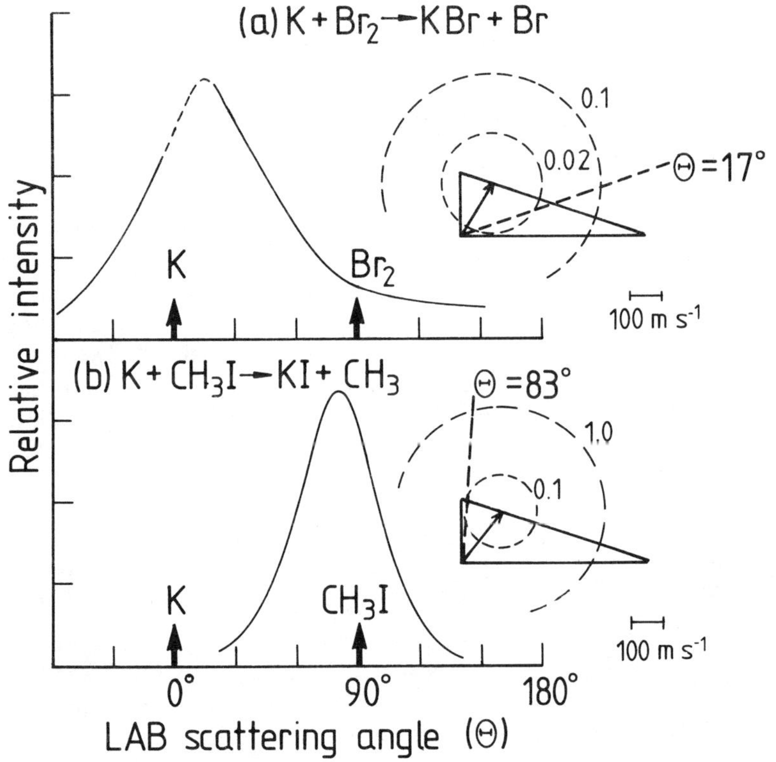

Figure 7.7. Reactive scattering in early molecular beam studies using effusive sources. The Newton diagrams are drawn for the most probable reagent velocities. The concentric circles show the length of u'_{KX} *with the* fractional *yield of product translational energy as shown*

Early *reactive* scattering measurements revealed a second major difference between reactions (7.23) and (7.24)[1, 2, 19, 20]. *Figure 7.7* compares LAB angular distributions for the salt molecules produced in these two reactions. The reactant beams for the experiments were of the effusive type and therefore contained molecules with a wide range of velocities. The Newton diagrams are drawn for the most probable velocities in each beam. Despite the limitations in the experiments, they did show clearly that the KBr from reaction (7.23) was scattered mainly at LAB angles less than that specifying the direction of $\boldsymbol{C}$, whereas the KI from reaction (7.24) was predominantly found at values of $\Theta > \Theta_{CM}$. Although, in the absence of velocity selection and analysis, the LAB distributions could not be properly transformed into CM angular distributions, it was plain that the CM distributions could not be symmetric about $\theta = 90$ degrees. This proved that these reactions both proceed *directly*. However, their detailed dynamics must differ since in $K + Br_2$ collisions the

molecular product scatters 'forward', along the same general direction as the incident K atoms, whereas when KI is formed in $K + CH_3I$ collisions it bounces back, roughly contrary to the direction of the incident atom beam. These two reactions exemplify *stripping* and *rebound* dynamics.

Reaction (7.25) was one of the first proceeding via a long-lived ($\geqslant 10^{-11}$ s) collision complex to be investigated in crossed-beam experiments[21]. It was studied using thermal beams, but the differential surface ionization detector incorporated a mass filter, so that the scattering of all four species could be resolved. *Figure 7.8* shows the LAB intensity distributions and the Newton diagram. The reactively scattered Rb has a bimodal distribution. Because of the $\sin\theta$ factor in equation (7.18), it is reasonable to assume (but see below in Section 7.4.2) that the observed LAB peaks correspond to peaks in the CM distribution at $\theta = 0$ and 180 degrees. Consequently, the most probable CM recoil energy could be estimated by observing where lines drawn at $\Theta = 95$ degrees and $\Theta = -15$ degrees (corresponding to the Rb peaks in the LAB distribution) cut the vector $\boldsymbol{v}$, or its extension. This procedure gave the most probable recoil energy ε'_{mp} as between 4 and 8 kJ mol^{-1}.

In the observed distribution, the peak at $\Theta = 95$ degrees had appreciably greater intensity than that at $\Theta = -15$ degrees. However, this symmetry is entirely associated with the effect of the $(v'_{AB}/u'_{AB})^2$ Jacobian factor in equation

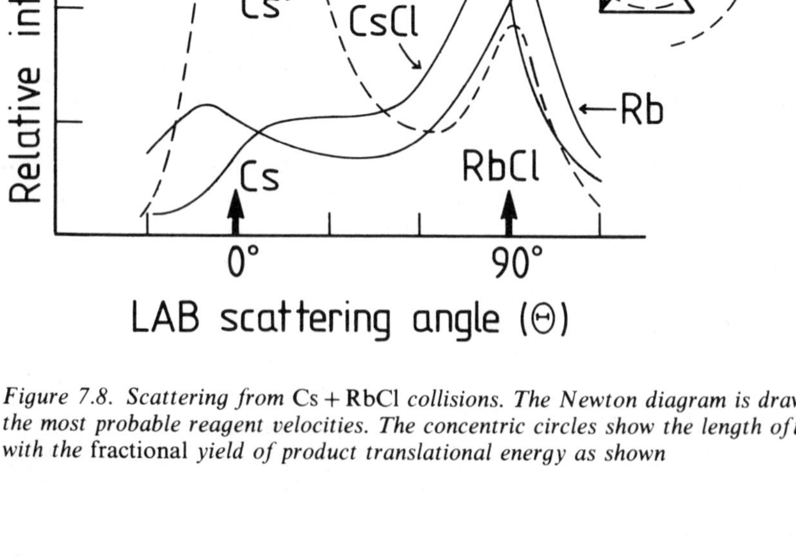

Figure 7.8. Scattering from Cs + RbCl *collisions. The Newton diagram is drawn for the most probable reagent velocities. The concentric circles show the length of* $\boldsymbol{u}'_{CsCl}$ *with the* fractional *yield of product translational energy as shown*

(7.16). In CM coordinates, the Rb angular distribution—and that of CsCl—are symmetric about $\theta = 90$ degrees. The distribution of Cs atoms scattered without reaction offers further proof of the formation of collision complexes. The peak at $\Theta = 90$ degrees results from the breakdown of complexes to Cs + RbCl, rather than to Rb + CsCl, in this nearly thermoneutral system.

Herschbach and his co-workers[21] carried out a quantitative analysis of the scattering from Cs + RbCl collisions. They assumed, as is reasonable in reactions proceeding by way of a complex, that the angular and relative velocity distributions of products in CM coordinates were uncorrelated, so that

$$\sigma_r(v', \theta) = f(\theta)f(v') \tag{7.27}$$

Using simple forms for $f(\theta)$ and $f(v')$, each with only one adjustable parameter, the observed distributions could be matched remarkably well. It was found that the angular function $f(\theta)$ was strongly peaked at $\theta = 0$ and 180 degrees and that the average recoil energy was appreciably greater than the collision energy, i.e., $\varepsilon'_T > \varepsilon_T$.

By considering the constraints imposed by angular momentum conservation, Herschbach[21] was able to establish a remarkably detailed picture of reaction (7.25). The strong peaking at the poles of the CM scattering distributions proved that the CsClRb complex both forms and fragments with orbital momentum much larger than the rotational momentum of either RbCl or CsCl. This means that complexes must be created with high probability in collisions with impact parameters as high as 0.8—0.9 nm (corresponding to a collision cross-section of ~ 2—3 nm^2). This, in turn, is consistent with a model in which there are long-range dipole–induced dipole forces between the reagents, and any encounters in which the centrifugal barrier is surmounted lead to complex formation.

7.4.2 Experiments with velocity selection and velocity analysis

When the results of primitive experiments on reaction (7.25) were analysed it was assumed that the distributions of v' and θ were uncorrelated. The results of primitive experiments on *direct* reactions have been analysed in a similar manner. For these reactions, the assumptions that $f(\theta)$ and $f(v')$ are uncorrelated and that neither depends strongly on the collision energy are less justifiable. Furthermore, the kinematic analysis has frequently been based on a Newton diagram drawn for the most probable velocities in the crossed beams, although this diagram only corresponds to the most probable Newton diagram *for the reactive collisions*, if there is zero threshold energy for reaction, and if the total reaction cross section varies only slowly with collision energy[5]. As indicated earlier, this is true for reaction (7.25). Furthermore, as it proceeds via a collision complex, this system retains no memory of its detailed mode of formation, so the v' and θ distributions only depend weakly on ε_T.

A simplified analysis also led to a correct estimate of the product translational excitation in reaction (7.23) between K atoms and Br_2. The

reasons for this are not difficult to identify. The very large cross-section shows that this reaction, like (7.25), must occur with high, and similar, probabilities at most thermal collision energies. Furthermore, the spread of collision energies, even in experiments with thermal effusive sources, is small compared with the energy released in the reaction. Consequently, it is reasonable (though no more than that) to assume that the angular and velocity distributions of the products are independent of ε_T and not *strongly* correlated.

Application of the simple analysis to the results of primitive experiments on the $K + Br_2$ reaction is illustrated on the Newton diagram in *Figure 7.7.* The dashed line drawn at $\Theta = 17$ degrees corresponding to the observed peak in the LAB angular distribution, cuts v at $\theta = 0$ degrees and indicates a most probable value of the final translational energy of about 4 kJ mol^{-1}. The great majority of the energy released during the reaction ($\sim$ 180 kJ mol^{-1}) must cause *internal* excitation of the products.

The polar diagram in *Figure 7.9* shows the results of more sophisticated experiments in which ε_T was defined using a supersonic atom beam and the LAB velocities of the reactively scattered KBr were measured at several LAB angles[6]. Contours join points of equal KBr flux reaching the detector. The LAB distributions are transformed to CM distributions using equation (7.16). *Figure 7.9*(b) shows that nearly all the KBr is scattered at angles $\theta < 90$ degrees; the peak flux at $\theta = 0$ degrees corresponds to a recoil energy of about 6 kJ mol^{-1}. These conclusions are in broad agreement with those deduced from the earlier primitive experiments. However, the later experiments do reveal a long tail in the distribution of final translational energies.

A simple analysis of primitive beam experiments led to an underestimate of the translational energy released in the reaction between K atoms and CH_3I. As *Figure 7.7*(b) shows, the distribution of KI reaches a peak at $\Theta = 83$ degrees. A line drawn on the 'average' Newton diagram at this angle intersects $\boldsymbol{v}$ at $\theta = 180$ degrees and a value u'_{KI} which corresponds to $\varepsilon'_T \sim 16$ kJ mol^{-1}. This implies that, once again, the reaction energy is channelled mainly into internal excitation of the products, but experiments with better resolution[22] have modified this conclusion. The results of such experiments are shown in *Figure 7.10*. KI is, indeed, scattered predominantly at angles $\theta > 90$ degrees; but the relative translational energy of the products is an appreciable fraction of the total available energy: very few encounters lead to values of ε'_T less than 25 per cent of the total energy and, in a high proportion, more than half the energy released appears as translational excitation. In all, there are three ways in which rebound reactions, like (7.24), differ from stripping reactions, such as (7.23): (1) their reaction cross-sections are at least an order-of-magnitude smaller; (2) the molecular product is scattered mostly into the backward hemisphere; (3) there is usually less internal excitation of the products.

Theoretical models of alkali-metal atom reactions have as their cornerstone the 'electron-jump' mechanism originally proposed, almost 40 years ago, to explain the very rapid rates observed in diffusion-flame experiments on these reactions[23]. The proposal is that as M approaches the molecular reagent there comes a point at which the difference between the ionization potential (IP) of

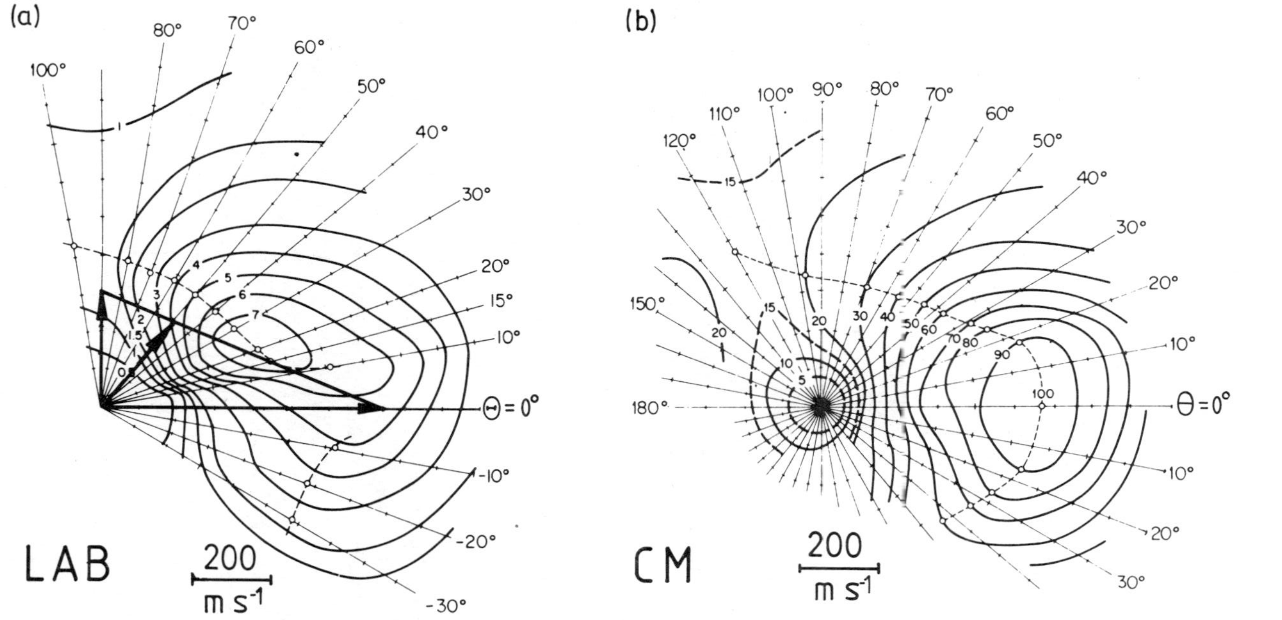

Figure 7.9. Flux of reactively scattered **KBr** *from* $K + Br_2 \rightarrow KBr + Br$. *Contour lines joint points of equal flux as a function of* (a) LAB *velocity* (v'_{KBr}) *and scattering angle* (Θ), *and* (b) CM *velocity* (u'_{KBr}) *and scattering angle* (θ) [*reproduced, with permission from Grice, R., in* Molecular Beams Physical and Chemical Applications, *Ed. Lawley, K. P. John Wiley, New York* (1975)]

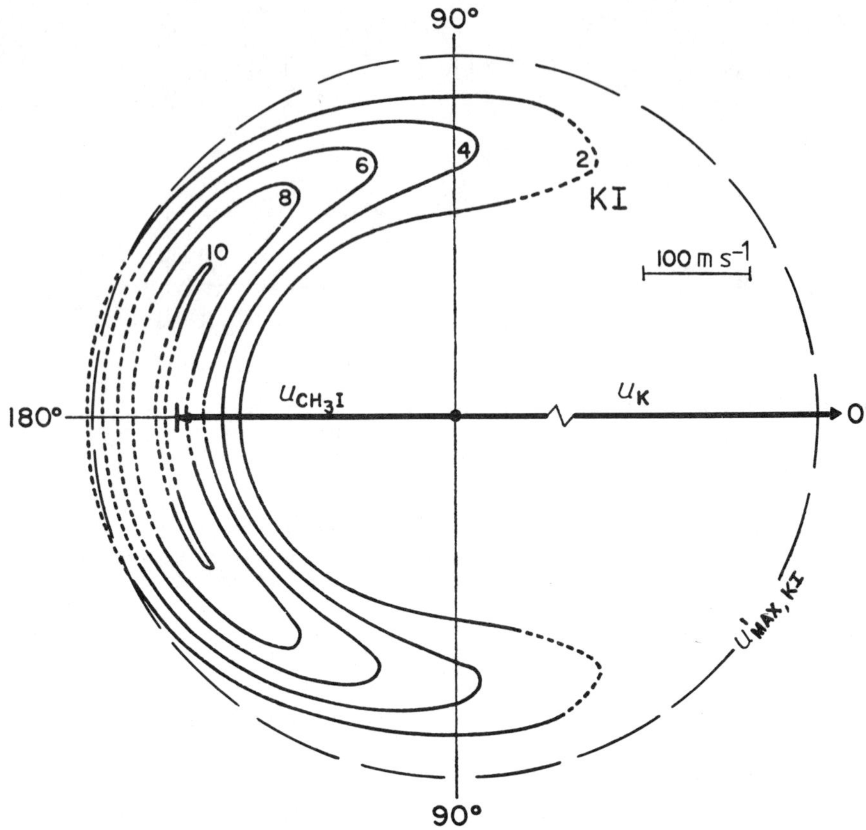

Figure 7.10. Flux of reactively scattered KI *from* $K + CH_3I \rightarrow KI + CH_3$ *as a function of* CM *velocity and scattering angle* [*reproduced, with permission, from Bernstein, R. B., and Rulis, A. M.* Faraday Discussions of the Chemical Society, 55, 293 (1973)]

M and the vertical electron affinity (EA_V) of the molecule is equal to the Coulombic attraction between two oppositely charged particles at that separation (r_c), i.e.,

$$\varepsilon^2/(4\pi\varepsilon_0)r_c = IP - EA_V \tag{7.28}$$

An electron is then transferred 'instantaneously' from M to the molecule (cf. *Figure 2.6*), the molecule undergoing a vertical transition to an ionic state, as illustrated in *Figure 7.11*.

Different values of the total cross-section for reactions occurring by an 'electron-jump' mechanism can be related—at least, approximately—to the values of r_c predicted by equation (7.28). For Br_2, EA_V is approximately 1.4 eV ($\sim$ 140 kJ mol^{-1}), reaction occurs for all collisions with impact parameters $\leqslant$ 0.7 nm, and $S_r \sim 2$ nm^2. On the other hand, EA_V is negative for CH_3I; the electron jump only occurs once the reagents are much closer

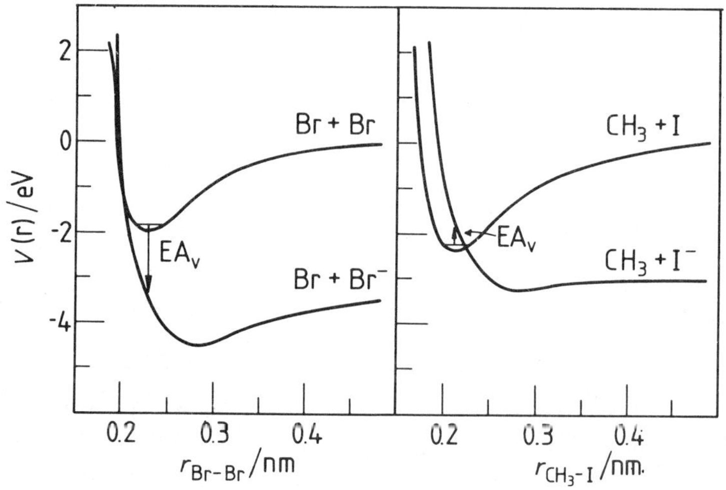

Figure 7.11. Potential curves showing vertical electron affinities (EA_V) *for* Br_2 *and* CH_3I [*adapted, with permission from Grice, R.,* Molecular Scattering: Physical and Chemical Applications, *Ed. Lawley, K. P. John Wiley, New York* (1975)]

together and the reaction cross-section is much less.

The wide variety of dynamics exhibited by reactions of alkali-metal atoms can be explained in terms of the electronic structure and energy of the negative ion formed when the electron is transferred. In some cases, for example $SnCl_4$ and $HgBr_2$, the electron enters a non-bonding σ orbital or a weakly antibonding π^* orbital and the ion is stable with respect to dissociation. Such reactions proceed via a collision complex. In Br_2 and CH_3I, however, the transferred electron enters an antibonding σ^* orbital and the molecular ions are formed above their dissociation limits, as shown in *Figure 7.11*. With Br_2^-, there is only modest repulsion between the separating fragments, Br^- and K^+ are attracted by Coulombic forces, and KBr is formed with high internal excitation. The Br product is a virtual 'spectator' in this process and, like the K^+ ion, it proceeds with little change in direction; hence KBr is scattered forward in the CM frame-of-reference. With CH_3I^-, rather more energy is released as repulsion between CH_3 and I^-. Furthermore, the K^+ ion is already close, so there is less mixed energy release than in $K + Br_2$ collisions. Finally, there is a more strongly preferred orientation for this reaction with the K atom approaching from the I end of the CH_3I molecule. These factors combine to cause the KI to rebound back along the direction from which the K atoms approached.

Spectacular evidence for the importance of orientation effects has been obtained from reactive scattering experiments in which symmetric-top molecules are partially oriented in the scattering region using a combination of electric fields[24]. With CH_3I, the probability of reaction is less when the I end of the molecules is oriented away from the approaching K (or Rb) atom beam than when it faces towards the K (or Rb) atoms. This finding is consistent with

the previous model of this reaction—and what one might expect on a simple consideration of steric factors. With CF_3I, however, the two orientations lead to roughly the same amount of reaction. Moreover, if the I end is closest to the incoming K atoms, KI is scattered *backward*, whereas *forward* scattering is observed when the orientation is reversed.

Two factors may contribute to the difference in reactivity of oriented CH_3I and CF_3I. First, there is the fact that the CH_3 group shields the I atom from attack from the rear more effectively than does CF_3. This is because the partially oriented molecules precess about the total angular momentum vector in the scattering region. In CH_3I, the precessional cone is very wide and the period of the motion is comparable with the duration of an intermolecular collision. The CF_3I molecule, however, precesses more slowly over a narrower cone and the I atom is less shielded.

The different dynamical behaviour may also be the result of differences in the electron-jump mechanism for the two reactions. The vertical electron affinity for CF_3I is probably larger than that for CH_3I allowing the electron to be transferred at larger separations. As with CH_3I, there will be repulsion between CF_3 and I^- with the result that the KI will be scattered along the direction to which the I atom originally pointed. If the electron jumps at sufficiently large distances, the reaction probability at either end of the molecule will be similar, as is found experimentally.

Full-scale trajectory calculations have been carried out on the $K + Br_2$ and $K + CH_3I$ reactions, and their analogues[25]. They essentially support the qualitative models that have been described in the preceding few paragraphs. One feature which has not been brought out so far is migration of the electron between the two separating fragments of the molecular reagent. This appears to be important in reactions like $K + Br_2$ [26]. Following its transfer, the electron is located on the Br atom closest to the approaching K^+ ion. Later, however, the K^+ ion may get closer to the second Br and the electron can then switch to this species. Consequently, a major fraction of the KBr which is scattered sharply forward appears to be formed in collisions in which the K^+ ion, having initially been attracted by one Br^-, skips by this species and makes off with the other.

7.4.3 Reactions of alkali-metal dimers[6]

When alkali-metal atoms (M) expand through a supersonic nozzle, a high proportion of dimer molecules (M_2) are formed as the gas cools. Typically, dimers constitute about 25 per cent of the species in the beam after the expansion is complete. Furthermore, the remaining atoms can be removed by deflecting them out of the beam with a magnetic field. The dimer molecules which remain have low internal energies and a well-defined translational energy.

The bonds in alkali-metal dimers are rather weak, varying from 109 kJ mol^{-1} for Li_2 to 42 kJ mol^{-1} for Cs_2. As alkali-metal atoms form strong ionic bonds with electronegative atoms, the dimers undergo a variety of

highly exoergic reactions. Only two of the several types of reaction which have already been studied in crossed-beam experiments will be considered here.

The reactions of halogen atoms (X = Cl, Br, I) with alkali-metal dimers (M_2 = Na_2, K_2, Rb_2, Cs_2) are all very exoergic, e.g.,

$$K_2(^1\Sigma_g^+) + Cl(3^2P) \rightarrow KCl(^1\Sigma^+) + K(4^2S);\ \Delta H_0^\circ = -372\ \text{kJ mol}^{-1} \quad (7.29a)$$

The reactions give rise to strong chemiluminescence[27], particularly from the resonant ^2P state of the atoms, e.g.,

$$K_2(^1\Sigma_g^+) + Cl(3^2P) \rightarrow KCl(^1\Sigma^+) + K^*(4^2P);\quad \Delta H_0^\circ = -216\ \text{kJ mol}^{-1} \quad (7.29b)$$

but emission has been observed from higher atomic states right up to the energetically accessible limit. The total reaction cross-section for Cl, Br + K_2 is large, ~ 0.5—1.0 nm^2, and that for reaction (7.29b) producing $K^*(4^2P)$ atoms is of comparable magnitude.

The observation of efficient chemiluminescence from the X + M_2 reactions is especially interesting as it was predicted many years ago. The arguments were quite simple and based on the correlations between the electronic states of the reactants, the transitory ionic intermediate (the reactions proceed directly), and the products. Because halogen atoms have a ^2P ground state, the X + M_2 reagents interact to yield three distinct potential hypersurfaces, two of $^2A''$ and one of $^2A'$ symmetry. The products of reaction (7.29a) correlate with a single $^2A'$ state, whereas those of (7.29b) correlate with two $^2A''$ and one $^2A'$ hypersurfaces. Consequently, there are twice as many adiabatic pathways for (7.29b) as for (7.29a). The reaction leading to ground-state products doubtless occurs via an electron jump mechanism involving the lowest $^2A'$ state of an ionic intermediate formed from $X^-(^1S_0) + M_2^+(^2\Sigma_g^+)$. The large cross-section for reaction (7.29b) suggests that it proceeds by a similar mechanism involving an excited ionic intermediate.

The model described in the previous paragraph explains the large reaction cross-section and the highly efficient production of alkali-metal atoms in their excited resonant state. However, the detection of emission from many other atomic states indicates that a model based on electronically adiabatic collisions is oversimplified. Apparently these systems can transfer easily from one hypersurface to another—at least, when these are of the same symmetry type (i.e., $^2A'$ or $^2A''$). The relative chemiluminescence intensities have been compared with the 'prior', statistical, expectation[28]. The surprisal associated with the observed distributions is small, further demonstrating the ease with which these systems can move between the many ionic and covalent potential energy hypersurfaces which exist in an energetically accessible range.

Many reactions of alkali-metal dimers with halogen-containing molecules have been studied by molecular beam techniques. Only those with halogen and interhalogen diatomic molecules will be considered here. These reactions also involve electron transfer, and are even more exothermic than the X + M_2 reactions. Although there is no orbital degeneracy associated with either

reagent, there are several distinct reaction pathways energetically available:

$$M_2 + XY \rightarrow M + MX + Y \quad (7.30a)$$

$$\rightarrow MX + MY \quad (7.30b)$$

$$\rightarrow M^* + MX + Y \quad (7.30c)$$

$$\rightarrow M^+ + MX + Y^- \quad (7.30d)$$

$$\rightarrow M_2X^+ + MY^- \quad (7.30e)$$

$$\rightarrow M^+ + MXY^- \quad (7.30f)$$

$$\rightarrow MX + MY^* \quad (7.30g)$$

For $M = K$ and $X = Y = Br$, $\Delta H_0^\circ = -260$ kJ mol^{-1} for reaction (7.30a) and -520 kJ mol^{-1} for reaction (7.30b).

Reactions (7.30a) to (7.30g) have all been identified in crossed-beam experiments and a very detailed picture of the dynamics of reactive collisions between alkali-metal dimers and diatomic halogens has been developed[6, 27, 29—31]. The total reaction cross-sections are in the range 1—2 nm^2, and the product KX is scattered predominantly in the forward direction relative to the incident K_2 beam. This asymmetry in the case of homonuclear halogens (X = Y) implies that (7.30b) cannot be the dominant reaction pathway. Otherwise, the need to conserve momentum would require symmetrical scattering about $\theta = 90$ degrees, whatever the dynamical mechanism. For the $K_2 + Br_2$ reaction, it has been estimated that the cross-section for reaction (7.30a) is about three times that for (7.30b)[30]. The cross-sections for the chemiluminescent and chemi-ionization channels (7.30c) to (7.30g) are much smaller ($\leqslant 0.1$ nm^2).

The $M_2 + XY$ reactions should resemble the prototype $M + X_2$ reactions in regard to the electron jump mechanism. Following transfer of an electron from M_2 to XY at a separation of about 0.7—0.8 nm, these systems should follow stripping dynamics. The vertical transitions following the electron jump produce M_2^+ close to its equilibrium separation but the electron added to XY enters a strongly antibonding σ^* orbital. Consequently, the product moving in the forward direction is intially $M_2^+X^-$ whereas Y is ejected backwards. However, $M_2^+X^-$ has only a very transient existence. It rapidly breaks up, but the MX formed continues to move in the forward direction. In a small fraction of encounters, it appears that $M_2^+X^-$ can reach configurations, with the X^- located between the alkali metal species, where the ground and excited state potentials approach. This allows the transient to dissociate non-adiabatically yielding M*, according to equation (7.30c).

Channels (7.30b), (7.30d) and (7.30f) all require the transfer of a second electron during the reactive encounter. Molecular orbital theory has been applied to predict what geometries of the system might favour this. For example, chemi-ionization is likely to be preferred in collisions where the system approaches a linear MMXY configuration. The final path (7.30g), producing an electronically excited molecule, does not require a second

electron transfer in the usual way, since the emitting state is homopolar. The charge distribution is rearranged when a photon is emitted and the molecule falls to its ionic ground state. It appears that reaction (7.30g) initially follows the dynamics of (7.30a) but the MX molecule acts as a third-body allowing recombination into the MY excited state.

This discussion of the $M_2 + XY$ reactions provides a prime example of how scattering experiments can lead to a very detailed understanding of the molecular dynamics in reactive collisions. In fact, our understanding of these systems is enhanced by studies of collisions between pairs of alkali-metal halide molecules[32]. In these, the dynamics are dominated by the existence of a deep potential well on the surface corresponding to the quadrupolar alkali-metal halide dimer, $Y^- \overset{M'^+}{\underset{M^+}{}} X^-$. The reactions proceed by way of a collision complex. They have large cross-sections ($\bar{S}_r \sim 1$—2 nm^2) and, together with the $M_2 + X_2$ reactions, serve as prototypes of facile four-centre reactions proceeding via strongly ionic intermediate states.

7.4.4 Reactions of alkaline earth metal atoms

The ionization potentials of the alkaline earth elements Ca, Sr and Ba are less than 6 eV; that of barium, which has been used most as a reagent in molecular beam experiments, is only 5.2 eV. On these grounds, their reactions with molecules containing electronegative atoms might be expected to resemble those of the alkali-metal atoms. On the other hand, the ground states of the alkaline earth metals are $^1S_0$, rather than $^2S_{\frac{1}{2}}$, and potentially two valence electrons are available for transfer—although the second ionization potential is much higher than the first (for Ba, it is 10.0 eV). These factors lead to some similarities with the reactions of alkali-metal dimers.

The ionization potentials of the alkaline earth metal atoms and their salts are sufficiently high to prevent their detection by surface ionization, but they are ionized with unusual efficiency by electron bombardment. An important property is that the salts have low-lying, bound, electronic states, leading to discrete absorption spectra at visible wavelengths. This has allowed products in defined vibrational levels of the electronic ground state to be detected via laser-induced resonance fluorescence[15—17].

A number of reactions of alkaline earth metal atoms with halogen-containing molecules have been studied in standard scattering experiments. The results, for example, for the reactions with diatomic halogens[33, 34] closely follow those for the corresponding reactions with alkali-metal atoms. Thus, the reactions

$$Ba + X_2 \, (= Cl_2, Br_2) \rightarrow BaX + H; \qquad (7.31)$$

have large cross-sections (0.1—1.0 nm^2), lead to strong forward scattering of the M′X product, and deposit only about 10—20 per cent of the available energy into the relative translational motion of the products. Clearly, the reactions proceed by an electron-jump stripping mechanism, like the alkali-

metal atom plus halogen molecule reactions. Although chemiluminescence from electronically excited $BaCl_2$ has been observed in experiments where a beam of Ba atoms was fired into Cl_2 scattering gas, the crossed-beam experiments indicate that dihalide is produced in < 5 per cent of reactive collisions[35].

The vibrational distributions of BaX molecules formed in the family of reactions between Ba atoms and hydrogen halides,

$$\mathrm{Ba + HX\,(= F, Cl, Br, I) \rightarrow BaX + H;} \qquad (7.32a, b, c, d)$$

$$\Delta H_0^\circ = -46, -39, -51, -112\ \mathrm{kJ\ mol^{-1}}$$

have been measured using the laser induced resonance fluorescence technique[15, 16]. The kinematics of the reactions are similar to those of reaction (7.1). However much energy is released into relative translation of the products, the BaX molecules must be scattered close to the ***C*** vector, the spread of BaX LAB velocities must be small, and the product molecules remain in the scattering zone—which is illuminated by the laser—much longer than the scattered H atoms. These factors facilitate the use and interpretation of resonance fluorescence measurements.

Zare and his co-workers[16] note that their experiments, in contrast to standard reactive scattering measurements, yield relative number densities of products rather than product fluxes. In calculating the densities, they assume that all molecules remain within the 'observation zone' for the same time, irrespective of their internal state, or of the incident velocities in the reactive collisions in which they are formed (these two factors determine, via the energy balance equation, the final translational excitation of a pair of product species). This assumption is also made when interpreting the spectra from infrared chemiluminescence experiments under conditions of arrested relaxation (Section 6.2.1). There the assumption is justified, because reaction does not occur under single collision conditions. In molecular beam experiments, it is an approximation[5], even though the favourable kinematics in the Ba + HX reactions ensure that no great error ensues.

The BaX vibrational distributions produced in reactions (7.32a, b, c, d) are shown in *Figure 7.12*. The overall yields are low; corresponding to $\langle f_{V'} \rangle = 0.12, 0.28, 0.36$ and 0.18 of the available energy along the sequence: Ba + HF, HCl, HBr and HI. The populations in individual rotational levels were not resolved, but the yields of rotational energy in BaF and BaCl could be estimated from the 'rotational envelopes' of individual bands. The yields were quite high ($\langle f_{R'} \rangle = 0.13$ for BaF and 0.18 for BaCl). No doubt, this is because the reduced mass of the separating products is very small, so a large part of the initial angular momentum has to become final rotational momentum of BaX. Nevertheless, certainly for reactions (7.32a) and (7.32b), and probably for (7.32c) and (7.32d), $\langle f_{T'} \rangle$ exceeds 0.5.

From the experimental results and the theoretical prediction that the Ba + XH mass combination could lead to considerable 'mixed' energy release and hence high vibrational excitation, it is evident that these reactions must

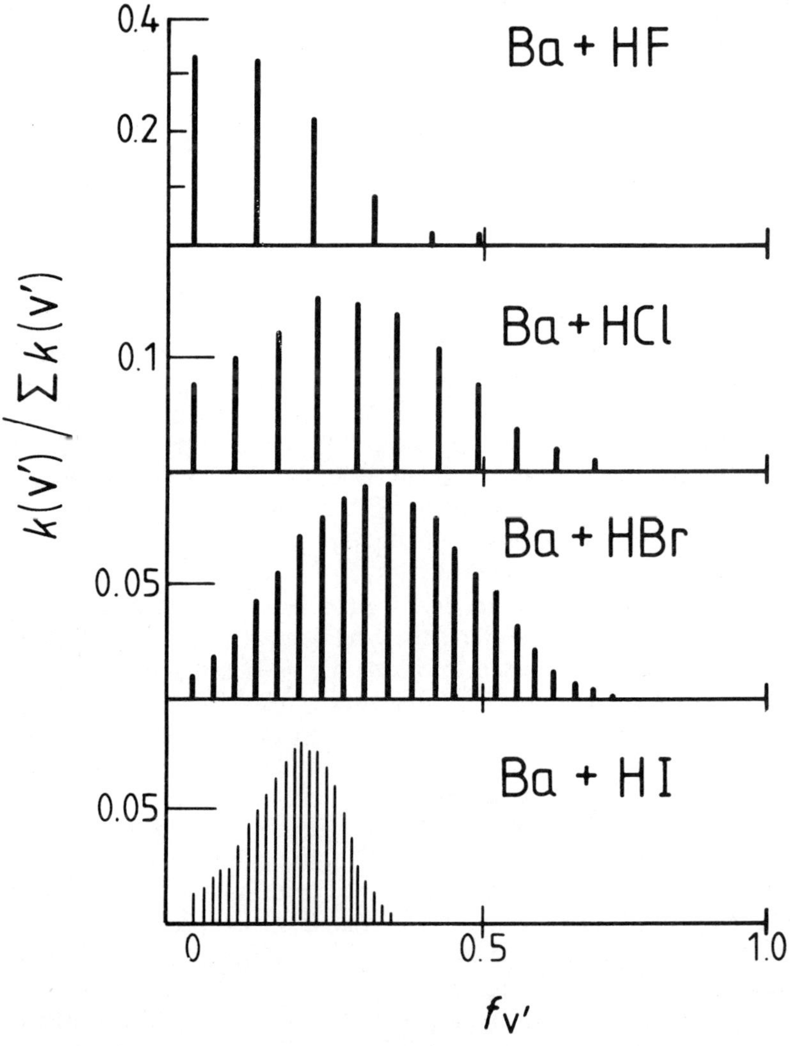

Figure 7.12. Relative vibrational populations of barium halides formed in the reactions Ba + HX → BaX + H [*adapted, with permission, from Cruse, H. W., Dagdigian, P. J., and Zare, R. N.*, Faraday Discussions of the Chemical Society, **55**, 277 (1973)]

proceed across rather repulsive potential hypersurfaces. The reasons for this are not difficult to find. The electron affinities of the HX molecules do not favour the early transfer of an electron from an approaching Ba atom. By the time the electron is transferred, the opportunity for much attractive energy release through the Coulombic attraction of Ba^+ and Cl^- has been lost. Furthermore, the transferred electron enters a strongly antibonding σ^* orbital on HX. Consequently, HX is formed well above its dissociation energy and the energy is released almost entirely as relative translational motion. This

interpretation is very similar to that given earlier to explain the moderate yields of internal energy in the reactions of alkali-metal atoms with CH_3I.

7.5 Reactions of non-metal species

Differences between beam *experiments* on metal and non-metal reactions have been emphasized so far. In particular, it has been pointed out how their low ionization potentials allow alkali metals and their salts to be detected via surface ionization. Such differences are not the most fundamental, nor the most interesting. A second more basic effect of metals' low ionization potentials is that they frequently form strongly ionic bonds. Consequently, the rearrangement of electronic structure which accompanies any chemical reaction requires the more or less complete transfer of an electron during the course of any reaction producing a metal salt. We have seen that, when the non-metallic reagent has a favourable electron affinity, the electron transfer can take place very early in the reactive encounter, whilst the reagents are still far apart, and before their electron clouds overlap to any appreciable extent. Then a simplified, but useful, description of the reaction dynamics can be given in terms of the forces between (1) the two parts of the fragmenting negative ion, and (2) the Coulombic attraction between the ions which come together to form the salt molecule.

The ultrasimple electron transfer model is less satisfactory for metal reactions, such as those of $K + CH_3I$, $K + HX$ and $Ba + HX$, where the electron is transferred at quite small reagent separations. In such cases, the charge clouds on the reagents already overlap and one should take account of three-centre interactions in any wholly satisfactory description of the electronic state through which the system passes. These reactions serve as a halfway house between reactions characterized by early electron transfer, such as that between K atoms and Br_2, and reactions of non-metallic species. In interpreting the latter, it is helpful to consider the form and energy of the three-centre molecular orbitals that can be formed as the reaction proceeds. Much use has been made of reasoning based not only on semi-quantitative molecular orbital theory but also on empirical information gathered from spectroscopy and photochemistry. Examples of this approach are given below, especially in Sections (7.5.2) and (7.5.4).

7.5.1 The $H + H_2$ and $F + H_2$ reactions revisited

The reactions between H and H_2 and between $F + H_2$—along with their isotopic relations—were featured prominently in Chapters 5 and 6. This was because their electronic simplicity allows their potential energy hypersurfaces to be calculated accurately, so they provide the most promising meeting ground for full scale theory and detailed experiment. Examples of both systems, i.e.,

$$D + H_2 \rightarrow HD + H \tag{7.32}$$

and

$$F + D_2 \rightarrow DF + D \tag{7.33}$$

have recently been studied in crossed molecular beam experiments[36–38], providing further data for comparison with predictions about the collision dynamics of these reactions.

Crossed-beam experiments on reactions (7.32) and (7.33) are not easy. However, the interpretation of the observed reactive scattering is simplified, because the internal states of the molecular products are widely spaced in energy. The salt molecules produced in the reactions of metal atoms possess a densely packed manifold of vibrational and rotational levels. Usually, the spread in the incident collision energy and that in the final rotational energy exceed the separation of neighbouring vibrational levels, with the result that any 'structure' in the spectrum of final CM velocities is completely obscured. Furthermore, the product molecules are frequently scattered over a fairly wide range of CM scattering angles. The nett result is that the distributions of reactively scattered product over LAB angles (Θ) and final LAB velocities (v') are continuous; there is no evidence for the existence of several peaks associated with the production of molecules in different quantum states.

In contrast, reactions (7.32) and (7.33) produce molecules with widely spaced rovibrational levels. Moreover, the potential hypersurfaces for these reactions ensure that, at collision energies close to threshold, reaction can only occur in near collinear collisions of low impact parameter. This limits the yield of rotational energy [for reaction (7.33), this is confirmed by infrared chemiluminescence measurements, as reported in Section 6.3.1], and the molecular product rebounds into a rather narrow cone around $\theta = 180$ degrees. If the collision energy is well-defined, the spectrum of final relative velocities consists of a number of narrow bands; each corresponding to the formation of the molecular product in a different vibrational level, with the width of a band determined by the distribution over rotational levels within that vibrational state. These 'spectra' of recoil energy or velocities have been predicted for several reactions from the results of infrared chemiluminescence measurements of the detailed distribution of a reaction product over internal states[39]. A straightforward inversion procedure, based on the conservation of energy, provides the corresponding translational energy spectrum.

The $D + H_2$ reaction has been studied in a crossed-beam experiment[36] using extensive liquid helium cryo-pumping. The D atoms were produced by thermal dissociation at 3000 K and this beam crossed an H_2 beam from a nozzle source at 77 K. Neither beam was velocity-selected, but the D atom beam was mechanically chopped. Scattered HD was observed using a quadrupole mass spectrometer. A major difficulty was to differentiate between the reactively scattered HD and any HD impurity in the D atom beam which was scattered elastically by collisions with H_2. Fortunately, the kinematics for these two processes are quite different, as the Newton diagram in *Figure 7.13* shows.

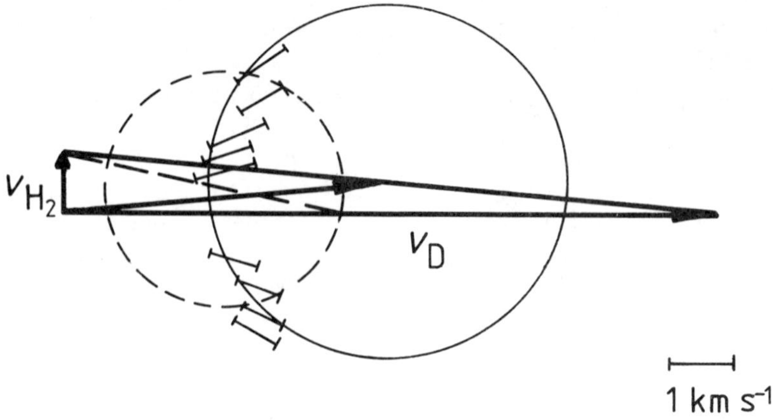

Figure 7.13. Newton diagram for $D + H_2 \rightarrow HD + H$. *Symbols* (|—|) *represent the observed spread of* LAB *velocities at different* Θ. *The full circle indicates the locus of reactively scattered* HD *velocity vectors with all the energy released in translational motion of the products. The dashed circle indicates where the velocity vectors for elastically scattered* HD *impurity would terminate* [*adapted, with permission, from Geddes, J., Krause, H. F., and Fite, W. L.,* Journal of Chemical Physics, **56**, 3298 (1972)]

The D atom velocity in the Newton diagram is chosen to represent a collision energy of 32 kJ mol^{-1} (0.33 eV), which is the threshold energy for reaction (7.32). The full circle represents the locus of CM velocities for reactively scattered HD, assuming no internal excitation. The dashed circle shows the distribution of CM velocities for elastically scattered HD. A number of pieces of evidence show that the observed HD was predominantly reaction product rather than impurity. The most important is the result of the velocity analysis. Measuring the phase lag between the chopping of the D atom beam and the detected signal gave the LAB velocities indicated on *Figure 7.13*. It is plain that these are consistent with backward reactive scattering of HD with little or no internal excitation, but not with elastic scattering. A low yield of internal product energy is to be expected. The total energy is sufficient to produce HD(v′ = 1) in only a very small fraction of the reactive collisions, whilst angular momentum restrictions ensure that HD is formed predominantly in low rotational levels.

The transformation from the scattering distribution over LAB angles Θ to that over CM angles θ is simplified by the narrow spread of CM final velocities, and by the fact that most *reactive* collisions would occur at energies fairly close to ε_T^0, the threshold energy, even though there was a wide spread of LAB velocities in the thermal D atom beam. The differential reaction cross-section which was deduced could be fitted to the simple expression

$$\sigma_r(v, \theta) = S_r(v) \cos^2[1.35(180 - \theta)] \quad (7.34)$$

The sharply peaked backward scattering which this formula implies is in good agreement with three-dimensional quantum scattering calculations on the H $+H_2$ reaction[40].

Absolute values of the differential reaction cross-sections were found by comparing the signals from reactively scattered HD with those from elastically scattered D atoms. The elastic scattering cross-sections could be calculated from earlier beam measurements on $H + H_2$ collisions[41]. Integration over all angles gave the total reaction cross-section. Assuming that the excitation function varied in the manner predicted by simple collision theory led to the result:

$$S_r(\varepsilon_T) = (9.0 \pm 4.5) \times 10^{-3}[1 - (\varepsilon_T^0/\varepsilon_T)]\ \text{nm}^2 \qquad (7.35)$$

Integration of this expression over a thermal distribution of velocities with $T = 1400$ K (the temperature which approximately describes the distribution of relative velocities in the beam experiment) gave a thermal rate constant: $k(T = 1400\ \text{K}) = 3.2 \times 10^{-12}\ \text{cm}^3\ \text{molecule}^{-1}\ \text{s}^{-1}$. This value is, remarkably, within a factor of two of that obtained by extrapolating the thermal rate data from lower temperatures (*see Table* 5.8).

In contrast to the $D + H_2$ reaction, the reaction between F atoms and D_2 is appreciably exoergic and has only a small activation energy. It has been studied using a velocity-selected atom beam and a D_2 nozzle beam[36, 37]. Fluorine atoms were obtained by dissociating F_2 at 675 K; the F_2 bond is so weak that, even at this moderate temperature, dissociation is almost complete at low pressure ($\sim 100\ \text{N m}^{-2}$ or 1 Torr). The distribution of LAB velocities was measured at several LAB angles, and for several well-defined collision energies. The Newton diagram for $\varepsilon_T = 7.0\ \text{kJ mol}^{-1}$ is drawn in *Figure 7.14*.

The molecular product of the $F + D_2$ reaction, like the HD formed in reaction (7.32), has widely spaced internal states, but now several vibrational levels are energetically accessible. The highest of these is DF (v′ = 4) and, as *Figure 7.14* shows, the relative velocity vectors associated with the formation of molecules in this state must terminate close to the origin of the CM coordinates. Production of DF in lower v′ levels leads to higher recoil energies, represented by the larger circles superimposed on the Newton diagram. *Figure 7.14* indicates the range of LAB angles over which DF(v′) molecules would be found *if* they were formed without any rotational excitation and *if* they rebounded into a narrow cone of CM angles defined by $165 \leqslant \theta \leqslant 180$ degrees. If these conditions were fulfilled, the maximum signals from molecules formed in different vibrational states would appear at different LAB angles. Furthermore, the peaks in intensity due to DF (v′ = 4) and, to a lesser extent, DF(v′ = 3), could be clearly resolved.

At low collision energies, the simplified picture of the scattering which is illustrated in *Figure 7.14* is fairly satisfactory. The peak in the LAB angular distribution due to DF (v′ = 4) can be clearly distinguished in *Figure 7.15* and was even sharper when the collision energy was reduced to $3.4\ \text{kJ mol}^{-1}$. The

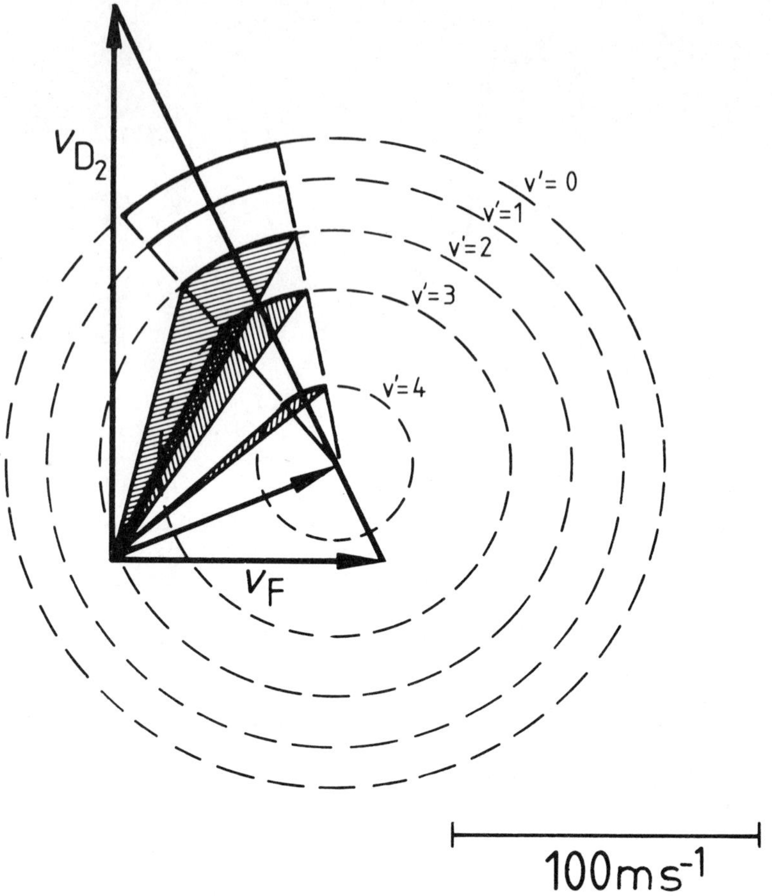

Figure 7.14. Newton diagram for $F + D_2 \rightarrow DF + D$. *The circles indicate the recoil velocities of* DF(v', j' = 0). *The hatched area encloses the range of* LAB *angles into which* DF(v' = 4, 3 *and* 2, j' = 0) *would be scattered, if they rebounded into a cone defined by* $165° \leqslant \theta \leqslant 180°$

model holds particularly well for DF (v' = 4), because when these molecules are produced in low-energy collisions there is little energy left over for rotational excitation. Consequently, DF (v' = 4) is formed in near-collinear collisions with very small impact parameters, and therefore *must* be scattered at angles close to $\theta = 180$ degrees. The formation of DF(v' < 4) is not so constrained by energy and momentum conservation and these molecules are formed over a wider range of θ. A second peak due to DF (v' = 3) molecules is visible, but the weaker scattering into lower v' levels can only be identified as a result of velocity analysis.

An interesting result was obtained as the collision energy was increased. The kinematic controls, which at lower energies require DF(v' = 4) to scatter sharply backward, become progressively weaker. DF(v' = 4) molecules were

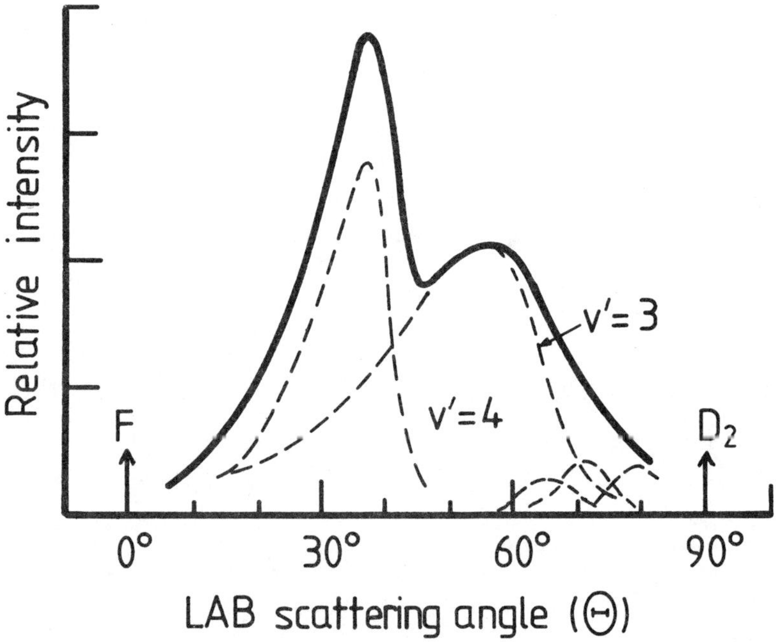

Figure 7.15. Scattering of DF *product from* $F + D_2$ *collisions at an energy of approximately* 7.0 kJ mol^{-1}. *The full line shows the total product intensity; the dashed lines those associated with molecules in particular* v′ *levels as deduced from the velocity analysis [adapted, with permission, from Lee, Y. T. in* Physics of Electronic and Atomic Collisions, *Eds. Govers T. R., and de Heer, F. J., North-Holland, Amsterdam* (1971)]

found over an increasingly wider range of θ and eventually the 'structure' in the LAB angular scattering disappeared.

The experiments which have been reviewed in this and earlier sections have provided an extremely detailed picture of the reactive collisions which occur between H atoms and H_2, and between F atoms and H_2. Most aspects of the dynamics of these two prototypes of atom-transfer reactions can be understood in terms of their calculated potential energy hypersurfaces and quasiclassical models of their dynamics. Classical dynamics are inadequate, however, at energies close to threshold. Threshold energies exist not only for the overall occurrence of reaction, but also for the formation of products in specified internal states. It is likely that future experiments will be designed to probe these threshold regions, in order to test the accuracy of quantum mechanical calculations on the reaction dynamics.

7.5.2 Direct atom-transfer reactions

An important development in the interpretation of molecular beam experi-

ments has been the increasing use of semiquantitative molecular orbital theory to explain the distributions of reaction products over angle and relative velocity. In this section, this approach is illustrated by applying it to two families of atom-transfer reactions: the reactions of H atoms with diatomic halogens (XY) and those of the type, $X + YZ \rightarrow XY + Z$, where X, Y and Z are all halogen atoms. The electronic and nuclear structures of both these systems can be discussed in terms of orbital correlation rules—which were established by Walsh a quarter of a century ago[42].

For any collinear three-atom system, the molecular orbitals can be assigned as σ or π, according to whether λ, the quantum number for the component of angular momentum whose vector is directed along the internuclear axis, is 0 or 1 (we shall not be concerned with orbitals for which $\lambda > 1$). For σ orbitals, there is no plane containing the internuclear axis through which the orbital wavefunction changes sign, whereas there is one such plane for π orbitals. For a bent structure of C_s symmetry, the only symmetry which remains is reflection through the plane containing the three nuclei; orbitals are assigned as a' or a'', depending on whether they remain unchanged or change sign on undergoing this operation.

Figure 7.16 shows the orbital correlations for the **HAB** systems (where A and B are any second-row elements) and also provides a pictorial representation of the nature of some orbitals for linear and 90 degree bent geometries, according to Walsh. It can be seen that the energies of the three low-lying orbitals which have σ symmetry for collinear HAB, either remain the same or increase, as HAB is bent and the orbitals take on a' symmetry. The degeneracy of the two bonding π orbitals and that of the antibonding π^* orbitals is split as the symmetry is lowered. The orbitals which are built from p orbitals perpendicular to the molecular plane assume a'' character and therefore remain of essentially the same energy. The in-plane bonding π orbital, like the bonding σ orbitals, has its energy raised. Consequently, HCN, with ten valency electrons which can just fill the lowest five molecular orbitals, is linear in its electronic ground state.

In contrast to the behaviour of the bonding orbitals, the π^* anti-bonding orbital which takes on a' character in C_s symmetry has its energy steeply lowered as the HAB angle is decreased. Because of this, the species HCO, HNO, HO_2 and HOCl, which have respectively 11, 12, 13 and 14 valency electrons, are all strongly bent in their electronic ground state. The addition of further electrons requires occupation of the σ^*/a'^* orbital whose energy rises steeply as the HAB angle is lowered. With 16 electrons present, there seems little doubt that a linear structure should provide the lowest energy. The $H + XY$ case (X and Y representing halogens) corresponds to 15 valency electrons, and the preferred geometry cannot be confidently predicted on the basis of semiquantitative arguments. For $H + F_2$, *a priori* calculations[43] on the electronic energy indicate that the lowest barrier to reaction is on the surface for collinear HFF, but whether this is so for other systems may depend on the absolute and relative electronegativities of X and Y.

Molecular beam experiments incorporating velocity analysis have been

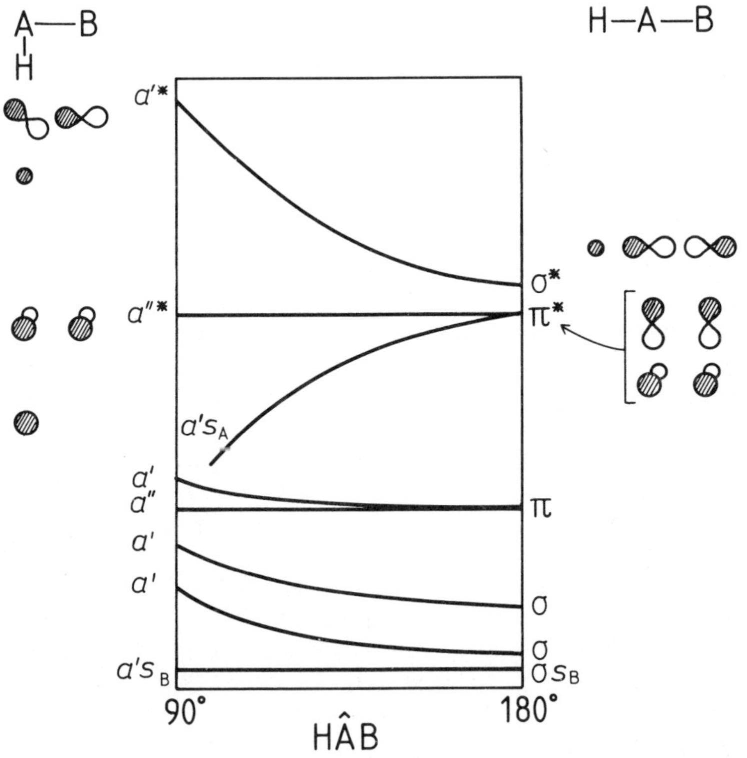

Figure 7.16. Correlation of molecular orbitals for linear and bent conformations of HAB

performed on the reactions of D atoms with Cl_2, Br_2, I_2, ICl and IBr[44]. The D atoms were generated by thermal dissociation of D_2 and the halogen beam came from a supersonic source. These experiments measured distributions of the products' translational energy, and complement infrared chemiluminescence measurements of the products' internal state distribution from the same, or related, reactions. The two sets of results are consistent. However, it has to be appreciated that the experiments cover quite different ranges of collision energy; the scattering studies were carried out at an average collision energy of roughly 40 kJ mol^{-1}, well in excess of the value in the chemiluminescence experiments.

Figure 7.17(a) shows the usual contour line representation of reactively scattered HCl from

$$H + Cl_2 \rightarrow HCl + Cl \tag{7.36}$$

The HCl is scattered almost exclusively into the backward ($\theta > 90$ degrees) hemisphere. The maximum is at $\theta = 180$ degrees with a velocity corresponding to about 40 per cent of the total available energy going into the translational recoil of the products. These observations indicate that a collinear HClCl

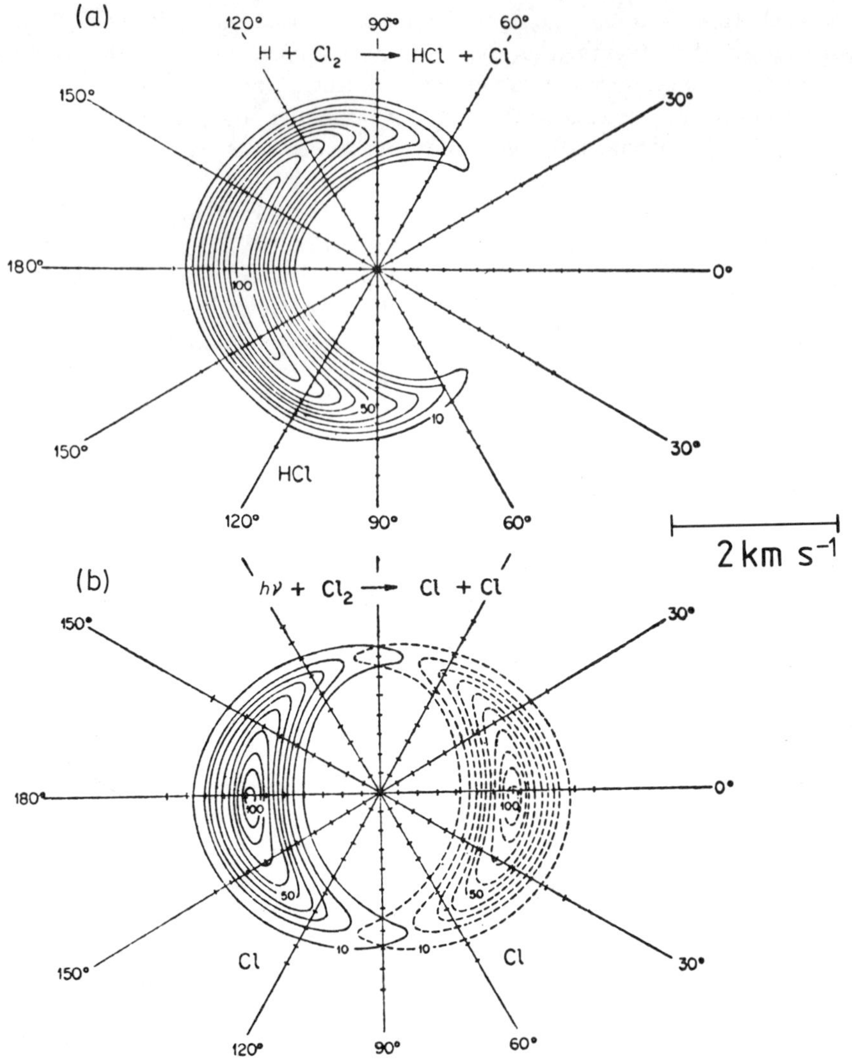

Figure 7.17. Comparison of polar contour maps for (a) HCl *reactively scattered in* $H + Cl_2$ *collisions, and* (b) Cl *atoms scattered in the photodissociation of* Cl_2 [*reproduced, with permission, from Herschbach, D. R.*, Faraday Discussion of the Chemical Society, **55**, 233 (1973)]

geometry is favoured for reaction and that there are strong repulsive forces between the separating products.

Reaction (7.36) can proceed adiabatically across a $^2A'$ potential hypersurface with the highest-energy electron occupying the σ^*/a' molecular orbital whose energy, according to the correlations illustrated in *Figure 7.16*, rises steeply as the HAB angle is lowered. The scattering results indicate that this effect is important in HClCl so that collinear and nearly collinear collisions are

most likely to lead to reaction. Herschbach has carried the molecular orbital approach further[4]. Arguing that the σ^*/a'^* orbital corresponds, in the HClCl case, primarily to bonding overlap of the 1*s* orbital located on the H atom with the σ^* orbital of Cl_2, he suggests that the antibonding nodal plane of the three-centre molecular orbital is roughly midway between the two Cl atoms. An analogy is then drawn between the translational energy release in reactive collisions, during which the HClCl σ^*/a' orbital becomes singly occupied, and that in photodissociation where an electron is promoted to the $Cl_2\sigma^*$ orbital. The contour maps in *Figure 7.17* show that the angular and velocity distributions of HCl formed in reaction (7.36) and of Cl atoms created by laser photodissociation of Cl_2 at 347 nm[45] are very close, and a number of similar correlations have been found.

Along the sequence Cl_2, Br_2, I_2, the peak in the HX distribution moves from $\theta = 180$ to $\theta = 135$ to $\theta = 105$ degrees and $\langle f_{T'} \rangle$ falls, indicating greater product internal excitation. These changes probably reflect a lowering in the energy barrier to reaction (reflected in the decrease in activation energy along the series $H + Cl_2$, Br_2, I_2), a parallel increase in the 'attractiveness' of the surface, and the increasing incidence of reaction in collisions with large impact parameters. The shift in the peak of the angular distribution may also be an indication that as the electronegativity of the X and Y atoms decreases a bent geometry for HXY is likely to be preferred. This is consistent with Walsh's rules.

The H + ICl reaction provides a striking illustration of how differences in the electronegativity of the X and Y atoms can cause differences in the reactivity at the two ends of a reactant molecule. The experiments show that HI + Cl are formed at least 3 or 4 times more frequently than HCl + I, despite a considerable energetic bias towards the second pair of products (the dissociation energy of HCl is 134 kJ mol^{-1} greater than that of HI). However, Walsh's rules predict that the HXY conformation will have a lower energy than HYX, where X is less electronegative than Y. The reason for this is that, whereas the bonding orbitals are predominantly built from orbitals in Y, the more electronegative atom, the XY antibonding orbitals are constructed largely from orbitals on X. Consequently, the conformation HXY favours reaction more than HYX, because it optimizes the bonding overlap between the H atom 1*s* orbital and the *p* orbital located on the central atom, thereby lowering the energy of the σ^*/a'^* orbital.

Walsh has also discussed the electronic structure and molecular geometry of systems comprised of three non-hydride atoms. The XYZ systems, where all three atoms are halogens, occupy a position in this family like that of HXY in the HAB family. Three halogen atoms possess a total of 21 valency electrons and the highest energy electron occupies an orbital which, in collinear XYZ, is built primarily by out-of-phase overlap of p_z orbitals on X and Z with the 2*s* orbital on the central atom. Systems with 20 valency electrons, like Cl_2O, are strongly bent, whereas those with 22, such as I_3^-, are linear. Again Walsh's rules suggest that the most stable species will be those with I, the least electronegative of the halogens, in the central position.

It was molecular beam experiments on the X + YZ reactions that first took reactive scattering studies out of the alkali-metal-only age, and measurements have now been made on many of these systems. They show a rich variety of reaction dynamics. This is because the triatomic species XYZ have electronic energies roughly comparable with those of the corresponding separated species X + YZ or XY + Z. The fact that some triatomic halogens possess some stability can be inferred from the exceptional efficiency of diatomic halogens as third-bodies for halogen atom recombination via the bound-complex mechanism (*see* Section 7.8.2). For example, these measurements indicate a dissociation energy for I_3 of about 20 kJ mol^{-1}.

On the basis of Walsh's rules, 'stable' XYZ is expected to be linear, or nearly so, because the outermost electron occupies an orbital whose antibonding character rises steeply as the XYZ angle is decreased. Furthermore, complexes containing I atoms are more stable than those including Cl and Br because of easier charge-transfer stabilization. Where XYZ contains different atoms, the most favourable configuration has the least electronegative atom in the Y position.

The existence of short-range attractive forces in most of the X + YZ systems is confirmed by the results of reactive scattering experiments. For most of these reactions, the cross-section is found to lie between 0.01 and 0.1 nm^2. For F + ICl, the major reaction[6] is

$$F + ICl \rightarrow FI + Cl \tag{7.37}$$

and the scattering appears to be symmetric about $\theta = 90$ degrees, although this conclusion needs confirmation in experiments incorporating velocity analysis. These observations are consistent with the formation of an FICl collision complex which can survive for longer than a rotational period, and which is favoured over FClI because of the central position of the I atom.

When the difference between the electronegativities of the central and end atoms is less than in FICl, the stability of the intermediate is reduced. Where the lifetime is only of the order of a rotational period, the symmetry of the angular distribution about $\theta = 90$ degrees is lost. Where the lifetime is of the order of half a rotational period, the molecular product scatters predominantly into the forward hemisphere. This result is found, for example, for the reaction[46, 47]

$$Cl + BrI \rightarrow ClBr + I \tag{7.38}$$

Reactive scattering of ICl from Cl + BrI collisions occurs over a wider range of angles than that of ClBr, indicating that ClIBr is rather more stable than ClBrI.

Triatomic halogens become least stable when the most electronegative atom is positioned centrally. This situation is epitomized by BrClI. The cross-section for the reaction

$$Br + ClI \rightarrow BrCl + I \tag{7.39}$$

is about five times smaller than for other X + YZ reactions that have been studied[46]. Moreover, BrCl is scattered backward. This result, signifying a

rebound mechanism, demonstrates that the halogen atom plus halogen molecule family of reactions is capable of exhibiting the full range of dynamical behaviour and that the differences between related reactions can be rationalized by means of simple molecular orbital theory.

7.5.3 Reactions proceeding via collision complexes

The determination of CM differential reaction cross-sections in molecular beam experiments provides the crucial test of whether a reaction proceeds directly or via a collision complex. If the variation of the differential reaction cross-section—or, more simply, the distribution of products over CM scattering given in Section 3.3 Because $l' \approx l$, the products must separate in a must proceed via a complex which can survive for at least several rotational periods. As was pointed out in Section 7.4.1, the halogen atom exchange reaction (7.25) between Cs atoms and RbCl was one of the first reactions which was shown to involve a collision complex. When the LAB scattering intensities for CsCl (or Rb) were transformed to the corresponding CM quantities, $f(\theta)$ was found to be symmetrical around $\theta = 90$ degrees. The distribution was, however, far from isotropic; $f(\theta)$ peaked sharply at $\theta = 90$ and 180 degrees. In addition, the reaction was found to have a large total cross-section, roughly 2 nm^2.

To explain the experimental observations on reaction (7.25), it was proposed that a complex held together by largely ionic forces could be formed in any collisions which surmounted the centrifugal barrier on the effective potential. This barrier would be at large reagent separations, there being strong long-range attraction between the highly ionic RbCl and the highly polarizable Cs atom. As a result reaction could take place in collisions with high impact parameters ($b \lesssim 0.8$ nm) and, on average, the orbital angular momentum (l) would far exceed the rotational momentum of the reacting RbCl molecule. The strong polar peaking in $f(\theta)$ was evidence that the *final* orbital momentum (l') was very much greater than the rotational momentum of CsCl, because similar effects influence the break-up of the complex. The strong peaking at $\theta = 0$ and 180 degrees can then be understood in terms of the discussion of two-body scattering given in Section 3.3. Because $l' \approx l$, the products must separate in a plane close to that in which the reagents came together to form the complex. Consequently, the $(1/\sin\theta)$ factor in equation (3.29) plays an important role in how $f(\theta)$ varies with θ.

In theoretical treatments of reactions which proceed via collision complexes[48, 49], it is usual to separate any discussion of the bimolecular formation of the complex from the later stage of their unimolecular decay. At sufficiently high total pressures, the complex could be stabilized by collision, and an association reaction occur. Under single collision conditions, this cannot happen but no direct measure is obtained of the lifetime of the collision complex. In mass spectrometric studies of the breakdown of primary ions[50], 'metastable' species, whose lifetime is comparable with the flight time through the mass spectrometer, are well known. There have not, however, been any

molecular beam studies of reactions, where the complex lasts sufficiently long ($\geqslant 10^{-5}$ s), either for the scattering pattern to be distorted by passage of the complexes out of the scattering zone, or for it to show up as a single entity at the detector.

Because the collision complex lasts for times in excess of a rotational period, it is usually assumed that its energy is completely randomized long before the complex breaks up and that its behaviour is determined solely by its total energy (ε) and total angular momentum (J). Discussion of the reaction dynamics concentrates on the unimolecular process, and this is usually analysed by means of the statistically based theories of reaction that were introduced in Chapter 4. However, it was stressed there—and has been re-emphasized at other points in this book—that RRKM and related theories of unimolecular reactions assume statistical sharing of energy in the phase space of the energized reagent and at the transition state, but make no mention of the products. The *final* distribution of energy may depend critically on the 'later' regions of the potential hypersurface connecting the transition state to the separated reaction products. For the purposes of discussion, the unimolecular process can itself be subdivided: into the part in which an energized collision complex decays into the transition state, and into the much more rapid subsequent stage in which the products separate.

The results of experiments on the reactions:

$$CH_3 + CF_3 \rightarrow (CH_3 - CF_3)^{\dagger} \rightarrow CH_2 = CF_2 + HF \qquad (7.40)$$

and

$$CH_2 = CHCl + h\nu \rightarrow (CH_2 = CHCl)^{\dagger} \rightarrow CH \equiv CH + HCl \qquad (7.41)$$

were described in Section 6.3.4. In both reactions, the vibrational excitation of the hydrogen halide was greater than it would be if energy was shared statistically between all the degrees of freedom of the reaction products. This is not surprising. The reactions in which the products of (7.40) and (7.41) combine both have high activation energies, so a considerable fraction of the total energy liberated in reaction (7.40) or (7.41) is released after the system has passed through the transition state for the dissociation and as the products separate. There is no reason to suppose that this energy will be partitioned evenly among the products' degrees of freedom.

Reactions (7.40) and (7.41) are examples of unimolecular processes which occur via 'tight' transition states. The Cs + RbCl reaction illustrates the other extreme of behaviour. It appears that fragmentation of the CsRbCl complex is only restricted by the requirement to overcome the centrifugal maximum in the effective potential for CsCl + Rb at large separation. A velocity analysis of CsCl (or Rb) scattered from reaction (7.25) has not been carried out. It was pointed out in Section 4.6.3 that the distribution of product recoil energies from such a reaction would not correspond to the distribution of translational energies along the reaction coordinate (ε_x^+) because the energy of the centrifugal barrier, as well as ε_x^+ becomes final relative translational energy.

The reactions passing through a collision complex which have been most thoroughly studied using molecular beams are those between halogen atoms

(F or Cl) and olefins or substituted olefins[51—53]. Addition of the atom creates an energized alkyl, or substituted alkyl, radical which can subsequently eliminate a more weakly bound halogen atom, e.g.,

$$Cl + CH_2 = CHBr \rightarrow (C_2H_3BrCl)^{\dagger} \rightarrow CH_2 = CHCl + Br \qquad (7.42)$$

an H atom, e.g.,

$$F + C_2H_4 \rightarrow (C_2H_4F)^{\dagger} \rightarrow C_2H_3F + H \qquad (7.43)$$

or a methyl radical, e.g.,

$$F + \text{i-}C_4H_8 \rightarrow (C_4H_8F)^{\dagger} \rightarrow C_3H_5F + CH_3 \qquad (7.44)$$

Unimolecular decomposition of these radicals seems to fall into a category between the extremes represented by reaction (7.25)—i.e., reaction via a very 'loose' transition state—and (7.40)—i.e., reaction via a 'tight' transition state, with potential energy much higher than that of the separated products. Thus, kinetic data on the reverse addition reactions indicate that the exit channel barriers in reactions like (7.42) to (7.44) vary from almost zero for Cl or Br elimination, to about 12.5 kJ mol^{-1} for H elimination to approximately 33 kJ mol^{-1} for loss of CH_3.

Reactions of both Cl and F atoms have been studied in crossed molecular beam experiments incorporating velocity analysis. The apparatus which is shown in *Figure 7.2* is that used on reactions of F atoms. In these experiments, the collision energy could be selected, and scattering studies on reaction (7.43) have been performed at collision energies between 9.0 and 50 kJ mol^{-1}.

The reactions of Cl with all of $CH_2{=}CHBr$, $CH_3CH{=}CHBr$ and $CH_2{=}CBr{-}CH_3$ have quite large cross-section (0.2—0.35 nm^2) at thermal collision energies and scatter products symmetrically about $\theta = 90$ degrees. The angular distributions peak fairly strongly at $\theta = 0$ and 180 degrees, indicating that the final angular momentum is once again largely orbital angular momentum. Surprisingly, the products of

$$Cl + CH_2{=}CH{-}CH_2Br \rightarrow ClCH_2CH{=}CH_2 + Br \qquad (7.45)$$

are not scattered symmetrically. Peaks are found at $\theta = 0$ and 180 degrees but the forward scattering is about three times more intense than the backward scattering. This indicates that reaction proceeds by way of an osculating complex with an average lifetime somewhat shorter than a rotational period ($\leqslant 10^{-12}$ s).

The product angular distributions from the reactions of F atoms with butenes are almost isotropic. On the other hand, from reaction (7.43) the products show definite sideways peaking; that is the intensity of reactive scattering is greatest at angles close to $\theta = 90$ degrees. This observation suggests a detailed picture of how this reaction occurs, and again illustrates how unusual dynamical features arise from the requirement to conserve total angular momentum when one reaction product is an H atom, so that the reduced mass of the separating products (μ'), and hence the final orbital momentum, is small.

The proposed mechanism for the $F + C_2H_4$ reaction is portrayed in *Figure 7.18*. The F atom attacks along a direction roughly perpendicular to the plane of the C_2H_4 molecule. As the reaction cross-section is quite large (0.022 nm^2), the initial orbital momentum is the main component in determining the total angular momentum of the complex, i.e., $\boldsymbol{J} \approx \boldsymbol{l}$, and the complex rotates in a plane close to that containing the three heavy atoms. Eventually an H atom is ejected in a direction nearly perpendicular to the plane of the product molecule, leading to a sideways peaked angular distribution. The exit impact parameter (b'), as well as μ', must be small, so $\boldsymbol{j}' \approx \boldsymbol{J} \approx \boldsymbol{l}$, and hence C_2H_3F is formed with its rotation strongly polarized.

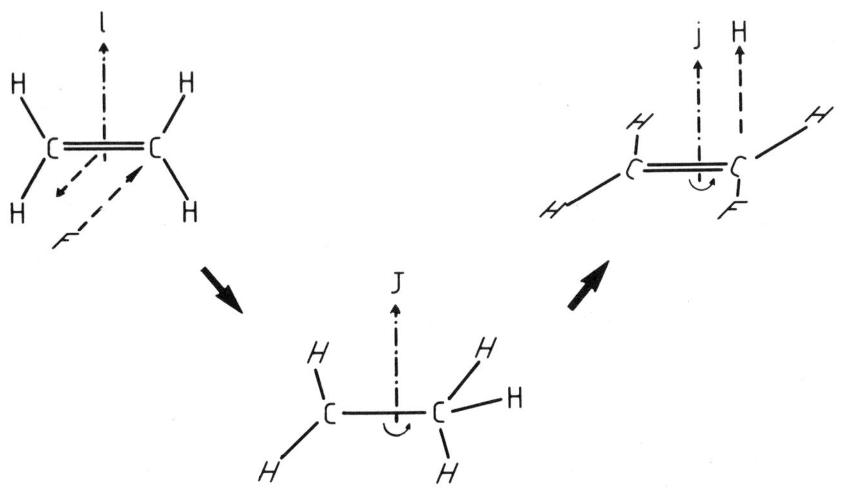

Figure 7.18. Mechanism of the reaction $F + C_2H_4 \rightarrow C_2H_3F + H$ *illustrating the relationship between the angular momentum and the preferred orientation of the product separation*

A good deal of interest has been aroused by measurements of the product translational energy distributions for the addition–elimination reactions of halogen atoms with olefinic molecules. The $\varepsilon_{T'}$ distribution from reaction (7.42) is given in *Figure 7.19*(a).The second curve was calculated by the method outlined in Section 4.6.3 for a complex decomposing via a very loose transition state. The approach is based on RRKM theory but recognizes that there will be a centrifugal barrier in the exit channel which the system must surmount if the complex is to separate to products. The final translational energy of an individual pair of separating products will exceed the energy of motion along the reaction coordinate in the transition state by the energy of the centrifugal barrier. The predictions of this model, incorporating angular momentum effects and therefore designated (RRKM + AM), are in good agreement with the observations for reaction (7.42). As the addition reactions of Br atoms to olefins occur with little or no activation energy, the assumptions of the (RRKM + AM) model, especially the absence of any redistribution of energy

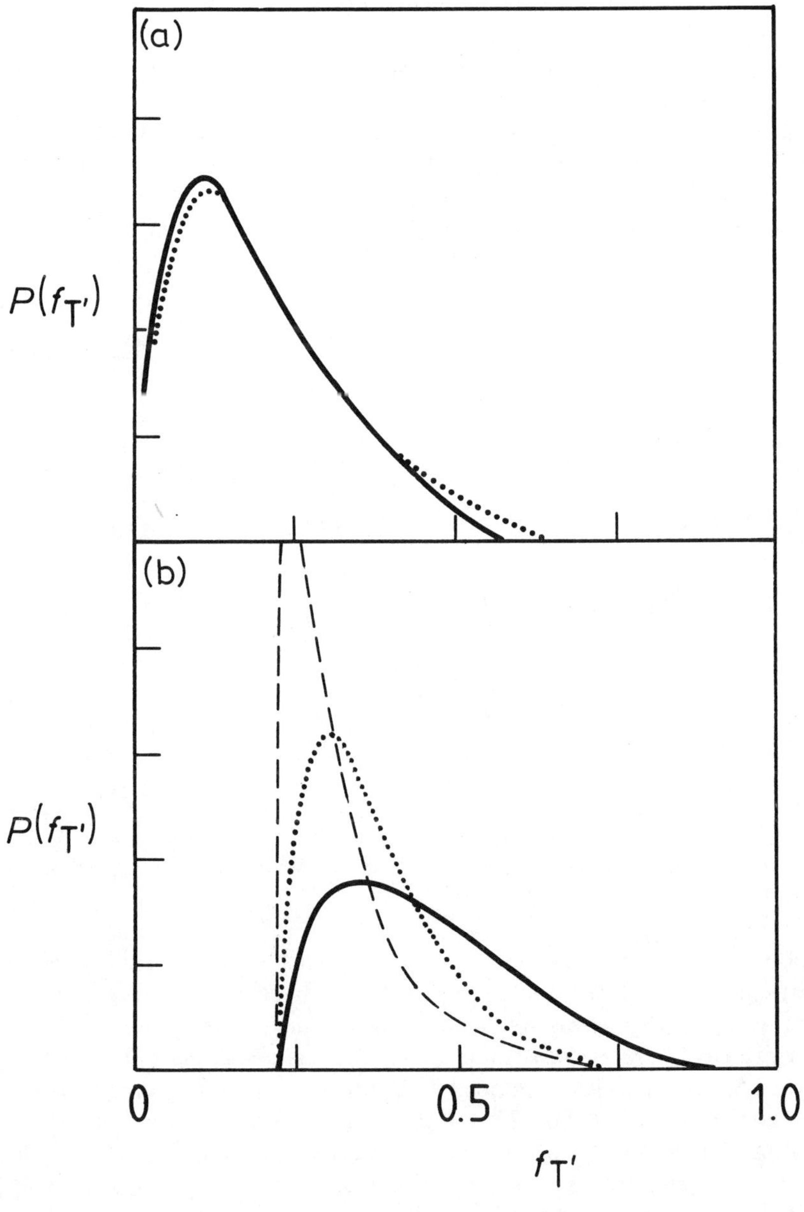

Figure 7.19. Distributions of product translational energies from (a) $Cl + CH_2{=}CHBr \rightarrow CH_2{=}CHCl + Br$, *and* (b) $F + i{-}C_4H_8 \rightarrow C_3H_5F + CH_3$. *The full lines show the experimental distributions, the dotted lines those calculated assuming statistical sharing of the energy, but taking account of angular momentum effects. The dashed line in* (b) *shows the result of neglecting angular momentum effects in the statistical calculations*

once the transition is traversed, seem appropriate, and the agreement between experiment and theory is satisfactory.

The second panel of *Figure 7.19* shows a similar comparison of experimental and calculated recoil energy distributions for reaction (7.44). There is now a clear discrepancy between experiment and theory, and similar differences are found for all the F atom plus olefin reactions eliminating CH_3 or H, which have been studied by Lee's group[51]. Of course, the dynamics of these decompositions involve a transition state determined by a 'real' potential barrier rather than a centrifugal barrier, and considerably more energy is released after one of these systems passes through the transition state than in (7.42). The theoretical curve plotted in *Figure 7.19*(b) is calculated according to phase-space theory (*see* Section 4.6.2), except that it is assumed that energy released after passage through the transition state is channelled into relative translational motion. This amount can be varied but this alone does not allow the gap between experiment and theory to be closed.

The discrepancy described in the previous paragraph has led to a lively debate[54,55], which is not yet concluded[49,53]. There can be no doubt that the experimental observations do not *prove* that energy is not distributed statistically in the transition state, although they may be a strong indirect indication of this. The unknown factor is the dynamical behaviour as the products separate from the transition state. One view is that, especially when an H atom is ejected, there can be little coupling between the relative translation motion and the other degrees of freedom. Comparison with other, dynamically less complex, systems would suggest that this is likely to be the case if the release of energy in this region can be classified as repulsive. If the energy released in this post-transition state region does become relative translational energy in the products, then the failure of phase-space calculations to match the observed $\varepsilon_{T'}$ distribution does strongly indicate a non-statistical distribution of energies in the transition state itself.

Greater internal excitation of the products could be explained if their structure in the transition state was still somewhat different from their final equilibrium structure. This would correspond to a 'late', attractive, release of energy. It would cause the products to recoil with less translational energy than when this effect was absent. It is harder to explain the long high-energy tail of the $\varepsilon_{T'}$ distribution. It seems unlikely that the separating products could make a 'secondary encounter' allowing energy to be transferred from internal modes to relative translation. This seems especially improbable when one of the products is an H atom, since then the products will separate rapidly.

The latest beam experiments[53] on the $F + C_2H_4$ reaction have explored the dependence of the recoil energy distribution on the incident energies. At all the collision energies used ($9.0 \leqslant \varepsilon_T \leqslant 50$ kJ mol^{-1}), the $\varepsilon_{T'}$ distribution was broad, with the average product translational energy essentially constant at 50 per cent of the total available energy. The results could not be matched either by phase-space theory or by calculations based on Marcus's theory invoking a 'tight' transition state. Further recent evidence on the non-statistical sharing of energy in the products of reaction (7.43) comes from observations on the

infrared chemiluminescence from the vinyl fluoride produced[56]. These show that the populations in excited vibrational levels are 'non-statistical'.

In conclusion, it can be stated that there is now strong circumstantial evidence that when some chemically activated species unimolecularly decompose the distribution of energy among the modes in the transition state is not governed solely by statistical factors. Here is another area of reaction dynamics where further experimental and theoretical efforts will undoubtedly be made.

7.5.4 Four-centre and six-centre reactions

Four-centre reactions of the general type

$$AB + CD \rightarrow AC + BD \tag{7.46}$$

have not featured prominently in this volume. It was pointed out in Section 5.3.5 that these reactions usually have high activation energies and occur at an appreciable rate only at high temperatures. In studies at low temperatures, chemical change according to equation (7.46) may be observed, but it can be difficult to establish that reaction occurs via a single elementary process and not via a more complex mechanism, possibly one aided by surface catalysis.

In this chapter, two families of facile four-centre reactions have been introduced: the reactions between alkali-metal dimers and halogen-containing molecules and those between two alkali-metal halides. Both proceed via highly ionic electronic states. One further class of facile four-centre reaction has been discovered in molecular beam experiments[57]. These reactions involve an alkali-metal halide and a halogen molecule, e.g.,

$$CsI + Cl_2 \rightarrow CsCl + ICl \tag{7.47}$$

The angular and velocity distributions of the products of this reaction are symmetrical about $\theta = 90$ degrees, demonstrating that reaction proceeds via a collision complex. It appears that once again ionic interactions are important, the proposal being that the collision complex corresponds to the alkali-metal trihalide molecule, $Cs^+(ClICl)^-$. With the least electronegative halogen in the central position, the linear trihalide ion is stable, according to arguments outlined in Section 7.5.2, and the complex survives until it reaches a sufficiently linear structure for Cs^+ to strongly polarize the triatomic species and pull off a Cl^- ion.

The reaction between H_2 and I_2,

$$H_2 + I_2 \rightarrow 2HI; \quad \Delta H_0^\circ = -8.2 \text{ kJ mol}^{-1} \tag{7.48}$$

provides the most celebrated case where a reaction originally thought to proceed in a single four-centre process has been found to involve a more complex mechanism. Comparison of thermal and photochemical rate data[58] showed conclusively that below 800 K, reaction proceeds via the steps:

$$I_2 \leftrightarrows 2I; \quad \Delta H_0^\circ = +149.0 \text{ kJ mol}^{-1} \tag{7.49a}$$

$$2I + H_2 \leftrightarrows 2HI; \quad \Delta H_0^\circ = -157.2 \text{ kJ mol}^{-1} \tag{7.49b}$$

The energy requirements for two HI molecules to react have been explored in molecular beam experiments. Accelerated beams of HI were directed into a long chamber containing DI at low pressure[59]. The mean collision energy was varied from 84 to 456 kJ mol^{-1}, but no HD could be detected from any of the experiments. This result indicates that, for the reverse of reaction (7.49b)—or the reverse of (7.48)—to proceed, vibrational (or possibly rotational) excitation of the HI is necessary.

The existence of a high energy barrier for reaction (7.48), and by implication for similar four-centre reactions which cannot proceed by way of a strongly ionic intermediate state, can be discussed in terms of molecular orbital correlations[60]. It is helpful, though not essential, to concentrate on the simplest system of this kind, namely

$$H_2 + D_2 \rightleftarrows 2HD \tag{7.50}$$

Figure 7.20 presents the orbital correlations for this reaction proceeding via a sequence of rectangular → square → rectangular geometries (i.e., $D_{2h} \rightarrow D_{4h} \rightarrow D_{2h}$ symmetries). The symbols indicate the symmetry of the four-centre orbitals with respect to reflection through two orthogonal planes which are both perpendicular to the plane containing the atoms. The in-phase combinations of the two σ_g orbitals on the 'separated' products are symmetric (s) with respect to both operations, and hence correlate directly with one another. However, there is no such correlation for the out-of-phase combinations of the σ_g orbitals: these have opposite symmetries with respect to reflection in each of the symmetry planes.

The qualitative arguments that reactions like (7.50) and, possibly (7.48), cannot proceed through rather symmetrical transition states of low energy is supported, in the case of (7.50), by full-scale quantum mechanical calculations

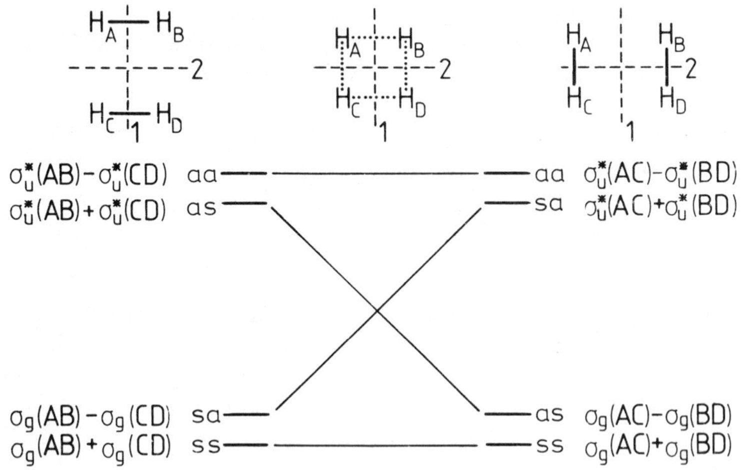

Figure 7.20. Diagram showing the correlation of molecular orbitals for the molecular hydrogen-exchange reaction proceeding via a square transition state

on the electronic energy of H_4 [61–63]. Fairly extensive calculations have failed to discover a geometry which could serve as the transition state for reaction (7.50) with an energy lower than 456 kJ mol^{-1}, although T- and Y-shaped structures have not yet been investigated[64]. On the other hand, shock tube experiments determine an energy of activation of 169 kJ mol^{-1} for this reaction[65], and experiments involving selective excitation of the reagents' vibration indicate that it may be vibrational energy rather than translational energy which is needed to promote the reaction[66]. At the present time, there is an unresolved and interesting discrepancy between theory and experiment for this reaction, which challenges researchers in both camps.

The controversy about whether covalently bonded molecules can react in a concerted four-centre process has recently been given a fresh dimension, as the result of a series of beautiful new crossed-beam experiments. In these studies, a comparison is made between the reactions of a closed-shell molecule (Br_2 or HI) with either normal monomeric halogen molecules, e.g., Cl_2, or the dimer, $(Cl_2)_2$, formed in the expansion from a supersonic nozzle source and held together by van der Waals forces[67,68]. Experiments on collisions of Cl_2 monomers with Br_2 or HI were carried out using heated nozzle sources for both beams[67]. The chlorine beam was 'seeded' with helium to provide collision energies up to 125 kJ mol^{-1}. Even at the highest collision energies, no evidence was found for formation of BrCl or ICl in the reactions:

$$Cl_2 + Br_2 \rightarrow 2BrCl \tag{7.51}$$

$$Cl_2 + HI \rightarrow HCl + ICl \tag{7.52}$$

It was estimated that the reaction cross-sections are less than 10^{-4} nm^2.

Dimeric $(Cl_2)_2$ could be generated by operating the nozzle source at high pressure and unheated[68]. Under these conditions, up to 10 per cent of the species in the beam were dimers. In these experiments, the Br_2 beam was seeded with helium, thereby providing a variable collision energy up to about 105 kJ mol^{-1}. Cross-sections for the reactions of $(Cl_2)_2$ with Br_2 and HI were found to be large and four pathways could be distinguished:

$$Br_2 + (Cl_2)_2 \rightarrow 2BrCl + Cl_2 \tag{7.53a}$$

$$Br_2 + (Cl_2)_2 \rightarrow (BrCl.Cl_2) + BrCl \tag{7.53b}$$

$$Br_2 + (Cl_2)_2 \rightarrow (Br_2.Cl_2) + Cl_2 \tag{7.53c}$$

$$Br_2 + (Cl_2)_2 \rightarrow Br_2 + 2Cl_2 \tag{7.53d}$$

At low collision energies, the 'molecule-transfer' reaction (7.53c) predominated but was essentially replaced at higher collision energies by collision-induced dissociation, i.e., reaction (7.53d). The $(Br_2.Cl_2)$ product of reaction (7.53c) was scattered backward and both this and the distribution of recoil energies were analogous to the rebound behaviour found for many 'atom-transfer' reactions.

Reactions (7.53a) and (7.53b) are rather more interesting since, as they are clearly single collision processes, they must proceed via a six-centre cyclic

intermediate. The distributions of scattering angle and relative velocity for the products indicate that these reactions proceed via a common collision complex. This first breaks up to yield a BrCl . . . Cl_2 . . . ClBr chain held together by van der Waals forces. Usually, the two weak bonds break in succession so that reaction (7.53a) occurs; in a small fraction of cases, the second bond does not give way and the products of reaction are (BrCl . . Cl_2) + ClBr.

The existence of a low-energy pathway for reaction via a six-centre intermediate, where there is no comparable path for the four-centre process, can be understood in terms of molecular orbital correlations. For $Cl_2 + Br_2$, the orbital correlations are similar to those already described for $H_2 + D_2$ and the four-centre reaction is therefore predicted to have a high activation energy. On the other hand, orbital correlations for a six-centre intermediate, along the lines suggested for cycloaddition reactions by Hoffman and Woodward[69] give no indication of a high potential barrier to reaction, in agreement with results of the molecular beam experiments. It is interesting to note that a termolecular exchange reaction via H_6 may be possible and, indeed, preferred to reaction via a four-centre transition state[70]. The development of this totally new area of chemical kinetics has scarcely started, and its growth can be eagerly awaited.

References

Section 7.1

1 HERSCHBACH, D. R., in *Molecular Beams*, Ed. Ross, J., John Wiley, New York (1966), Vol. X in the series *Advances in Chemical Physics*, 319
2 KINSEY, J. L., in *International Review of Science. Physical Chemistry*, series 1, Vol. 9, *Chemical Kinetics*, Ed. Polanyi, J., Butterworths, London (1972), Chap. 6
3 FLUENDY, M. A. D., and LAWLEY, K. P., *Chemical Applications of Molecular Beam Scattering*, Chapman and Hall, London (1973)
4 HERSCHBACH, D. R., *Faraday Discussions of the Chemical Society*, **55**, 233 (1973)
5 FARRAR, J. M., and LEE Y. T., *Annual Review of Physical Chemistry*, **25**, 357 (1974)
6 GRICE, R., in *Molecular Scattering: Physical and Chemical Applications*, Ed. Lawley, K. P., John Wiley, New York (1975), Vol. XXX in the series *Advances in Chemical Physics*, 247

Section 7.2

7 BRADLEY, J. N., *Fast Reactions*, Oxford University Press, Oxford (1975)
8 AMDUR, I., and JORDAN, J. E., in *Molecular Beams*, Ed. Ross, J., John Wiley, New York (1966), Vol. X in the series *Advances in Chemical Physics*, 29
9 HERMAN, Z., and WOLFGANG, R., in *Ion–Molecule Reactions*, Ed. Franklin, J. L., Plenum Press, New York (1972)
10 ODIORNE, T. J., BROOKS, P. R., and KASPER, J. V. V., *Journal of Chemical Physics*, **55**, 1980 (1971)
11 DOEMENY, L. J., van ITALLIE, F. J., and MARTIN, R. M., *Chemical Physics Letters*, **4**, 302 (1969)
12 CHOU, M. S., CRIM, F. F., and FISK, G. A., *Chemical Physics Letters*, **20**, 464 (1973)
13 FREUND, S. M., FISK, G. A., HERSCHBACH, D. R., and KLEMPERER, W., *Journal of Chemical Physics*, **54**, 2510 (1971)
14 BENNEWITZ, H. G., HAERTEN, R., and MULLER, G., *Chemical Physics Letters*, **12**, 335 (1971)
15 SCHULTZ, A., CRUSE, H. W., and ZARE, R. N., *Journal of Chemical Physics*, **57**, 1354 (1972)

16 CRUSE, H. W., DAGDIGIAN, P. J., and ZARE, R. N., *Faraday Discussions of the Chemical Society,* **55,** 277 (1973)
17 SMITH, G. P., and ZARE, R. N., *Journal of Chemical Physics,* **64,** 2632 (1976)

Section 7.4

18 GREENE, E. F., MOURSUND, A. L., and ROSS, J., in *Molecular Beams,* Ed., Ross, J., John Wiley, New York (1966), Vol. X in the series *Advances in Chemical Physics,* 135
19 WILSON, K. R., KWEI, G. H., NORRIS, J. A., HERM, R. R., BIRELY, J. H., and HERSCHBACH, D. R., *Journal of Chemical Physics,* **41,** 1154 (1964)
20 HERSCHBACH, D. R., *Discussions of the Faraday Society,* **33,** 149 (1962)
21 MILLER, W. B., SAFRON, S. A., and HERSCHBACH, D. R., *Discussions of the Faraday Society,* **44,** 108 (1967)
22 BERNSTEIN, R. B., and RULIS, A. M., *Faraday Discussions of the Chemical Society,* **55,** 293 (1973)
23 POLANYI, M., *Atomic Reactions,* Williams and Norgate, London (1932)
24 BROOKS, P. R., *Faraday Discussions of the Chemical Society,* **55,** 299 (1973)
25 PORTER, R. N., *Annual Review of Physical Chemistry,* **25,** 317 (1974)
26 KUNTZ, P. J., MOK, M. H., and POLANYI, J. C., *Journal of Chemical Physics,* **50,** 4623 (1969)
27 STRUVE, W. S., KRENOS, J. R., McFADDEN, D. L., and HERSCHBACH, D. R., *Journal of Chemical Physics,* **62,** 404 (1975)
28 FAIST, M. B., and LEVINE, R. D., *Chemical Physics Letters,* **47,** 5 (1977)
29 FOREMAN, P. B., KENDALL, G. M., and GRICE, R., *Molecular Physics,* **23,** 127 (1972)
30 WHITEHEAD, J. C., HARDIN, D. R., and GRICE, R., *Molecular Physics,* **25,** 515 (1973)
31 LIN, S. M., WHITEHEAD, J. C., and GRICE, R., *Molecular Physics,* **27,** 741 (1974)
32 KING, D. L., and HERSCHBACH, D. R., *Faraday Discussions of the Chemical Society,* **55,** 331 (1973)
33 LIN, S.-M., MIMS, C. A., and HERM, R. R., *Journal of Chemical Physics,* **58,** 327 (1973)
34 MIMS, C. A., LIN, S.-M., and HERM, R. R., *Journal of Chemical Physics,* **58,** 1983 (1973)
35 JOHAN, C. D., and ZARE, R. N., *Chemical Physics Letters,* **9,** 65 (1971)

Section 7.5

36 GEDDES, J., KRAUSE, H. F., and FITE, W. L., *Journal of Chemical Physics,* **56,** 3298 (1972) and **59,** 566 (1973)
37 SCHAFER, T. P., SISKA, P. E., PARSON, J. M., TULLY, F. P., WONG, Y. C., and LEE, Y. T., *Journal of Chemical Physics,* **53,** 3385 (1970)
38 LEE, Y. T., in *Physics of Electronic and Atomic Collisions,* Eds. Govers, T. R., and de Heer, F. J., North-Holland, Amsterdam, (1971), p. 357
39 ANLAUF, K. G., CHARTERS, P. E., HORNE, D. S., MACDONALD, R. G., MAYLOTTE, D. H., POLANYI, J. C., SKRLAC, W. J., TARDY, D. C., and WOODALL, K. B., *Journal of Chemical Physics,* **53,** 4091 (1970)
40 SCHATZ, G. C., and KUPPERMANN, A., *Journal of Chemical Physics,* **65,** 4668 (1976)
41 HARRISON, H., *Journal of Chemical Physics,* **37,** 1164 (1962)
42 WALSH, A. D., *Journal of the Chemical Society,* 2288 (1953)
43 O'NEIL, S. V., PEARSON, P. K., SCHAEFER III, H. F., and BENDER, C. F., *Journal of Chemical Physics,* **58,** 1126 (1973)
44 McDONALD, J. D., LeBRETON, P. R., LEE, Y. T., and HERSCHBACH, D. R., *Journal of Chemical Physics,* **56,** 769 (1972)
45 BUSCH, G. E., MAHONEY, R. T., MORSE, R. I., and WILSON, K. R., *Journal of Chemical Physics,* **51,** 449 (1969)
46 LEE, Y. T., LeBRETON, P. R., McDONALD, J. D., and HERSCHBACH, D. R., *Journal of Chemical Physics,* **51,** 455 (1969)
47 LOESCH, H. J., and BECK, D., *Berichte der Bunsengesellschaft für Physikalische Chemie,* **75,** 736 (1971)
48 SAFRON, S. A., WEINSTEIN, N. D., HERSCHBACH, D. R., and TULLY, J. C., *Chemical Physics Letters,* **12,** 564 (1972)
49 MARCUS, R. A., *Journal of Chemical Physics,* **62,** 1372 (1975)

50 COOKS, R. G., BEYNON, J. H., CAPRIOLI, R. A., and LESTER, G. R., *Metastable Ions*, Elsevier, Amsterdam (1973)

51 PARSON, J. M., SHOBATAKE, K., LEE, Y. T., and RICE, S. A., *Faraday Discussions of the Chemical Society*, **55,** 344 (1973)

52 CHEUNG, J. T., McDONALD, J. D., and HERSCHBACH, D. R., *Faraday Discussions of the Chemical Society*, **55,** 377 (1973)

53 FARRAR, J. M., and LEE, Y. T., *Journal of Chemical Physics*, **65,** 1414 (1976)

54 MARCUS, R. A., *Faraday Discussions of the Chemical Society*, **55,** 379, 381 (1973)

55 PARSON, J. M., SHOBATAKE, K., LEE, Y. T., and RICE, S. A., *Faraday Discussions of the Chemical Society*, **55,** 380 (1973)

56 MOEHLMANN, J. G., GLEAVES, J. T., HUDGENS, J. W., and McDONALD, J. D., *Journal of Chemical Physics*, **60,** 4790 (1974)

57 KING, D. L., and HERSCHBACH, D. R., *Faraday Discussions of the Chemical Society*, **55,** 331 (1973)

58 SULLIVAN, J. H., *Journal of Chemical Physics*, **46;** 73 (1967)

59 JAFFE, S. B., and ANDERSON, J. B., *Journal of Chemical Physics*, **49,** 2859 (1968)

60 HOFFMAN, R., *Journal of Chemical Physics*, **49,** 3739 (1968)

61 CONROY, H., and MALLI, G., *Journal of Chemical Physics*, **50,** 5049 (1969)

62 WILSON, Jr., C. W., and GODDARD, Jr., W. A., *Journal of Chemical Physics*, **51,** 716 (1969)

63 RUBINSTEIN, M., and SHAVITT, I., *Journal of Chemical Physics*, **51,** 2014 (1969)

64 GIMRAC, B. M., *Journal of Chemical Physics*, **53,** 1623 (1970)

65 KERN, R. D., and NIKA, G. G., *Journal of Physical Chemistry*, **75,** 1615 (1971)

66 BAUER, S. H., LEDERMAN, D. M., RESLER, E. L., and FISHER, E. R., *International Journal of Chemical Kinetics*, **5,** 93 (1973)

67 KING, D. L., DIXON, D. A., and HERSCHBACH, D. R., *Journal of the American Chemical Society*, **96,** 3328 (1974)

68 DIXON, D. A., and HERSCHBACH, D. R., *Journal of the American Chemical Society*, **97,** 6268 (1975)

69 HOFFMAN, R., and WOODWARD, R. B., *Journal of the American Chemical Society*, **87,** 2046 (1965)

70 WRIGHT, J. S., *Chemical Physics Letters*, **6,** 476 (1976)

Note added in proof (see p. 309).

Experiments which utilise the intensity, monochromaticity, and polarization of laser radiation threaten the truth of the statements in the first paragraph of this Chapter; *see*, for example, APT, J., and PRITCHARD, D. E., *Physical Review Letters*, **37,** 91 (1976); CAMPBELL, J. D., YU, M. H., MANGIR, M., and WITTIG, C., *Journal of Chemical Physics*, **69,** 3854 (1978)

Note added in proof (*see* p. 319). For further use of lasers, *see* KARNY, Z., and ZARE, R. N., *Journal of Chemical Physics*, **68,** 3360 (1978); DISPERT, H. H., GEIS, M. W., and BROOKS, P. R., *Journal of Chemical Physics*, **70,** 5317 (1979).

Appendix 1

Units, conversion factors and fundamental constants

A1.1 Basic units

Physical quantity	*Name of unit*	*Symbol*
length	metre	m
mass	kilogramme	kg
time	second	s
electric current	ampere	A
temperature	kelvin	K
amount of substance	mole	mol

A1.2 Derived units

Physical quantity	*Name of unit*	*Symbol*	*Definition*
energy	joule	J	$kg\ m^2\ s^{-2}$
force	newton	N	$kg\ m\ s^{-2}$
electric charge	coulomb	C	A s
potential difference	volt	V	$J\ A^{-1}\ s^{-1}$
frequency	hertz	Hz	s^{-1}
pressure	pascal	Pa	$N\ m^{-2}$

A1.3 Prefixes for SI units

Prefix	*Symbol*	*Factor*
mega	M	10^6
kilo	k	10^3
(centi	c	10^{-1})[a]
milli	m	10^{-3}
micro	μ	10^{-6}
nano	n	10^{-9}

[a] Not strictly allowable in the SI system, but cm and cm^3 are still widely used..

A1.4 Conversion factors

Physical quantity	*Unit*	*Symbol*	*SI equivalent*
length	ångstrom	Å	10^{-10} m
	atomic unit (≡ Bohr radius)	a.u.	5.292×10^{-11} m
volume	litre	l	10^{-3} m^3
	cubic centimetre	cm^3	10^{-6} m^3
mass	atomic unit	a.m.u.	1.6605×10^{-27} kg
temperature	degree Centigrade	°C	K − 273.15
energy	electron-volt (≡ 96.49 kJ mol^{-1})	eV	1.602×10^{-19} J
	erg	erg	10^{-7} J
	atomic unit	Hartree	4.359×10^{-18} J
	calorie	cal	4.184 J
force	dyne	dyne	10^{-5} N
electric charge	esu	esu	3.333×10^{-10} C
frequency	wavenumber	cm^{-1}	3.336×10^{-11} Hz
pressure	standard atmosphere	atm	1.013×10^{5} Pa
	torr	Torr	1.33×10^{2} Pa

For an ideal gas at 298 K,
1 atm ≡ 2.463×10^{19} molecule cm^{-3}; 1 Torr ≡ 3.241×10^{16} molecule cm^{-3}

A1.5 Fundamental constants

Constant	*Symbol*	*Value*
gas constant	R	8.314 J K^{-1} mol^{-1}
Boltzmann's constant	k_B	1.381×10^{-23} J K^{-1}
Avogadro's number	N_A	6.022×10^{23} mol^{-1}
Planck's constant	h	6.626×10^{-34} J s
speed of light in vacuum	c	2.998×10^{8} m s^{-1}
charge on electron	e	1.602×10^{-19} C
mass of electron	m_e	9.109×10^{-31} kg
mass of proton	m_p	1.672×10^{-27} kg
permittivity of a vacuum	ε_0	8.854×10^{-12} F m^{-1}

Appendix 2

Symbols

This Appendix lists and defines the principal symbols used in this book. A few symbols have more than one meaning but the correct interpretation will be clear from the context. Many of the symbols which are used in only one section are not listed here.

A	pre-exponential or A-factor in Arrhenius rate equation (1.7)
$A_{n,m}$	Einstein spontaneous emission coefficient for transition from level n to level m
α	polarizability of atom or molecule
B_v	rotational constant of molecule in specified vibrational level
b	impact parameter
β	collisional efficiency of third-body gas: also, parameter in Morse potential, equation (2.23)
$\mathbf{C}$	vector of centre-of-mass in a two-body collision
C	constant in the attractive part, $-Cr^{-n}$, of an intermolecular potential
χ	total angular deflection in a collision
D_0, D_e	dissociation energy from zero-point level (or from minimum) of the potential
d	hard-sphere collision diameter
E, ε	energy on molar (molecular) scale
E_{act}	activation energy defined by equation (1.8)
ΔE	difference between internal energy of products and reagents
ΔE_0, $\Delta \varepsilon_0$	energy difference between lowest states of products and of reagents
$\Delta E_0^{\ddagger}$, $\Delta \varepsilon_0^{\ddagger}$	energy difference between lowest state of transition state and reagents
ε_T^0	threshold translation energy for bimolecular reaction
ε^0	critical internal energy for unimolecular reaction
ε_T, $\varepsilon_{T'}$	relative translational energy of reagents (products)

$\varepsilon_{n}, \varepsilon_{n'}$	energy of reagent (product) internal state
ε^{+}	energy in transition state related to lowest level in the transition state
ε_{x}	energy of motion along reaction coordinate
$\varepsilon_{J}, \varepsilon_{J}^{+}$	energy of reagents (in transition state) in adiabatic modes
ε_{zp}	zero-point energy
ε_{LJ}	'well' depth in Lennard-Jones potential
$g_{n}, g_{n'}$	degeneracy of reagent (product) internal state
$\mathscr{H}$	the Hamiltonian operator in quantum mechanics
H, H_{rel}	the classical Hamiltonian for total (relative) motions of the system
ΔH_0°	standard enthalpy change for a reaction at 0 K
I	moment of inertia; also, intensity of beam of particles or photons
$I(n')$	surprisal associated with formation of species in specified state, equation (4.126)
$\mathbf{J}$, J	vector (quantum number) associated with total angular momentum
$\mathbf{j}$, j, $\mathbf{j}'$, j′	vector (quantum number) associated with rotational angular momentum of reagent (product)
K_c	equilibrium constant in terms of concentrations (molecule cm^{-3})
$k(T), k_f(T), k_r(T)$	thermal rate constant (for forward, i.e., left-to-right, and reverse directions)
$k(n' \mid n; T)$	detailed rate constant for reaction connecting reagents and products in specified states
$k(n' \mid ; T), k(\mid n'; T)$	partially detailed rate constants for reaction to (from) specified internal states
$k(\varepsilon), k_J(\varepsilon)$	rate constants for reagents with specified energy (and angular momentum)
$k^{\infty}(T), k^{0}(T)$	rate constant of pressure dependent reaction in the limit of high (low) total pressure
k_{rad}	first-order rate constant for radiative decay
k	quantum number associated with relative motion, defined by equation (4.51a)
κ	factor allowing for difference between reactive and total flux through S^*
κ_{qu}	quantum correction allowing for non-classical motion along the reaction coordinate
$\boldsymbol{l}, l, \boldsymbol{l}', l'$	vector (quantum numbers) associated with orbital angular momentum of reagents (products)
$\lambda_{v'}, \lambda_{v'}^{\circ}$	parameters from surprisal analysis, equation (4.128)
M, m	total mass of system (mass of individual species)
M	'third-body gas'

M(M′)	in Chapter 7, alkali-metal (alkaline-earth metal atom)
μ	reduced mass
N_n	concentration (molecule cm^{-3}) in specified state
n_n	total number of molecules in specified state
$N^*(\varepsilon^+)$	total number of active mode states at $\varepsilon \leqslant \varepsilon^+$
n, n′	represents set of quantum numbers identifying reagent (product) internal state
ν	frequency (Hz)
$\tilde{\nu}_e$	frequency (cm^{-1}) of small displacements about r_e
$\tilde{\nu}_e x_e$	anharmonicity constant (cm^{-1})
$\tilde{\nu}_x^*$	imaginary frequency (cm^{-1}) from normal mode analysis in transition state
Ω, ω	solid angles in laboratory (LAB) and centre-of-mass (CM) coordinates
ω	collisional frequency in unimolecular rate theory, $= Z[M]$
P	probability, especially steric factor in equation (3.61) for $k(T)$
$P_r(b)$	opacity function
$P(n'), P^\circ(n')$	observed (*a priori*) probability of finding reaction product in specified state, equation (4.126)
p	momentum
ϕ	quantum yield
Q_{int}, Q_T, etc	partition function for specified motion or states
q	position coordinate
r	distance between two species, usually atoms
r_e	equilibrium internuclear separation
$\rho(\varepsilon)$	density of energy states, per unit volume where translational states are included
ΔS_c°	difference between entropy of products and reagents, with 1 molecule cm^{-3} as standard state
S_{act}	entropy of activation defined by equation (1.19)
$S(v), \sigma(v)$	total (differential) scattering cross-section
$S_r(v), \sigma_r(v)$	total (differential) reaction cross-section
S^*	the critical $(6N-1)$-dimensional surface in phase space dividing reagents from products
s	number of active degrees of freedom in unimolecular rate theory
T	temperature
$T_{v'}$	temperature-like parameter arising from surprisal analysis, equation (4.133)
τ, τ_{rad}	lifetime (radiative lifetime)
$d\tau$	elemental volume of phase space
Θ, θ	scattering angles in laboratory (LAB) and centre-of-mass (CM) coordinates

$\mathbf{u}_A$	velocity of A in CM system
V	total volume
$\mathscr{V}$	electronic potential energy
$\Delta\mathscr{V}$	difference between potential energy of products and reagents
$\mathscr{V}^*$	maximum potential energy along path of minimum energy
$\mathscr{V}_{eff}$	'effective' potential, including centrifugal term
$\mathbf{v}$, v	relative velocity (magnitude of relative velocity)
$\mathbf{v}_A$	velocity of A in LAB system
v, v′	vibrational quantum number of reagent (product)
x	reaction coordinate, i.e., along reaction path of minimum energy
x^*	value of x at the transition state
x_e	value of x at reagent equilibrium separation
x°	in unimolecular rate theory, $= \varepsilon^\circ/k_B T$
$Z(T)$	rate constant for bimolecular collisions
z	number of charges on an ion

Appendix 3

Statistical mechanics

The basic postulate of statistical mechanics, the definition of the partition function for a molecular system, and the justification for writing the total partition function as a product of partition functions associated with different separable motions are summarized at the start of Section (4.1). The important relationships are

$$\frac{n_i}{n} = \frac{g_i \exp(-\varepsilon_i/k_B T)}{\sum g_i \exp(-\varepsilon_i/k_B T)}$$

therefore

$$\frac{N_i}{N} \equiv \frac{g_i \exp(-\varepsilon_i/k_B T)}{Q_{total}} \qquad \text{(A3.1)}$$

$$\varepsilon_i = \varepsilon_T + \varepsilon_R + \varepsilon_V + \varepsilon_{El} \ldots \qquad \text{(A3.2)}$$

$$Q_{total} = Q_T Q_R Q_V Q_{El} \ldots \qquad \text{(A3.3)}$$

where the symbols are defined in Appendix 2.

In this Appendix, formulae for different separable partition functions are briefly derived and examples are given of their values for a few atoms and molecules.

A3.1 **Translation** (one-dimension)

Solution of the wave equation for a particle of mass m moving in a 'one-dimensional' box of length l, yields the expression

$$\varepsilon_n = n^2 h^2/8ml^2 \qquad \text{(A3.4)}$$

for the energy of the non-degenerate level with quantum number n. Therefore

$$Q_{T,1-D} = \sum_{n=1}^{\infty} \exp(-n^2 h^2/8ml^2 k_B T) \qquad \text{(A3.5)}$$

When the energy levels are closely packed, i.e., their typical separation $\ll k_B T$, as is the case for atoms or molecules with l an appropriate distance for free translation in the gas phase, equation (A3.5) can be replaced, without loss of

accuracy, by

$$Q_{T,1-D} = \int_0^\infty \exp(-n^2h^2/8ml^2k_BT)\,dn$$

$$= \tfrac{1}{2}\left(\frac{\pi 8ml^2k_BT}{h^2}\right)^{\frac{1}{2}}$$

$$= (2\pi mk_BT/h^2)^{\frac{1}{2}}l \qquad \text{(A3.6)}$$

A3.2 Translation (three-dimensions)

Because the free motions in three directions (x, y, z) are quite independent, the total translational energy is given by

$$\varepsilon_T = \varepsilon_{T,x} + \varepsilon_{T,y} + \varepsilon_{T,z} \qquad \text{(A3.7)}$$

so

$$Q_T = Q_{T,x}Q_{T,y}Q_{T,z}$$

$$= (2\pi mk_BT/h^2)^{\frac{3}{2}}V \qquad \text{(A3.8a)}$$

or

$$(Q_T/V) = (2\pi mk_BT/h^2)^{\frac{3}{2}} \qquad \text{(A3.8b)}$$

Alternatively, the expression for (Q_T/V) can be derived *via* the phase integral (*see* equation 4.6). The total energy is $p^2/2m$ and $d\tau = 4\pi p^2 dp$, i.e., the required volume element in phase space includes all those translational states whose momentum has a magnitude between p and $p + dp$. Then

$$(Q_T/V) = (1/h^3)\int_0^\infty 4\pi p^2 \exp(-p^2/2mk_BT)\,dp$$

$$= (2\pi mk_BT/h^2)^{\frac{3}{2}} \qquad \text{(A3.9)}$$

It is useful to note that, making the substitution $\varepsilon_T = p^2/2m$, equation (A3.9) can be written

$$(Q_T/V) = \int_0^\infty \left\{\left(2^{\frac{1}{2}}\hbar^3\pi^2\right)^{-1} m^{\frac{3}{2}}\varepsilon_T^{\frac{1}{2}}\right\}\exp(-\varepsilon_T/k_BT)\,d\varepsilon_T \qquad \text{(A3.10)}$$

showing that the density of translational states per unit volume is

$$\rho(\varepsilon_T) = (2^{\frac{1}{2}}\hbar^3\pi^2)^{-1}\, m^{\frac{3}{2}}\varepsilon_T^{\frac{1}{2}} \qquad \text{(A3.11)}$$

A3.3 Rotation (linear molecules)

For a rigid linear rotor with moment of inertia I, the energy of the $(2j + 1)$-degenerate level with quantum number j is given by

$$\varepsilon_j = j(j+1)\hbar^2/2I \tag{A3.12}$$

Therefore,

$$Q_R = \sum_{j=0}^{\infty} (2j+1)\exp[-j(j+1)\hbar^2/2Ik_B T] \tag{A3.13}$$

Generally, the summation over j can be replaced by an integration, so

$$Q_R = \int_0^{\infty} (2j+1)\exp[-j(j+1)\hbar^2/2Ik_B T]\,dj$$

$$= (2k_B T/\hbar^2)I$$

or

$$= (k_B T/hcB) \tag{A3.14}$$

where B is the rotational constant in cm^{-1}.

For real, i.e., non-rigid, molecules, equation (A3.14) is usually a very good approximation. In a thermal sample at room temperature, only the lowest vibrational level will be appreciably populated and Q_R should be evaluated using the moment of inertia, or rotational constant for this state. For certain applications, $Q_{R,v}$ is required, where $v > 0$, and then the appropriate value of I_v or B_v should be used. Use of the classically based expression causes some loss of accuracy for some diatomic hydrides at low temperatures, and a direct summation according to equation (A3.13) may be desirable. Finally, it may be necessary to take account of effects which arise from the presence of identical nuclei. Thus, for homonuclear diatomic molecules, the expressions in equation (A3.14) should be divided by a symmetry factor of 2.

A3.4 **Rotation** (non-linear molecules)

Based on the same assumptions as in the previous section, it is possible to show that for spherical top molecules (all three moments of inertia, I_A, I_B and I_C, the same) and symmetric top molecules ($I_A < I_B = I_C$ or $I_A = I_B < I_C$)

$$Q_R = \pi^{\frac{1}{2}}(2k_B T/\hbar^2)^{\frac{3}{2}}(I_A I_B I_C)^{\frac{1}{2}} \tag{A3.15}$$

The same formula holds approximately for asymmetric tops ($I_A \neq I_B \neq I_C$).

A3.5 Internal rotation

If two symmetric top groups in a molecule are *free* to rotate counter to one another, the partition function for this one-dimensional rotation is given by

$$Q_{IR} = \pi^{\frac{1}{2}}(2k_B T/\hbar^2)^{\frac{1}{2}} I^{\frac{1}{2}} \tag{A3.16}$$

If the axis around which the internal rotation takes place coincides with principal axes of rotation for the two groups, $I = I_1 I_2/(I_1 + I_2)$ where I_1 and I_2 are the individual group moments of inertia around the common axis.

Otherwise, evaluation of I is more difficult. Frequently, motions of this kind are partially restricted, i.e., there is a potential barrier to rotation. In this situation, it is hard to calculate the partition function directly and it may be better to rely on thermodynamic information and the equations in Section (A3.8) below.

A3.6 Vibration

For a single harmonic oscillator of frequency $\tilde{\nu}$ cm^{-1}, the energy of the non-degenerate level with quantum number v, relative to $(r_e) = 0$, is given by

$$\varepsilon_v = (v + \tfrac{1}{2})hc\tilde{\nu} \qquad \text{(A3.17a)}$$

Relative to the zero-point level (v = 0),

$$\varepsilon_v = vhc\tilde{\nu} \qquad \text{(A3.17b)}$$

Partition functions based on each of these expressions can be useful. The required summation is straightforward and yields

$$Q_V = \{1 - \exp(-hc\tilde{\nu}/k_B T)\}^{-1} \qquad \text{(A3.18a)}$$

or

$$Q_V = \exp(-hc\tilde{\nu}/2k_B T)/\{1 - \exp(-hc\tilde{\nu}/k_B T)\} \qquad \text{(A3.18b)}$$

If the s normal modes of a polyatomic molecule are treated as separable and harmonic, one derives

$$Q_V = \prod_s \{1 - \exp(-hc\tilde{\nu}_s/k_B T)\}^{-1} \qquad \text{(A3.19a)}$$

or

$$Q_V = \prod_s [\exp(-hc\tilde{\nu}_s/2k_B T)/\{1 - \exp(-hc\tilde{\nu}_s/k_B T)\}] \qquad \text{(A3.19b)}$$

The expression for the density of vibrational levels in a polyatomic molecule in the classical limit is important in unimolecular reaction rate theory and is derived in Section (4.5.4).

A3.7 Electronic states

The separation of electronic energy states is generally $\gg k_B T$. Consequently, the electronic partition function,

$$Q_{El} = \sum g_{El} \exp(-\varepsilon_{El}/k_B T) \qquad \text{(A3.20)}$$

is frequently unity. Q_{El} may differ from unity when the electronic ground state is not a singlet. If this state only involves spin degeneracy, e.g., $H(^2S_{\frac{1}{2}})$, Q_{El} is an integer. When L for an atom or Λ for a linear molecule is non-zero, it may be necessary to evaluate equation (A3.20) over the spin–orbit components.

Examples are given below for $O(^3P)$, $Cl(^2P)$, $OH(^2\Pi)$ and $NO(^2\Pi)$. It is important to remember that for an atom, each spin–orbit component is (2J + 1)-degenerate. For a linear molecule, however, each component is only doubly degenerate since Ω can only be directed in one or other direction along the internuclear axis.

A3.8 Internal energy and entropy

The link between statistical mechanics and classical thermodynamics is embodied in the equations relating thermodynamic quantities to the partition function. For internal energy and entropy, these are

$$E = RT^2(\partial \ln Q_{\text{total}}/\partial T)_V \tag{A3.21}$$

$$S - R\{1 + \ln(Q_{\text{total}}/N_A)\} + E/T \tag{A3.22}$$

For simple molecules, these equations allow one to calculate thermodynamic properties of a molecule from its observed microscopic properties. For larger molecules, the equations can be employed in the other direction; that is, the partition function can be calculated from thermodynamic observations, using the equation:

$$\begin{aligned}\ln Q_{\text{int}} &= \ln (Q_{\text{El}} Q_{\text{V}} Q_{\text{R}}) \\ &= 3.665 + \frac{S^\circ(T)}{R} - \frac{[H^\circ(T) - H^\circ(0\,\text{K})]}{RT} \\ &\quad - 1.5 \ln (\text{m/a.m.u.}) - 2.5 \ln (T/\text{K})\end{aligned} \tag{A3.23}$$

Table A3.1 Values of partition functions at 298 K

Species	*Electronic ground state*	Q_{El}	$(Q_V - 1)$[a]	Q_R	Q_T	Q_{total}
H	^2S	2	–	–	9.8 (23)[b]	1.95 (24)
O	^3P	6.74	–	–	6.2 (25)	4.21 (26)
Cl	^2P	4.03	–	–	2.1 (26)	8.31 (26)
H_2	$^1\Sigma_g^+$	1	6.2(−10)	{7.69}[c]	2.8 (24)	2.12 (25)
HCl	$^1\Sigma^+$	1	5.4 (−7)	19.9	2.2 (26)	4.27 (27)
O_2	$^3\Sigma$	3	4.9 (−4)	72.1	1.8 (26)	3.82 (28)
OH	$^2\Pi$	3.02	1.5 (−8)	11.2[d]	6.8 (25)	2.50 (27)
NO	$^2\Pi$	3.12	1.0 (−4)	122.	1.6 (26)	6.11 (28)
CO_2	$^1\Sigma_g^+$	1	8.6 (−2)	266.	2.8 (26)	8.21 (28)
NO_2	$^2A_1$	2	3.0 (−2)	2.22 (3)	3.0 (26)	1.39 (30)
HNO_3		1	Q_{int} = 3.6(4)[e]		4.9 (26)	1.77 (31)
C_2H_6	$^1A_{1g}$	1	Q_{int} = 1.21(3)[e]		1.6 (26)	1.95 (29)

[a] Q_V evaluated according to equation (A3.18a) or (A3.19a)
[b] 9.8 (23) ≡ 9.8×10^{23}
[c] Evaluated by direct summation, includes degeneracies associated with nuclear spin.
[d] For accurate work, account should be taken of the coupling between rotational and electronic motions.
[e] Q_{int} evaluated using equation (A3.23).

Reaction Index

In this index, all species with one or more unpaired electrons in their electronic ground state, e.g., NO, O_2, $O(^1D)$, are classified as *radicals*; species with paired electrons in their electronic ground states are classified as *molecules*.

Subject Index

Entries in italics are for terms which are used frequently in the text. The page numbers with these entries give the place where the term is defined or first introduced.